Digital Telephony

WILEY SERIES IN TELECOMMUNICATIONS

Donald L. Schilling, Editor
City College of New York

Digital Telephony
Second Edition

John Bellamy
Vice President of Engineering
AMBIT Systems Inc
Carrollton, Texas

A Wiley-Interscience Publication
JOHN WILEY & SONS, INC.
New York · Chichester · Brisbane · Toronto · Singapore

Copyright © 1991 John Wiley & Sons, Inc.

Library of Congress Cataloging-in-Publication Data:

Bellamy, John, 1941-
 Digital telephony / John Bellamy. — 2nd ed.
 p. cm. — (Wiley series in telecommunications)
 "A Wiley-Interscience publication."
 Includes bibliographical references and index.

 1. Digital telephone systems. I. Title. II. Series.
TK5103.7.B44 1990
621.385—dc20 90-37906
ISBN 0-471-62056-4 CIP

Printed in the United States of America

10 9 8 7 6 5 4 3 2 1

*The first edition of Digital Telephony was dedicated
to my family as a token repayment for their indulgence.
The fact that they supported me a second time
(when they knew beforehand what was involved)
more than warrants repetition:*

To Judy, Michelle, Joy, Kelly, and Cary

CONTENTS

PREFACE

When the first edition of *Digital Telephony* was written (1980–1981) fiber optic transmission was given little more coverage than to say it represented an exciting new technology for transmission systems. In 1988 AT&T incurred a 6.7 billion dollar write off for analog radio systems being replaced by optical fiber transmission systems. If the dramatic emergence of fiber can cause AT&T to incur its first yearly loss in 100 years it seems acceptable that I write a new edition to expand upon the above understatement (or as one colleague commented: "to at least remove it").

In addition to fiber optic transmission, other new or more extensively covered topics in this edition are: Integrated Services Digital Network (ISDN), Synchronous Optical Network (SONET), Common Channel Signaling (CCS), the Intelligent Network, Cellular Radio Systems, low rate voice encoding, Extended Superframe (ESF), Digital Cross-Connect Systems (DCS), and Trellis Coded Modulation (TCM). All sections have been updated with newer references and examples of recent implementations. The only chapter to survive unscathed is the last chapter—that deals with traffic theory. (Mathematics is more enduring than communications systems technology.)

As mentioned in the preface of the first edition, the term "digital telephony" refers to the use of digital technology in the message path of voice communications networks. In this case the term "digital" refers to a method of encoding the signal—that is, a form of modulation. Hence digital telephony implies voice transmission and switching applications, not data communications. However, one of the most attractive features of voice digitization is that the encoded signals (pulses) are divorced from the analog waveforms of the source. Thus the digital transmission and switching equipment of a voice network is inherently capable of servicing any traffic of a digital nature, particularly data traffic. It is this property of a digital telephone network that is exploited to provide ISDN services.

This book provides an introduction to all aspects of digital communications with an emphasis on voice applications: voice digitization, digital transmission, digital switching, network synchronization, network control, and network analysis. For the most part, the book is an out-growth of the author's work, study, and teaching in the areas listed. Because telephony in the United States

has been the province of the common carriers who have traditionally trained their own engineers, publicly available textbooks covering much of the material presented here have been unavailable. In preparing this book the author relied heavily on articles in the *AT&T Technical Journal*, the *IEEE Transactions on Communications*, and the proceedings of technical conferences.

It should be emphasized that this is not a technical book in the traditional, analytical sense of communications theory. Since numerous books covering communications theory are already available, this book stresses the application and operational aspects of communications system design. Some basic theory is presented in both qualitative and, when appropriate, quantitive terms. The purpose of this volume, however, is to introduce terminology and explain how the various systems work and how applications influence implementations. In most cases the concepts are supported by citing example implementations in the telephone network. The examples are mostly taken from the U.S. network, although some examples pertaining to international (CCITT) standards are also included.

This book is primarily intended for graduate electrical engineers as either a textbook or a reference. The electrical engineering student is most capable of appreciating occasional references to theory and its influence on the practice. However, because analytical rigor is waived in favor of operational descriptions, less technically oriented readers should have no difficulty understanding the principles.

Most of the material in this book has been used for graduate courses in electrical engineering at Southern Methodist University and for corporate telecommunications managers covering telecommunications, data communications, and specifically, digital telephony. There is too much material for students with no previous exposure to the terminology, practice, and systems concepts of telecommunications. Chapter 6 (covering digital radio and modulation) and Chapter 10 (covering traffic analysis) can be omitted and still provide a coherent, full semester of study in the other areas. A complete list of topics is as follows:

Chapter 1: Overview of public analog telephone network in the United States followed by a brief review of the introduction of digital technology into the network.
Chapter 2: A detailed discussion of the basic advantages and disadvantages of digital implementations for voice communications.
Chapter 3: Descriptions and comparisons of the most common voice digitization algorithms.
Chapter 4: Fundamentals of digital wire-line transmission and multiplexing.
Chapter 5: Basic concepts and operations of digital switching machines.
Chapter 6: Digital modulation and radio fundamentals.
Chapter 7: Network synchronization, control, and management requirements and implementations.
Chapter 8: Fiber optic transmission systems and SONET.
Chapter 9: High-level descriptions of integrated digital networks including ISDN.

Chapter 10: Fundamentals of traffic analysis for designing networks.

The appendixes cover the derivation of equations, PCM voice coding relationships, fundamentals of digital communications theory, and traffic tables.

JOHN BELLAMY

Coppell, Texas
July 1990

ACKNOWLEDGMENTS

In preparing this second edition, I am most indebted to Bob Hicks and David Clifton of Rockwell International who made the corporate library available to me for research. I also must thank the library staff: Wanda Fox, Diana Wright and Sharon Bulls for their assistance.

I also want to thank Larkin Crutcher and Steve Bootman of DSC Corporation and Bob Cubbage of Rockwell International for reviewing Chapter 8 and offering numerous helpful suggestions regarding fiber optic systems and SONET. A large portion of the material on fiber optic transmission systems was obtained from *Optical Fiber Telecommunications II* edited by Stewart Miller and Ivan Kaminow.

J. B.

1

BACKGROUND AND TERMINOLOGY

Beginning in the 1960s, telecommunications in the United States began undergoing radical changes in several different areas. First, the conventional analog telephone network was being called upon to provide many new and different services, most of which emanated from the data processing industry. Second, the marketplace and the regulatory agencies stimulated competition in both old and new areas of traditionally monopolistic services. Third, digital technology emerged to implement many of the fundamental transmission and switching functions within the U.S. telephone network and other networks around the world. The main purpose of this book is to describe the design, application, and operational aspects of this new digital equipment. As background, the technology of the analog telephone network is reviewed to provide a framework for the introduction of digital equipment.

It must be emphasized that the introduction of digital technology into the telephone network was motivated by desires to improve the quality, add new features, and reduce the costs of conventional voice services. Digitization of the network did not arise from the needs of the data processing industry for better data transmission services. Indeed, most of the digital technology introduced into the network was initially inaccessible to data traffic, except through analog channels. Of course, a digital network is a natural environment for data communications services. As more and more of the network became digitized more and more support for direct use of the facilities became available for data applications. However, it was not until facilities of the Integrated Services Digital Network (ISDN) became available that end-to-end, switched digital channels could be used for both voice and data.

As a point of historical reference, Figure 1.1 has been included to show that the idea of integrated voice and data is not new. This figure depicts a concept of a German inventor named Phillip Reis [1] to add voice communications to the prevailing means of electrical communications at the time—the telegraph. Reis developed the equipment in the 1860's and died in 1874—2 years before Alexander Graham Bell received his patent for the telephone. As indicated, the figure implies alternate use of the wires for voice or data communications (i.e.,

1

Figure 1.1. Back to the future: the first integrated voice/data communication system.

integrated transmission). Reis actually used the telegraph attachment to signal information pertaining to voice tests—an indication of inadequate voice quality.

To implement simultaneous voice and telegraph communications the telephone in Figure in 1.1 would have to have been digital. Because of technology limitations at the time such an implementation was impossible and telephone systems necessarily evolved with analog technology. One hundred years later the situation changed significantly. Telephone equipment developers and service providers had an abundance of new technology—they were challenged with how to make effective use of it.

For the most part the following discussions describe individual systems and technical reasons for transitions from conventional analog equipment to seemingly less natural digital counterparts. Thus one purpose of this book is to describe how digital technology improves and expands the capabilities of various subsystems within voice telephone networks. Another purpose of the book is to describe the ultimate benefits derived when an entire network is implemented with digital techniques. A great degree of synergism exists when individual systems are designed into one cohesive network utilizing digital implementations throughout. The synergistic effect benefits conventional voice services and other existing or yet to arise uses for the public network.

Most of the equipment descriptions and design examples presented in this

book come from material authored by engineers at AT&T Laboratories and other suppliers for the public telephone network. The basic principles, however, are by no means unique to the public telephone network. The concepts and implementation examples are applicable to any communications network: public or private, voice or data. An inherent attribute of a digital network is that it can, to a large extent, be designed independently of its application.

Terminals, Transmission, and Switching

The three basic elements of a communications network are terminals, transmission systems, and switches. The first part of this chapter provides an overview of these elements as implemented in analog telephone networks. Then, the last part of this chapter provides a brief overview of digital implementations *within* the analog network. Following a detailed discussion of the motivation for digital implementations in Chapter 2, the next four chapters describe the operation and design of the basic elements of a digital voice network. Chapter 3 discusses digital voice terminals and the most common algorithms used to convert analog voice signals into digital bit streams. Chapter 4 presents the basics of digital transmission systems. Fundamentals of digital switching follow in Chapter 5. Basic digital modulation techniques and their application to point-to-point digital microwave systems are described in Chapter 6. A discussion of various synchronization and control considerations for digital networks is provided in Chapter 7. Chapter 8 describes fiber optic transmission systems and the synchronous multiplexing standard (SONET).

The main emphasis of the first eight chapters involves circuit switching as traditionally implemented for voice telephone networks. A circuit-switched network is one that assigns a complete end-to-end connection in response to each request for service. Each connection, with its associated network facilities, is held for the duration of the call. Chapter 9 describes a different type of network, commonly referred to as a packet-switched network, that is particularly suited to servicing data traffic. Chapter 9 then discusses basic considerations to be made when designing a network to accommodate both voice and data traffic: The Integrated Services Digital Network (ISDN). The last chapter presents the basics of traffic theory: the fundamental mathematics for analyzing and predicting telecommunications network performance.

1.1 TELECOMMUNICATIONS STANDARD ORGANIZATIONS

Prior to the breakup of the Bell System on January 1, 1984 telecommunications standards in North America were essentially established by the dominant equipment designer and supplier to the Bell System: Bell Telephone Laboratories and Western Electric. Independent telephone companies, that provided local service to the 20% of the country not covered by the Bell System, relied on the U.S. Independent Telephone Association (USITA) (now referred to as the

U.S. Telephone Association: USTA) to formulate and disseminate standards, particularly for interconnecting with the Bell System.

In anticipation of the divestiture of the Bell Operating Companies (BOCs) from AT&T, the Exchange Carriers Standards Association (ECSA) was formed in 1983 as a nonprofit trade association to represent the interests of all exchange carriers (BOCs and independents). In February of 1984 ECSA sponsored the establishment of the T1 standards committee to formulate new interconnection standards for the U.S. national network. The T1 committee is accredited by the American National Standards Institute (ANSI) to ensure that standards approvals follow principles of openness. Thus T1 committee standards are designated as ANSI T1.nnn-date. (T1 stands for Telecommunications standards entity number 1.) Table 1.1 lists the major subcommittees within T1 and the respective responsibilities.

Other organizations in North America that establish standards related to telecommunications are the Electronic Industries Association (EIA), the Institute of Electrical and Electronic Engineers (IEEE), and Bell Communications Research (Bellcore). Bellcore is an organization chartered to establish standards and qualify equipment for the BOCs. The IEEE is most known for its data communications standards listed in Table 1.2 but has also established numerous standards for measuring and characterizing telecommunications equipment.

Most of the world outside of North America relies on two international telecommunications standards committees established under the auspices of the International Telecommunication Union (ITU). These committees are the International Telegraph and Telephone Consultative Committee (CCITT) and the International Radio Consultative Committee (CCIR). CCITT establishes recommendations for telephone, telegraph, and data transmission circuits and equipment. CCIR is concerned with coordinating the use of the radio spectrum. In the United States, use of the radio spectrum is controlled by the Federal Communications Commission (FCC).

The primary source for CCITT recommendations is the Conference of European Posts and Telecommunications Administrations (CEPT). North American standards and CEPT standards have often been incompatible in the past. North American standards established by the Bell System were therefore

TABLE 1.1 T1 Standards Subcommittees

Subcommittee	Responsibility
T1E1	Carrier to customer installation interfaces
T1M1	Internetwork operations, administration, maintenance and provision
T1Q1	Performance
T1S1	Services, architecture, and signaling
T1X1	Transmission
T1Y1	Specialized subjects

TABLE 1.2 IEEE Data Communications Standards

802.1	Higher Layer (Protocol) Interface Standard
802.2	Logical Link Control Standard (LLC)
802.3	CSMA/CD Local Area Network (Ethernet)
802.4	Token Bus Local Area Network (Arcnet)
802.5	Token Ring Local Area Network (IBM Ring)
802.6	DQDB Metropolitan Area Network (QPSX)
802.7	Broadband
802.8	Optical Fiber
802.9	Integrated Voice and Data LAN Interfaces

incorporated into CCITT recommendations as a subset. Due to the need for more international compatibility the T1 subcommittees and CEPT now work more closely together to establish joint standards. A major example of the joint effort is the SONET standard for synchronous fiber transmission systems described in Chapter 8.

The international Standards Organization (ISO) is an organization with standards activities in a wide range of subject matters—some of which involve telecommunications. Technical subcommittees within ISO work closely with CCITT study groups in formulating CCITT recommendations—particularly those related to ISDN protocols that adhere, as much as possible, to the ISO data communications standard for Open Systems Interconnection (OSI) Reference Model.

1.2 THE ANALOG NETWORK HIERARCHY

Because the analog telephone networks of the world evolved over a period of almost 100 years a great amount of diversity in equipment implementations also developed. It is a remarkable achievement that vast networks, like the U.S. network, could accommodate the myriad of equipment types and function properly. In 1980* in the United States alone, there were 181 million telephones [2], almost all of which could directly dial any public telephone number and have a good quality connection. This achievement was made possible by standardized interfaces and well-defined functional hierarchies. As newer digital equipment was installed it necessarily adhered to the standardized practices of the analog network. The fact that the equipment was implemented with digital technology was transparent to the rest of the network.

*Although there is no specific date at which digital technology took over from analog technology, 1980 is significant in that it represents the time frame in which fiber optics emerged to displace analog radios for intercity transmission—the last stronghold of analog technology in the U.S. public network.

1.2.1 Bell System Hierarchy

Alexander Graham Bell invented the first practical telephone in 1876. It soon became apparent, however, that the telephone was of little use without some means of changing connections on an "as needed" basis. Thus the first switching office was established in New Haven, Connecticut only 2 years later. This switching office, and others following, was located at a central point in a service area and provided switched connections for all subscribers in the area. Because of their locations in the serivce areas, these switching offices are often referred to as central offices.

As telephone usage grew and subscribers desired longer distance connections, it became necessary to interconnect the individual service areas with trunks between the central offices. Again, switches were needed to interconnect these offices, and a second level of switching evolved. Continued demand for even longer distance connections, along with improved long-distance transmission facilities, stimulated even more levels of switching. In this manner the public telephone network in the United States evolved to a total of five levels. These levels are listed in Table 1.3.

At the lowest level of the network are class 5 switching offices, also called central offices (CO) or end offices (EO). The next level of the network is composed of class 4 toll offices. The toll network of the Bell System contained three more levels of switching: primary centers, sectional centers, and regional centers.

To illustrate the structure and motivation for hierarchical networks a symbolic, three-level example is shown in Figure 1.2. In contrast, Figure 1.3

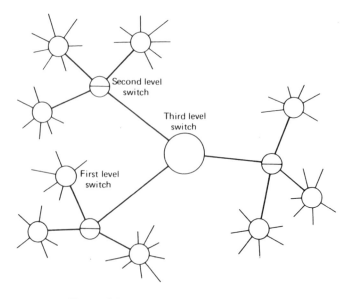

Figure 1.2. Three level switching hierarchy.

TABLE 1.3 Public Network Hierarchy of the Bell System (1982) [3]

Switch Class	Functional Designation	No. in Bell System	No. in Independents	Total
1	Regional center	10	0	10
2	Sectional center	52	0	67
3	Primary center	148	20	168
4.	Toll center	508	425	933
5	End office	9803	9000	18,803

depicts a different network structure for interconnecting all of the first-level switches: a fully connected mesh structure. Obviously, the hierarchical network requires more switching nodes but achieves significant savings in the number of trunks: the transmission links between switching offices. Determination of the total number of trunk circuits in either network is necessarily a function of the amount of traffic between each pair of switching nodes. (Chapter 10 provides the mathematics for determining the number of trunk circuits.) As a first approximation, the trunk costs of a mesh can be determined as the total number of connections (trunk groups) N_c between switching offices:

$$N_c = N(N - 1)/2 \qquad (1.1)$$

where N is the number of nodes.

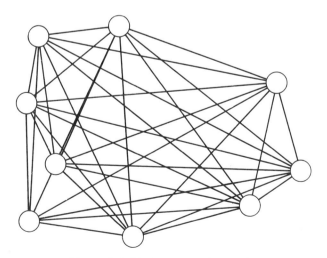

Figure 1.3. Mesh connected network.

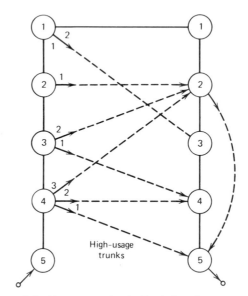

Figure 1.4. Alternate routing in North American network.

Thus the mesh network of Figure 1.3 has 36 connections as compared to 12 connections in Figure 1.2. In the case of fiber optic transmission the cost comparison of the trunks is almost exactly 3:1 because a single fiber system can provide more capacity than is needed between any two switches.

A less obvious difference between the networks of Figures 1.2 and 1.3 involves the method of establishing connections between two offices. In the hierarchical network there is one and only one route between any two switching nodes. In the mesh network most connections would be established on the direct route between the two offices. However, if the direct route is unavailable (because of a traffic overload or an equipment failure) and the first-level switches can provide trunk-to-trunk connections (called tandem switching functions) the mesh network provides many alternatives for establishing connections between any two nodes. Hence the reliability of a network architecture must be considered in addition to just the costs. In general, neither a pure mesh nor a purely hierarchical network is desirable.

Taking these factors into account Figure 1.4 depicts alternate routing as implemented in the former Bell System. As indicated, the basic backbone hierarchical network is augmented with high usage trunks. High-usage trunks are used for direct connections between switching offices with high volumes of interoffice traffic. Normally, traffic between two such offices is routed through the direct trunks. If the direct trunks are busy (which may happen frequently if they are highly utilized), the backbone hierarchical network is still available for alternate routing.

Traffic was always routed through the lowest available level of the network.

This procedure not only used fewer network facilities but also implied better circuit quality because of shorter paths and fewer switching points. Figure 1.4 shows the basic order of selection for alternate routes. The direct interoffice trunks are depicted as dashed lines, while the backbone, hierarchical network is shown with solid lines.

In addition to the high-usage trunks, the backbone network was also augmented with additional switching facilities called tandem switches. These switches were employed at the lowest levels of the network and provided switching between end offices. Tandem switches are not part of the toll network, as indicated in Figure 1.5, but are part of what is referred to as an exchange area. Generally speaking, an exchange area is an area within which all calls are considered to be local calls (i.e., toll free).

In general terms, any switching machine in a path between two end offices provides a tandem switching function. Thus toll switches also provide tandem switching functions. Within the public telephone network, however, the term "tandem" refers specifically to intermediate switching within the exchange area.

The basic function of a tandem office is to interconnect those central offices within an exchange area having insufficent interoffice traffic volumes to justify direct trunks. Tandem offices also provide alternate routes for exchange area calls that get blocked on direct routes between end offices. Although Figure 1.5 depicts tandem offices as being physically distinct from end offices and toll offices, tandem switches were often colocated with either or both types. Operationally, exchange area switching and toll network switching were always separated. The primary reason for the separation was to simplify tandem switching by avoiding billing and network routing. A toll switch must measure call durations for billing purposes but a tandem switch does not. Prior to the introduction of computer controlled switching, billing functions were a significant consideration. The operational separation also implied that toll-connecting

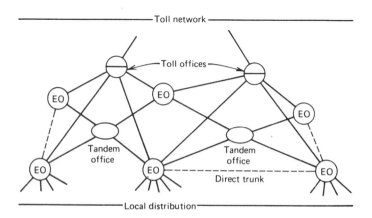

Figure 1.5. Exchange area network.

trunk groups were separate from tandem trunk groups. The flexibility of computer-controlled switching has eliminated the need for the separation.

The separation of exchange facilities from toll facilities had an important effect on the transmission and switching equipment utilized in the respective applications. Exchange area connections were usually short and only involved a few switching offices. Toll connections, on the other hand, could involve numerous switching offices with relatively long transmission links between them. Thus, for comparable end-to-end quality, individual exchange area equipment need not provide as much quality as did toll network counterparts.

1.2.2 Postdivestituture U.S. Network

In the decade of the 1980s the structure of the public telephone network in the United States changed significantly as a result of changes in the technology and the regulatory environment. The main technological changes were (1) extensive deployment of very large digital switching machines, (2) the adaptation of computer-controlled switches to provide multiple switching functions in one machine (e.g., the integration of end office, tandem, and toll switching functions), and (3) the deployment of fiber optic transmission systems that could carry very large cross sections of traffic. Notice that all three of these technological developments suggest a network with fewer, larger switching offices than before.

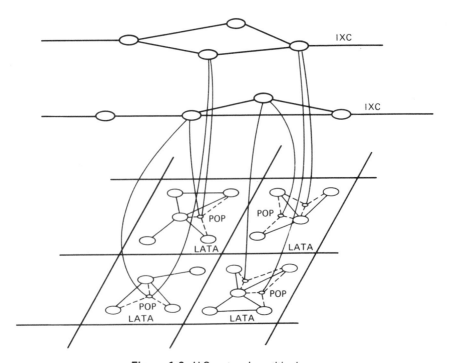

Figure 1.6. U.S. network partitioning.

These technological influences on the network topology are discussed more fully in Chapters 8 and 9.

The most dramatic and immediate effect on the network occurred on January 1, 1984 when the breakup of AT&T officially took effect. Because the breakup involved divestiture of the Bell Operating Companies (BOCs) from AT&T the network itself became partitioned at a new level. The new partitioning is shown in Figure 1.6, which depicts AT&T as one of several competing long distance carriers referred to as interexchange carriers (IXCs) and local access and transport areas (LATAs), which are the exclusive domain of local exchange carriers (LECs). In addition to AT&T the other two main IXCs are MCI and U.S. Sprint. The LECs include 23 BOCs and the former independent telephone companies like GTE, Contel, United Telecommunications, and some 1500 mostly small-town telephone companies.

The number of LATAs in the United States was initially 164 but the number changes slightly as adjustments in service boundaries are sometimes made. Because a LATA entails an area that includes many exchange areas LECs complete toll calls that traverse different exchange areas within one LATA. (The BOCs assumed responsibility for 20% of the #4ESS toll switches [4].) The IXCs are not allowed to carry intra-LATA traffic. Similarly, an LEC is not allowed to carry traffic between two LATAs even when both LATAs might be service areas of a single BOC. Only an IXC is allowed to carry inter-LATA traffic. To ensure that these service partitions are adhered to each IXC interfaces with a LATA at a single point in the LATA referred to as a point of presence (POP). IXC equipment at a POP can be a switching office or merely a junction for collecting traffic carried elsewhere to be switched.

A major aspect of the modified final judgment (MFJ) that specified the divestiture was the condition of equal access, which meant that an LEC (specifically a BOC) was to treat all IXCs equally in regards to exchange access. The conditions of equal access meant that access to all end offices in a LATA would be equal in type, quality, and price for all IXCs. The LATA network structure [5] established to accomplish equal access is shown in Figure 1.7.

The design of the LATA network for intra-LATA traffic is left to the discretion of the LECs. Thus intra-LATA connections can involve multiple switching offices between end offices. However, connections between an EO and a POP can involve at most one intermediate switching office referred to as an Access Tandem (AT). An LEC can establish more than one AT in a LATA but must provide tandem inter-LATA connecting (TIC) trunk groups from each AT to the POP of each IXC. If the LEC so chooses, it can directly connect an EO to a POP with a direct inter-LATA connecting (DIC) trunk group. Whether EO access occurs via an AT or a DIC the same circuit quality must be provided (3 dB of loss in both cases).

With respect to the previous Bell System hierarchy, an AT takes the place of a class 4 toll switch. However, billing functions, which were formerly performed in class 4 switches, are now performed within the IXC network. Although Figure 1.6 shows access tandem and basic tandem switching functions as being distinct,

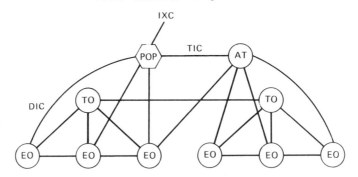

POP: Point of Pressence
AT: Access Tandem
TO: Tandem Office
EO: End Office
TIC: Tandem Inter-LATA Connecting
DIC: Direct Inter-LATA Connecting
TC TC: Tandem Connecting

Figure 1.7. LATA hierarchy and access architecture.

access tandem functions can be integrated into regular tandem switches if the tandem switch can provide AT features. Foremost among these features are providing proper signal levels and quality on the trunks, the ability to forward automatic number identification (ANI) information to an IXC for billing, and the ability to route calls to different IXC POPs depending on presubscription or per-call three-digit carrier designations.

1.2.3 Switching Systems

In general terms the equipment associated with any particular switching machine can be categorized as providing one of the following functions:

1. Signaling
2. Control
3. Switching.

The basic function of the signaling equipment is to monitor the activity of the incoming lines and forward appropriate status or control information to the control element of the switch. Signaling equipment is also used to place control signals onto outgoing lines under direction of the switch control element.

The control element processes incoming signaling information and sets up connections accordingly. The switching function itself is provided by a switching matrix: an array of selectable crosspoints used to complete connections between input lines and output lines. These basic constituents of a switching machine are shown in Figure 1.8.

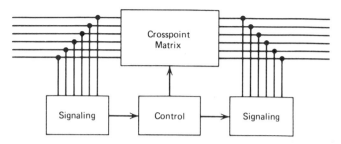

Figure 1.8. Switching system components.

Electromechanical Switching

Prior to the introduction of digital electronic switching machines in the late 1970s switching offices in North America, and around the world, were equipped with one of two basic types of electromechanical switches: step-by-step* and crossbar. As shown in Figure 1.9, cross-points of a step-by-step switch are wiper contacts that move in direct response to dial pulses. As the pulses of the first digit enter the switch they immediately "step" the vertical wiper to a horizontal row corresponding to the first digit. After the proper row is selected, the wiper is rotated across another set of contacts until an idle line to the next stage of switching is located. The next set of dial pulses, representing the second digit, then "steps" the second stage in like manner. The process continues through however many stages are needed for a particular switch size.

As the name implies, a step-by-step switch uses *direct progressive control*: successive segments of a path through the switch are established as each digit is dialed. With progressive control, the control elements of the switch are integrated into the switching matrix. This feature is very useful for implementing a variety of switch sizes and allowing relatively easy expansion. A progressive control switch, however, has a number of significant limitations:

1. A call may be blocked even though an appropriate path through the switch exits but is not attempted because an unfortunate path gets selected in an early stage.

2. Alternate routing for outgoing trunks is not possible. That is, the outgoing line is directly selected by incoming dial pulses and can not be substituted.

3. Signaling schemes other than dial pulses (e.g., push-button tone signaling) are not directly usable.

4. Number translation is impossible.

In contrast to a step-by-step switch, a crossbar switch is one that uses centralized, common control for switch path selection. As digits are dialed, the

*A step-by-step switch is also referred to as a "Strowger switch" in honor of its inventor Almon B. Strowger.

SLEEVE BANK

SLEEVE WIPER

LINE BANK

LINE WIPER

WIPER CORDS

VERTICAL WIPER

VERTICAL COMMUTATOR (USED IN LINE FINDERS)

Figure 1.9. Step-by-step switching element. (Copyright 1977, Bell Telephone Laboratories. Reprinted by permission.)

control element of the switch receives the entire address before processing it. When an appropriate path through the switch is determined (which may involve number translation or alternate routing), the control element transfers the necessary information in the form of control signals to the switching matrix to establish the connection. The fundamental feature, and advantage, of a common control switch is that the control function implementation is separate from the switch implementation. Common control crossbar systems introduced the ability to assign logical addresses (telephone numbers) independently of physical line numbers.

The crosspoints of a crossbar switch (Figure 1.10) are mechanical contacts with magnets to set up and hold a connection. The term "crossbar" arises from the use of crossing horizontal and vertical bars to initially select the contacts. Once established, the switching contacts are held by electromagnets energized with direct current passing through the established circuit. When the circuit is opened, the loss of current causes the crosspoints to be released automatically.

Because of the operational limitations of progressive control, step-by-step switches were used primarily in smaller class 5 switching offices. Crossbar switches, on the other hand, were used predominantly in metropolitan areas and within the toll network. In some cases step-by-step switches were augmented with common control by receiving the digits into special control equipment. After processing the request, the control equipment generated pulses that set up a connection as if the switch was receiving dial pulses directly.

Figure 1.10. Crossbar switching element. (Copyright 1977, Bell Telephone Laboratories, Reprinted by permission.)

Stored Program Control

Step-by-step and crossbar switching systems use electromechanical components for both the switching matrix and the control elements. In some cases the electromechanical control elements in these switches represent rudimentary forms of special purpose digital computers. The hardwired electromechanical logic, however, has limited capabilities and is virtually impossible to modify.

A major milestone for telephony was established in 1965 when the Bell System installed its first computer-controlled switching system: the No. 1 ESS.* This switching system uses a stored program digital computer for its control functions. The stored program control (SPC) feature of the No. 1 ESS allowed the introduction of new features such as abbreviated dialing, call forwarding, call waiting, and three-way calling.

Stored program control not only provides significant advantages to end users, but also simplifies many administrative and maintenance tasks for the

*Computer controlled PBXs were avilable before 1965. The No. 1 ESS represents the first instance of computer control in the public network hierarchy.

operating companies. A large part of line administration that formerly required many manual modifications (mainframe cross-connects) can now be accomplished with changes in computer data tables of an SPC switch. Furthermore, physical line numbers are easily independent of the logical (directory) line numbers, thus making number changes easy. Some of the other benefits of SPC to an operating company are: automated record keeping, lower blocking probabilities, generation of traffic statistics, automated call tracing, and message unit accounting (per call charges as opposed to flat rate billing for unlimited local calling).

The switching matrix of the No. 1 ESS (and also the No. 2 ESS, No. 3 ESS, and No. 1A ESS) is implemented with electromechanical reed relays. Thus the term ESS, which stands for electronic switching system, refers in general to computer-controlled switching and not to the nature of the switching matrix itself. Similarly, the EAX terminology of GTE stands for electronic automated exchange and does not necessarily imply anything about the nature of the matrix. However, the No. 4 ESS, which was first installed in 1976, is a high-capacity toll switch using computer control and digital electronics for its switching matrix. Thus the No. 4 ESS is "electronic" in its control and its switching matrix. Furthermore, the DMS switches of Northern Telecom, the No. 5 EAX of GTE, and the No. 5 ESS of AT&T also utilize digital logic circuits for the crosspoint matrix.

Private Branch Exchanges

In the United States the term "private branch exchange" (PBX) refers generically to any switching system owned or leased by a business or organization to provide both internal switching functions and access to the public network. Thus a PBX in the United States may use either manual or automatic control. The term "PABX" is also used in the United States, and particularly in other countries, to refer specifically to automatically controlled PBXs.

The historical development of PBX systems has followed closely that of switches in the public network. PBXs with computerized control became available in 1963 (before the No. 1 ESS) when AT&T's No. 101 ESS was first installed. Since that time a large number of independent manufacturers have developed computer controlled PBXs. In fact, the PBX market has long been one of the most competitive and innovative businesses in all telecommunications.

The use of computer control for PBXs has proved to be even more rewarding for users than computer control in the public network. Not only are customized calling features (abbreviated dialing, etc.) usually provided, but numerous facilities for cost management are also available. Some of the more useful commonplace features in a PBX are the following:

1. Accounting summaries by individual employee or department.
2. Multiple classes of service with priorities and access restrictions to area codes, WATS lines, and so on.

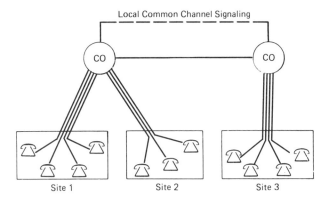

Figure 1.11. Centrex service to multiple sites.

3. Least-cost routing to automatically select tie lines, foreign exchange circuits, WATS, DDD, and so forth.
4. Automatic callback when circuits are available.
5. Traffic monitoring and analysis to determine the utilization of existing cicuits or to ascertain blocking probabilities and network cost effectiveness.

Centrex

Many of the features initially provided by PBXs are also offered by operating companies as centrex features. As indicated in Figure 1.11 centrex is a business customer service offering supported by switching equipment in the central office. Every telephone or data device at the customer premises has a dedicated channel to the switch in the central office. Originally, each channel implied a dedicated pair of wires. It is now more common to use multiplexing techniques (described in Section 1.2.5) to reduce the transmission costs. Nevertheless, from the central office point of view each centrex extension has a dedicated appearance at the CO with a unique public network telephone number. A software partition in the central office treats the centrex lines as a closed user group to provide the following basic features:

1. Direct dialing to centrex extensions from the public network.
2. Station-to-station calling using extension numbers as opposed to full, seven-digit public numbers.
3. Common voice user features such as forwarding, transfer, call waiting, pick up, and three-way calling.
4. Multiple sites with transparent numbering plans and features. With city-wide centrex the sites can be supported from multiple offices interconnected by common channel signaling (CCS) described in Section 1.2.10.

5. Centralized attendant/message desk with call origination information for informed processing of forwarded calls.
6. High availability because CO equipment and directly connected analog phones are powered at the CO with backup power sources.
7. Virtually unlimited growth.

Figure 1.12. Multipair cable. (Copyright 1977, Bell Telephone Laboratories. Reprinted by permission.)

1.2.4 Transmission Systems

Functionally, the communications channels between switching systems are referred to as trunks. These channels are implemented with a variety of facilities including: pairs of wires, coaxial cable, point-to-point microwave radio links, and optical fibers.

Open Wire

A classical picture of the telephone network in the past consisted of telephone poles with crossarms and glass insulators used to support uninsulated open-wire pairs. Except in rural environments, the open wire has been replaced with multipair cable systems or fiber. The main advantage of an open-wire pair is its relatively low attenuation (a few hundredths of a decibel per mile at voice frequencies). Hence, open wire is particularly useful for long rural customer loops. The main disadvantages are having to separate the wires with crossarms to prevent shorting, and the need for large amounts of copper. (A single open-wire strand has a diameter that is five times the diameter of a typical strand in a multipair cable. Thus open wire uses roughly 25 times as much copper as does cable.) As a result of maintenance problems and low electronics costs, open wire in rural environments has been mostly replaced with cable systems using amplifiers to offset attenuation on long loops.

Paired Cable

In response to overcrowded crossarms and high maintenance costs, multipair cable systems were introduced as far back as 1883. Today a single cable may contain anywhere from 6 to 2700 wire pairs. Figure 1.12 shows the structure of a typical cable. When telephone poles are used, a single cable can provide all the circuits required on the route—thereby eliminating the need for crossarms. More recently the preferred means of cable distribution is to bury it directly in the ground (buried cable) or use underground conduit (underground cable).

Table 1.4 lists the most common wire sizes to be found within paired cable systems. The lower-gauge (higher-diameter) systems are used for longer dis-

TABLE 1.4 Wire Gauge and Resistance of Common Paired Cable[a]

Gauge	Diameter (in.)	DC Resistance (Ω/1000 ft)
26	0.016	40.81
24	0.020	25.67
22	0.025	16.14
19	0.036	8.051

[a]Note that the loop resistance of a pair is twice the resistance of a single wire given in the table.

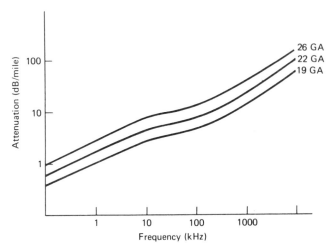

Figure 1.13. Attenuation versus frequency of common gauges of paired cable.

tances where signal attenuation and DC resistance can become limiting factors. Figure 1.13 shows attenuation curves [6] for the common gauges of paired cable as a function of frequency. An important point to notice in Figure 1.13 is that the cable pairs are capable of carrying much higher frequencies than required by a voice signal (approximately 3 kHz).

In the past, the exchange areas of the telephone network used paired cable almost exclusively for short-haul interoffice transmission. Up until the introduction of multiplexing techniques, described later in this chapter, each voice circuit (trunk) was carried on a separate pair of wires. Because of the dramatic drop in electronics costs interoffice transmission systems now use multiplexing to carry multiple channels on a single pair of wires or fiber.

Two-Wire versus Four-Wire

All wire-line transmission in the telephone network is based on transmission through pairs of wires. As shown in Figure 1.14, transmission through a single wire (with a ground return) is possible and has been used in the past. However, the resulting circuit is too noisy for customer acceptance. Instead, balanced pairs of wires as shown in Figure 1.15 are used with signals propagating as a voltage

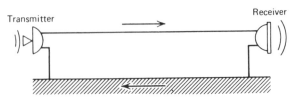

Figure 1.14. Single-wire transmission with ground return.

Figure 1.15. Two-wire transmission.

difference between the two wires. The electrical current produced by the difference signal flowing through the wires in opposite directions is called a "metallic current." In contrast, induced noise or interference is coupled equally into both wires of a pair and propagates along the pair in one direction. Current propagating in the same direction in both wires is referred to as common mode or longitudinal current. Longitudinal currents are not coupled into a circuit output unless there is an imbalance in the wires that converts some of the longitudinal signal (noise or interference) into a difference signal. Thus the use of a pair of wires for each circuit provides much better circuit quality than does single-wire transmission. Some older switching systems used single-wire (un-balanced) transmission to minimize the number of contacts. Unbalanced circuits were only feasible in small switches where noise and crosstalk could be controlled.

Virtually all subscriber loops in the telephone network are implemented with a single pair of wires.* The single pair provides for both directions of transmission. If users on both ends of a connection talk simultaneously, their conversations are superimposed on the wire pair and can be heard at the opposite ends. In contrast, wire-line transmission over longer distances, as between switching offices, usually involves two pairs of wires: one pair for each direction. Longer distance transmission requires amplification and most often involves multiplexing. These operations are implemented most easily if the two directions of transmission are isolated from each other. Thus interoffice trunks typically use two pairs of wires and are referred to as four-wire systems. The use of two-wire pairs does not necessarily imply the use of twice as much copper as a two-wire circuit. Four-wire systems are frequently used with some form of multiplexing to provide multiple channels in one direction on one wire pair. Thus, a net savings in copper results.

Sometimes the bandwidth of a single pair of wires is separated into two subbands that are used for the two directions of travel. These systems are referred to as "derived four-wire systems." Hence, the term "four-wire" has evolved to imply separate channels for each direction of transmission—even when wires may not be involved. For example, radio systems that necessarily use separate channels for each direction are also referred to as "four-wire systems."

The use of four-wire transmission had a direct impact on the switching systems of the toll network. Since toll network circuits were four-wire, the switches were designed to separately connect both directions of transmission.

*It is now commonplace to use fiber for the feeder portion of a subscriber loop but the drop to a residence is a single pair per telephone.

Hence, two paths through the switch were needed for each connection. A two-wire switch, as used in older analog end offices, required only one path through the switch for each connection.

Two-Wire to Four-Wire Conversion

At some point in a long distance connection it was always necessary to convert from two-wire transmission of local loops to four-wire transmission on long-distance trunks. Usually the conversion occured at the trunk interface of the end office switch. Sometimes, however, two-wire transmission was used for interoffice trunks so that the conversion occurred at the next switching office.

The interconnection of two-wire and four-wire facilities for a typical connection is shown in Figure 1.16. The basic conversion function is provided by hybrid circuits that couple the two directions of transmission as shown. Hybrid circuits have been traditionally implemented with specially interconnected transformers. More recently, however, electronic hybrids have been developed. Ideally a hybrid should couple all energy on the incoming branch of the four-wire circuit into the two-wire circuit, and none of the incoming four-wire signal should be transferred to the outgoing four-wire branch.

When the impedance matching network Z exactly matches the impedance of the two-wire circuit, near perfect isolation of the two four-wire branches can be realized. Impedance matching used to be a time-consuming, manual process and was therefore not commonly used. Furthermore, the two-wire circuits were usually switched connections so the impedance that had to be matched would change with each connection. For these reasons the impedances of two-wire lines connected to hybrids were rarely matched. The effect of an impedance mismatch is to cause an echo—the power level of which is related to the degree of mismatch. The effect of echoes on voice quality and the means of controlling them is discussed later in this chapter.

Loading Coils

The attenuation curves shown in Figure 1.13 indicate that the higher frequencies of the voice spectrum (up to 3 kHz) experience more attenuation than the lower frequencies. This frequency dependent attenuation distorts the voice signal and is referred to as amplitude distortion. Amplitude distortion becomes most significant on long cable pairs where the attenuation difference is greatest.

Figure 1.16. Interconnection of two-wire and four-wire circuits.

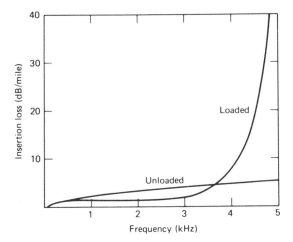

Figure 1.17. Effect of loading on 24-gauge cable pair.

The usual method of combating amplitude distortion on intermediate length (3–15 miles) wire pairs is to insert artificial inductance into the lines. The extra inductance comes from loading coils that are inserted at 3000, 4500, or 6000 ft intervals. Figure 1.17 shows the effect of loading coils on a 24-gauge loop. Notice that the voiceband response up to 3 kHz is greatly improved, but the effect on higher frequencies is devastating.

Prior to the introduction of wire line and fiber carrier systems loading coils were used extensively on exchange area interoffice trunks. Loading coils are also used on the longer, typically rural, subscriber loops. Here too, carrier systems have displaced most of the single pairs of wires being used on long routes.

1.2.5 Pair-Gain Systems

Providing service to rural subscribers has always been an expensive proposition because of the length of the routes involved and the small number of households to support the cost of both the initial installation and the maintenance. In the past, a common means of reducing the costs was to use party lines, which involved sharing of a wire pair among multiple households. A party line is particularly useful for satisfying new service requests on routes with no spare pairs but is obviously objectionable due to the lack of privacy and the lack of availability of the line.

A pair-gain system is an alternate approach to sharing pairs of wires that is much more acceptable to the users. This section describes two basic types of pair-gain systems: concentrators (remote switches) and multiplexers (carrier systems).

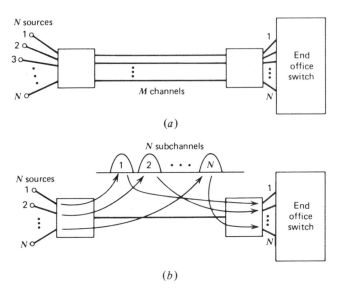

Figure 1.18. Pair gain systems: concentration and multiplexing. (a) Concentration ($n > M$). (b) Multiplexing.

Concentration

The first form of a pair-gain system in Figure 1.18 depicts a basic line concentration system. When viewed from the station set end of the system, a pair-gain system provides concentration by switching some number of active stations to a smaller number of shared output lines. At the other end of the system, deconcentration (expansion) occurs by switching from the shared lines to individual inputs of the switching office corresponding to the active stations. Expanding the traffic back to the original number of stations ensures that the system is operationally transparent to both the switch and the user. Notice that a definition of which end provides concentration and which end provides expansion is dependent on the point of view.

Since a concentrator is incapable of simultaneously connecting all stations it services, a certain amount of blocking is necessarily introduced by concentration. When the activity of individual stations is low enough, significant amounts of concentration can be achieved with acceptable blocking probabilities. For example, 40 stations that are each active only 7.5% of the time can be concentrated onto 10 lines with a blocking probability of 0.001.* This is an acceptable degradation in service since an equally active called station is busy 75 times as often.

Notice that a concentration system requires the transfer of control informa-

*A discussion of traffic analysis is provided in Chapter 10—from which this result can be obtained.

tion between the concentrator/expander switch terminals. When one end of the system establishes a new connection to one of the shared lines, the other end must be informed to set up the appropriate reverse connection.

Multiplexing

As shown in Figure 1.13 the inherent bandwidth of a typical wire pair is considerably greater than that needed for a single voice signal. Thus, multiplexing can be used to carry multiple voice channels on a single pair of wires. The increase in attenuation implied by the higher frequencies is offset by amplifiers in the multiplex equipment and at periodic points in the transmission lines. The particular multiplexing technique shown in Figure 1.18b is a frequency division multiplex system. Another form of multiplexing, time division multiplexing of digital voice signals, is also used in pair-gain systems and is discussed later in this chapter.

As shown in Figure 1.18b, there is a one-to-one relationship between the customer lines and the subchannels of the multiplexer. Thus, unlike the concentration system, there is no possibility of blocking in a multiplexing type of pair-gain system. Also, there is no need to transfer switching information since the same one-to-one relationship defines the correspondence between customer lines at one end and switching office lines at the other end. A major drawback of multiplexing pair-gain systems is that the subchannels are highly underutilized if the sources are relatively inactive. In these situations a combination of concentration and multiplexing can be used. A digital subscriber loop multiplexer described in a later section is an example of such a system.

1.2.6 Modulation/Multiplexing

The introduction of cable systems into the transmission plant to increase the circuit packing density of open wire is one instance of multiplexing in the telephone network. This form of multiplexing, referred to as "space division multiplexing," involves nothing more than bundling more than one pair of wires into a single cable. The telephone network uses two other forms of multiplexing, both of which use electronics to pack more than one voice circuit into the bandwidth of a single transmission medium. Analog frequency division multiplexing (FDM) has been used extensively in the point-to-point microwave radios of the network and to a much lesser degree on some coaxial cable and wire line systems. Digital time division multiplexing (TDM) is now the dominant form of multiplexing used in the U.S. network and is discussed briefly in Section 1.3.2 and extensively in Chapters 4, 7, and 8.

Frequency Division Multiplexing

As indicated in Figure 1.18b, an FDM system divides the available bandwidth of the transmission medium into a number of narrower bands or subchannels. Individual voice signals are inserted into the subchannels by amplitude modulating appropriately selected carrier frequencies. As a compromise be-

tween realizing the largest number of voice channels in a multiplex system and maintaining acceptable voice fidelity, the telephone companies have established 4 kHz as the standard bandwidth of a voice circuit.* If both sidebands produced by amplitude modulation are used (as in N1 or N2 carrier systems on paired cable), the subchannel bandwidth is 8 kHz, and the corresponding carrier frequencies lie in the middle of each subchannel. Since double-sideband modulation is wasteful of bandwidth, single-sideband (SSB) modulation is used whenever the extra terminal costs are justified. The carrier frequencies for single-sideband systems lie at either the upper or lower edge of the corresponding subchannel, depending on whether the lower or upper sideband is selected. The A5 channel bank multiplexer of AT&T uses lower-sideband modulation.

FDM Hierarchy

In order to standardize the equipment in the various broadband transmission systems of the network, the Bell System established an FDM hierarchy as provided in Table 1.5. Each level of the hierarchy is implemented using a set of standard FDM modules. The multiplex equipment is independent of particular broadband transmission media.

All multiplex equipment in the FDM hierarchy use single-sideband modulation. Thus, every voice circuit requires approximately 4 kHz of bandwidth. The lowest-level building block in the hierarchy is a channel group consisting of 12 voice channels. A channel group multiplex uses a total bandwidth of 48 kHz. Figure 1.19 shows a block diagram of an A5 channel group multiplexer, the most common A-type channel bank used for first-level multiplexing. Twelve modulators using 12 separate carriers generate 12 double-sideband signals as indicated. Each channel is then bandpass filtered to select only the lower sideband of each double-sideband signal. The composite multiplex signal is produced by superposing the filter outputs. Demultiplex equipment in a receiving terminal uses the same basic processing in reverse order.

TABLE 1.5 FDM Hierarchy of the Bell Network

Multiplex Level	Number of Voice Circuits	Formation	Frequency Band (kHz)
Voice channel	1		0–4
Group	12	12 voice circuits	60–108
Supergroup	60	5 groups	312–552
Mastergroup	600	10 supergroups	564–3,084
Mastergroup Mux	1,200–3,600	Various	312,564–17,548
Jumbogroup	3,600	6 mastergroups	564–17,548
Jumbogroup Mux	10,800	3 jumbogroups	3,000–60,000

*Actually, the usable bandwidth of a voice channel is closer to 3 kHz owing to guard bands needed by FDM separation filters.

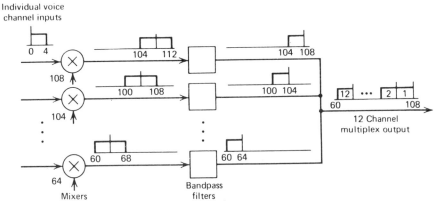

Figure 1.19. A5 channel bank multiplexer.

Notice that a sideband separation filter not only removes the upper sideband, but also restricts the bandwidth of the retained signal: the lower sideband. These filters therefore represent a basic point in the telephone network that defines the bandwidth of a voice circuit. Since frequency division multiplexing was used on all long-haul analog circuits, long-distance connections provide somewhat less than 4 kHz of bandwidth. (The loading coils discussed previously also produce similar bandwidth limitations into a voice circuit.)

As indicated in Table 1.5, the second level of the FDM hierarchy is a 60 channel multiplex referred to as a supergroup. Figure 1.20 shows the basic implementation of an LMX group bank which multiplexes five first-level channel groups. The resulting 60 channel multiplex output is identical to that obtained when the channels are individually translated into 4 kHz bands from 312 to 552 kHz. Direct translation requires 60 separate single-sideband systems with 60 distinct carriers. The LMX group bank, however, uses only five single-sideband systems plus five lower-level modules. Thus two-stage multiplexing, as implied by the LMX group bank, requires more total equipment but achieves economy through the use of common building blocks. (Engineers at AT&T Laboratories also developed a multiplexer that generates a supergroup directly [7].)

Because a second-level multiplexer packs individual first-level signals together without guard bands, the carrier frequencies and bandpass filters in the LMX group bank must be maintained with high accuracy. Higher-level multiplexers do not pack the lower-level signals as close together. Notice that a mastergroup, for example, does not provide one voice channel for every 4 kHz of bandwidth. It is not practical to maintain the tight spacing between the wider bandwidth signals at higher frequencies. Furthermore, higher-level multiplex signals include pilot tones to monitor transmission link quality and aid in carrier recovery.

1.2.7 Wideband Transmission Media

Wire pairs within multipair cables have usable bandwidths that range from a little under 1 mHz up to about 4 mHz depending on the length, the wire gauge, and the type of insulation used on the pairs. Multiplexed transmissions on these wire pairs consequently have capacities that range from 24 channels (on analog N3 or digital T1 carrier systems) up to 96 channels (on digital T2 carrier systems). In contrast, an analog L5E coaxial cable system can carry 13,200 voice channels. Experimental optical fiber systems have been developed that carry 30,000 voice channels with much larger capacities possible when higher speed supporting electronics become available. Because optical transmitters (LEDs and lasers) function best in a pulsed mode of operation fiber systems are primarily digital in nature. Digital fiber optic transmission systems are described in Chapter 8. This section discusses coaxial cable and point-to-point microwave radio systems.

Coaxial Cable

Coaxial cable systems were used predominantly to satisfy long-haul requirements of the toll network. The first commercial system was installed in 1941 for transmission of 480 voice circuits over a 200 mile stretch between Minneapolis, Minnesota and Stevens Points, Wisconsin [7]. To combat attenuation, repeater amplifiers were installed at 5.5 mile intervals. Considering the maximum capacity of 12 voice circuits on open-wire or cable at the time, the introduction of "coax" was a significant development. Since that time coaxial cable capacity has been steadily increased by (1) using larger diameter cables (0.375 in.) to reduce attenuation, (2) decreasing the distance between repeaters, and (3) improving the noise figure, linearity, and bandwidth of the repeater amplifiers.

A summary of the analog coaxial cable systems used in the Bell System is provided in Table 1.6. Notice that each system reserves one pair of tubes as

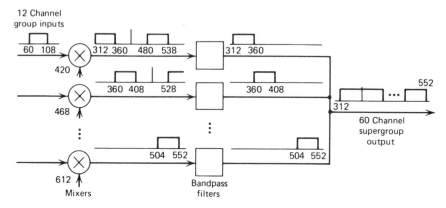

Figure 1.20. LMX group bank multiplexer.

TABLE 1.6 Coaxial Cable Systems in the Bell Network

System Designation	Pairs per System[a]	Signal Designation	Repeater Spacing (miles)	Capacity per Pair	Total Capacity
L1	3/6	Mastergroup	8	600	1,800
L3	5/6	Mastergroup Mux	4	1,860	9,300
L4	9/10	Jumbogroup	2	3,600	32,400
L5	10/11	Jumbogroup Mux	1	10,800	108,·00

[a]The number of pairs are shown as working/total.

spares in the event of failure—a particularly important consideration since each tube carries a high volume of traffic. Because of wider bandwidths and greater repeater spacings optical fibers have replaced coax as the wideband transmission media of choice.

Microwave Radio

Much of the impetus for terrestial microwave radio systems came from the need to distribute television signals nationwide. However, as the volume of long-distance traffic increased steadily, radio systems also proved to be the most economical means of distributing voice circuits in the long-distance network. Beginning in 1948, when the first system was installed between New York and Boston, the number of microwave radio systems grew to supply 60% of the voice circuit miles in the U.S. toll network in 1980 [3]. It was just a few years after that that optical fibers began to take over for high-density interoffice routes. Radio systems require line-of-sight transmission with repeater spacings typically 26 miles apart. The major advantage of radio systems is that a continuous right-of-way is not required—only small plots of land spaced 20–30 miles apart for towers and equipment shelters. A major cost of guided transmission, for example, wire pairs, coax, or fiber, is the right-of-way costs. In many metropolitan areas, microwave routes are heavily congested and cannot be expanded with presently allocated frequency bands. Thus coaxial cable systems and fiber optics are, sometimes, the only alternatives for high-capacity transmission.

The frequency bands allocated by the FCC for common carrier use in the United States are listed in Table 1.7. Of these bands, 4 and 6 GHz have been the most popular. The 2-GHz band has not been used extensively because the relatively narrow allocated channel bandwidths do not permit implementation of economical numbers of voice circuits. The basic drawback of the 11-GHz band is its vulnerability to rain attenuation. However, 11-GHz radios are used in some short-haul applications.

The microwave radio systems of the Bell network are listed in Table 1.8. Notice that each radio system is designed to carry one of the multiplex hierarchies described previously. All of these radios except the AR-6A use low-index frequency modulation (FM) of the signal generated by the FDM

TABLE 1.7 Microwave Frequencies Allocated for Common Carrier use in the United States

Band (MHz)	Total Bandwidth (MHz)	Channel Bandwidths (MHz)
2,110–2,130	20	3.5
2,160–2,180	20	3.5
3,700–4,200	500	20
5,925–6,425	500	30
10,700–11,700	1000	40,20

multiplexer equipment. Thus, the FM radios transmit the SSB-FDM signal as a baseband signal with a bandwidth as indicated in Table 1.5. FM modulation was chosen to permit the use of nonlinear power amplifiers in the transmitters and to take advantage of FM signal-to-noise ratio performance.

Examination of Tables 1.7 and 1.8 indicates that the bandwidth utilized per voice circuit in TD-2 radios is 13.3 kHz and 14.3 kHz in TH-3 radios. The indicated channel densities have been achieved through continual refinement of the FM equipment. Although small increases in circuit capacity may be possible, the state of the art has approached the theoretical maximum efficiency for FM. Hence FM will always require a significant increase in the 4 kHz bandwidth of the individual single-sideband voice circuits. In contrast, the single sideband AR-6A radio provides 6000 voice circuits in the 30-MHz channels at 6 GHz.

Since many voice circuits are carried by each radio channel, microwave systems usually include extra equipment and extra channels to maintain service despite outages that may result from any of the following:

1. Atmospheric-induced multipath fading
2. Equipment failures
3. Maintenance.

TABLE 1.8 Bell System Analog Microwave Radios

System	Band (GHz)	Voice Circuits	Application
TD-2	4	600–1500	Long haul
TD-3	4	1200	Long haul
TH-1	6	1800	Short/long haul
TH-3	6	2100	Short/long haul
TM-1	6	600–900	Short/long haul
TJ	11	600	Short haul
TL-1	11	240	Short haul
TL-2	11	600–900	Short haul
AR-6A	6	6000	Long haul (SSB)

On some routes, the most frequent source of outage in a microwave radio system arises from multipath fading. Figure 1.21 depicts a simple model of a multipath environment arising as a result of atmospheric refraction. As indicated, the model involves two rays: a primary ray and a delayed secondary ray. If the secondary ray arrives out of phase with respect to the primary ray, the composite signal is effectively attenuated. The amount of attenuation is dependent on both the magnitude and the phase of the secondary ray. Quite often only nominal amounts of fading occur and can be accommodated by excess signal power in the transmitter, called a fade margin. In some instances, however, the received signal is effectively reduced to zero, which implies that the channel is temporarily out of service.

Frequency Diversity

Fortunately, exceptionally deep fades normally affect only one channel (carrier frequency) at a time. Thus, a backup channel including a spare transmitter and a spare receiver can be used to carry the traffic of a faded primary channel. Selection of and switching to the spare channel is performed automatically without a loss of service. This remedy for multipath fading is referred to as frequency diversity. Notice that frequency diversity also provides hardware backup for equipment failures.

A fully loaded TD-3 radio system uses 12 channels: 10 main channels and two backup channels for protection. This is referred to variously as 2 for 10, 10 by 2, or 10×2 protection switching. Some short-haul systems use one-for-one protection switching because it is simpler to implement. However, since only half of the required bandwidth is actually carrying traffic new systems with one-for-one protection are only allowed in uncongested environments.

Except in maintenance situations, protection switching must be automatic in order to maintain service continuity. A typical objective is to restore service within 30 msec to minimize noticeable effects in the message traffic. A more critical requirement is to restore service before the loss of signal is interpreted by some signaling schemes as a circuit disconnect. Inadvertent disconnects occur if the outage lasts for more than 1 to 2 sec.

Figure 1.21. Two-ray model of multipath propagation.

Space Diversity

Since deep fades only occur when a secondary ray arrives exactly out of phase with respect to a primary ray, it is unlikely that two paths of different lengths experience fading simultaneously. Figure 1.27 depicts a technique, called space diversity, using different path lengths to provide protection against multipath fading. As indicated, a single transmitter irradiates two receive antennas separated by some distance on the tower. Although the path length difference may be less than a meter, this difference is adequate at microwave frequencies, which have wavelengths on the order of tenths of meters.

Rain is another atmospherically based source of microwave fading. As already mentioned, rain attenuation is a concern mostly in higher-frequency radios (11 GHz and above). Unfortunately neither frequency diversity (at the high frequencies) nor space diversity provides any protection against rain fades.

Satellites

Following the April 1965 launch of the Soviet Union's Molniya domestic communications satellite and the first international communications satellite, INTELSAT I, the use of satellites for international telephone traffic grew phenomenally. The 1970s and early 1980s also produced significant use of satellites in the United States for television program distribution and for corporate voice and data networks. The first domestic system in North America was the Canadian Anik A in 1972 followed by Western Union's Westar system for U.S. service in 1974 [8].

In one sense a satellite system is a microwave radio system with only one repeater: the transponder in outer space. In fact, some satellite systems use the same 4- and 6-GHz frequency bands used by terrestrial microwave radios. In another sense, however, the broadcast nature of the down link offers additional opportunites for new services not available from point-to-point terrestrial systems. Distribution of network television programming is one application particularly suited to the broadcast nature of satellites. Direct broadcast satellites (DBS) to receive only home receivers have been successfully introduced in numerous countries around the world [9]. In the United States, the large installed base of TV cable systems has seemingly precluded commercial viability of DBS.*

Another application that is well suited to satellites is communications with mobile stations. The INMARSAT system, for example, began supporting the maritime industry in 1982 and was augmented with digital Skyphone [10] for aeronautical telephone service in 1989.

One drawback to satellite communications is the inherent propagation delay of the long transmission path. For a stationary satellite, this delay (not including ground links) is 250 msec up and down. A complete circuit with satellite links for both directions of travel therefore implies greater than a one-half second round-

*A DBS system is designed to use receiver antennae that are less than a meter in diameter. The home satellite receiver systems that are common in the United States use 3- to 5-m antennae to receive (intercept) commercial program distribution.

Figure 1.22. Space diversity.

trip propagation time. Delays of this magnitude are noticeable in a voice conversation but not prohibitive. The effects of the propagation delays can be alleviated somewhat by pairing each satellite circuit with a ground based circuit in the opposite direction. Thus, the round trip delay involves only one satellite link. In this manner, existing terrestial circuits can be used advantageously while demands for more capacity are economically satisfied by new satellite systems.

Long transmission delays can have significant, almost debilitating, effects on some data communications circuits. The most popular form of error control for data communications involves retransmission of errored data blocks. In the simpler data communications protocols only one block at a time is transmitted; the source waits for a response from the opposite end of the circuit before sending the next block. Thus, transmission delays much greater than a block transmission time lead to low throughputs and inefficient channel utilizations.

As is the case with point-to-point microwave radio, fiber optic transmission has displaced the use of satellites for high-density, domestic and international telephone communications. Hence the use of satellites is primarily directed at thin-route telephone and data applications, mobile systems, and broadcast applications of any type.

1.2.8 Transmission Impairments

One of the most difficult aspects of designing an analog telephone network is determining how to allocate transmission impairments to individual subsystems within the network. Using subjective evaluations by listeners, certain objectives for end-to-end transmission quality can be established in a relatively straightforward manner [11]. After tempering the goals with economic feasibility, the end-to-end objectives can be established. However, considering the myriad of equipment types and connection combinations in a network, designing the individual network elements to meet these objectives in all cases is a complex problem. A great deal of credit is due the Bell System for having developed a nationwide analog network with the level of consistent performance it had.

The major factors to be considered in establishing transmission objectives for an analog network are: signal attenuation, noise, interference, crosstalk, distortion, echos, singing, and various modulation and carrier related imperfections.

Signal Attenuation
Subjective listening tests have shown that the preferred acoustic-to-acoustic loss [12] in a telephone connection should be in the neighborhood of 8 dB. A study

of local telephone connections [13] demonstrates that the typical local call had only 0.6 dB more loss than ideal. Surveys of the toll network [14] indicate that the average toll connection had an additional 6.7 dB of loss. This same survey also showed that the standard deviation of loss in toll connections was 4 dB (most of which was attributable to the local loops).

Since trunks within the toll network used amplifiers to offset transmission losses it would have been straightforward to design these trunks with zero decibel nominal insertion loss. However, as discussed later, echo and singing considerations dictated a need for certain minimum levels of net loss in most analog trunk circuits.

Interference

Noise and interference are both characterized as unwanted electrical energy fluctuating in an unpredictable manner. Interference is usually more structured than noise since it arises as unwanted coupling from just a few signals in the network. If the interference is intelligible, or nearly so, it is referred to as crosstalk.* Some of the major sources of crosstalk are: coupling between wire pairs in a cable, the effects of nonlinear components on FDM signals, inadequate filtering or carrier offsets in FDM equipment, and intersymbol interference in time division multiplex equipment.

Crosstalk, particularly if intelligible, is one of the most disturbing and undesirable imperfections that can occur in a communications network. Crosstalk in analog systems is particularly difficult to control since voice signal power levels vary considerably (i.e., across a dynamic range of 40 dB). The absolute level of crosstalk energy from a high-level signal must be small compared to a desired low-level signal. In fact, crosstalk is most noticeable during speech pauses when the power level of the desired signal is zero.

Two of the most common forms of crosstalk of concern to communications engineers are "near-end crosstalk" (NEXT) and "far-end crosstalk" (FEXT). Near-end crosstalk refers to coupling from a transmitter into a receiver at a common location. Often this form of crosstalk is most troublesome because of a large difference in power levels between the transmitted and received signals. Far-end crosstalk refers to unwanted coupling into a received signal from a transmitter at a distant location. Both forms of crosstalk are illustrated in Figure 1.23.

Noise

The most common form of noise analyzed in communications systems is white noise with a Gaussian (normal) distribution of amplitude values. This type of noise is both easy to analyze and easy to find since it arises as thermal noise in all electrical components. Battery systems used to power customer loops are also a

*Crosstalk is also used to characterize signal-like interferences in nonvoice networks. For example, crosstalk in a data circuit would refer to an interfering signal being coupled in from another similar data circuit.

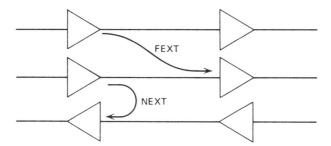

Figure 1.23. Near-end and far-end crosstalk.

source of this type of noise. White noise is truly random in the sense that a sample at any instant in time is completely uncorrelated to a sample taken at any other instant in time. The other most common forms of noise in the telephone network are impulse noise and quantization noise in digital voice terminals (Chapter 3). Impulse noise can occur from switching transients in electromechcanical switching offices or from rotary dial telephones. Step-by-step switches are the most frequent culprits. More modern electromechanical switches which use glass encapsulated reed relays for cross-points produce much less noise. Whereas white noise is usually quantified in terms of average power, impulse noise is usually measured in terms of so many impulses per second. Impulse noise is usually of less concern to voice quality than background (white noise. However, impulse noise tends to be the greatest concern in a data communications circuit.

The power level of any disturbing signal, noise or interference, is easily measured with a root mean square voltmeter. However, disturbances at some frequencies within the passband of a voice signal are subjectively more annoying than others. Thus, more useful measurements of noise or interference power in a speech network take into account the subjective effects of the noise as well as the power level. The two most common such measurements in telephony use: a C-message weighting curve or a psophometric weighting curve as shown in Figure 1.24. These curves essentially represent filters that weight the frequency spectrum of noise according to its annoyance effect to a listener. C-message weighting represents the response of the 500 type telephone set. As far as perceived voice quality is concerned, only the noise that gets passed by the telephone set is important. Notice that disturbances between 1 and 2 kHz are most perceptible. C-message weighting is used in North America while psophometric weighting is the European (CCITT) standard.

A standard noise reference used by telephone engineers is 1 pW, which is 10^{-12} W or -90 dBm (dBm is power in decibels relative to a milliwatt). Noise measured relative to this reference is expressed as so many decibels above the reference (dBrn). Thus, a noise level of 30 dBrn corresponds to -60 dBm or 10^{-9} W of power. If the readings are made using C-message weighting, the power level is expressed by the abbreviation dBrnC. Similarly, psophometrically

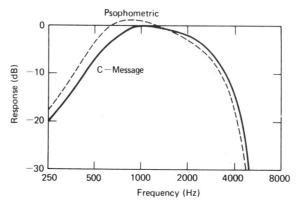

Figure 1.24. C-message and psophometic weighting.

weighted picowatts are expressed by the abbreviation pWp. The relationships between various noise power measurements are given in Table 1.9.

The quality of an analog voice circuit is usually not specified in terms of the classical signal-to-noise ratio. The reason is that relatively low levels of noise or interference are noticeable during pauses in speech when there is no signal. On the other hand, high levels of noise can occur during speech and be unnoticeable. Thus, absolute levels of noise are more relevant than signal-to-noise ratios for specifying voice quality.* The objectives for maximum noise levels in the analog AT&T network are 28 dBrnC for connections up to 60 miles in length and 34 dBrnC for 1000 mile circuits.†

TABLE 1.9 Relationships between Various Noise Measurements

	To Convert	
From	To	
dBm	dBrn	Add 90 dB
dBm 3 kHz flat	dBrnC	Add 88 dB
dBm 3 kHz flat	dBp	Add 87.5 dB
dBrn 3 kHz flat	dBrnC	Substract 2 dB
dBC	dBp	Subtract 0.5 dB
pW 3 kHz flat	pWp	Multiply by 0.562

*It is a common practice in the industry to specify the quality of a voice circuit in terms of a test tone to noise ratio. However, the test tone must be at a specific power level so the ratio, in fact, specifies absolute noise power.

†These noise power values are related to a particular point in a circuit called a "zero-transmission-level point," discussed later.

Distortion

In a previous section signal attenuations were considered with the tacit assumption that the received waveforms were identical in shape to the source waveforms but merely scaled down in amplitude. Actually, the received waveforms generally contain certain distortions not attributable to external disturbances such as noise and interference but which can be attributed to internal characteristics of the channel itself. In contrast to noise and interference, distortion is deterministic; it is repeated every time the same signal is sent through the same path in the network. Thus distortions can be controlled or compensated for once the nature of the distortion is understood.

There are many different types and sources of distortion within the telephone network. The telephone companies have minimized those types of distortion that most affect the subjective quality of speech. Later on they also became concerned with distortion effects on data transmission. Some distortions arise from nonlinearities in the network such as: carbon microphones, saturating voice-frequency amplifiers, and unmatched compandors (Chapter 3). Other distortions are linear in nature and are usually characterized in the frequency domain as either amplitude distortion or phase distortion.

Amplitude distortion refers to attenuating some frequencies in the voice spectrum more than others. The loading coils discussed earlier represent one means of eliminating amplitude distortion on long voice-frequency wire pairs. Amplitude distortion is also introduced by spectrum limiting filters in FDM equipment. Ideally these filters should uniformly pass all voiceband frequencies up to 4 kHz and reject all others. Practical designs, however, imply the need for gradual attenuation "roll-offs" beginning at about 3 kHz. Figure 1.25 shows the attenuation vs. frequency response of a typical analog toll connection.

Phase distortion is related to the delay characteristics of the transmission

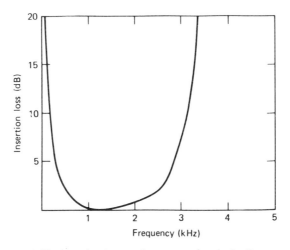

Figure 1.25. Insertion loss vs. frequency of typical toll connection.

medium. Ideally a transmission system should delay all frequency components in a signal uniformly so the proper phase relationships exist at the receiving terminal. If individual frequency components experience differing delays, the time domain representation at the output becomes distorted because superposition of the frequency terms is altered at the output. For reasons not discussed here the delay of an individual frequency components is usually referred to as its envelope delay. For a good explanation of envelope delay see reference [15].

Uniform envelope delay relates to a phase response that is directly proportional to frequency. Thus systems with uniform envelope delay are also referred to as linear phase systems. Any deviation from a linear phase characteristic is referred to as phase distortion. The perceptual effects of phase distortion to a voice signal, however, are small. Thus only minimal attention need be given to the phase response of a voice network. The phase response and corresponding envelope delay of a typical toll connection is shown in Figure 1.26.

In addition to the distortions just mentioned, carrier systems sometimes introduce other frequency-related distortions such as frequency offsets, jitter, phase hits, and signal dropouts. The effects of these imperfections and phase distortion are adequately controlled for voice traffic but present difficulties for high-rate voiceband data traffic.

Echos and Singing

Echos and singing both occur as a result of transmitted signals being coupled into a return path and fed back to the respective sources. The most common cause of the coupling is an impedance mismatch at a four-wire to two-wire hybrid. As shown in Figure 1.27, mismatches cause signals in the incoming branch of a four-wire circuit to get coupled into the outgoing branch and return to the source. It is impractical to provide good impedance matches at this point

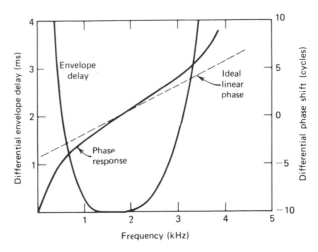

Figure 1.26. Envelope delay and phase response of typical toll connection.

Figure 1.27. Generation of echos at two-wire to four-wire interface.

for all possible connections since the switched network involves many different local loops and trunks—each with its own characteristic impedance.

If only one reflection occurs, the situation is referred to as "talker echo." If a second reflection occurs, "listener echo" results. When the returning signal is repeatedly coupled back into the forward path to produce oscillations singing occurs. Basically, singing results if the loop gain at some frequency is greater than unity. If the loop gain is only slightly less than unity, a near-sing condition causes damped oscillations. Singing and near-singing conditions have a disturbing effect on both the talker and the listener. Talker echo is usually the most noticeable and troublesome.

The degree of echo annoyance experienced by a talker is dependent on both the magnitude of the returning signal and the amount of delay involved [16, 17]. On short connections the delay is small enough that the echo merely appears to the talker as natural coupling into his ear. In fact, a telephone is purposely designed to couple some speech energy (called sidetone) into the earpiece. Otherwise, the telephone seems dead to a talker. Near-instantaneous echos merely add to the sidetone and go unnoticed. As the roundtrip delay increases, however, it becomes necessary to increasingly attenuate the echos to eliminate the annoyance to a talker. Hence, long-distance circuits require significant attenuation to minimize echo annoyance. Fortunately, an echo experiences twice as much attenuation as does the signal since it traverses twice the distance, Intermediate-length connections are typically designed with 2 to 6 dB of path attenuation depending on the delay. All transmission links within the Bell System were designed with specific amounts of net loss call via net loss (VNL) that depended on the length of the link and the position in the hierarchy [17]. In general, the VNL network design established end-to-end attenuation in proportion to the length of the circuit.

Connections that produce more than 45 msec of roundtrip delay (representing 1800 miles of wire) require more attenuation for echo control than can be tolerated in the forward path. In these cases one of two types of devices is used to control the echo: an echo suppressor or an echo canceler.

As shown in Figure 1.28 an echo suppressor operates on four-wire circuits by measuring the speech power in each leg and inserting a large amount of loss (35 dB typically) in the opposite leg when the power level exceeds a threshold. Thus, a returning echo is essentially blocked by the high level of attenuation.

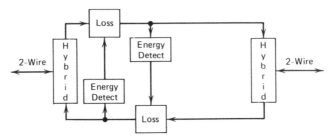

Figure 1.28. Echo suppressor.

Notice that an echo suppressor converts a full-duplex circuit into a half-duplex circuit with energy sensing being the means of turning the line around.

One drawback of echo suppressors for voice circuits is that they may clip beginning portions of speech segments. If a party at one end of a connection begins talking at the tail end of the other party's speech, the echo suppressor does not have time to reverse directions. Echo suppressors with good performance are able to reverse directions in 2 to 5 msec [16]. When fast release of the attenuation is used split echo suppressors are necessary for good performance. A split echo suppressor is one that separates the echo control of each leg so the loss insertion of each leg is closest to the point of echo occurrence.

A second, and much preferred, form of echo control is echo cancellation [18]. As shown in Figure 1.29 an echo canceler operates by simulating the echo path to subtract a properly delayed and attenuated copy of a transmitted signal from the receive signal to remove (cancel) echo components. Thus echo cancellation requires training to determine how much delay and attenuation is needed to simulate the echo characteristics of the circuit. Notice in Figure 1.29 that echos are canceled close to the source so that delays in the echo canceler are minimized. The important feature of an echo canceler is that it maintains a full-duplex circuit so clipping does not occur. Satellite circuits with greater than 500 msec of roundtrip delay require echo cancelers for acceptable performance [19].

Another method of echo control involves impedance matching of at the

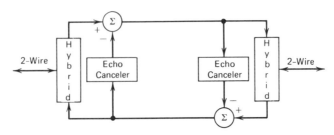

Figure 1.29. Echo canceler.

hybrids to reduce the magnitude of the echo. Some electronic hybrids provide dynamic balancing to automatically eliminate echoes. In fact, a common way of implementing the impedance matching is to build a echo canceler with near zero delay. These circuits eliminate, or greatly reduce, echoes occurring at the associated hybrid but do not eliminate echoes that may occur elsewhere in the network. For a detailed discussion of all types of echo control see reference [20].

In general, the procedures used to control echos also control singing. On some fairly short connections, however, no echo control is necessary, and singing may become a problem. Singing can also occur in idle circuits and still be a problem if it overloads amplifiers or causes intermodulation in frequency division multiplex equipment.

1.2.9 Power Levels

As indicated in previous paragraphs, voice signal power in long-distance connection needs to be rigidly controlled. The delivered signal power must be high enough to be clearly audible but, at the same time, not be so strong that circuit instabilities such as echo and singing result.

To maintain rigid control on the end-to-end power level of a circuit involving a variety of transmission systems, telephone companies necessarily control the net attenuation and amplification of each transmission system. These systems are designed for a certain amount of net loss (VNL) as described previously.

To administer the net loss of transmission links, the transmission levels of various points in a transmission system are specified in terms of a reference point. CCITT recommendations call this point the zero-relative-level point and the North American term is a zero-transmission-level point (0-TLP). The reference point may not exist as an accessible point but has long been considered to be at the sending end terminal of a two-wire switch. In North America the sending end of a four-wire switch is defined to be a -2-dB TLP. Hence, a 0-dB TLP is only a hypothetical point on a four-wire circuit. Nevertheless, it is useful in relating the signal level at one point in the circuit to the signal level at another point in the circuit.

If a 0-dBm (1-mW) test tone is applied at a 0-TLP, the power level at any other point in the circuit is determined directly (in decibels referred to 1 mW) as the TLP value at that point. It should be emphasized, however, that TLP values do not specify power levels—only the gain or loss at a point relative to the reference point.

Signal power levels or noise power levels are not normally expressed in terms of local measured values. Instead, powers are expressed in terms of their value at the 0-TLP. For example, if an absolute noise power of 100 pW (20 dBrn or -70 dBm) is measured at a -6-dB TLP, it is expressed as 26 dBrn0. (The "0" indicates that the specification is relative to the 0-TLP.) If noise power is measured with C-message weighting, it is designated as so many dBrnC0. Similarly, psophometric weighted noise is commonly expressed in units of picowatts psophometrically weighted (dBm0p or pWp0).

Figure 1.30. TLP values for Example 1.1.

Example 1.1. Using Figure 1.30 for reference determine each of the following: (1) The signal power to be applied at point B to determine if points A and C are at the proper levels. (2) The amount of gain (loss) a signal experiences when propagating from A to C. (3) The amount of noise that would be measured at C if 27 dbrnC of absolute noise is measured at B and no additional noise occurs on the B to C link.

Solution. (1) Because point B is a -13-dB TLP, the proper test tone level is -13 dBm (0.05 mW). (2) Because the TLP values drop by 2 dBm there is 2-dB net loss from A to C. (3) An absolute measurement of 27 dBrnC at B is 40 dBrnC0. This is also 40 dBrnC0 at C. The absolute noise power measured at C would be $40 - 4 = 36$ dBrnC.

To put signal and noise powers in perspective, a survey of voice signals in the Bell System [21] indicated the average speech signal has -16 dBm0 of power. In comparison the noise power objective for a 1000 mile analog circuit was 34 dBrnC0 (-56 dBmC0).

1.2.10 Signaling

The signaling functions of a telephone network refer to the means for transferring network related control information between the various terminals, switching nodes, and users of the network. There are two basic aspects of any signaling system: specially encoded electrical waveforms (signals), and how these waveforms should be interpreted. The most common control signals to a user are dial tone, ringback, and busy tone. These signals are referred to as call progress tones and may never change. The signaling procedures used internally to the network are not constrained by user convention and are often changed to suit particular characteristics of transmission and switching systems. As a result the analog network used a wide variety of signaling schemes to transfer control information between switching offices.

Signaling Functions

Signaling functions can be broadly categorized as belonging to one of two types: supervisory or information bearing. Supervisory signals convey status or control of network elements. The most obvious examples are: request for service (off-hook), ready to receive address (dial tone), call alerting (ringing), call termination (on-hook), request for an operator (hook flash), called party ringing (ringback), and network or called party busy tones. Information bearing signals

include: called party address, calling party address, and toll charges. In addition to call related signaling functions, switching nodes communicate between themselves and network control centers to provide certain functions related to network management. Network related signals may convey status such as maintenance test signals, all trunks busy, equipment failures, or they may contain information related to routing and flow control. Chapter 7 discuses some of the basic considerations of network management for routing and flow control.

In-Channel Signaling

Signals are transmitted with one of two basic techniques: in-channel signaling or common channel signaling. In-channel signaling (sometimes referred to as "per trunk signaling") uses the same transmission facilities or channel for signaling as for voice. Common channel signaling, as discussed in the next section, uses one channel for all signaling functions of a group of voice channels. In the past most signaling systems in the telephone network were the in-channel variety.

In-channel signaling systems can be further subdivided into in-band and out-of-band techniques. In-band systems transmit the signaling information in the same band of frequencies used by the voice signal. The main advantage of in-band signaling is that it can be used on any transmission medium. The main disadvantage arises from a need to eliminate mutual interference between the signaling waveforms and a user's speech. The most prevalent example of in-band signaling is single frequency (SF) signaling, which uses a 2600-Hz tone as an on-hook signal for interoffice trunks. Although normal speech rarely produces a pure 2600-Hz signal, inadvertent disconnects have occurred as a result of user-generated signals. Two other common examples of in-band signaling are addressing by dual tone multifrequency (DTMF) signals from push-button telephones or multifrequency (MF) signaling between switching offices.

In-channel but out-of-band signaling uses the same facilities as the voice channel but a different portion of the frequency band. Thus out-of-band signaling represents a form of frequency division multiplexing within a single voice circuit. The most common instance of out-of-band signaling is direct current signaling as used on most customer loops. With this form of signaling, the central office recognizes the off-hook condition by the flow of direct current in the line. Other commonly used loop signals are dial pulses generated by a rotary dial at a rate of 10 pulses per second and a 20-Hz ringing voltage from the central office. All of these signals use lower frequencies than those generated in speech. Thus there is no possibility of one being mistaken for the other. The major disadvantage of out-of-band signaling is its dependence on the transmission system. For example, single sideband carrier systems filter out the very low frequencies associated with each voice channel. Thus, the on-hook/off-hook signal must be converted to something like SF signaling for FDM transmission. Out-of-band signaling is also implemented with frequencies above the cut-off frequency of voice separation filters but below the 4 kHz limit of a channel. CCITT recommends the use of 3825 Hz for this purpose.

Common Channel Signaling

Instead of sending signaling information over the same facilities that carries the message traffic (voice signals), common channel signaling (CCS) uses a dedicated data link between the stored program control elements (computers) of switching systems. Figure 1.31 depicts such a data link between two switching offices. Notice that the per-trunk signaling equipment associated with the trunks has been eliminated. The data link sends messages that identify specific trunks and events related to the trunks. Thus the type of CCS shown in Figure 1.30 is referred to as "channel associated" common channel signaling. The following are the main advantages of common channel signaling:

1. Only one set of signaling facilities is needed for each associated trunk group instead of separate facilities for each individual circuit.
2. A single dedicated control channel allows transfer of information such as address digits directly between the control elements (computers) of switching offices. In-channel systems, on the other hand, must have the control information switched from the common control equipment of the originating office onto the outgoing channel, and then the receiving office must switch the incoming control information from the voice channel into its common control equipment. The simpler procedure for transferring information directly between switch processors is one of the main motivations for CCS.
3. Since separate channels are used for voice and control there is no chance of mutual interference.
4. Since the control channel of a common channel system is inaccessible to users, a major means for fraudulent use of the network is eliminated.
5. Connections involving multiple switching offices can be set up more rapidly since forwarding of control information from one office can overlap a circuit set up through the node. With in-channel systems the associated circuit must first be established before the control information can be transferred across it. The CCS data link can also operate at a much higher data rate than common analog network signaling schemes—implying even faster connection set up.

Figure 1.31. Trunk group associated common channel signaling.

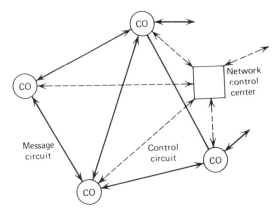

Figure 1.32. Disassociated common channel signaling network.

6. The channel used for common channel signaling does not have to be associated with any particular trunk group. In fact, the control information can be routed to a centralized control facility where requests are processed, and from which switching offices receive their connection control information. Figure 1.32 depicts a common channel signaling network that is disassociated from the message network structure. One advantage of centralized control is its ability to process requests with knowledge of networkwide traffic conditions. Centralized control is also attractive for managing multiple switching offices that are too small to warrant call processing facilities of their own. The transition from in-channel signaling to disassociated common channel signaling at the network level is analogous to the lower level transition from direct progressive control switches (step-by-step) to common control switches.

The following are the major disadvantages of common channel signaling:

1. Control information pertaining to an established circuit, such as a disconnect, must be relayed from one node to the next in a store-and-forward fashion. An in-channel disconnect signal, on the other hand, automatically propagates through the network enabling all nodes involved in the connection to simultaneously process the disconnect and release the associated facilities.

2. If one node in a common channel system fails to relay the disconnect information properly, facilities downstream from the disconnect will not be released. Thus, a high degree of reliability is required for the common channel—both in terms of physical facilities (duplication) and in terms of error control for the data link.

3. Since the control information traverses a separate path from the voice signal, there is no automatic test of the voice circuit as when the voice

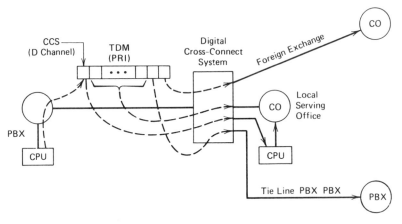

Figure 1.33. TDM link with multiple destinations for the channels.

channel is used to transfer control information. CCIS systems include special provisions for testing a voice circuit when it is set up.

4. In some instances all trunks in a group do not logically terminate at the same switch. Figure 1.33 depicts a multiplexed transmission link from a PBX to the public network. As indicated, some of the channels compose a trunk group to the local CO but other channels may represent tie lines to other PBXs or foreign exchange circuits to other COs. The digital cross-connect system (DCS) is a specialized switching system (described in Chapter 5) that routes individual circuits within a trunk group to individual locations. If CCS is implemented on the trunk group (as in an ISDN primary rate interface described in Chapter 9) the signaling information for the latter two trunk groups terminates at the CO. The CO must then forward the information to the respective destinations (assuming those destinations are equipped with CCS). All-in-all these latter two cases are easier to handle if the signaling information accompanies the message channels as it does automatically with in-channel signaling.

As a final note, it should be pointed out that some signaling functions originating or terminating with an end user inherently require in-channel implementations. For example, dial tone, ringback, and busy tones must be in-channel signals to get to the user. Furthermore, a user sometimes needs the capability to access certain control elements in the network that are associated with an established connection. The most practical means of accomplishing this is to use in-channel signals. For example, a request for operator assistance on an existing connection typically uses an in-channel hook flash to alert the operator. In addition, data terminals wishing to disable echo suppressors in dial-up connections need to send special tones that get recognized by echo suppressors

in the circuit. In similar fashion automatic originating facsimile equipment generate 1100 Hz beeps to allow automatic recognition of the originating equipment type by the destination.

1.2.11 Analog Interfaces

The design, implementation, and maintenance of any large and complex system require partitioning of the system into subsystems. Associated with each subsystem is an interface that defines the inputs and outputs independent of internal implementations. Well-established interfaces are a fundamental requirement to maintain compatibility between old and new equipment. Within the U.S. telephone network standardized interfaces are particularly necessary to support competition among equipment suppliers and service providers in almost all facets of the network. The principal analog interfaces used in the periphery of the network are subscriber loops, loop-start trunks, ground-start trunks, direct-inward-dial trunks, and E&M trunks.

Subscriber Loop Interface

The most common interface in the network involves the two-wire connection of individual telephone lines to end office switches: the subscriber loop interface. Because of the nature of industry standard telephones (2500 station sets) and the electromechanical switches to which they were connected, this interface has a number of characteristics that are difficult to satisfy with modern integrated circuit technology. The fundamental characteristics of this interface are as follows:

1. *Battery*: Application of dc power to the loop (48 V normally) to enable dc signaling and provide bias current for carbon microphones.
2. *Overvoltage Protection*: Protection of equipment and personnel from lightning strikes and power line induction or shorts.
3. *Ringing*: Application of a 20-Hz signal at 86-V rms for ringer excitation. Typical cadence is 2 sec on and 4 sec off.
4. *Supervision*: Detection of off-hook/on-hook by flow/no-flow of dc current.
5. *Test*: Access to the line to test in either direction: toward the subscriber or back into the switch.

In the case of a digital end office, two more functions are necessarily required: two-wire to four-wire conversion (*hybrid*) and analog-to-digital *coding* (and digital-to-analog decoding). Taken together, these functions are referred to as BORSCHT. The basic implementation of a BORSCHT interface is described in the context of digital switching in Chapter 5.

Loop-Start Trunks

A loop-start (LS) trunk is a two-wire connection between switches (usually between a central office and CPE equipment: a PBX). From an operational

point of view, an LS trunk is identical to a subscriber loop. Thus an LS interface in a PBX emulates a telephone by closing the loop to draw current for call origination and recognizing ringing voltages for incoming calls. To send address information the PBX interface generally waits for a few seconds and assumes that dial tone is present before sending DTMF tones or generating dial pulses by interrrupting loop current. Some PBXs provide dial tone detection so faulty equipment or connections are more easily recognized and addressing can be sent as soon as the other end is ready.

One significant difficulty with two-way LS trunks arises when both ends of the line seize the line at the same time (or nearly the same time). Because both ends of the line think they are originating a call the line becomes hung. This situation is commonly referred to as glare. If the PBX detects dial tone before sending digits it will recognize the glare condition by timing out on the wait for dial tone and can then generate a disconnect to release the glare condition but drop the incoming call. More commonly, the PBX blindly sends the address digits and connects the originating PBX station to the line. Generally this means the incoming call gets connected to the wrong station. For this reason, LS trunks are normally used only as one-way trunks: either one-way incoming or one-way outgoing.

Ground-Start Trunks

The aforementioned problem with glare on two-way LS trunks can be largely resolved by augmenting the call origination process to use Ground-Start (GS) procedures [22]. When originating a call the end office applies a ground potential to the tip lead of the tip and ring pair and waits for the PBX to acknowledge the seizure by drawing loop current. When the PBX originates a call it first applies ground to the ring lead and closes the loop waiting for loop current. (The CO does not apply battery during an idle state as it does in an LS interface.) The CO acknowledges the connect request by applying battery to the tip/ring pair and momentarily applying ground to the tip. (Maintaining low noise circuits requires removal of all paths to ground during the connected state.) A GS protocol prevents simultaneous seizures unless the originations occur within a few hundred milliseconds of each other. In contrast, an LS protocol allows multiple seizures to occur with windows up to 4 sec (the silent interval between ring bursts). Moreover, a glare condition on GS trunks can be recognized by the interface equipment so it can be resolved by redirecting the calls to different trunk circuits.

Another advantage of GS trunks is the ability of the CO to signal network disconnects to the PBX (the CO removes battery). With LS trunks the network does not generally provide disconnect signaling so the PBX must rely on the end user to hang up. (A situation that often produces hung trunks with data connections.) Furthermore, when the CO places an incoming call that eventually gets abandoned, because no one answers, a CO immediately signals the abandonment by removing ground from the tip lead. With LS trunks, abandoned calls can be recognized only by the absence of ring voltage, which can take 6 sec.

Direct-Inward-Dial Trunks

Direct-inward-dial (DID) trunks are particularly simple two-wire trunk interfaces because they are always one-way trunks: incoming only with respect to a PBX. As implied by the name they allow a serving CO to forward the extension number of incoming calls so a PBX can immediately route the call to a destination without going through an attendant. In contrast to LS and GS trunks, the PBX end of a DID trunk provides battery voltage so the CO can signal an incoming call by merely closing the loop to draw current. After the PBX reverses battery momentarily (winks) to signify it is ready to receive digits, the CO either generates dial pulses or DTMF tones to send the extension number (two, three, or four digits). After the designated station answers, the PBX reverses battery again to signify the connected state and holds that state for the duration of the call. DID trunks are also referred to as "loop reverse-battery supervision" runks with variations in the signaling protocol depending the type of CO [22].

E&M Trunks

As indicated in Figure 1.34, an E&M trunk is actually defined as an interface to a transmission system and not a transmission system itself. The interface in Figure 1.34 has a four-wire voice path, an E lead with an associated return lead (SG), and an M lead with an associated return (SB). Thus there are eight wires in this interface (referred to as a type II E&M interface). Other types of E&M interfaces are defined with as few as four wires [23] (a two-wire voice path, an E lead, and an M lead with ground returns).

In any type of E&M interface, supervision signaling is always conveyed on the E and M leads and not on the voice pair (or pairs). The PBX signifies off-hook by closing the M-SB loop to draw current while the transmission equipment indicates off-hook by closing the E-SG loop to draw current. How the transmission equipment conveys the supervision is a function of the transmission link. A variety of timing protocols are defined for the start of address signaling, which can be in-channel DTMF tones or dial pulses generated by momentary opens on the respective E and M leads.

Figure 1.34. Type II E&M interface.

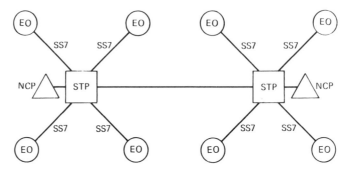

Figure 1.35. CCS network/intelligent network platform.

Although E&M signaling is formally defined as just an interface they are often used (with up to four pairs of wires) as direct connections between PBXs. Because of the requirement for multiple pairs such applications usually occur when the PBXs are located within a single building or campus complex. The availability of external control leads also leads to E&M interfaces being used for special applications such as paging systems where the M lead can be used to turn on the loudspeaker.

1.2.12 The Intelligent Network

The first common channel signaling facilities of AT&T were installed in the toll network between a No. 4A crossbar switch in Madison, Wisconsin and a No. 4 ESS switch in Chicago, Illinois in 1976 [24]. The 2400-bps CCS data links between SPC switching offices replaced in-channel SF/MF signaling facilities to reduce costs and significantly speed up long distance call connect times. As more and more of the toll switchs were implemented with CCS capabilities a CCS network evolved that not only improved the performance of the existing network but also established a platform for the introduction of new features. As indicated in Figure 1.35 the CCS network became a disassociated packet switching network that separated control of the network from the switching machines themselves.

The packet switching nodes of the CCS network are referred to as signal transfer points (STPs). The network control points (NCPs) of Figure 1.35 initially represented network database servers that defined how to route calls, verify credit cards, or process special services such as 800 numbers. The same basic structure is also being installed within the LATAs to extend CCS features all the way to the end offices. The communications protocol used between CCS entities is CCITT recommendation: Signaling System No. 7 (SS7).

As the capabilities of the NCPs are expanded from being data base servers to processing service requests the concept of an intelligent network (IN) [25] begins to take shape. In its ultimate form advanced services of the network are

executed in the NCPs [or service control points (SCPs) as they are called in the IN] and not in the switching machines themselves. The switching machines provide only the switching fabric—connection commands come from an SCP. The fundamental motivations behind developing the IN are to

1. Enable the deployment of network-based features such as city-wide Centrex mentioned previously.
2. Allow the introduction of new features without having to change the hardware or software of switching vendors.
3. Speed up the design/deployment cycle of new features because only the SCP software needs upgrading.
4. Allow customizing of services by the operating companies according to the needs of their customers.

The obvious disadvantage of such centralized control is the vulnerability of the network to a failure in the CCS network or the SCP. For this reason basic service will most surely always remain with the switching machines. In this case, software in the switching machines recognizes special service situations as software triggers to involve an SCP. Leaving basic service processing in the switching machines also reduces the load on the SS7 links and the processing speed in the SCPs.

1.2.13 Dynamic Nonhierarchical Routing

The alternate routing procedures shown in Figure 1.4 for the hierarchical network of the Bell System were necessarily simple and straightforward because of the original reliance on mechanical switching machines. Such simplicity, however, leads to the following limitations:

1. Even though the topology of the network might allow a large number of routing alternatives, constraints on how the routes are selected limit the actual choices to a relatively small number.
2. The routing patterns cannot be changed according to the time of day or network traffic patterns.
3. The process of progressive setup precludes retracing a path to a previous switching node and trying a new, untried route when a previous choice leads to a node that is blocked.
4. A completed connection may involve a large number of intermediate facilities using a lot of network resources for one connection.

All of these deficiencies result from using relatively simple decision logic in each node with no knowledge of the network status as a whole. The introduction of SPC switching machines and CCS changes the situation so that more efficient routing procedures are possible. Along these lines, AT&T completed the

deployment of dynamic nonhierarchical routing (DNHR) in 1987 [26, 27]. Specific features of DNHR are as follows:

1. Routing tables in the NCPs list all possible two-link routes in the order of cost so use of network resources are minimized.
2. Routing is dynamic to take advantage of traffic noncoincidence.
3. Blocking at an intermediate node produces "crankback," which allows selection of untried routes from the originating node.
4. Routes that produce too much delay for the echo cancelers in the circuit are excluded.

Reference [27] reports that DNHR reduced blocking on Thanksgiving day from 34% in 1986 to 3% in 1987. After installation of DNHR Mother's day traffic saw three to four times as many first attempt completions as before. The implementation of DNHR demonstrates that a network architecture is as much a function of the software in the network as of the physical topology. It also demonstrates that more centralized control of a network is more vulnerable to failures. On January 15, 1990 an obscure bug in the CCS software of the No. 4 ESS switches in AT&T's long distance network prevented completion of many calls during a 9-h period [28]. Ironically, the bug occurred in software intended to isolate the network from a node failure.

1.2.14 Cellular Radio Telephone System

Prior to 1983 mobile telephone users in the United States were restricted to using the services of radio common carriers (RCCs) who had limited bandwidth and limited resources available for services. The service environment was necessarily one of limited availability and extreme congestion. October 1983 marks a significant date for mobile telephony as the time when commercial cellular mobile telephone service started in Chicago, Illinois. As indicated in Figure 1.36 a cellular system [29] consists of a number of radio base stations connected to a (digital) switching office referred to as the mobile telephone switching office (MTSO). As a mobile subscriber moves from one cell to another the MTSO automatically switches connections with the respective base stations to maintain a continuous connection with the public switched network. The basic advantages of the cellular architecture are as follows:

1. The ability to reuse allocated channel frequencies within a serving area. Using a combination of antenna directivity in the base stations and signal attenuation from distant cells the same channel can be reused in one of every seven cells [30].
2. Reduced transmit power requirements for the mobiles. The power savings advantage for automobiles is primarily one of reducing the cost of the transmitter. For hand-held units the power savings is important for

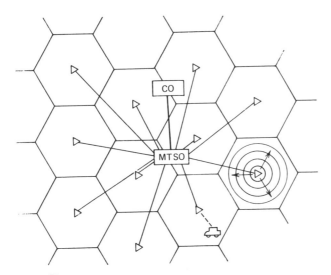

Figure 1.36. Cellular mobile telephone topology.

battery size and time between recharges. In fact, widespread acceptance and use of personal radio systems requires much lower transmit power levels and consequently smaller cells [31].

3. Reduced occurrences of multipath propagation. Shorter distances imply less chance of signal reflections causing multipath signal degradation.

4. Expandability. A system can be installed with comparatively large cells to minimize start-up costs. After service requirements grow and revenue is being received the capacity of the system can be expanded by subdividing the congested cells.

5. Reliability. Because the cells actually have significant overlapping coverage if one cell fails neighboring cells can provide service until repairs are made.

As of 1988 the FCC in the United States had defined 728 mobile service areas referred to as cellular geographic service areas (CGSAs). Each CGSA is allocated 832 radio channels, which are equally divided between two competing service providers: one a wireline carrier and the other a nonwireline carrier. The wireline carrier provides local telephone service in the area but competes with the nonwireline carrier (RCC) for cellular service. To ensure effective competition, the wireline carrier must not use facilities that are shared with local telephone service. Specifically, the MTSOs must be separate from the local switching offices. Thus, both types of carriers must backhaul all traffic to their respective MTSOs. The service providers can have more than one MTSO in an area but cannot interconnect them by the switched public network. (They are typically interconnected by leased private lines or digital microwave.) This is just

one example of how optimum use of technology conflicts with stimulating competition in telecommunications. For other examples see reference [32].

When a mobile unit first activates itself it scans the channels to determine which idle channel has the strongest signal. Using that channel the unit registers with the system to identify itself and place calls. After the initialization process, the network continually monitors signals from the mobile and controls it to switch channels when necessary. To complete calls to a mobile, the cellular network pages for the designated subscriber beginning with its home cell. (Unless the cellular control center already knows where the subscriber is located.) When a subscriber crosses a MTSO boundary, in addition to a cell boundary, a common channel signaling network is used to transfer the call to the new cell and the new MTSO. The connection to the public network is unchanged. Thus, the original MTSO then becomes a tandem node to the public network.

1.2.15 Voiceband Data Transmission

The primary concern of telephone companies, voice service, is sometimes referred to as POTS: plain old telephone service. In addition to POTS the telephone companies also provide a number of special services such as program distribution of radio and television, telephoto, teletype, facsimile, and data transmission. In most cases, these services are supported by adapting the sources to use the ubiquitous voice grade telephone channel. An exception is television distribution, which was an initial stimulus for the nationwide microwave radio network. To a certain extent, wideband FDM voice signals were designed to conform to the network TV distribution facilities. Although nationwide TV is now distributed primarily by satellites fiber is providing more and more TV distribution services for local applications.

One significant aspect of a voiceband telephone channel in regards to nonvoice applications is the absence of dc and low-frequency transmission. Although the metallic circuit provided in customer loops passes frequencies down to and including dc, most of the rest of the network does not. The equipments that are primarily responsible for blocking the low frequencies are transformers in two-wire to four-wire hybrids, FDM separation filters, and digital voice encoders that purposely block low frequencies to avoid 60-Hz hum that might arise from power line interference. Because baseband signals from facsimile and most data applications have frequency content down to dc, these signals must be modulated for voiceband transmission.

A second consideration when using the telephone network for data is the bandwidth restriction of approximately 3 kHz imposed by FDM separation filters, loading coils, and band-limiting filters in digital voice terminals (Chapter 3). The main implication of a restricted bandwidth is a limitation on the signaling rate or baud rate, which in turn directly relates to the data rate. A common signaling rate is 2400 symbols per second using carriers between 1700 and 1800 Hz. Symbol rates of 4800 symbols per second are also used in lower-sideband modems with a carrier at 2850 Hz.

As mentioned, user acceptance of voice quality does not require stringent control of the phase response (envelope delay) of the channel. High-speed data transmission, however, requires comparatively tight tolerance on the phase response to prevent intersymbol interference. Thus voice channels used for high-speed data transmission require special treatment. The special treatment is available at additional monthly rates from the telephone companies in the form of C-type conditioning for leased lines. C-type conditioning is available in several different grades that provide various amounts of control of both phase and amplitude distortion. Comparable functions are also available in newer modems in the form of equalizers.

Most medium rate (synchronous) modems include fixed equalization circuitry designed to compensate for the phase distortion in a typical connection. For higher data rates, automatically adjustable equalization is needed. An automatic equalizer first undergoes a training sequence in which the characteristics of the transmission channel are determined by measuring the response to known test signals. Then, equalizing circuitry in the receiver of each modem is adjusted (and continually readjusted during transmission) to provide compensation for the amplitude and phase distortions in the channel. High-speed transmission over dial-up lines requires automatic equalization in the modems since the channel characteristics change with each connection.

Another form of conditioning, referred to as a D-type conditioning, provides lower levels of noise and harmonic distortion than normally provided in leased lines, even with C-type conditioning. D-type conditioning usually does not involve special treatment of any particular line. Instead, the telephone company tests a number of different circuits until one with suitable quality is found. Sometimes this means avoiding a cable that includes pairs from a noisy switching office. Unlike voice, which is reasonably tolerant of impulse noise, a data circuit is more susceptible to impulse noise than to the normal background (white) noise. The elimination of older equipment (e.g., step-by-step switches) has eliminated much of the impulse noise in the network.

Even when the best leased lines available use both types of conditioning, some residual amounts of noise and distortion always exist. The remaining imperfections generally preclude any data rate higher than 9.6 or 19.2 kbps. One common way of obtaining 9600 bps is to signal at 2400 symbols per second and encode 4 bits into one of 16 symbols during each signal interval. Dial-up connections are generally limited to 4800 or 9600 bps because they necessarily pass through switching equipment, present different characteristics to equalize each time they are set up, and typically contain more noise than leased lines.

Another important consideration for data transmission over long-distance circuits is the effect of echo suppressors. As mentioned in Section 1.2.8, an echo suppressor blocks the signal in a return path when the corresponding forward path of a four-wire circuit is active. Thus, operative echo suppressors effectively preclude a full-duplex operation. Even in a half-duplex mode of operation the echo suppressors must be given 100 msec of deactivation time to reverse the direction of propagation. For these reasons the common carriers have provided

a means of disabling the echo suppressors by using an in-channel control signal from the data terminals. Echo suppressors are disabled by transmitting a pure tone between 2010 and 2240 Hz for 400 msec. The echo suppressors remain disabled as long as there is continuous energy in the channel. Thus, the modems can switch to signaling frequencies and begin full-duplex data transmission after the suppressors are first disabled. If the energy in the channel is removed for 100 msec, the echo suppressors are reactivated. Hence, rapid line turnaround is required for half-duplex operations.

1.3 THE INTRODUCTION OF DIGITS

Voice digitization and transmission first became feasible in the late 1950s when solid state electronics became available. In 1962 Bell System personnel established the first commercial use of digital transmission when they began operating a T1 carrier system for use as a trunk group in a Chicago area exchange [33]. Since that time a whole family of T-carrier systems (T1, T1C, T1D, T2, T3, T4) has been developed—all of which involve time division multiplexing of digitized voice signals.

The world's first commercially designed digital microwave radio system was established in Japan by Nippon Electric Company (NEC) in 1968 [34]. In the early 1970s digital microwave systems began to appear in the United States for specialized data transmission services. The first digital microwave link in the U.S. public telephone network was supplied by NEC of Japan for a New York Telephone link between Brooklyn and North Staten Island in 1972 [34]. Digital microwave systems have since been developed and installed by several U.S. manufacturers for use in intermediate length toll and exchange area circuits.

The Bell System's first commercial use of digital fiber optic transmission occurred in September of 1980 on a short-haul route between Atlanta and Smyrna, Georgia [35]. Three years later the first long-haul system between New York and Washington DC was put into service.

In addition to transmission systems, digital technology has proven to be equally useful for implementing switching functions. The first country to use digital switching in the public telephone network was France in 1970 [36]. The first application of digital switching in the public network of the United States occurred in early 1976 when the Bell System began operating its No. 4ESS [37] in a class 3 toll office in Chicago. Two months after that Continental Telephone Company began operation in Ridgecrest, California of an IMA2 digital toll switch supplied by TRW-Vidar [38]. The first digital end office switch in the United States became operational in July of 1977 in the small town of Richmond Hill, Georgia [39]. This switch was supplied by Stromberg-Carlson for Coastal Utilities, Inc., a small independent telephone company.

1.3.1 Voice Digitization

The basic voice coding algorithm used in T-carrier systems, and most other digital voice equipment in telephone networks around the world, is shown in Figure 1.37. The first step in the digitization process is to periodically sample the waveform. As discussed at length in Chapter 3, all of the information needed to reconstruct the original waveform is contained in the samples—if the samples occur at an 8 kHz rate. The second step in the digitization process involves quantization: identifying which amplitude interval of a group of adjacent intervals a sample value falls into. In essence the quantization process replaces each continuously variable amplitude sample with a discrete value located at the middle of the appropriate quantization interval. Since the quantized samples have discrete levels, they represent a multiple level digital signal.

For transmission purposes the discrete amplitude samples are converted to a binary code word. (For illustrative purposes only, Figure 1.37 shows 4 bit code words.) The binary codes are then transmitted as binary pulses. At the receiving end of a digital transmission line the binary data stream is recovered, and the discrete sample values are reconstructed. Then a low-pass filter is used to "interpolate" between sample values and recreate the original waveform. If no transmission errors have occurred, the output waveform is identical to the input waveform except for a small amount of quantization distortion: the difference between a sample value and its discrete representation. By having a large number of quantization intervals (and hence enough bits in a code word to encode them), the quantization intervals can be small enough to effectively eliminate quantization effects.

It is worth noting that the bandwidth requirements of the digital signal increase as a result of the binary encoding process. If the discrete, multiple-amplitude samples are transmitted directly, the bandwidth requirements are theoretically identical to the bandwidth of the original signal. When each discrete sample is represented by a number of individual binary pulses, the signal bandwidth increases accordingly. The two-level pulses, however, are much less vulnerable to transmission impairments than are the multiple-amplitude pulses (or the underlying analog signal).

1.3.2 Time Division Multiplexing

Basically, time division multiplexing (TDM) involves nothing more than sharing a transmission medium by establishing a sequence of time slots during which individual sources can transmit signals. Thus the entire bandwidth of the facility is periodically available to each user for a restricted time interval. In contrast, FDM systems assign a restricted bandwidth to each user for all time. Normally, all time slots of a TDM system are of equal length. Also, each subchannel is usually assigned a time slot with a common repetition period called a frame

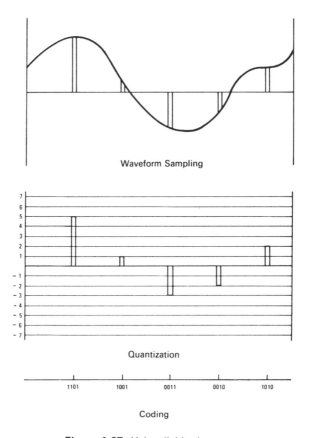

Figure 1.37. Voice digitization process.

interval. This form of time division multiplexing (as shown in Figure 1.38) is sometimes referred to as synchronous time division multiplexing to specifically imply that each subchannel is assigned a certain amount of transmission capacity determined by the time slot duration and the repetition rate. In contrast, another form of TDM (referred to as "statistical or asynchronous time division multiplexing") is described in Chapter 9. With this second form of multiplexing, subchannel rates are allowed to vary according to the individual needs of the sources. All of the backbone digital links of the public telephone network (T-carrier and digital microwave) use a synchronous variety of TDM.

Time division multiplexing is normally associated only with digital transmission links. Although TDM can be used conceivably for analog signals by interleaving samples from each signal, the individual samples are usually too sensitive to all varieties of transmission impairments. In contrast, time division switching of analog signals is more feasible than analog TDM transmission because noise and distortion of the switching equipment is more controllable. As discussed in Chapter 5, analog TDM techniques have been used in some PBXs.

Figure 1.38. Time division multiplexing.

T-Carrier Systems

The volume of interoffice telephone traffic in the United States has been growing more rapidly than local traffic. This rapid growth puts severe strain on interoffice transmission facilities that are designed for lower traffic volumes. Telephone companies are then faced with the necessary task of expanding the number of interoffice circuits. T-carrier systems evolved primarily as a cost effective means for interoffice transmission: both for initial installations and for relief of crowded interoffice cable pairs.

Despite the need to convert the voice signals to a digital format at one end of a T1 line and back to analog at the other, the combined conversion and multiplexing cost of a digital TDM terminal is lower than the cost of a comparable analog FDM terminal. The first T-carrier systems were designed specifically for exchange area trunks at distances between 10 and 50 miles.

A T-carrier systems consists of terminal equipment at each end of a line and a number of regenerative repeaters at intermediate points in the line. The function of each regenerative repeater is to restore the digital bit stream to its original form before transmission impairments obliterate the identity of the digital pulses. The line itself, including the regenerative repeaters, is referred to as a span line. The terminal equipment is referred to as D-type (digital) channel banks, which come in numerous versions. The transmission lines are wire pairs using 16 to 26 gauge cable. A block diagram of a T-carrier system is shown in Figure 1.39.

The first T1 systems used D1A channel banks for interfacing, converting, and multiplexing 24 analog circuits. A channel bank at each end of a span line provides interfacing for both directions of transmission. Incoming analog

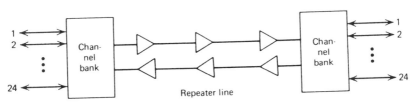

Figure 1.39. T1-carrier system.

signals are time division multiplexed and digitized for transmission. When received at the other end of the line, the incoming bit stream is decoded into analog samples, demultiplexed, and filtered to reconstruct the original signals. Each individual TDM channel is assigned 8 bits per time slot. Thus, there are $(24)(8) = 192$ bits of information in a frame. One additional bit is added to each frame to identify the frame boundaries, thereby producing a total of 193 bits in a frame. Since the frame interval is $125 \mu sec$, the basic T1 line rate becomes 1.544 Mbps. This line rate has been established as the fundamental standard for digital transmission in North America and Japan. The standard is referred to as a DS1 signal (for digital signal 1).

A similar standard of 2.048 Mbps has been established by CCITT for most of the rest of the world. This standard evolved from a T1-like system that provides 32 channels at the same rate as the North American channels. Only 30 of the channels in the CCITT standard, however, are used for voice. The other two are used for frame synchronization and signaling. Signaling and control information for T1 systems are inserted into each voice channel (or transmitted separately by CCIS facilities). Digital signaling and control techniques for both systems are discussed in Chapter 7.

The greatly increased attenuation of a wire pair at the frequencies of a DS1 signal (772 kHz center frequency) mandate the use of amplification at intermediate points of T1 span line. In contrast to an analog signal, however, a digital signal can not only be amplified but it can also be detected and regenerated. That is, as long as a pulse can be detected it can be restored to its original form and relayed to the next line segment. For this reason T1 repeaters are referred to as regenerative repeaters. The basic functions of these repeaters are

1. Equalization
2. Clock recovery
3. Pulse detection
4. Transmission.

Equalization is required because the wire pairs introduce certain amounts of both phase and amplitude distortion that cause intersymbol interference if uncompensated. Clock recovery is required for two basic purposes: first, to establish a timing signal to sample the incoming pulses; second, to transmit outgoing pulses at the same rate as at the input to the line.

Regenerative repeaters are normally spaced every 6000 ft in a T1 span line. This distance was chosen as a matter of convenience for converting existing voice frequency cables to T-carrier lines. Interoffice voice frequency cables typically use loading coils that are spaced at 6000 ft intervals. Since these coils are located at convenient access points (manholes) and must be removed for high-frequency transmission, it was only natural that the 6000 ft interval be chosen. One general exception is that the first regenerative repeater is typically spaced 3000 ft from a central office. The shorter spacing of this line segment is

needed to maintain a relatively strong signal in the presence of impulse noise generated by some switching machines.

The operating experience of T1 systems was so favorable that they were continually upgraded and expanded. One of the initial improvements produced T1C systems which provide higher transmission rates over 22-gauge cable. A T1C line operates at 3.152 Mbps for 48 voice channels: twice as many as a T1 system.

Another level of digital transmission became available in 1972 when the T2 system was introduced. This system is designed for toll network connections. In contrast, T1 systems were originally designed only for exchange area transmission. The T2 system provides for 96 voice channels at distances up to 500 miles. The line rate is 6.312 Mbps which is referred to as a DS2 standard. The transmission media is special low capacitance 22-gauge cable. By using separate cables for each direction of transmission and the specially developed cables, T2 systems can use repeater spacings up to 14,800 ft in low-noise environments.

TDM Hierarchy

In a manner analogous to the FDM hierarchy, AT&T established a digital TDM hierarchy that has become the standard for North America. Starting with a DS1 signal as a fundamental building block, all other levels are implemented as a combination of some number of lower-level signals. The designation of the higher-level digital multiplexers reflects the respective input and output levels. For example, an M12 multiplexer combines four DS1 signals to form a single DS2 signal. Table 1.10 lists the various multiplex levels, their bit rates, and the transmission media used for each. Notice that the bit rate of a high-level multiplex signal is slightly higher than the combined rates of the lower-level inputs. The excess rate is included for certain control and synchronization functions discussed in Chapter 7. A similar digital hierarchy has also been

TABLE 1.10 Digital TDM Signals of North America and Japan

Digital Signal Number	Number of Voice Circuits	Multiplexer Designation	Bit Rate (Mbps)	Transmission Media
DS1	24	D channel bank (24 analog inputs)	1.544	T1 paired cable
DS1C	48	M1C (2 DS1 inputs)	3.152	T1C paired cable
DS2	96	M12 (4 DS1 inputs)	6.312	T2 paired cable
DS3	672	M13 (28 DS1 inputs)	44.736	Radio Fiber
DS4	4032	M34 (6 DS3 inputs)	274.176	T4M coax WT4 waveguide Radio

TABLE 1.11 Digital TDM Hierarchy of CCITT

Level Number	Number of Voice Circuits	Multiplexer Designation	Bit Rate (Mbps)
CEPT1	30		2.048
CEPT2	120	M12	8.448
CEPT3	480	M23	34.368
CEPT4	1920	M34	139.264
CEPT5	7680	M45	565.148

established by CCITT as an international standard. As shown in Table 1.11, this hierarchy is similar to the North American standard but involves different numbers of voice circuits at all levels.

Digital Pair-Gain Systems

Following the successful introduction of T1 systems for interoffice trunks, most major manufacturers of telephone equipment developed digital TDM systems for local distribution. These systems are most applicable to long rural loops where the cost of the electronics is offset by the savings in wire pairs. No matter what the distance is, unexpected growth can be most economically accommodated by adding electronics, instead of wire, to produce a pair-gain system. The possibility of trailer parks and apartment houses springing up almost overnight causes nightmares in the minds of cable plant forecasters. Pair-gain systems, however, provide networking alternative to dispel those nightmares.

Digital pair-gain systems are also useful as alternatives to switching offices in small communities. Small communities are often serviced by small automatic switching systems normally unattended and remotely controlled from a larger switching office nearby. These small community switches are referred to as community dial offices (CDOs). A CDO typically provides only limited service features to the customers and often requires considerable maintenance. Because digital pair-gain systems lower transmission costs for moderate-sized groups of subscribers, they are a viable alternative to a CDO: stations in the small community are serviced from the central office by way of pair-gain systems. A fundamental consideration in choosing between pair-gain systems and remote switching involves the traffice volumes and calling patterns within the small community. The basic techniques of analyzing traffic patterns and determining trunk group sizes are provided in Chapter 10.

The first two digital pair-gain systems used in the Bell System are the subscriber loop multiplex (SLM) system [40, 41] and, its successor: the subscriber loop carrier (SLC-40) system [40, 42]. Although these systems used a form of voice digitization (delta modulation) different from that used in T-carrier systems (pulse code modulation), they both used standard T1 repeaters for digital transmission at 1.544 Mbps. Both systems also converted the digitized

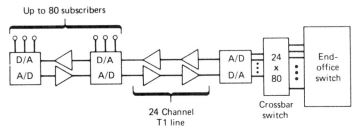

Figure 1.40. Subscriber loop multiplexer (SLM).

voice signals back into individual analog interfaces at the end-office switch to achieve system transparency. Figures 1.40 and 1.41 show block diagrams of these systems. Notice that the SLM system provided both concentration and multiplexing (80 subscribers for 24 channels) while the SLC-40 was strictly a multiplexer (40 subscribers assigned in a one-to-one manner to 40 channels).

The SLM and SLC-40 systems used delta modulation voice coding because it is simpler than pulse code modulation as used in T1 systems and was therefore less costly to implement on a per channel basis—a desirable feature for modular system implementations. The original T1 systems, on the other hand, minimized electronics costs by using common encoders and decoders which precluded implementation of less than 24 channels. (An unnecessary feature in an interoffice application.) By the late 1970s low cost, integrated circuit implementations of standard pulse code modulation became available which led the way to the first (1979) installation of the SLC-96: a subscriber carrier system using voice coding that was compatible with T1 systems and the emerging digital end office switching machines [43].

The SLC-96 system (which is functionally equivalent to four T1 lines) can interface directly with a digital end office and not be demultiplexed into 24 distinct analog interfaces. Thus this capability, which is referred to as integrated digital loop carrier (IDLC), greatly reduces the prove-in distance where digital carrier becomes less expensive than separate subscriber pairs. Subsequent enhancements of the SLC-96 system include use of fiber transmission, support for local switching functions in the remote module, and configurability of analog interfaces from the central office [44].

Figure 1.41. Subscriber loop carrier (SLC-40).

Figure 1.42. Data under voice.

1.3.3 Data under Voice

After the technology of T-carrier systems had been established, AT&T began offering leased digital transmission services for data communications. This service, known as Dataphone Digital Service (DDS), uses T1 transmission links with special terminals (channel banks) that provide direct access to the digital line. An initial drawback of DDS service arose because T-carrier systems were originally used only for exchange area and short toll network trunks. Without some form of long-distance digital transmission, the digital circuits in separate exchange areas could not be interconnected. AT&T's response to long-distance digital transmission was the development of a special radio terminal called the 1A radio digital terminal (1A-RDT) [45, 46]. This terminal encoded one DS1 signal (1.544 Mbps) into less than 500 kHz of bandwidth. As shown in Figure 1.42, a signal of this bandwidth can be inserted below the lowest frequency of a mastergroup multiplex (Table 1.5). Since this frequency band is normally unused in TD or TH radio systems, the DS1 signal could be added to existing analog routes without displacing any voice channels. The use of frequencies below those used for voice signals leads to the designation "data under voice" (DUV).

It is important to point out that data under voice represented a special development specifically intended for data transmission and not for voice services. In fact, DUV was used only to provide long-distance digital transmission facilities for Digital Dataphone Service. Although the DDS network uses common technology (such as T1 lines) and common facilities (such as TD radios) of the voice network, the system is functionally separate from the voice network. That is, the voice network does not use DDS circuits, nor does the DDS network use digital circuits within the voice network.

1.3.4 Digital Microwave Radio

In contrast to data under voice systems, which piggyback 1.544 Mbps onto an analog radio for data services, a common carrier digital microwave system uses digital modulation exclusively to transmit and receive higher-level digital multiplex signals for voice traffic. Digital radios use the same frequency bands

allocated for analog radios as listed in Table 1.7. Thus a primary design requirement of a digital radio is that it must confine its radiated power to prevent excessive interference into adjacent, possibly analog, channels. Moreover, the FCC stipulated that the digital radios had to be capable of providing roughly the same number of voice circuits as existing analog radios. Table 1.12 lists the minimum number of voice circuits and the resulting bit rate that must be provided by digital radios in each of the common carrier microwave bands [47].

Despite the design constraints imposed by compatibility requirements with analog radios, digital radios have proven to be more economical than analog radios in several applications. Because of lower terminal costs (multiplexing and demultiplexing), digital radio systems are generally less expensive than analog radio systems for distances up to about 300 miles, and on longer routes requiring channel access (drop-and-insert) at intermediate points in the route [48]. The major impetus for digital radio in the United States has been the emergence of digital toll switches like the No. 4ESS. The interconnection of digital radio signals to a digital switch avoids costly lower level multiplex equipment.

1.3.5 Fiber Optic Transmission

Of all the new technology introduced into the telephone network fiber has certainly had the most profound effect. Prior to its emergence, transmission engineers would have considered the combination of extremely wide bandwidth, extremely low attenuation, and immunity from interference as something akin to perpetual motion. Low attenuation allows long repeater spacings which equates to low maintenance costs: the reason for wholesale replacement of long-haul analog radio systems. On December 5, 1986 AT&T completed the last section of a transcontinental fiber system [49]. By the end of the decade virtually all of the high-density routes were converted to fiber. Radio systems are and will continue to be used for carrying low-density traffic where right-of-way costs are

TABLE 1.12 Minimum Voice Circuit Requirements of Digital Radios in the United States

Frequency Band (MHz)	Minimum Number of Circuits	Equivalent Number of DS1 Signals	Resultant Bit Rate (Mbps)[a]	Channel BW (MHz)
2,110–2,130	96	4	6.144	3.5
2,160–2,180	96	4	6.144	3.5
3,700–4,200	1152	48	73.7	20
5,925–6,425	1152	48	73.7	30
10,700–11,700	1152	48	73.7	40

[a]The actual bit rate is usually slightly greater owing to extra framing and synchronization bits. Furthermore, most radio systems provide significantly more voice circuits than the minimum.

a dominant consideration. However, because these routes are relatively short spurs off of digital fiber arteries, interface costs imply digital implementations—thus completing the economic foundation for the all-digital network.

As discussed in Chapter 8, an optical fiber is not an inherently digital transmission system. However, the interface electronics (drivers and receivers) and the application of connecting to digital switches have stimulated their use of digital transmission. Furthermore, considering the amount of bandwidth available there is little incentive to use analog transmission to conserve bandwidth as in the case of microwave radios.

1.3.6 Digital Switching

The original research into digital switching at Bell Laboratories was reported by Earle Vaughan in 1959 [50]. Laboratory models were developed to demonstrate the concept of integrating digital time division multiplex transmission systems with time division switching systems. Unfortunately the necessary solid state electronics had not matured sufficiently at that time so commercial development of digital switching was not pursued, and development of the No. 1 ESS continued along the lines of space division electromechanical technology. Almost 10 years later, however, Bell Labs began development of a digital toll switch, the No. 4 ESS.

When placed in service in January of 1976, the No. 4 ESS provided several new capabilities for the toll network. First, it was the first toll switch to be designed for stored program control at the outset.* Second, its capacity was three times that of the prevailing electromechanical switch at the time: the No. 4A cross bar. The larger capacity of the No. 4 ESS meant that many metropolitan areas could consolidate toll traffic into one switch instead of several. Third, the digital time division design of the No. 4 ESS allowed direct connection to digital T-carrier lines. This last feature illustrates the particular attraction of digital switching for toll and tandem offices of the network. By 1976, when the first No. 4 ESS was installed, it was clear that digital transmission was dominating the exchange area and shorter intertoll trunks. Thus significant economies and improvements in transmission quality resulted by eliminating channel banks at the interface between digital trunks and a switching system.

The early development of digital end-office switches in the United States was undertaken by independent equipment manufacturers with the first system being placed in service in 1977. These systems were primarily designed for the smaller switching offices of the independent telephone companies. Digital switches are particularly attractive to rural telephone companies because they can provide significant copper savings when directly connected to digital pair gain transmission systems. The first large digital end office switching system to be introduced into the North American network was the DMS-100 provided by

*Stored program control was first implemented in the toll network beginning in 1969 by retrofitting crossbar switches [51].

TABLE 1.13 Digital Central Office Switching Systems of North America

Manufacturer	Designation	Date of Introduction	Application	Line Size
AT&T	4 ESS	1976	Toll	107,000
AT&T	5 ESS	1983	Local	100,000
CIT-Alcatel	E 10-five	1982	Local	100,000
GTE	3 EAX	1978	Toll/tandem	60,000
GTE	5 EAX	1982	Local	145,000
LM Ericsson	AXE10	1978	Local/toll	200,000
NEC	NEAX-61	1979	Local/toll	80,000
ITT	System 1210	1978	Local/toll	26,000
Northern Telecom	DMS-10	1977	Local	7,000
Northern Telecom	DMS-100	1979	Local	100,000
Northern Telecom	DMS-200	1978	Toll	60,000
Siemens	EWSD	1981	Local	200,000
Stromberg Carlson	DCO	1977	Local	32,000
Vidar	ITS4	1977	Toll/tandem	7,000
Vidar	ITS4/5	1978	Local/toll	12,768

Northern Telecom. Table 1.13 lists digital switching machines available for the North American network.

The switching systems listed in Table 1.13 pertain only to equipment used in the public network. A vast array of digital switching equipment is also available for private exchanges. The first commercially available digital switch to be used in the United States was an Automatic Call Distributor† (ACD) supplied by Collins/Rockwell in 1974.

The functional essence of a digital time division switching matrix is illustrated in Figure 1.43. As indicated, all inputs are time division multiplex links. These links may represent digital pair-gain systems, T-carrier interoffice trunks, or the outputs of colocated channel banks used to interface analog lines to the digital switch. In any event the switching matrix itself is designed to service only TDM input links.

Basically, the switching matrix is required to transfer information arriving in a specified time slot (channel) on an incoming TDM link to a specified time slot on an outgoing TDM link. Since an arbitrary connection involves two different physical links and two different time slots, the switching process requires spatial translation (space switching) and time translation (time switching). Thus the basic operation is sometimes referred to as two-dimensional switching. Space switching is achieved with conventional digital logic selector (or multiplexer)

†An automatic call distributor is a switching machine designed to connect incoming requests to any one of a number of available attendants. If all attendants are busy, calls are put on hold and automatically switched to attendants as they become available. ACDs are commonly used for operator assistance, airline reservations, service organizations, or any other application where requests can be serviced by any one of a number of attendants.

Figure 1.43. Digital time division switching operation.

circuits, and time switching is achieved by temporarily storing information in a digital memory or register circuit.

For convenience the inputs to the switching matrix of Figure 1.43 are all shown on the left, and the outputs are on the right. Since the links are inherently four-wire, each input time slot is paired with the same time slot on a corresponding output link. Hence, a full-duplex circuit also requires a "connection" in the reverse direction which is achieved by transferring information from time slot 17 of link N to time slot 3 of link 1 in the example shown. Operational and implementation details of a large range of digital time division switches are provided in Chapter 5.

1.3.7 Digital Network Evolution

The evolution of the analog telephone network into one that is mostly digital is summarized in Figure 1.44. The process began in the 1960s (a) with T1 systems being installed on relatively short haul interoffice trunks within the exchange areas. Next, in the early 1970s (b), digital transmission was introduced into the toll network with T2 systems for relatively short routes between toll offices. It was in the late 1970s (c) that digitization really began to take over. T1 coverage expanded greatly, digital loop carrier (DLC) systems came into use,* and digital switches became available at all levels of the network: PBXs (DPBXs), end offices (DEOs), tandem offices (DTOs), and toll offices (DTOs). Moreover, microwave digital radios (MDRs) proved to be advantageous to use in both the exchange areas and the shorter toll network routes due to low interface costs to digital switches. Thus the late 1970s produced a number of integrated islands where digital switches within a region were interconnected by digital transmission links but there was little digital connectivity between the islands. (Data under voice was installed as an overbuild to analog routes for limited capacity data services.) A fully integrated and interconnected digital network became a reality in the early 1980s (d) when fiber optic transmission emerged as the technology of choice for high-density long-haul routes. Digital connectivity to

*Digital subscriber carrier systems were actually introduced in the early 1970s but these systems utilized a voice digitization technology (delta modulation) that was incompatible with the rest of the network and therefore did not figure into the integrated network. These early digital loop carrier systems are now obsolete.

Figure 1.44. Digital network evolution.

customer premises equipment also occurred in this time frame as T1 became the preferred voice trunk interface for large PBXs.

End-to-end digital connectivity for voice or data services became a reality in the late 1980s (e) with the introduction of ISDN basic rate (ISDN BR, 2B + D) and ISDN primary rate (ISDN PR, 23B + D) digital connections to the customer. In addition, fiber technology became much more ubiquitous as DS3 rate systems virtually eliminated T2 systems in the toll network and fiber-based systems became the preferred technology for loop carrier and feeder systems— even at relatively short distances.

REFERENCES

1 S. P. Thompson, *Phillip Reis: inventor of the telephone*, Sylvanus P. Thompson, London, New York, 1883.

2 R. Smith and W. Sonnerville, "Record '81 Telco Construction Set at $23.5 Billion, Up 5.6%," *Telephone Engineer and Management*, Jan 15, 1981, p 55.

3 Technical Staff, Bell Laboratories, *Engineering and Operations in the Bell System*, 2nd Ed., Bell Telephone Laboratories Inc., 1983.

4 F. T. Andrews, "Divestiture: A Record of Technical Achievement," *IEEE Communications Magazine*, Dec 1985, pp 54–58.

5 D. M. Mardon, "Plan for access—tandem and trunking," *Telephone Engineer & Management*, Nov 15, 1983, pp 140–142.

6 L. Jachimowicz, J. Olszewski, and I. Kolodny, "Transmission Properties of Filled Thermoplastic Insulated and Jacketed Telephone Cables at Voice and Carrier Frequencies," *IEEE Transactions on Communications*, March 1973, pp 203–209.

7 Technical Staff, Bell Laboratories, *Engineering and Operations in the Bell System*, Bell Telephone Laboratories Inc., 1977.

8 W. L. Pritchard, "The History and Future of Commercial Satellite Communications," *IEEE Communications Magazine*, May 1984, pp 22–37.

9 P. Kullstam, ed., "Special Issue—Satellite Communications," *IEEE Communications Magazine*, March 1984.

10 I. Boyd, C. B. Southcott, D. P. Crowe, and P. J. Bolingbroke, "Spech Codec for the Skyphone Aeronautical Telephone Service," *British Telecommunications Engineering*, July 1986, pp 83–91.

11 A. J. Aikens and D. A. Lewinski, "Evaluation of Message Circuit Noise," *Bell System Technical Journal*, July 1970, pp 879–909.

12 J. L. Sullivan, "A Laboratory System for Measuring Loudness Loss of Telephone Connections," *Bell System Technical Journal*, October 1971, pp 2663–2739.

13 W. K. Macadam, "A Basis for Transmission Performance Objective in a Telephone Communications System," *AIEE Transactions on Communications and Electronics*, May 1978, pp 205–209.

14 F. P. Duffy and T. W. Thatcher, Jr., "Analog Transmission Performance on the Switched Telecommunications Network," *Bell System Technical Journal*, April 1971, pp 1311–1347.

15 W. R. Bennett and J. R. Davey, *Data Transmission*, McGraw-Hill, New York, 1965.

16 Members of Technical Staff, Bell Telephone Laboratories, *Transmission Systems for Communications*, 4th ed., Western Electric Company, 1971.

17 H. R. Huntley, "Transmission Design of Intertoll Telephone Trunks," *Bell System Technical Journal*, September 1953, pp 1019–1036.

18 K. Murano, S. Unagami, and F. Amano, "Echo Cancellation and Applications," *IEEE Communications Magazine*, January 1990, pp 49–55.

19 D. L. Duttweiler and Y. S. Chen, "A Single-Chip VLSI Echo Cancellor," *Bell Systems Technical Journal*, February 1980, pp 149–160.

20 H. G. Suyderhoud, M. Onufry, and S. J. Campanella, "Echo Control in Telephone Communications," *National Telecommunications Conference Record*, 1976, pp 8.1-1–8.1-5.

21 F. T. Andrews, Jr. and R. W. Hatch, "National Telephone Network Transmission Planning in the American Telephone and Telegraph Company," *IEEE Transactions on Communications Technology*, June 1971, pp 302–314.

22 *Compatibility Information for Telephone Exchange Service, Bellcore Technical Advisory TA-NPL-000912*, Bell Communications Research, Issue 1, Feburary 1989.

23 *Notes on THE NETWORK*, American Telephone and Telegraph Company, Network Planning Division, AT&T, New York, 1980, pp 44–58.

24 "Special Issue: Stored Program Control Network," *Bell System Technical Journal*, September 1982.

25 J. S. Doyle and C. S. McMahon, "The Intelligent Network Concept," *IEEE Transactions on Communications*, December 1988, pp 1296–1301.

26 J. M. Akinpelu, "The Overload Performance of Engineered Networks with non-hierarchical and Hierarchical Routing," *Bell System Technical Journal*, September 1984, pp 1261–1281.

27 G. R. Ash and E. Oberer, "Dynamic Routine in the AT&T Network—Improved Service Quality at Lower Cost," *IEEE Globecom Proceedings*, 1989, pp 9.1.1–9.1.6.

28 C. Mason, "Software Problem Cripples AT&T Long-Distance Network," *Telephony*, January 22, 1990, pp 10–11.

29 K. W. Strom, "On The Road with AUTOPLEX System 1000," *AT&T Technical Journal*, Vol. 3, No. 3, 1988, pp 42–51.

30 "Special Issue on Advanced Mobile Phone Service (AMPS)," *Bell System Technical Journal*, January 1979.

31 D. C. Cox, "Portable Digital Radio Communications—An Approach to Tetherless Access," *IEEE Communications Magazine*, July 1989, pp 30–40.

32 I. Dorros, "The New Fugure—Back to Technology," *IEEE Communications Magazine*, January 1987, pp 57–62.

33 K. E. Fultz and D. B. Penick, "The T1 Carrier System," *Bell System Technical Journal*, September 1965, pp 1405–1451.

34 "Supply Record on NEC PCM Microwave Equipment," Nippon Electric Company, Tokyo, May 1974.

35 M. Schwartz, "Optical Fiber Transmission—From Conception to Prominence in 20 Years," *IEEE Communications Magazine*, May 1984, pp 38–48.

36 J. B. Jacob, "The New Approaches and Solutions Offered by the Digital ESS.E 10 to Deal with Switching, Operation, and Maintenance Functions," *International Communications Conference Proceedings*, 1979, pp 46.6-1–46.6-5.

37 "Special Issue: No. 4 ESS," *Bell System Technical Journal,* July/August 1981.

38 J. C. McDonald and J. R. Baichtal, "A New Integrated Digital Switching System," *National Telecommunications Conference Record,* 1976, p 3.2-1–3.2-5.

39 "First Class 5 Digital Central Office goes into Service in Georgia," *Telephony,* August 1, 1977, pp 24–25.

40 F. T. Andrews, Jr., "Customer Lines Go Electronic," *Gell Labs Record,* February 1972, pp 59–64.

41 M. T. Manfred, G. A. Nelson, and C. H. Sharpless, "Digital Loop Carrier Systems," *Bell System Technical Journal,* April 1978, pp 1129–1142.

42 S. J. Brolin, G. E. Harrington, and G. R. Leopold, "The Subscriber Loop Carrier System Gets Even Better," *Bell Labs Record,* May 1977, pp 119–122.

43 S. Brolin, Y. S. Cho, W. P. Michaud, and D. H. Williamson, "Inside the New Digital Subscriber Loop System," *Bell Labs Record,* April 1980, pp 110–116.

44 G. E. Harrington and R. W. Wyndrum, "A 'smart' SLC carrier meets the changing nature of services in the evolving digital loop," *Bell Labs Record,* September 1985, pp 13–17.

45 R. C. Prime and L. L. Sheets, "The 1A Radio Digital System Makes 'Data Under Voice' a Reality," *Bell Labs Record,* December 1973, pp 335–339.

46 R. R. Grady and J. W. Knapp, "1A Radio Digital Terminals Put 'Data Under Voice'," *Bell Labs Record,* May 1974, pp 161–166.

47 FCC *Memorandum Opinion and Order* released January 29, 1975 modifying FCC *Report and Order* of Docket 19311 released Septeber 27, 1974.

48 W. S. Chaskin, "Digital Transmission Economics," *Digital Microwave Transmission Engineering Symposium,* Rockwell International, September 1978.

49 R. C. Hansen, "AT&T's Digital Network Evolution," *IEEE International Conference on Communications,* 1987, pp 19B.1.1–1.1.5.

50 H. E. Vaughan, "Research Model for Time-Separation Integrated Communication," *Bell System Technical Journal,* July 1959, pp 909–932.

51 B. T. Fought and C. J. Funk, "Electronic Translator System for Toll Switching," *IEEE Transactions on Communications,* June 1970, p 168.

PROBLEMS

1.1 Express 33 dBrnC of noise in terms of picowatts and milliwatts of absolute power.

1.2 Twenty-seven picowatts of noise with a flat spectrum between 0 and 3 kHz is how many dBrnC?

1.3 A value of 30 dBrnCO is how many picowatts of absolute noise power at a −3-dB TLP?

1.4 An idle channel noise power measurement of 21 dBrnC occurs at a −7-dB TLP. Express the noise power of this measurement in dBrnCO and determine what power measurement this noise would produce at another point in the circuit that is designated as a −2-dB TLP.

1.5 A transmission link has 14 dBrn of absolute noise power at a −13-dB TLP input testpoint and 27 dBrn of absolute noise power at a −3-dB TLP output testpoint. How much absolute noise is added on the transmission link?

1.6 A transmission link with a −2-dB TLP at the transmit end and a −4-dB TLP at the receive end is part of a voice circuit that produces the following idle channel noise measurements: 18 dBrnC0 at the transmit end and 20 dBrnC0 at the receive end. What is the contribution (in pWp0) of the transmission link toward the total absolute power at the receive end?

2

WHY DIGITAL?

The first chapter provides an overview of the existing analog telephone network and a brief introduction to the use of digital transmission and switching. This chapter discusses the basic technical advantages of digital implementations that stimulated the replacement of almost all of the analog technology.

2.1 ADVANTAGES OF DIGITAL VOICE NETWORKS

A list of technical features of digital communications network is provided in Table 2.1. These features are listed in the order that the author considers to be their relative importance for general telephony. In particular applications, however, certain considerations may be more or less significant. For instance, the last item, ease of encryption, is a dominant feature favoring digital networks for the military.

Most of the features of digital voice networks listed in Table 2.1 and discussed in the following paragraphs pertains to advantages of digital transmission or switching relative to analog counterparts. In some instances, however, the features pertain only to all-digital networks. Encryption, for example, is practical and generally useful only if the secure form of the message is

TABLE 2.1 Technical Advantages of Digital Communications Networks

1. Ease of multiplexing
2. Ease of signaling
3. Use of modern technology
4. Integration of transmission and switching
5. Signal regeneration
6. Performance monitorability
7. Accommodation of other services
8. Operability at low signal-to-noise/interference ratios
9. Ease of encryption

established at the source and translated back into the clear only at the destination. Thus an end-to-end digital system that operates with no knowledge of the nature of the traffic (i.e., provides transparent transmission) is a requirement for digital encryption applications. For similar reasons end-to-end digital transmission is needed for data related applications. When a network consists of a mixture of analog and digital equipment, universal use of the network for services such as data transmission dictates conformance to the least common denominator of the network: the analog channel.

2.1.1 Ease of Multiplexing

As mentioned in Chapter 1, digital techniques were first applied to general telephony in interoffice T-carrier (time division multiplex) systems. In essence these systems trade electronics costs at the ends of a transmission path for the cost of multiple pairs of wires between them. (A trade that is more cost effective every year.) Although frequency division multiplexing of analog signals has also been used in the past to reduce cable costs, FDM equipment is much more expensive than TDM equipment—even when the cost of digitization is included. After voice signals have been the digitized, TDM equipment costs are quite small by comparison. Since digitization occurs only at the first level of the TDM hierarchy, high-level digital TDM is even more economical than high-level FDM counterparts.

It should be pointed out that time division multiplexing of analog signals is also very simple and does not require digitization of the sample values. The drawback of analog TDM lies in the vulnerability of narrow analog pulses to noise, distortion, crosstalk, and intersymbol interference. These degradations can not be removed by regeneration as in a digital system. Hence, analog TDM is not feasible except for noiseless, distortion-free environments.* In essence the ability to regenerate a signal, even at the expense of greater bandwidth, is almost a requirement for TDM transmission.

2.1.2 Ease of Signaling

Control information (on hook/off hook, address digits, coin deposits, etc.) is inherently digital and, hence, readily incorporated into a digital transmission system. One means of incorporating control information into a digital transmission link involves time division multiplexing the control as a separate but easily identifiable control channel. Another approach involves inserting special control codes into the message channel and having digital logic in the receiving terminals decode that control information. In either case, as far as the transmission system is concerned, control information is indistinguishable from message traffic.

*Analog time division multiplexing has been used in a few telephone applications. Farinon's Subscriber Radio System, for example, uses pulse width modulated TDM [1]. As discussed in Chapter 5 some older PBXs also used analog TDM.

In contrast, analog transmission systems require special attention for control signaling. Many analog transmission systems present unique and sometimes difficult environments for inserting control information. An unfortunate result was that many varieties of control signal formats and procedures evolved. The control formats depend on the nature of both the transmission system and its terminal equipment. In some interfaces between network subsystems control information must be converted from one format to another. Signaling on analog links therefore represented a significant administrative and financial burden to the operating telephone companies.

The move to common channel signaling removes most of the signaling costs associated with interoffice trunks but does not change the situation for individual subscriber lines, which must carry signaling on the same facility as the message channel. The use of digital subscriber lines (DSLs) reduces the signaling costs associated with analog subscriber lines, which helps offset the higher cost of a DSL and a digital telephone. DSLs are a fundamental aspect of ISDN as described in Chapter 9.

In summary, digital systems allow control information to be inserted into and extracted from a message stream independently of the nature of transmission medium (cable, fiber, microwave, satellite, etc.). Thus the signaling equipment can (and should) be designed separately from the transmission system. It then follows that control functions and formats can be modified independently of the transmission subsystem. Conversely, digital transmission systems can be upgraded without impacting control functions at either end of the link.

2.1.3 Use of Modern Technology

A multiplexer or switching matrix for time division digital signals is implemented with the same basic circuits used to build digital computers: logic gates and memory. The basic crosspoint of a digital switch is nothing more than an "AND" gate with one logic input assigned to the message signal and the other inputs used for control (crosspoint selection). Thus the dramatic developments of digital integrated circuit technology for computer logic circuits and memory are applicable directly to digital transmission and switching systems. In fact, many standard circuits developed for use in computers are directly usable in a switching matrix or multiplexer. Figure 2.1 shows the basic implementation of a 16 channel, bit-interleaved, digital time division multiplexer using common digital logic circuits. As indicated, the multiplexing function involves nothing more than cyclically sampling the 16 input data streams. Such an operation assumes all of the data streams are synchronized to each other. As discussed in Chapter 7, the process of synchronizing the data streams requires logic circuitry that is much more complicated than that shown. Nevertheless, the implementation of a TDM multiplexer is much less expensive than analog FDM.

Even greater advantages of modern technology have been achieved by using large-scale integrated (LSI) circuits designed specifically for telecommunications functions such as voice encoding/decoding codecs, multiplexers, switching

Figure 2.1. Sixteen to one TDM multiplexer.

matrices, and special purpose and general purpose digital signal processors (DSPs). Digital signal processing functions are described in Section 2.2.

The relative low cost and high performance of digital circuits allows digital implementations to be used in some applications that are prohibitively expensive when implemented with comparable analog components. Completely nonblocking switches, for example, are not practical with conventional analog implementations, except in small sizes. In a modern digital switch the cost of the switching matrix itself is relatively insignificant. Thus, for medium-size applications, the size of the switch matrix can be increased to provide nonblocking operations, if desired. The automatic call distributer developed by Collins-Rockwell [2] is an early example of a digital switch operating in an analog environment. A digital implementation was chosen largely because it could economically provide a nonblocking operation.

The benefits of modern device technology are not confined to digital circuits alone. Analog integrated circuits have also progressed significantly—allowing traditional analog implementations to improve appreciably. One of the primary requirements of an analog component, however, is that it be linear. It appears, if only because of research and development emphasis, that fast digital components are easier to manufacture than linear analog counterparts. In addition, digital implementations appear to have an inherent functional advantage over analog implementations. This advantage is derived from the relative ease with which digital signals can be multiplexed. A major limitation with the full use of LSI components results from limited availability of external connections to the device. With time division multiplex techniques, a single physical pin can be

used for multiple channel access into the device. Thus the same technique used to reduce costs in transmission systems can also be used within a local module to minimize the interconnections and maximize the utilization of very large scale integration. In the end, a "switch on a chip" is possible only if a great number of channels can be multiplexed onto a relatively small number of external connections.

The technological development to have the most significant impact on the telephone network is certainly fiber optic transmission. Although fibers themselves do not favor digital transmission over analog transmission, the interface electronics to a fiber system function primarily in an on–off (nonlinear) mode of operation. Thus digital transmission dominates fiber applications although research into analog fiber technology is significant—particularly for video signals.

2.1.4 Integration of Transmission and Switching

Traditionally the analog transmission and switching systems of telephone networks were designed and administered by functionally independent organizations. In the operating telephone companies, these two equipment classes are referred to as outside plant and inside plant, respectively. These equipments necessarily provide standardized interfaces, but, other than that, transmission equipment is functionally independent of switching equipment.

When time division multiplexing of digital voice signals was introduced into the exchange area and communications engineers began considering digital switching, it became apparent that time division multiplexing operations were very similar to time division switching functions. In fact, as described in later chapters, the first stages of digital switches generate first level TDM signals by nature—even when interfaced to analog transmission links. Thus the multiplexing operations of a transmission system can be easily integrated into the switching equipment.

The basic advantage of integrating the two systems is shown in Figure 2.2. The demultiplexing equipment (channel banks) at the switching offices is unnecessary, and first stage switching equipment is eliminated. If both ends of the digital TDM trunks are integrated into a digital switch, the channel banks at both ends of the trunk are eliminated. In a totally integrated network voice signals are digitized at or near the source and remain digitized until delivered to their destination. Furthermore, all interoffice trunks and internal links of a switching system carry TDM signals exclusively. Thus first level multiplexing and demultiplexing is nonexistent except at the periphery of the network. Although integration of DS1 signals into switching machines is commonplace, integration of higher level signals is complicated by higher level multiplexing formats (pulse stuffing) described in Chapter 7. A newer multiplexing format (SONET) described in Chapter 8 is much more amenable to direct interconnection into a switching system.

Integration of transmission and switching functions not only eliminates much

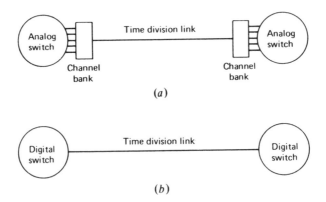

Figure 2.2. Integration of transmission and switching. (a) Nonintegrated transmission and switching. (b) Integrated time time division switching and transmission.

equipment, but it also greatly improves end-to-end voice quality. By eliminating multiple analog-to-digital and digital-to-analog conversions and by using low error rate transmission links, voice quality is determined only by the encoding process. In summary, the implementation benefits of a fully integrated digital network are as follows:

1. Long-distance voice quality is identical to local voice quality in all aspects of noise, signal level, and distortion.
2. Since digital circuits are inherently four-wire, echos are eliminated, and true full-duplex, four-wire digital circuits are available.
3. Cable entrance requirements and mainframe distribution of wire pairs is greatly reduced because all trunks are implemented as subchannels of a TDM signal.

2.1.5 Signal Regeneration

As described more fully in the next chapter, the representation of voice (or any analog signal) in a digital format involves converting the continuous analog waveform into a sequence of discrete sample values. Each discrete sample value is represented by some number of binary digits of information. When transmitted, each binary digit is represented by only one of two possible signal values (e.g., a pulse versus no pulse or a positive pulse versus a negative pulse). The receiver's job is to decide which discrete values were transmitted and represent the message as a sequence of binary encoded discrete message samples. If only small amounts of noise, interference, or distortion are impressed upon the signal during transmission, the binary data in the receiver are identical to the binary sequence generated during the digitization or encoding process. As shown in Figure 2.3, the transmission process, despite the existence of certain imper-

fections, does not alter the essential nature of the information. Of course, if the imperfections cause sufficient changes in the signal, detection errors occur and the binary data in the receiver does not represent the original data exactly.

A fundamental attribute of a digital system is that the probability of transmission errors can be made arbitrarily small by inserting regenerative repeaters at intermediate points in the transmission link. If spaced close enough together, these intermediate nodes detect and regenerate the digital signals before channel-induced degradations become large enough to cause decision errors. As demonstrated in Chapter 4, the end-to-end error rate can be made arbitrarily small by inserting a sufficient number of regeneration nodes in the transmission link.

The most direct benefit of the regeneration process is the ability to localize the effects of signal degradations. As long as the degradations on any particular regenerated segment of a transmission link do not cause errors, their effects are eliminated. In contrast, signal impairments in analog transmission accumulate from one segment to the next. Individual subsystems of a large analog network must be designed with tight controls on the transmission performance to provide acceptable end-to-end quality. An individual subsystem of a digital network, on the other hand, need only be designed to ensure a certain minimum error rate—usually a readily realizable goal.

When an all-digital network is designed with enough regeneration points to effectively eliminate channel errors, the overall transmission quality of the network is determined by the digitization process and not by the transmission systems. The analog-to-digital conversion process inherently introduces a loss of signal fidelity since the continuous analog source waveform can only be represented by discrete sample values. By establishing enough discrete levels, however, the analog waveform can be represented with as little conversion error as desired. The increased resolution requires more bits and consequently more bandwidth for transmission. Hence, a digital transmission system readily provides a trade-off between transmission quality and bandwidth. (A similar trade-off exists for frequency-modulated analog signals.)

2.1.6 Performance Monitorability

An additional benefit of the source independent signal structure in a digital transmission system is that the quality of the received signal can be ascertained with no knowledge of the nature of the traffic. The transmission link is designed

Figure 2.3. Signal regeneration in a digital repeatered line.

to produce well defined pulses with discrete levels. Any deviation in the receive signal, other than nominal amounts planned for in the design, represents a degradation in transmission quality. In general, analog systems can not be monitored or tested for quality while in service since the transmitted signal structure is unknown. FDM multiplex signals typically include pilot signals to measure channel continuity and power levels. The power level of a pilot is an effective means of estimating the signal-to-noise ratio—only in a fixed noise environment. Hence, noise and distortion is sometimes determined by measuring the energy level in an unused message slot or at the edge of the signal passband. In neither case, however, is the quality of an in-service channel being measured directly.

One common method of measuring the quality of a digital transmission link is to add parity or CRC bits to the message stream. The redundancy introduced to the data stream by the parity bits enables digital logic circuits in a receiver to readily ascertain channel error rates. If the error rate exceeds some nominal value, the transmission link is degraded.

Another technique for measuring in-service transmission quality is used in T-carrier lines. This technique involves monitoring certain redundancies in the signal waveform itself. When the redundancy pattern at the receiver deviates from normal, decision errors have occurred. A complete description of the line coding format used in T-carrier systems is provided in Chapter 4. Other methods of measuring transmission quality in digital systems are discussed in Chapters 4 and 6.

2.1.7 Accommodation of Other Services

It was previously pointed out that a digital transmission system readily accommodates control (signaling) information. This fact is representative of a fundamental aspect of digital transmission: any digitally encoded message (whether inherently digital or converted from analog) presents a common signal format to the transmission system. Thus the transmission system need provide no special attention to individual services and can, in fact, be totally indifferent to the nature of the traffic it carries.

In an analog network the transmission standard is the 4-kHz voice circuit. All special services such as data or facsimile must be transformed to "look like voice." In particular, data signals must be converted to an analog format through the use of modems.

The standard analog channel is necessarily optimized for voice quality. In so doing, certain transmission characteristics (such as the phase response and impulse noise) have received less attention than more noticeable voice quality impairments. Some less-emphasized considerations, phase distortion in particular, are critical for high-rate data services. Use of the analog network for nonvoice services may require special compensation for various analog transmission impairments. If the analog channel is too poor, it may be unusable for a particular application. In contrast, the main parameter of quality in a digital

system is the error rate. Low error rate channels are readily obtainable. When desired, the effects of channel errors can be effectively eliminated with error control procedures implemented by the user.

An additional benefit of the common transmission format is that traffic from different types of sources can be intermixed in a single transmission medium without mutual interference. The use of a common transmission medium for analog signals is sometimes complicated because individual services require differing levels of quality. For example, television signals, which require greater transmission quality than voice signals, are not usually combined with FDM voice channels in a wideband analog transmission system [3].

Although telephone companies are primary concerned with voice services (POTS), the rapid growth in data communications stimulated increased attention to data transmission performance requirements. The inherent advantages of digital systems for data communications will help to stimulate even more growth in nonvoice services as digital channels become accessible through ISDN.

2.1.8 Operability at Low Signal-to-Noise/Interference Ratios

Noise and interference in an analog voice network becomes most apparent during speech pauses when the signal amplitude is low. Relatively small amounts of noise occurring during a speech pause can be quite annoying to a listener. The same levels of noise or interference are virtually unnoticeable when speech is present. Hence it is the absolute noise level of an idle channel that determines analog speech quality. Subjective evaluations of voice quality [4, 5] have led to maximum noise level standards of 28 dBrnC0 (-62 dBm0) for short-haul systems and 34 dBrnC0 (-56 dBm0) for long-haul systems. For comparison, the power level of an active talker is typically -16 dBm0. Thus representative end-to-end, signal-to-noise ratios in analog networks are 46 and 40 dB for short- and long-haul systems, respectively. Signal-to-noise ratios on individual transmission systems are necessarily higher.

In a digital system speech pauses are encoded with a particular data pattern and transmitted at the same power level as active speech. Because signal regeneration virtually eliminates all noise arising in the transmission medium, idle channel noise is determined by the encoding process and not the transmission link. Thus speech pauses do not determine maximum noise levels as they do in an analog system. As discussed in Chapter 4, digital transmission links provide virtually error-free performance at signal-to-noise ratios of 15 to 25 dB, depending on the type of line coding or modulation used.

The ability of a digital transmission system to reject crosstalk is sometimes more significant than its ability to operate in relatively high levels of random noise. One of the most troublesome considerations in the design and maintanance of an analog network is the need to eliminate crosstalk between conversations. The problem is most acute during pauses on one channel while an interfering channel is at maximum power. At these times relatively low-level

crosstalk may be noticeable. The crosstalk is particularly undesirable if it is intelligible and therefore violates someone's privacy. Again, speech pauses do not produce low-amplitude signals on digital transmission links. The transmission links maintain a constant amplitude digital signal. Thus, low levels of crosstalk are eliminated by the regeneration process in a digital repeater or receiver. Even if the crosstalk is of sufficient amplitude to cause detection errors, the effects appear as random noise and, as such, are unintelligible.

Considering the fact that a digital system typically needs a greater bandwidth than a comparable analog system and that wider bandwidths imply greater crosstalk and noise levels, the ability to operate at lower SNRs may be as much a requirement of a digital system as an advantage.

2.1.9 Ease of Encryption

Although most telephone users have little need for voice encryption, the ease with which a digital bit stream can be scrambled and unscrambled [6] means that a digital network provides an extra bonus for users with sensitive conversations. In contrast, analog voice is much more difficult to encrypt and is generally not nearly as secure as digitally encrypted voice. For a comparison of common analog voice encryption techniques see references [7], [8], and [9]. As mentioned previously, it is the encryptability of digital voice that has attracted military interest.

2.2 DIGITAL SIGNAL PROCESSING

The preceding paragraphs emphasize the advantages of digital technology in implementing the transmission and switching systems of a network. Another significant application of digital technology is the area of signal processing. Basically, signal processing refers to an operation on a signal to enhance or transform its characteristics. Digital signal processing can be applied to either analog or digital waveforms. Amplification, equalization, modulation, and filtering are common examples of signal processing functions.

Digital signal processing (DSP) refers to the use of digital logic and arithmetic circuits to implement signal processing functions on digitized signal waveforms. Sometimes analog signals are converted to digital representations for the express purpose of processing them digitally. Then the digital representations of the processed signals are converted back to analog. These operations are illustrated in Figure 2.4, where a sinewave corrupted by noise is digitally filtered to remove the noise. The main advantages of digitally processing signal waveforms are listed in Table 2.2.

It is important to point out that digital signal processing in this context refers to the technology used to condition, manipulated, or otherwise transform a signal waveform (a digitized representation thereof). In another context signal processing refers to the interpretation of control signals in a network by the

Figure 2.4. Digital signal processing of an analog signal.

TABLE 2.2 Digital Signal Processing Features

Reproducibility
 The immunity of digital circuits to small imperfections and parasitic elements implies that circuits can be produced with consistent operational characteristics without fine adjustments or aging tolerances.

Programmability
 A single basic structure can be used for a variety of signal types and applications by merely changing an algorithmic or parametric specification in a digital memory.

Time sharing
 A single digital signal processing circuit can be used for multiple signals by storing temporary results of each process in random access memory and processing each signal in a cyclic (time divided) fashion.

Automatic test
 Since the inputs and outputs of a digital signal processing circuit are digital data, tests can be performed routinely by comparing test responses to data patterns stored in memory.

Versatility
 Because of the decision making capabilities of digital logic, digital signal processing can perform many functions that are impossible or impractical with analog implementations.

control processors of switching systems. In the latter case the logical interpretation of a control code is processed and not the underlying signal waveforms.

2.2.1 DSP Applications

The following four sections identify applications of digital signal processing that represent either lower cost solutions to functions that have been traditionally implemented with analog technology or are functions that have no practical implementation with analog technology.

Echo Cancellers
The cost and performance of DSP echo cancellers have improved to the point that they can be justified for almost all long-distance circuits—thereby providing full-duplex circuits (no echo suppression) and no artificial attenuation (no via net loss). The inherent delays and adaptation required for an echo canceller

imply DSP implementations are significantly less expensive than analog implementations.

Tone Receivers

Detection of DTMF, MF, SF, or other analog tones is easily and economically realized by converting the analog signals to digital representations for the explicit purpose of detecting the tone. Of course, a DSP implementation is even more economical when the tones are already digitized, which is the case with a digital switch. The programmability feature of a DSP circuit is particularly useful for tone receivers because one hardware implementation can be used for multiple functions by selecting different filter options (programs) depending on the application [10].

High-Speed Modems

Reliable operation (low bit error rates) of high speed (9600 bps and higher) voiceband modems over the switched telephone network requires sophisticated modulation techniques (described in Chapter 6) and sophisticated signal conditioning referred to as adaptive equalization. The most practical way to implement these functions is with DSP circuitry. References [11] and [12] describe the application of DSP to a 9600-bps and a 14,400-bps modem, respectively. Reference [13] describes the use of DSP for adaptive equalization of a 400-Mbps digital radio. Previous digital radios used analog adaptive equalizers because they were cheaper. Very high rate digital radios require more sophisticated equalizers, which are easier to implement with DSP.

Low Bit Rate Voice Encoding

The realization of low bit rate voice encoding algorithms described in Chapter 3 involves extensive numerical processing to remove redundancy in the digitized voice samples. DSP technology is the only economical means of implementing these algorithms on a real time basis. References [14], [15], and [16] describe DSP implementations for 32-kbps, 16-kbps, and 4.8-kbps voice coders, respectively.

2.3 DISADVANTAGES OF DIGITAL VOICE NETWORKS

The first part of this chapter discussed the basic technical advantages of digital networks. To balance the discussion this section reviews the basic technical disadvantages of digital implementations as listed in Table 2.3.

2.3.1 Increased Bandwidth

In the brief introduction to voice digitization presented in Chapter 1, it is pointed out that transmission of samples of an analog waveform requires no more bandwidth that the underlying waveform (at least in theory). The

bandwidth expansion comes when the samples are encoded into binary codes and transmitted with an individual pulse for each bit in the code. Thus a T1 system requires approximately eight times as much bandwidth as do 24 analog voice channels since each sample is represented by an 8 bit code word. In actual practice, a somewhat larger bandwidth is needed. Although more sophisticated digitization algorithms can be used to encode voice at a lower bit rate than that used on T1 systems (64 kbps) even the most sophisticated algorithms (described in Chapter 3) cannot provide standard telephone quality voice without at least a three to one bandwidth penalty.

In some portions of the present network, such as the local loops, the bandwidth increase does not represent much of a penalty since the inherent bandwidth is underutilized. In long-haul radio systems, however, bandwidth is at a premium, and digital systems are generally inefficient in terms of the number of voice channels provided. However, the ability of a digital system to overcome higher levels of noise and interference sometimes provides compensation for the bandwidth requirements—particularly in congested transmission environments where mutual interference can become a limiting consideration [3].

The bandwidth penalty imposed by voice digitization is directly dependent on the form of transmission coding or modulation used. With greater sophistication in the modulation/demodulation equipment, greater efficiency in terms of the bit rate in a given bandwidth is achievable. Basically, greater transmission efficiency is achieved by increasing the number of levels in the line code. With limited transmit power, however, the distances between discrete signal levels in the receiver are reduced dramatically. Thus, the transmitted signal is no longer as immune to noise and other imperfections as it is with lower information densities.

Using a combination of an advanced digital modulation technique (described in Chapter 6) and a lower rate digitization technique, point-to-point digital radios can provide voice channel efficiencies comparable to analog microwave systems. However, when lower rate digitization algorithms are used on internal equipment of the network it complicates the operation of the network because such equipment must be bypassed to provide data transparency.

2.3.2 Need for Time Synchronization

Whenever digital information is transmitted from one place to another a timing reference or "clock" is needed to control the transfer. The clock specifies when to

TABLE 2.3 Disadvantages of Digital Implementations

1. Increased bandwidth
2. Need for time synchronization
3. Topologically restricted multiplexing
4. Need for conference/extension bridges
5. Incompatibilities with analog facilities

sample the incoming signal to decide which data value was transmitted. The optimum sample times usually correspond to the middle of the transmitted pulses. Thus, for optimum detection, the sample clock must be synchronized to the pulse arrival times. In general, the generation of a local timing reference for detecting the digital signal is not difficult. Chapter 4 discusses some of the design considerations needed to establish proper sample clocking in the receiver of a digital transmission link.

More subtle problems arise, however, when a number of digital transmission links and switches are interconnected to form a network. Not only must the individual elements of the network maintain internal synchronization, but also certain networkwide synchronization procedures must be established before the individual subsystems can interoperate properly. Chapter 7 discusses these basic network synchronization requirements and implementations.

The need for some form of synchronization is not unique to digital networks. Single sideband FDM transmission systems present similar requirements for carrier synchronization in analog networks. In analog systems, however, the synchronization requirements are less critical by about two orders of magnitude [17].

2.3.3 Topologically Restricted Multiplexing

To the general public, the most apparent use of multiplexing is broadcast services for radio and television. In these systems the airspace is shared by using frequency division multiplexing of individual broadcast channels. With this system there are no operational restrictions to the geographic location of transmitters and receivers. As long as the transmitters confine their emissions to their assigned bandwidth and each receiver uses a sufficiently selective filter to pass only the desired channel, the network operates without mutual interference. Time division multiplexing, on the other hand, is not nearly as amenable to applications involving distributed sources and destinations. Since the time-of-arrival of data in a time slot is dependent on the distance of travel, distributed TDM systems require a guard time between time slots. FDM systems also require guardbands between the channels to achieve adequate channel separation. The width of the guardbands, however, is not dependent on the geographic location of the transmitters. In a TDM system the guard times must be increased as the geographic separation between transmitters increases. Furthermore, each time division source must duplicate the synchronization and time slot recognition logic needed to operate a TDM system. For these reasons, TDM has been used primarily in applications (e.g. interoffice trunks) where all of the information sources are centrally located and a single multiplexer controls the occurrence and assignment of time slots.

Time division multiple access satellites are examples of an application of TDM for distributed sources. These systems use sophisticated synchronization techniques so that each ground station times its transmission to arrive at the satellite at precisely defined times, allowing the use of small guard times between

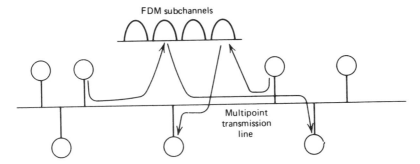

Figure 2.5. Frequency division multiplexing on distributed multipoint line.

time slots. Notice that this application involves only one destination, the satellite. If an application involves multiple distributed sources and destinations with transmission in more than one direction, larger guard times are unavoidable. Figure 2.5 shows such an application but uses frequency division multiplexing instead of time division multiplexing.

In addition to guard times, operation of a multipoint TDM network requires preamble overhead at the beginning of each transmission to allow receiving stations to synchronize to the individual sources. The guard time and preamble overhead are not too great a burden in a multisource, multidestination network like an Ethernet LAN because data are transferred in relatively large blocks. However, such overhead would be very significant in a multichannel TDM system in which delay considerations restrict the size of individual time slots. Time division multiplex loops (using a series of unidirectional transmission links) are a means of interconnecting distributed nodes with a TDM system using no guard times and very short preambles (framing patterns). TDM loops are described in Chapter 4.

2.3.4 Need for Conference/Extension Bridges

The process of combining multiple analog signals to form a conference call or function as multiple extensions on a single line can be accomplished by merely bridging the wire pairs together to superimpose all signals. Nowhere is this more convenient than when multiple extensions share a single two-wire line as indicated in Figure 2.6. When digitized voice signals are combined to form a conference the signals must either be converted to analog so they can be combined on two-wire analog bridges or the digital signals must be routed to a digital conference bridge, as shown in Figure 2.7. The digital bridge selectively adds the (four-wire) signals together (using digital signal processing) and routes separate sums back to the conferees as shown.

When conferencing is implemented in association with a switching system the need for a digital conference bridge is not much of a disadvantage and in fact can

Figure 2.6. Analog telephones connected to two-wire line.

significantly improve the quality of a conference by eliminating echoes and signal loss caused by power division. However, when digital extensions need to have their signals combined so multiple extensions can be active in a conversation, the need for a centralized bridge can be an onerous problem. Residential telephone wiring typically follows a daisy chain pattern as indicated in Figure 2.6. Thus the need to rewire all outlets and install a centralized conference box is a significant impediment to the deployment of digital station equipment in residential applications.

2.3.5 Incompatibilities with Analog Facilities

When digital equipment was first used in private and public telephone networks it necessarily provided standard analog interfaces to the rest of the network. Sometimes these interfaces represented a major cost of the digital subsystem. The foremost example of this situation arises in digital end offices. The standard analog subscriber loop interface described in Chapter 1 is particularly incompatible with electronic switching machines (analog or digital). Another aspect of digital switching that complicates its use in analog environments is the artificial delay inserted by a typical digital matrix. Both of these aspects of digital switching are discussed in Chapter 5.

One way to eliminate the problems with the analog interface is to use digital subscriber loops and digital telephones. Unfortunately, the overwhelming investment in the present loop plant for analog telephones complicates a widespread deployment of digital subscriber equipment. Most notable of the present practices that complicate a transition to digital loops are single wire

Figure 2.7. Use of conference bridge for digital telephones.

pairs, loading coils, bridged taps,* high-resistance or intermittent splices, and wiring gauge changes. The design of the standard ISDN digital subscriber loop described in Chapter 9 accommodates most of the above impediments but does so at significant expense.

REFERENCES

1 *SR-1536 Subscriber Radio, Technical Description*, Farinon SR Systems, Quebec, Canada, 1977.

2 R. J. Hirvela, "The Application of Computer Controlled PCM Switching to Automatic Call Distribution, *IEEE Communications Systems and Technology Conference*, Dallas, Texas, May 1974.

3 M. R. Aaron, "Digital Communications—The Silent (R)evolution?" *IEEE Communications Magazine*, January 1979, pp 16–26.

4 I. Nasell, "The 1962 Survey of Noise and Loss on Toll Connections," *Bell System Technical Journal*, March 1964, pp 697–718.

5 Bell Telephone Laboratories Members of Technical Staff, *Transmission Systems for Communications*, Western Electric Company, February 1970.

6 H. J. Hindin, "LSI-Based Data Encryption Discourages the Data Thief," *Electronics*, June 21, 1979, pp. 107–120.

7 N. S. Jayant, B. J. McDermott, S. W. Christensen, and A. M. Quinn, "A Comparison of Four Methods for Analog Speech Encryption," *International Communications Conference Record*, 1980, pp 16.6.1–16.6.5.

8 A. Gersho and R. Steele, Ed., "Special Issue on Encryption of Analog Signals," *IEEE Journal on Selected Areas in Communications*, May 1984.

9 H. J. Beker and F. Piper, "Speech Scrambling," *Speech Technology*, March/April 1987, pp 40–45.

10 A. Fukui and Y. Fujihashi, "A Single-Chip, 4-Channel MF/MFC/PB Receiver," *IEEE Globecom Conference Record*, 1987, pp 12.6.1–12.6.4.

11 P. Thomas, "The Design and Implementation of a CCITT V.32 Modem on a General Purpose Digital Signal Processing Microprocessor," *IEEE International Conference on Communications*, 1987, pp 12.5.1–12.5.4.

12 T. Kamitake, K. Uehara, M. Abe, and S. Kawamura, "Trellis Coding 14.4 kb/s Data Modem Implemented with a Single-Chip High-Speed Digital Signal Processor," *IEEE Globecom Conference Record*, 1987, pp 12.8.1–12.8.6.

13 H. Matsue, T. Shirato, and K. Watanabe, "256 QAM 400 Mb/s Microwave Radio System with DSP Fading Countermeasures," *IEEE International Conference on Communications*, 1988, pp 41.5.1–41.5.6.

14 J. L. So, "Implementation on an NIC (Nearly Instantaneous Companding) 32 kbps Transcoder Using the TMS320C25 Digital Signal Processor," *IEEE Globecom Conference Record*, 1988, pp 43.4.1–43.4.5.

15 J. L. Dixon, V. Varma, and D. W. Lin, "Software Considerations in the Design of a

*A bridged-tap is an unused pair of wires connected at some point to an in-use pair as another extension or for possible future reassignment of a cable pair.

16 kbps Speed Coder for a TDMA Portable Phone," *IEEE Globecom Conference Record*, 1988, pp 26.7.1–26.7.5.

16 K. Irie and S. Iai, "4.8 kbit/s Speech Codec Using Advanced Digital Signal Processors ((DSSP1," *IEEE Globecom Conference Record*, 1987, pp 20.4.1–20.4.5.

17 J. E. Abate, E. W. Butterline, R. A. Carley, P. Greendyk, A. M. Montenegro, C. D. Near, S. H. Richman, and G. P. Zampetti, "AT&T's New Approach to the Synchronization of Telecommunication Networks," *IEEE Communications Magazine*, April 1989, pp 35–45.

3

VOICE DIGITIZATION

Because of the interesting nature of the subject and its usefulness in a variety of applications, the field of voice digitization has been and continues to be an area of intense research. This research has produced many different types of voice digitization algorithms. The choice of a particular type is primarily dependent on the implementation cost and the performance requirements implied by the application. The algorithm chosen for T1 systems (companded PCM) provides excellent quality for all types of input signals (e.g., voice or data) at a moderate data rate (64 kbps) and a moderate cost. The algorithms used in the first generation digital PBXs, on the other hand, used different coding techniques (higher rate PCM or delta modulation) because a switching application is more sensitive to cost but less sensitive to quality or data rate. For example, the first digital PBX in the United States, introduced by Rolm Corporation in 1975, used uncompanded PCM at a data rate of 144 kbps because it was cheaper than companded PCM at the time [1]. Subsequent advantages derived from integrating transmission and switching and a dramatic drop in the cost of T1-compatible digital voice coders have made obsolete the use of "switching only" voice digitization algorithms. Because of high volume production, T1-compatible coder/decoder ICs (codecs) can be used in switching applications without a cost penalty.

Transmissions applications with strict bandwidth limits such as HF or digital cellular radio require much more sophisticated voice digitization algorithms to achieve data rates on the order of 8 to 16 kbps. As a help in reducing the data rate, the performance requirements of these applications are also relaxed as much as the application allows.

Another application for digitized voice is voice storage systems—either for recorded announcements or for voice messaging. Digital storage is particularly appropriate for recorded announcements because the playback quality does not deteriorate with time and individual announcements stored in memory or on a disc can be randomly accessed. Speech storage with limited memory is an example of an application that can use very low rate digitization algorithms with significant quality reductions. The Speak-and-Spell® learning machine developed by Texas Instruments, for example, stores words with a few hundred bits

per word—representing a data rate of about 850 bps [2]. In a similar example requiring better quality the same encoding algorithm (LPC) was used in the Voice Alert System of Chrysler automobiles to store 20 sec of speech (40 words) in 32,000 bits of read only memory—a data rate of 1600 bps [3].

The main reason voice messaging systems use digital storage is to have random access to the individual messages. To minimize storage (disk) space these systems typically use 16 to 32 kbps.

At the upper end of digital storage applications are high-fidelity recordings of voice and music. Many of the same advantages of digital transmission—as opposed to an analog transmission—also apply to digital recordings. Foremost among these advantages is the ability of defining the fidelity at the time of recording and maintaining the quality indefinitely by periodically copying (regenerating) the digitally stored information before inevitable deterioration produces bit errors. Thus a high-quality (high bit rate) digital recording of Bing Crosby, Elvis Presley, or Luciano Pavarotti (depending on your taste in music) can be saved for posterity. This feat could not be accomplished with analog recordings no matter how well cared for or preserved. As an example of high-fidelity audio recording, compact disc players [4] record two channels of audio at 705 kbps each.

Speech analysis and synthesis is another area of widespread research closely related to voice digitization. In fact, some of the lowest bit rate voice encoders and decoders use certain amounts of analysis and synthesis to digitally represent speech. In its ultimate form, however, analysis and synthesis have unique goals and applications fundamentally different from those of general voice digitization. Basically, goals of analysis and synthesis are to recognize words [5] or produce machine-generated speech (text to speech) [6].

One approach to analyzing speech is to process waveforms with the intent of recognizing speech phonemes—the basic units of speech from which spoken words are constructed. Once the phonemes have been identified, they are assigned individual codewords for storage or transmission. A synthesizer can then generate speech by recreating the combinations of phonemes. Analysis of this technique indicates that the information content of speech can be transmitted with a data rate of 50 bps [7]. It must be emphasized, however, that only the information content associated with the words themselves is transmitted—none of the more subjective qualities of speech such as naturalness, voice inflections, accents, and speaker recognizability. Thus such techniques, by themselves, are not applicable to general telephony, which customarily includes qualities other than the message content of spoken words.

Efficient encoding of facsimile images presents similar opportunities and limitations. Facsimile machines typically scan at 200 dots per inch, which implies there are 3.74 million bits of raw information on a 8.5 by 11 in. piece of paper. If the paper contains only recognizable text characters at 10 characters and 6 lines per inch the same information can be encoded as 5610 Ascii characters or 39,270 bits—a savings of almost 100 to 1. Besides being restricted to text-oriented messages, character-oriented encoding and decoding produces

the same output character font independent of the source (which could conceivably be hand written). Facsimile machines achieve one significant level of coding efficiency without sacrificing transparency by encoding white space into run length codes. Although this does not reduce the number of bits in a worst case (random dot) image, it greatly reduces the number of bits in the average image of interest. Similar processing is possible in voice applications by efficiently encoding silence. However, the voice problem is more complicated because real time voice requires reconstructing the temporal aspects of the source—restricting silence encoding to relatively large intervals.

Another level of speech analysis involves the actual recognition of spoken words. High levels of success have been achieved with the two main restrictions: that the system is trained on the speakers and the speakers are trained to speak with isolated words. As an example of one implementation [8] that tries to achieve the ultimate goal of speaker independence, continuous speech and large vocabularies, 71–96% recognition accuracy is possible depending on the level of the grammar specified. (A grammar defines which words are allowed to follow which words.)

Voice digitization techniques can be broadly categorized into two classes: those digitally encoding analog waveforms as faithfully as possible and those processing waveforms to encode only the perceptually significant aspects of speech and hearing processes. The first category is representative of the general problem of analog-to-digital and digital-to-analog conversions and is not restricted to speech digitization. The three most common techniques used to encode a voice waveform are pulse code modulation (PCM), differential PCM (DPCM), and delta modulation (DM). Except in special cases, digital telephony uses these techniques. Thus when studying the most common digital speech encoding techniques we are, in fact, investigating the more general realm of analog-to-digital conversion [9].

The second category of speech digitization is concerned primarily with producing very low data rate speech encoders and decoders for narrowband transmission systems or digital storage devices with limited capacity. A device from this special class of techniques is commonly referred to as a "vocoder" (voice coder). Although capable of producing intelligible speech, vocoder techniques generally produce unnatural or synthetic sounding speech. As such, low data rate vocoders do not provide adequate quality for general telephony. The following discussions emphasize waveform coding, but the basic principles and applications of vocoders are also included.

3.1 PULSE AMPLITUDE MODULATION

The first step in digitizing an analog waveform is to establish a set of discrete times at which the input signal waveform is sampled. Prevalent digitization techniques are based on the use of periodic, regularly spaced sample times. If the samples occur often enough, the original waveform can be completely recovered

Figure 3.1. Pulse amplitude modulation.

from the sample sequence using a low-pass filter to interpolate or "smooth out" between the sample values. These basic concepts are illustrated in Figure 3.1. A representative analog waveform is sampled at a constant sampling frequency $f_s = 1/T$ and reconstructed using a low-pass filter. Notice that the sampling process is equivalent to amplitude modulation of a constant-amplitude pulse train. Hence the technique represented in Figure 3.1 is usually referred to as a pulse amplitude modulation (PAM).

3.1.1 Nyquist Sampling Rate

A classical result in sampling systems was established in 1933 by Harry Nyquist when he derived the minimum sampling frequency required to extract all information in a continuous, time-varying waveform. This result—the Nyquist criterion—is defined by the relation

$$f_s > 2BW$$

where f_s = sampling frequency
BW = bandwidth of the input signal

The derivation of this result is indicated in Figure 3.2, which portrays the spectrum of the input signal and the resulting spectrum of the PAM pulse train. The PAM spectrum can be derived by observing that a continuous train of impulses has a frequency spectrum consisting of discrete terms at multiples of the sampling frequency. The input signal amplitude modulates these terms individually. Thus a double sideband spectrum is produced about each of the discrete frequency terms in the spectrum of the pulse train. The original signal waveform is recovered by a low-pass filter designed to remove all but the original signal spectrum. As shown in Figure 3.2, the reconstructive low-pass filter must have a cut-off frequency that lies between BW and $f_s - BW$. Hence, separation is only possible if $f_s - BW$ is greater than BW (i.e., if $f_s > 2BW$).

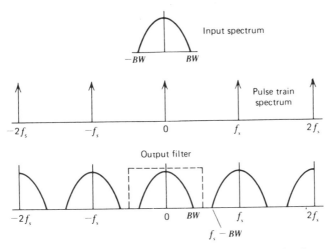

Figure 3.2. Spectrum of pulse amplitude modulated signal.

3.1.2 Foldover Distortion

If the input waveform of a PAM system is undersampled ($f_s < 2BW$), the original waveform cannot be recovered without distortion. As indicated in Figure 3.3, this output distortion arises because the frequency spectrum centered about the sampling frequency overlaps the original spectrum and cannot be separated from the original spectrum by filtering. Since it is a duplicate of the input spectrum "folded" back on top of the desired spectrum that causes the distortion, this type of sampling impairment is often referred to as "foldover distortion."

In essence, foldover distortion produces frequency components in the desired frequency band that did not exist in the original waveform. Thus another term for this impairment is "aliasing." Aliasing problems are not confined to speech digitization processes. The potential for aliasing is present in any sample data system. Motion picture taking, for example, is another sampling system that can produce aliasing. A common example occurs when filming moving stagecoaches in old Westerns. Often the sampling process is too slow to keep up with the

Figure 3.3. Foldover spectrum produced by undersampling an input.

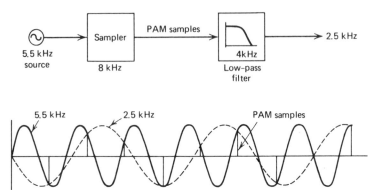

Figure 3.4. Aliasing of 5.5-kHz signal into a 2.5-kHz signal.

stagecoach wheel movements, and spurious rotations rates are produced. If the wheel rotates 355° between frames, it looks to the eye as if it has moved backward 5°.

Figure 3.4 demonstrates an aliasing process occurring in speech if a 5.5-kHz signal is sampled at an 8-kHz rate. Notice that the sample values are identical to those obtained from a 2.5-kHz input signal. Thus after the sampled signal passes through the 4-kHz output filter, a 2.5-kHz signal arises that did not come from the source. This example illustrates that the input must be bandlimited, *before sampling*, to remove frequency terms greater than $f_s/2$, even if these frequency terms are ignored (i.e., are inaudible) at the destination. Thus, a complete PAM system, shown in Figure 3.5, must include a bandlimiting filter before sampling to ensure that no spurious or source-related signals get "folded" back into the desired signal bandwidth. The input filter may also be designed to cut off very low frequencies to remove 60 cycle hum from power lines.

Figure 3.5 shows the signal being recovered by a sample-and-hold circuit that produces a staircase approximation to the sampled waveform. With use of the staircase approximation, the power level of the signal coming out of the reconstructive filter is nearly the same as the level of the sampled input signal.

Figure 3.5. End-to-end pulse amplitude modulated system.

Figure 3.6. Bandlimiting filter template designed to meet CCITT recommendations for PCM voice coders.

The response of the reconstructive filter, in this case, must be modified somewhat to account for the spectrum of the wider "staircase" samples. (The modification amounts to dividing the "flat" filter spectrum by the spectrum of the finite width pulse. See Appendix C.)

The bandlimiting and reconstructive filters shown in Figure 3.5 are implied to have ideal characteristics.* Since ideal filters are physically unrealizable, a practical implementation must consider the effects of nonideal implementations. Filters with realizable attenuation slopes at the band edge can be used if the input signal is slightly oversampled.

As indicated in Figure 3.2, when the sampling frequency f_s is somewhat greater than twice the bandwidth, the spectral bands are sufficiently separated from each other that filters with gradual roll-off characteristics can be used. As an example, sampled voice systems typically use bandlimiting filters with a 3-dB cutoff around 3.4 kHz and a sampling rate of 8 kHz. Thus the sampled signal is sufficiently attenuated at the overlap frequency of 4 kHz to adequately reduce the energy level of the foldover spectrum. Figure 3.6 shows a filter template

*An ideal filter is one with a frequency-independent time delay (linear phase), no attenuation in the passband (except as might be desired for pulse shaping), an arbitrarily steep cutoff, and infinite attenuation everywhere in the stopband.

designed to meet CCITT recommendations for out-of-band signal rejection in PCM voice coders. Notice that 14 dB of attenuation is provided at 4 kHz.

As mentioned in Chapter 1, the perceived quality of a voice signal is not greatly dependent upon the phase response of the channel (the relative delay of individual frequency components). For this reason the phase responses of the bandlimiting filters in the encoders and the smoothing filters in the decoders are not critical. Nonlinear phase responses in these filters, however, do impact high-rate voiceband data signals when digitized. Hence a somewhat paradoxical situation arises when voiceband data are transmitted over a T-carrier line: the process of converting the voiceband data signal (4800 bps typical) to a virtually error free 64-kbps digital signal causes distortion in the lower-rate data signal. However, because of regeneration the transmission process itself does not add to the signal degradation.

By interleaving the samples from multiple sources, PAM systems can be used to share a transmission facility in a time division multiplex manner. PAM systems are not generally useful over long distances, however, owing to the vulnerability of the individual pulses to noise, distortion, intersymbol interference, and crosstalk. Instead, for long-distance transmission the PAM samples are converted into a digital format, thereby allowing the use of regenerative repeaters to remove transmission imperfections before errors result. Time division PAM techniques have been used in some PBX systems described in Chapter 5.

3.2 PULSE CODE MODULATION

The preceding section describes pulse amplitude modulation, which uses discrete sample times with analog sample amplitudes to extract the information in a continuously varying analog signal. Pulse code modulation (PCM) is an extension of PAM wherein each analog sample value is quantized into a discrete value for representation as a digital code word. Thus, as shown in Figure 3.7, a PAM system can be converted into a PCM system by adding an analog-to-digital (A/D) converter at the source and a digital-to-analog (D/A) converter at the destination. Figure 3.8 depicts a typical quantization process in which a set of quantization intervals is associated in a one-to-one fashion with a binary code

Figure 3.7. Pulse code modulation.

Figure 3.8. Quantization of analog samples.

word. All sample values falling in a particular quantization interval are represented by a single discrete value located at the center of the quantization interval. In this manner the quantization process introduces a certain amount of error or distortion into the signal samples. This error, known as "quantization noise," is minimized by establishing a large number of small quantization intervals. Of course, as the number of quantization intervals increases, so must the number of bits increase to uniquely identify the quantization intervals.

3.2.1 Quantization Noise

A fundamental aspect of the design and development of an engineering project is the need for analytical measures of systems performance. Only then can a system be objectively measured and its cost effectiveness compared to alternate designs. One of the measures needed by a voice communication engineer is the quality of speech delivered to the listener. Measurements of speech quality are complicated by subjective attributes of speech as perceived by a typical listener. One subjective aspect of noise or distortion on a speech signal involves the frequency content, or spectrum, of the disturbance in conjunction with the power level. These effects of noise as a function of frequency are discussed in Chapter 1 with the introduction of C-message and psophometric weighting.

Successive quantization errors of a PCM encoder are generally assumed to be distributed randomly and uncorrelated to each other. Thus the cummulative effect of quantization errors in a PCM system can be treated as additive noise with a subjective effect that is similar to bandlimited white noise. Figure 3.9 shows the quantization noise as a function of signal amplitude for a coder with uniform quantization intervals. Notice that if the signal has enough time to change in amplitude by several quantization intervals, the quantization errors are independent. If the signal is oversampled, that is sampled much higher than the Nyquist rate, successive samples are likely to fall in the same interval causing a loss of independence in the quantization errors.

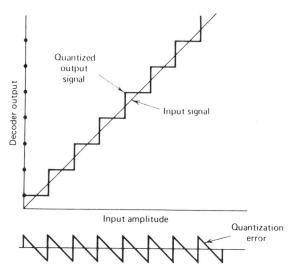

Figure 3.9. Quantization error as a function of amplitude over a range of quantization intervals.

The quantization error or distortion created by digitizing an analog signal is customarily expressed as an average noise power relative to the average signal power. Thus the signal-to-quantizing noise ratio SQR (also called a signal-to-distortion ratio or a signal-to-noise ratio) can be determined as

$$ \text{SQR} = \frac{E\{x^2(t)\}}{E\{[y(t) - x(t)]^2\}} \tag{3.1} $$

where $E\{\cdot\}$ = expectation or averaging
 $x(t)$ = the analog input signal
 $y(t)$ = the decoded output signal

In determining the expected value of the quantization noise, three observations are necessary:

1. The error $y(t) - x(t)$ is limited in amplitude to $q/2$ where q is the height of the quantization interval. (Decoded output samples are ideally positioned at the middle of a quantization interval.)
2. A sample value is equally likely to fall anywhere within a quantization interval—implying a uniform probability density of amplitude $1/q$.
3. Signal amplitudes are assumed to be confined to the maximum range of the coder. If a sample value exceeds the range of the highest quantization interval, overload distortion (also called peak limiting) occurs.

If we assume (for convenience) a resistance level of 1 Ω, the average quantization noise power is determined in Appendix A as

$$\text{quantization noise power} = \frac{q^2}{12} \tag{3.2}$$

If all quantization intervals have equal lengths (uniform quantization), the quantization noise is independent of the sample values and the signal-to-quantization ratio is determined as*

$$\text{SQR (dB)} = 10\log_{10}\left(\frac{v^2}{q^2/12}\right)$$

$$= 10.8 + 20\log_{10}\left(\frac{v}{q}\right) \tag{3.3}$$

where v is the rms amplitude of the input. In particular, for a sine wave input the signal-to-quantizing noise ratio produced by uniform quantization is

$$\text{SQR (dB)} = 10\log_{10}\left(\frac{A^2/2}{q^2/12}\right)$$

$$= 7.78 + 20\log_{10}(A/q) \tag{3.4}$$

where A is the peak amplitude of the sine wave.

Example 3.1. A sine wave with a 1 V maximum amplitude is to be digitized with a minimum signal-to-quantization noise ratio of 30 dB. How many uniformly spaced quantization intervals are needed, and how many bits are needed to encode each sample?

Solution. Using Equation 3.4, the maximum size of a quantization interval is determined as

$$q = (1)10^{-(30-7.78)/20}$$

$$= 0.078 \text{ V}$$

*The signal-to-quantization noise ratios commonly compare unfiltered decoder outputs to un-filtered quantization errors. In actual practice, the decoder output filter reduces the power level of both the signal and the noise. The noise power experiences a greater reduction than the signal power, since the uncorrelated noise samples have a wider spectrum than the correlated voice samples. Thus filtered signal-to-noise ratios are usually higher than the values calculated here by 1 to 2 dB.

Figure 3.10. Idle channel noise produced by mid-riser quantization.

Thus 13 quantization intervals are needed for each polarity for a total of 26 intervals in all. The number of bits required to encode each sample is determined as

$$N = \log_2(26)$$

$$= 4.7$$

$$= 5 \text{ bits per sample}$$

When measuring quantization noise power, the spectral content is often weighted in the same manner as noise in an analog circuit. Unfortunately, spectrally weighted noise measurements do not always reflect the true perceptual quality of a voice encoder/decoder. If the spectral distribution of the quantization noise more or less follows the spectral content of the speech waveform, the noise is masked by the speech and is much less noticeable than noise uncorrelated to the speech [10]. On the other hand, if the quantization process produces energy at voiceband frequencies other than those contained in particular sounds, they are more noticeable.

High-quality PCM encoders produce quantization noise that is evenly distributed across voice frequencies and independent of the encoded waveforms. Thus quantization noise ratios defined in Equation 3.4 are good measures of PCM performance. In some of the encoders discussed later (vocoders in particular) quantization noise power is not very useful. References [9], [11], and [12] describe other measures of encoder speech quality providing better correlations to quality as perceived by a listener.

3.2.2 Idle Channel Noise

Examination of Equations 3.3 and 3.4 reveals that the SQR is small for small sample values. In fact, as shown in Figure 3.10, the noise may actually be greater than the signal when sample values are in the first quantization interval. This effect is particularly bothersome during speech pauses and is known as idle channel noise. Figure 3.11 depicts one method of minimizing idle channel noise in PCM systems by establishing a quantization interval that straddles the origin.

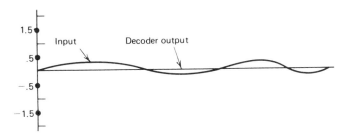

Figure 3.11. Elimination of idle channel noise by mid-tread quantization.

In this case all sample values in the central quantization interval are decoded as a constant zero output. PCM systems of this type use an odd number of quantization intervals since the encoding ranges of positive and negative signals are usually equal.

The quantization characteristics required to produce the output waveforms shown in Figures 3.10 and 3.11 are shown in Figures 3.12 and 3.13, respectively. The first characteristic (mid-riser) cannot produce a zero output level. The second characteristic (mid-tread) decodes low signals into constant, zero-level outputs. However, if the signal amplitude is comparable to the size of the quantization interval, or if a dc bias exists in the encoder, idle channel noise is a problem with mid-tread quantization also.

As mentioned in Chapter 1, noise occurring during speech pauses is more objectionable than noise with equivalent power levels during speech. Thus idle channel noise is specified in absolute terms separate from quantization noise which is specified relative to the signal level. For example, Bell system D3 channel bank specifications list the maximum idle channel noise as 23 dBrnC0 [13].

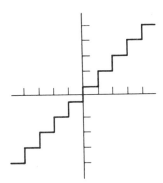

Figure 3.12. Mid-riser quantizer characteristic.

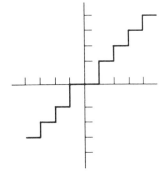

Figure 3.13. Mid-tread quantizer characteristic.

3.2.3 Uniformly Encoded PCM

An encoder using equal length quantization intervals for all samples produces code words linearly related to the analog sample values. That is, the numerical equivalent of each code word is proportional to the quantized sample value it represents. In this manner a uniform PCM system uses a conventional analog-to-digital converter to generate the binary sample codes. The number of bits required for each sample is determined by the maximum acceptable noise power. Minimum digitized voice quality requires a signal-to-noise ratio in excess of 26 dB [14]. For a uniform PCM system to achieve a signal-to-quantization ratio of 26 dB, Equation 3.4 indicates that $q_{max} = 0.123A$. For equal positive and negative signal excursions (encoding from $-A$ to A), this result indicates that just over 16 quantization intervals or 4 bits per sample are required.*

In addition to providing adequate quality for small signals, a telephone system must be capable of transmitting a large range of signal amplitudes referred to as dynamic range. Dynamic range (DR) is usually expressed in decibels as the ratio of the maximum amplitude signal to the minimum amplitude signal:

$$DR = 10 \log_{10}(P_{max}/P_{min})$$
$$= 20 \log_{10}(V_{max}/V_{min}) \tag{3.5}$$

A typical minimum dynamic range is 30 dB [14]. Thus signal values as large as 31 times A must be encoded without exceeding the range of quantiza-

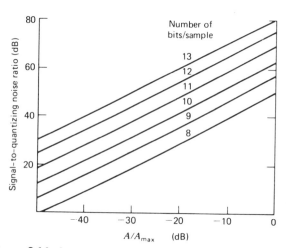

Figure 3.14. Signal-to-quantizing noise of uniform PCM coding.

*This SQR objective is for minimum acceptable performance and assumes all degradations occur in a single encoder. If additional signal impairments occur (such as multiple analog-to-digital conversions), the encoder must use more bits to provide noise margin for other elements in the network.

tion intervals. Assuming equally spaced quantization intervals for uniform coding, the total number of intervals is determined as 496, which requires 9 bit code words.*

The performance of an n-bit uniform PCM system is determined by observing that

$$q = \frac{2A_{max}}{2^n} \tag{3.6}$$

where A_{max} is the maximum (unoverloaded) amplitude.

Substituting Equation 3.6 into Equation 3.4 produces the PCM performance equation for uniform coding:

$$SQR = 1.76 + 6.02n + 20\log_{10}\left(\frac{A}{A_{max}}\right) \tag{3.7}$$

The first two terms of Equation 3.7 provide the SQR when encoding a full range sine wave. The last term indicates a loss in SQR when encoding a lower level signal. These relationships are presented in Figure 3.14 which shows the SQR of a uniform PCM system as a function of the number of bits per sample and the magnitude of an input sine wave.

Example 3.2. What is the minimum bit rate that a uniform PCM encoder must provide to encode a high fidelity audio signal with a dynamic range of 40 dB? Assume the fidelity requirements dictate passage of a 20-kHz bandwidth with a minimum signal-to-noise ratio of 50 dB. For simplicity, assume sinusoidal input signals.

Solution. To prevent foldover distortion, the sampling rate must be at least 40 kHz. Assuming an excess sampling factor comparable to that used in D-type channel banks (4000/3400), we choose a sampling rate of 48 kHz as a compromise for a practical bandlimiting filter. By observing that a full amplitude signal is encoded with an SQR of $40 + 50 = 90$ dB, we can use Equation 3.7 to determine the number of bits n required to encode each sample.

$$n = \frac{(90 - 1.76)}{6.02}$$

$$= 15 \text{ bits}$$

Thus the required bit rate is

$$(15)(48) = 720 \text{ kbps}$$

*This result is derived with the assumption of minimum performance requirements. Higher performance objectives (less quantization noise and greater dynamic range) require as many as 13 bits per sample for uniform PCM systems.

3.2.4 Companding

In a uniform PCM system the size of every quantization interval is determined by the SQR requirement of the lowest signal level to be encoded. Larger signals are also encoded with the same quantization interval. As indicated in Equation 3.7 and Figure 3.14, the SQR increases with the signal amplitude A. For example, a 26-dB SQR for small signals and a 30-dB dynamic range produces a 56-dB SQR for a maximum amplitude signal. In this manner a uniform PCM system provides unneeded quality for large signals. Moreover, the large signals are the least likely to occur. For these reasons the code space in a uniform PCM system is very inefficiently utilized.

A more efficient coding procedure is achieved if the quantization intervals are not uniform but allowed to increase with the sample value. When quantization intervals are directly proportional to the sample value, the SQR is constant for all signal levels. With this technique fewer bits per sample provide a specified SQR for small signals and an adequate dynamic range for large signals. When the quantization intervals are not uniform, a nonlinear relationship exists between the code words and the samples they represent.

Historically, the nonlinear function was first implemented on analog signals using nonlinear devices such as specially designed diodes [15]. The basic process is shown in Figure 3.15 where the analog input sample is first compressed and then quantized with uniform quantization intervals. The effect of the compression operation is shown in Figure 3.16. Notice that successively larger input signal intervals are compressed into constant length quantization intervals. Thus the larger the sample value, the more it is compressed before encoding. As shown in Figure 3.15, a nonuniform PCM decoder expands the compressed value using an inverse compression characteristic to recover the original sample value. The process of first compressing and then expanding a signal is referred to as "companding." When digitizing, companding amounts to assigning small quantization intervals to small samples and large quantization intervals to large samples.

Various compression–expansion characteristics can be chosen to implement a compandor. By increasing the amount of compression, we increase the dynamic range at the expense of the signal-to-noise ratio for large amplitude signals. One family of compression characteristics used in North America and Japan is the μ-law characteristic defined as

$$F_\mu(x) = \text{sgn}(x) \frac{\ln(1 + \mu|x|)}{\ln(1 + \mu)} \tag{3.8}$$

where x is the input signal amplitude $(-1 \leqslant x \leqslant 1)$, $\text{sgn}(x)$ is the polarity of x, and μ is a parameter used to define the amount of compression.

Because of the mathematical nature of the compression curve, companded PCM is sometimes referred to as log-PCM. A logarithm compression curve is ideal in the sense that quantization intervals and, hence, quantization noise are

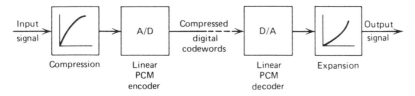

Figure 3.15. Companded PCM with analog compression and expansion.

proportional to the sample amplitude. The inverse or expansion characteristic for a μ-law compandor is defined as

$$F_\mu^{-1}(y) = \text{sgn}(y) \left(\frac{1}{\mu}\right) [(1 + \mu)^{|y|} - 1] \tag{3.9}$$

where y is the compressed value $= F_\mu(x)$ $(-1 \leqslant y \leqslant 1)$, $\text{sgn}(y)$ is the polarity of y, and μ is the companding parameter.

The first T-carrier systems in the United States used D1 channel banks [16], which approximated Equation 3.8 for $\mu = 100$. The compression and expansion functions were implemented with the specially biased diodes mentioned previously. Figure 3.17 depicts a block diagram of a D1 channel bank. Notice that the time division multiplexing and demultiplexing functions are implemented on analog PAM samples. Thus the companding and encoding/decoding functions were shared by all 24 voice channels. The ability to share this relatively expensive equipment was one of the reasons that PCM was originally chosen as

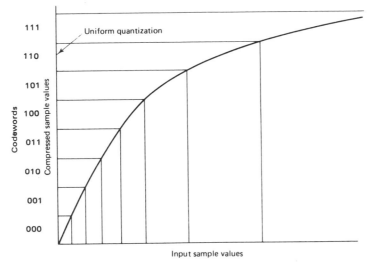

Figure 3.16. Typical compression characteristic.

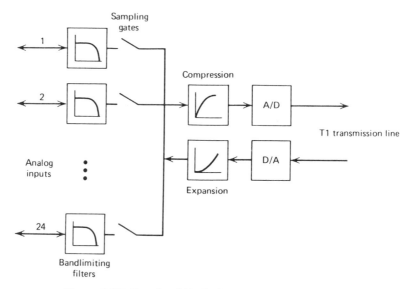

Figure 3.17. Functional block diagram of D1 channel bank.

the means of digitally encoding speech. Recent developments for large scale integrated circuit PCM codecs have diminished the need to share this equipment. Thus newer systems can use per channel codecs and provide more flexibility in implementing various system sizes. When most of the cost of a channel bank is in the common equipment, as in the original channel banks, less than fully implemented systems are overly expensive.

Each sample produced by a D1 channel bank was encoded into 7 bits: 1 polarity bit and 6 compressed magnitude bits. In addition, 1 signaling bit was added to each channel to produce an 8 bit code word for each time slot. Since the sampling rate is 8 kHz, a 64-kbps channel results. Even though the D1 channel banks have been superceded by newer channel banks utilizing a different coding format, the 64-kbps channel rate has been retained as a standard.

3.2.5 Easily Digitally Linearizable Coding

The success of the first T1 systems for interoffice **exchange** area transmission paved the way for further use of digital time division multiplex systems. As it became apparent that digital transmission was useful within the **toll network**, it also became clear that the form of PCM encoding used in the D1 channel banks was inadequate. In contrast to the exchange area, an end-to-end connection through the toll network could have conceivably involved as many as nine tandem connections.* Since digital switching was not in existence at the time that T-carrier systems for the toll network were being developed, each of these

*Nine tandem connections have never been known to occur.

tandem connections implied an extra digital-to-analog and analog-to-digital conversion. Thus the quality of each conversion had to be improved to maintain the desired end-to-end quality. The D2 channel bank [17] was therefore developed with improved digitized voice quality.

When the D2 channel bank was being developed digital switching was recognized as a coming technology—implying that channel banks would be paired on a dynamic basis as opposed to the one-to-one basis in T-carrier systems. Thus a greater degree of uniformity in companding characteristics would be required to allow pairing of channel banks on a nationwide basis. The main features incorporated into the D2 channel bank (and ensuing channel banks such as D3, D4, and D5) to achieve the improved quality and standardization are

1. Eight bits per PCM code word.
2. Incorporation of the companding functions directly into the encoder and decoder.
3. A new companding characteristic ($\mu255$).

The D1 channel banks use 1 bit per time slot for signaling and 7 bits for voice. Thus a signaling rate of 8 kbps was established, which was more than necessary for basic voice service. To provide a higher data rate for voice, signaling between D2, D3, D4, D5 channel banks is inserted into the least significant bit position of 8 bit code words in every sixth frame. Thus every sixth $\mu255$ PCM code word contains only 7 bits of voice information—implying that the effective number of bits per sample is actually $7\frac{5}{6}$ bits instead of 8. When common channel signaling is established, however, the associated T-carrier systems no longer need to carry signaling information on a per channel basis and a full 8 bits of voice can be transmitted in every time slot of every frame.

The compression and expansion characteristics of the D1 channel banks were implemented separately from the encoders and decoders. The D2 channel bank incorporates the companding operations into the encoders and decoders themselves. In these channel banks a resistor array is used to establish a sequence of nonuniformly spaced decision thresholds. A sample value is encoded by successively comparing the input value to the sequence of decision thresholds until the appropriate quantization interval is located. The digital output becomes whatever code is used to represent the particular quantization interval. (See Appendix B for a detailed description of the direct encoding procedure used in the D2 channel banks.) By incorporating the companding functions directly into the encoders and decoders, D2 channel banks avoid certain problems associated with parameter variability and temperature sensitivity of companding diodes in D1 channel banks.

The D2 channel banks also introduced improved performance in terms of the effect of channel errors on the decoded outputs. Of paramount concern in a PCM system is the effect of a channel error in the most significant bit position of

a code word. Bit errors in other positions of a code word are much less noticeable to a listener. A channel error in the most significant bit of a code word produced by a D1 channel bank causes an output error that is always equal to one half of the entire coding range of the coder. The D2 channel bank, on the other hand, uses a sign-magnitude coding format. With this format, a channel error in the polarity bit causes an output error that is equal to twice the sample magnitude (i.e. the polarity is inverted). In the worst case this error may correspond to the entire encoding range. Maximum amplitude samples are relatively rare, however, so most channel errors in a D2 coding format produce outputs with error magnitudes less than one-half the coding range. Thus, on average, the sign-magnitude coding format of the D2 and ensuing channel banks provide superior performance in the presence of channel errors [17].

In addition to a need for improved voice quality, it also became apparent that as more of the network began using digital techniques it would be necessary, or at least desirable, to implement many signal processing functions directly on digital signals and not convert them to an analog format for processing. Most signal processing functions (such as attenuating a signal or adding signals together) involve linear operations. Thus before processing a log-PCM voice signal it is necessary to convert the compressed transmission format into a linear (uniform) format.

To simplify the conversion process, the particular companding characteristic with $\mu = 255$ was chosen. This companding characteristic has the property of

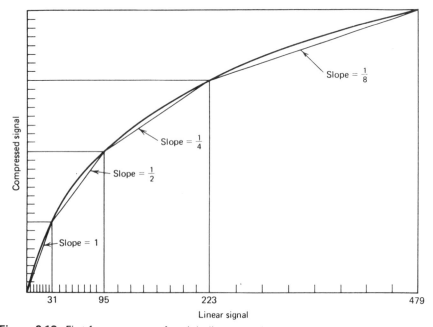

Figure 3.18. First four segments of straight line approximation to $\mu255$ compression curve.

being closely approximated by a set of eight straight-line segments also referred to as chords. Furthermore, the slope of each successive segment is exactly one-half the slope of the previous segment. The first four segments of a $\mu255$ approximation are shown in Figure 3.18. The overall result is that the larger quantization intervals have lengths that are binary multiples of all smaller quantization intervals. Because of this property a compressed code word is easily expanded into a uniform representation. Similarly, the uniform representation is easily converted into a compressed representation.* In fact, prevalent commercially available PCM codecs use digital compression of uniform code words instead of directly compressed encoding, as in the D2 channel banks. These techniques use a uniform encoder with a relatively large number of bits to cover the entire dynamic range of the signal. As described in Appendix B, the least significant bits of large sample values are discarded when compressing the code. The number of insignificant bits deleted is encoded into a special field included in the compressed code format. In this manner digital companding is analogous to expressing a number in scientific notation.

As shown in Figure 3.18, each major segment of the piecewise linear approximation is divided into equally sized quantization intervals. For 8 bit code words the number of quantization intervals per segment is 16. Thus an 8 bit $\mu255$ code word is composed of 1 polarity bit, 3 bits to identify a major segment, and 4 bits for identifying a quantizing interval within a segment. Table 3.1 lists the major segment endpoints, the quantization intervals, and the corresponding segment and quantization interval codes.

The quantization intervals and decoded sample values in Table 3.1 have been expressed in terms of a maximum amplitude signal of 8159 so that all segment endpoints and decoder outputs are integers. Notice that the quantizing step is doubled in each of eight successive linear segments. It is this property that facilitates the conversion to and from a uniform format. A complete encoding table is provided in Appendix B, along with a detailed description of the conversion process to and from uniform codes.

The straight-line approximation of the $\mu255$ companding curve is sometimes referred to as a 15 segment approximation. The 15 segments arise because, although there are eight positive segments and eight negative segments, the two segments nearest the origin are colinear and therefore can be considered as one. When viewed in this manner, the middle segment contains 31 quantization intervals with one segment straddling the origin (from -1 to $+1$ in Table 3.1). Code words for this middle quantization interval arise as a positive value less than $+1$ or a negative value greater than -1. There are, in effect, a positive zero and a negative zero. As represented in Table 3.1, these values are encoded as 00000000 and 10000000, respectively. However, $\mu255$ PCM channel banks invert all code words for transmission. The smaller amplitude signals, with mostly "0" segment codes, are most probable and would therefore cause less

*The inexorable advance of semiconductor technology has obviated much of the ingenuity that went into selecting EDL coding formats. Brute force, table look up conversion between codes using read only memories (ROMs) is now the most cost-effective approach.

than 50% pulses on the transmission line. The density of pulses is increased by inversion of the transmitted data, which improves the timing and clock recovery performance of the receiving circuitry in the regenerative repeaters. Thus the actual transmitted code words corresponding to a positive zero and a negative zero are respectively, 11111111 and 01111111, indicating strong timing content for the line signal of an idle channel.

TABLE 3.1 Encoding/Decoding Table for μ255 PCM[a]

Input Amplitude Range	Step Size	Segment Code S	Quantization Code Q	Code Value	Decoder Amplitude
0–1	1		0000	0	0
1–3			0001	1	2
3–5	2	000	0010	2	4
⋮			⋮	⋮	⋮
29–31			1111	15	30
31–35			0000	16	33
⋮	4	001	⋮	⋮	⋮
91–95			1111	31	93
95–103			0000	32	99
⋮	8	010	⋮	⋮	⋮
215–223			1111	47	219
223–239			0000	48	231
⋮	16	011	⋮	⋮	⋮
463–479			1111	63	471
479–511			0000	64	495
⋮	32	100	⋮	⋮	⋮
959–991			1111	79	975
991–1055			000	80	1023
⋮	64	101	⋮	⋮	⋮
1951–2015			1111	95	1983
2015–2143			0000	96	2079
⋮	128	110	⋮	⋮	⋮
3935–4063			1111	111	3999
4063–4319			0000	112	4191
⋮	256	111	⋮	⋮	⋮
7903–8159			1111	127	8031

[a]This table displays magnitude encoding only. Polarity bits are assigned as "0" for positive and "1" for negative. In transmission all bits are inverted.

In the interest of ensuring clock synchronization in the T1 repeaters, D2, D3, D4, and D5 channel banks alter the transmitted data in one other way. As indicated in Table 3.1, a maximum amplitude negative signal is all 1's, which would normally be converted to all 0's for transmission. Instead, for the all 0's code word only, the second least significant bit is set to 1 so that 00000010 is transmitted. In effect, an encoding error is produced to preclude an all 0's code

word. Fortunately, maximum amplitude samples are extremely unlikely so that no significant degradation occurs. (If the least significant bit were fored to a 1, a smaller decoding error would result. However, in every sixth frame this bit is "stolen" for signaling purposes and therefore is occasionally set to 0 independently of the code word. To ensure that "all-zero" code words are never transmitted, the second least significant bit is forced to a 1 when necessary.)

Example 3.3. Determine the sequence of code words for a $\mu255$ PCM encoded channel bank format representing a half maximum power 1-kHz digital signal.

Solution. Since the sampling rate of the standard $\mu255$ PCM channel bank is 8 kHz, a sequence of eight samples can be repeated in cyclic fashion to generate the 1-kHz waveform. For convenience the phases of the samples are chosen to begin at 22.5°. Thus the eight samples correspond to 22.5°, 67.5°, 112.5°, 157.5°, 202.5°, 247.5°, 292.5°, and 337.5°. With these phases, only two different magnitudes corresponding to 22.5° and 67.5° are required. The maximum amplitude of a half maximum power sine wave is (0.707) 8159 = 5768. Thus the two amplitudes contained in the sample sequence are

$$(5768)\sin(22.5°) = 2207$$

$$(5768)\sin(67.5°) = 5329$$

Using the encoding table in Appendix B, we determine the codes for these sample magnitudes to be 1100001 and 1110100, respectively. The sequence of eight samples can now be established as follows:

Polarity	Segment	Quantization	
0	1 1 0	0 0 0 1	22.5°
0	1 1 1	0 1 0 0	67.5°
0	1 1 1	0 1 0 0	112.5°
0	1 1 0	0 0 0 1	157.5°
1	1 1 0	0 0 0 1	202.5°
1	1 1 1	0 1 0 0	247.5°
1	1 1 1	0 1 0 0	292.5°
1	1 1 0	0 0 0 1	337.5°

Note. This sequence defines a 1-kHz test signal at a power level of 1 mW at the transmission level point (0dBm0). However, the actual transmitted data pattern is the complement of that provided above. Because only two amplitude samples are required to produce the test tone, this tone does not test all encoding/decoding circuitry. In general, a 1004-Hz tone is a better test tone since it is not harmonically related to a 8000-Hz sampling rate and will therefore exercise all encoder and decoder levels.

Performance of a μ255 PCM Encoder

As mentioned, the main motivation for changing the encoding algorithm of the D1 channel bank was to provide better speech quality for digital toll network transmission links. The signal-to-quantizing noise ratio for a maximum amplitude sine wave in the first segment of a $\mu255$ codec is determined easily from Equation 3.4 as

$$\text{SQR}(A = 31) = 7.78 + 20\log_{10}\left(\frac{31}{2}\right)$$

$$= 31.6\,\text{dB}$$

The signal-to-quantizing noise ratios for larger amplitude sinusoids are not as easy to calculate since the lengths of the quantization intervals vary with the sample size. Thus a general calculation of the quantizing noise power involves finding the expected value of the power of the quantization errors:

$$\text{noise power} = \frac{1}{12}\sum_{i=0}^{7} p_i q_i^2 \qquad (3.10)$$

where p_i = probability of a sample in the ith segment
$\quad q_i$ = quantization size for segment i
$\quad\ \ = 2^{i+1}$ for segmented $\mu255$ coding

Using Equation 3.10, we determine the signal-to-quantizing noise power for a full range sinusoid as

$$\text{SQR}(A = 8159) = 39.3\,\text{dB}$$

For comparison, if all quantization intervals had the maximum length of 256 as in the upper segment, Equation 3.4 provides an SQR of 37.8 dB. The difference of only 1.5 dB reflects the fact that a full scale sine wave spends 67% of the time in the upper segment where the quantization intervals are maximum (i.e., $p_7 = 0.67$). A voice signal, on the other hand, has a higher peak-to-average ratio than a sine wave. The average quantization error is smaller, but so is the average signal power. Hence the SQR is approximately the same.

The dynamic range of a segmented PCM encoder is determined as the signal power difference between a low-level signal occupying the entire range of the first segment and a high-level signal extending to the limits of the code. Thus the dynamic range of a segmented 255 coder is determined as

$$\text{DR}(A = 31 \text{ to } 8159) = 20\log_{10}(\tfrac{8159}{31}) = 48.4\,\text{dB}$$

In summary, an 8 bit $\mu255$ PCM codec provides a theoretical signal-to-quantizing noise ratio greater than 30 dB across a dynamic range of 48 dB. For

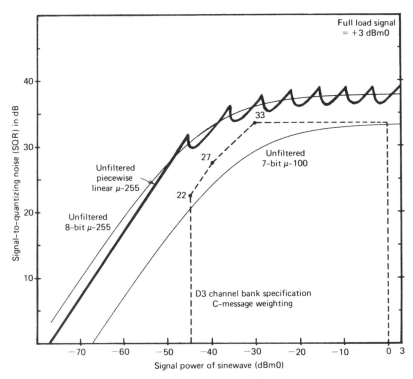

Figure 3.19. Signal-to-quantizing noise of μ-law coding with sine wave inputs.

comparison, Equation 3.4 or Figure 3.14 reveals that a uniform PCM encoder/decoder requires 13 bits for equivalent performance. (The extra quality of uniform encoding at high signal levels is unneeded.)

The theoretical performance of an 8 bit segmented $\mu255$ coder is shown in Figure 3.19 as a function of the amplitude of a sine wave input. Also shown is the theoretical performance of an unsegmented $\mu255$ coder and a 7 bit $\mu100$ coder used in the D1 channel bank. Notice that the 8 bit coders provide about 5 dB improvement over the 7 bit coder for high-level signals and even more improvement for low-level signals. The performance shown for the 8 bit coders does not include the effect of using only 7 bits for voice coding in every sixth frame. When this effect is included, the 8 bit coders lose 1.76 dB in performance.

The scalloped effect of the segmented coder occurs because the quantization intervals change abruptly at the segment endpoints instead of continuously as in analog companding. Also as shown in Figure 3.19, note the required performance of a D3 channel bank as specified by the Bell System [13]. This specification assumes all noise measurements are made using C-message weighting. C-message weighting reduces the effective noise level by 2 dB and therefore improves the SQR. Thus an ideal 8 bit $\mu255$ coder actually exceeds the specification by more than that shown in Figure 3.19. When the least significant

bit of every sixth frame is used for signaling, however, the SQR is reduced by a comparable 1.76 dB.

A-Law Companding

The companding characteristic recommended by CCITT for Europe and most of the rest of the world is referred to as an A-law characteristic. This characteristic has the same basic features and implementation advantages as does the μ-law characteristic. In particular, the A-law characteristic can also be well approximated by straight-line segments to facilitate direct or digital companding, and can be easily converted to and from a uniform format. The normalized A-law compression characteristic is defined as

$$F_A(x) = \text{sgn}(x)\left[\frac{A|x|}{1 + \ln(A)}\right] \qquad 0 \leqslant |x| \leqslant \frac{1}{A}$$

$$= \text{sgn}(x)\left[\frac{1 + \ln|Ax|}{1 + \ln(A)}\right] \qquad \frac{1}{A} \leqslant |x| \leqslant 1 \qquad (3.11)$$

The inverse or expansion characteristic is defined as

$$F_A^{-1}(y) = \text{sgn}(y)\frac{|y|[1 + \ln(A)]}{A} \qquad 0 \leqslant |y| \leqslant \frac{1}{1 + \ln(A)}$$

$$= \text{sgn}(y)\frac{(e^{|y|[1 + \ln(A)] - 1})}{A} \qquad \frac{1}{1 + \ln(A)} \leqslant |y| \leqslant 1 \qquad (3.12)$$

where $y = F_A(x)$ and $A = 87.6$.

Notice that the first portion of the A-law characteristic is linear by definition. The remaining portion of the characteristic $(1/A \leqslant |x| \leqslant 1)$ can be closely approximated by linear segments in a fashion similar to the μ-law approximation. In all, there are eight positive and eight negative segments. The first two segments of each polarity (four in all) are colinear and therefore are sometimes considered as one straight-line segment. Thus the segmented approximation of the A-law characteristic is sometimes referred to as a "13 segment approximation." For ease in describing the coding algorithms of the segmented companding characteristic, however, a 16 segment representation is used—just as in the case of the segmented μ-law characteristic.

The segment endpoints, quantization intervals, and corresponding codes for an 8 bit segmented A-law characteristic are shown in Table 3.2. The values are scaled to a maximum value of 4096 for integral representations. Figure 3.20 displays the theoretical performance of the A-law approximation where it is compared to the performance of a μ-law approximation presented in Figure 3.19. Notice that the A-law characteristic provides a slightly larger dynamic range. However the A-law characteristic is inferior to the μ-law characteristic in terms of small signal quality (idle channel noise). The difference in small signal

TABLE 3.2 Segmented *A*-Law Encoding/Decoding Table[a]

Input Amplitude Range	Step Size	Segment Code S	Quantization Code Q	Code Value	Decoder Amplitude
0–2			0000	0	1
2–4		000	0001	1	3
⋮			⋮	⋮	⋮
30–32	2		1111	15	31
32–34			0000	16	33
⋮		001	⋮	⋮	⋮
62–64			1111	31	63
64–68			0000	32	66
⋮	4	010	⋮	⋮	⋮
124–128			1111	47	126
128–136			0000	48	132
⋮	8	011	⋮	⋮	⋮
248–256			1111	63	252
256–272			0000	64	264
⋮	16	100	⋮	⋮	⋮
496–512			1111	79	504
512–544			0000	80	528
⋮	32	101	⋮	⋮	⋮
992–1024			1111	95	1008
1024–1088			0000	96	1056
⋮	64	110	⋮	⋮	⋮
1984–2048			1111	111	2016
2048–2176			0000	112	2112
⋮	128	111	⋮	⋮	⋮
3968–4096			1111	127	4032

[a]In transmission, every other bit is inverted.

performance occurs because the minimum step size of the *A*-law standard is 2/4096 whereas the minimum step size of μ-law is 2/8159. Furthermore, notice that the *A*-law approximation does not define a zero-level output for the first quantization interval (i.e., uses a mid-riser quantizer). However, the difference between mid-riser and mid-tread performance at 64 kbps is imperceptible [18].

3.2.6 Syllabic Companding

A significant attribute of the companding techniques described for PCM systems is that they instantaneously encompass the entire dynamic range of the coder on a sample-to-sample basis. Thus μ-law and *A*-law companding is sometimes referred to as instantaneous companding. However, because the power level of a speech signal remains fairly constant for 100 or more 8-kHz samples it is wasteful of code space to be able to encode very low level signals adjacent to very high level signals. In other words, it is unnecessary to allow for

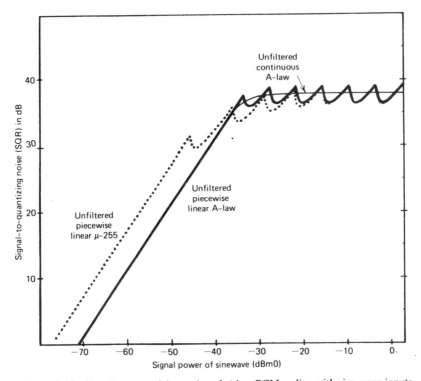

Figure 3.20. Signal-to-quantizing noise of *A*-law PCM coding with sine wave inputs.

instantaneous changes in the dynamic range of a signal. One technique of reducing the amount of transmission bandwidth allocated to dynamic range is to reduce the dynamic range of the signal at the source before encoding and then restore the original dynamic range of the signal at the receiver after decoding. When the adjustments to the dynamic range occur on a periodic basis that more or less corresponds to the rate of syllable generation, the technique is referred to as syllabic companding. Because the dynamic range adjustments occur only every 30 msec or so there is very little bandwidth needed to communicate the adjustments.

Syllabic companding was first developed for use on noisy analog circuits to improve the idle channel noise. As shown in Figure 3.21 the power level of low-level syllables is increased (compressing the dynamic range) for transmission but attenuated upon reception (expanding the dynamic range). The process of attenuating the received signal restores the low-power syllable to its original level, but attenuates any noise arising on the transmission link. Thus the receiver SNR is improved for low-level signals. The amount of amplification applied at the source is dependent on the short-term (syllabic) power level of the transmitted signal. Similarly, the compensating attenuation applied at the receiving terminal is determined from the short-term power level of the received signal.

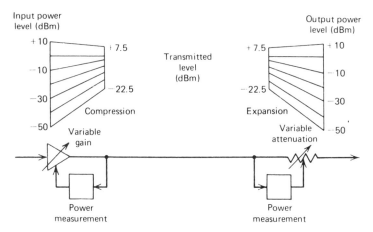

Figure 3.21. Syllabic companding of an analog signal.

Syllabic companding on digital systems provides the same basic improvement in signal-to-quantizing noise ratios as it does on noisy analog transmission links. When the digital encoders and decoders are considered as part of the transmission link, the process of amplifying low-level signals before encoding and attenuating them after decoding effectively reduces the quantization noise with no net change in signal level. In practice, syllabic companding as implemented in digitized voice terminals does not amplify the signal at the source and attenuate it at the destination. Instead, an equivalent process of controlling the step sizes in the encoder and decoder is used. As far as the transmitted bit stream is concerned, it makes no difference if the signal is amplified and encoded with fixed quantization or if the signal is unmodified but encoded with smaller quantization intervals. Thus syllabic compandors in digital voice terminals typically reduce the quantization intervals when encoding and decoding low-power syllables but increase the quantization intervals for high-power syllables.

Although syllabic companding can be used in conjunction with any type of voice coding, the technique has been applied most often to differential systems described in the following sections. In many of the applications, the adaption time has been reduced to 5 or 10 msec, which is somewhat shorter than the duration of a typical syllable (approximately 30 msec). The technique is still generally referred to as syllabic companding, however, to distinguish it from the instantaneous variety.

To adjust the step sizes in the decoder in synchronism with adjustments made in the encoder, some means must be established to communicate the step size information from the source to the destination. One method explicitly transmits the step size information as auxiliary information. A generally more useful approach is to derive the step size information from the transmitted bit stream. Thus, in the absence of channel errors, the decoder and the encoder operate on

the same information. This procedure is analogous to syllabic companded analog systems in which the receiver determines its attenuation requirements from the short-term power level of the received signal. In a digital system the bit stream is monitored for certain data patterns that indicate the power level of the signal being encoded. Indications of high power level initiate an increase in the step size, whereas indications of low levels cause a decrease.

Determining the step size information from the transmitted bit stream is generally better than explicitly transmitting the step sizes for the following reasons. Because there is no explicit transmission of step size, the transmission of sampled speech information is never interrupted, and the speech sample rate is equal to the transmission rate. Also, the bit stream does not have to be framed to identify step size information separately from the waveform coding. Furthermore, if the step size adjustments are made on a more gradual basis, the individual increments are small enough that occasional incorrect adjustments in the receiver caused by channel errors are not critical. However, on transmission links with very high error rates (one error in a hundred bits or so), better decoded voice quality can be obtained if the step size is transmitted explicitly and redundantly encoded for error correction [19].

3.2.7 Adaptive Gain Encoding

In the syllabic companding example of Figure 3.21 the dynamic range of a signal is reduced by a factor of 2 (in dBm). Thus, if an uncompanded signal has 36 dB of dynamic range the encoder sees only 18 dB. The 18-dB reduction implies three fewer bits are needed for dynamic range encoding. In the limit, if the power level of all input signals is adjusted to a single value, no bits of the encoder need to be allocated to dynamic range encoding. A process that adjusts all signals to a standard value is referred to as automatic gain control (AGC). AGC is traditionally used on carrier transmission systems to adjust all received signals to a standard value—thereby removing variations in propagation attentuation. AGC cannot be applied to a source voice signal with allowances for speech pauses when there is no signal present. Otherwise, idle channel noise would be amplified to the average level of active voice. Notice that with AGC there is no residual information in the power level of the encoded signal as there is in syllabic companding. To ascertain the original power level AGC must be augmented with adaptive gain encoding (AGE) as indicated in Figure 3.22.

There are two basic modes of operation for gain encoding depending on how gain factors are measured and to which speech segments the factors are applied. One mode of operation, as implied in Figure 3.22, involves measuring the power level of one segment of speech and using that information to establish a gain factor for ensuing speech segments. Obviously, this mode of operation relies on gradually changing power levels. This mode of operation is sometimes referred to as "backward estimation."

Another mode of operation involves measuring the power level of a speech segment and using the gain factor thus derived to adapt the encoder to the same

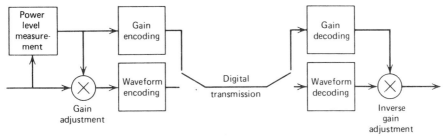

Figure 3.22. Adaptive gain encoding.

segment. This approach, referred to as "forward estimation," has the obvious advantage that the encoder and decoder use gain factors specifically related to the speech segments from which they are derived. The disadvantage is that each speech segment must be delayed while the gain factor is being determined. Although the availability of digital memory has made the cost of implementing the delay insignificant, the impact of the delay on echos and singing in a partially analog network must be considered.

Adaptive gain control with explicit transmission of gain factors is not without shortcomings. First, when the periodic gain information is inserted into the transmitted bit stream, some means of framing the bit stream into blocks is needed so gain information can be distinguished from waveform coding. Second, periodic insertion of gain information disrupts information flow causing higher transmitter clock rates that might be inconveniently related to the waveform sample clock. Third, correct reception of gain factors is usually critical to voice quality—indicating a need to redundantly encode gain information.

Reference [20] describes a modified form of PCM using forward estimation of gain factors that is referred to as "nearly instantaneously companded PCM." The need for transmitting speech segments in blocks is not a disadvantage in the application mentioned (mobile telephone) because repetitive bursts with error checking are used as a means of overcoming short-lived multipath fading. This system provides a bit rate reduction of 30% with respect to conventional PCM. Another example of the use of AGE is the subscriber loop multiplexer (SLM) system developed by Bell Labs [21]. The SLM system became obsolete when low-cost PCM codecs became available and the subscriber carrier systems could be integrated into digital end offices (with SLC 96 systems). All of the encoding algorithms described in the following sections use syllabic companding or AGE in some form to reduce the bit rate.

3.3 SPEECH REDUNDANCIES

As mentioned in the previous sections, a conventional PCM system encodes each sample of the input waveform independently from all other samples. Thus a PCM system is inherently capable of encoding an arbitrarily random waveform

whose maximum frequency component does not exceed one-half the sampling rate. Analyses of speech waveforms, however, indicate there is considerable redundancy from one sample to the next. In fact, as reported in reference [10] the correlation coefficient (a measure of predictability) between adjacent 8-kHz samples is generally 0.85 or higher. Hence the redundancy in conventional PCM codes suggests significant savings in transmission bandwidths are possible through more efficient coding techniques. All of the digitization techniques described in the rest of this chapter are tailored, in one degree or another, to the characteristics of speech signals with the intent of reducing the bit rate.

In addition to the correlation existing between adjacent samples of a speech waveform, several other levels of redundancy can be exploited to reduce encoded bit rates. Table 3.3 lists these redundancies. Not included are higher-level redundancies related to context dependent interpretations of speech sounds (phonemes), words, and sentences. These topics are not covered because techniques that analyze speech waveforms to extract only information content eliminate subjective qualities essential to general telephony.

TABLE 3.3 Speech Redundancies

Time domain redundancies
1. Nonuniform amplitude distributions
2. Sample-to-sample correlations
3. Cycle-to-cycle correlations (periodicity)
4. Pitch-interval to pitch-interval correlations
5. Inactivity factors (speech pauses)

Frequency domain redundancies
6. Nonuniform long term spectral densities
7. Sound specific short term spectral densities

3.3.1 Nonuniform Amplitude Distributions

As mentioned in the introduction to companding, lower amplitude sample values are more common than higher amplitude sample values. Most low-level samples occur as a result of speech pauses in a conversation. Beyond this, however, the power levels of active speech signals also tend to occur at the lower end of the encoding range. The companding procedures described in the previous section provide slightly inferior quality (i.e., lower signal-to-noise ratios) for small signals compared to large signals. Thus the average quality of PCM speech could be improved by further shortening of lower-level quantization intervals and increasing of upper-level quantization intervals. The amount of improvement realized by such a technique is minimal and probably would not justify the additional complexities, particularly if the linearizability of contemporary compandors was compromised. The most beneficial approach to processing signal amplitudes in order to reduce encoder bit rates involves some form of adaptive gain control as discussed earlier.

3.3.2 Sample-to-Sample Correlation

The high correlation factor of 0.85 mentioned in Section 3.3 indicates that any significant attempt to reduce transmission rates must exploit the correlation between adjacent samples. In fact, at 8-kHz sampling rates, significant correlations also exist for samples two to three samples apart. Naturally, samples become even more correlated if the sampling rate is increased.

The simplest way to exploit sample-to-sample redundancies in speech is to encode only the differences between adjacent samples. The difference measurements are then accumulated in a decoder to recover the signal. In essence these systems encode the slope or derivative of a signal at the source and recover the signal by integrating at the destination. Digitization algorithms of this type are discussed at length in later sections.

3.3.3 Cycle-to-Cycle Correlations

Although a speech signal requires the entire 300 to 3400 Hz bandwidth provided by a telephone channel, at any particular instant in time certain sounds may be composed of only a few frequencies within the band. When only a few underlying frequencies exist in a sound, the waveform exhibits strong correlations over numerous samples corresponding to several cycles of an oscillation. The cyclic nature of a voiced sound is evident in the time waveform shown in Figure 3.23. Encoders exploiting the cycle-to-cycle redundancies in speech are markedly more complicated than those concerned only with removing the redundancy in adjacent samples. In fact, these encoders more or less represent a transition from the relatively high-rate, natural-sounding waveform encoders to the relatively low-rate, synthetic-sounding vocoders.

3.3.4 Pitch Interval to Pitch Interval Correlations

Human speech sounds are often categorized as being generated in one of two basic ways. The first category of sounds encompasses "voiced" sounds that arise as a result of vibrations in the vocal cords. Each vibration allows a puff of air to flow from the lungs into the vocal tract. The interval between puffs of air exciting the vocal tract is referred to as the pitch interval or more simply, the rate of excitation is the pitch. Generally speaking, voiced sounds arise in the generation

Figure 3.23. Time waveform of voiced sound.

Figure 3.24. Time waveform of unvoiced sound.

of vowels and the latter portions of some consonants. An example of a time waveform for a voiced sound is shown in Figure 3.23.

The second category of sounds includes the fricatives or "unvoiced" sounds. Fricatives occur as a result of continuous air flowing from the lungs and passing through a vocal tract constricted at some point to generate air turbulence (friction). Unvoiced sounds correspond to certain consonants such as f, j, s, x. An example of a time waveform of an unvoiced sound is shown in Figure 3.24. Notice that an unvoiced sound has a much more random waveform than a voiced sound.

As indicated in Figure 3.23, a voiced sound not only exhibits the cycle-to-cycle redundancies mentioned in Section 3.3.3, but the waveform also displays a longer term repetitive pattern corresponding to the duration of a pitch interval. Thus one of the most efficient ways of encoding the voiced portions of speech is to encode one pitch interval waveform and use that encoding as a template for each successive pitch interval in the same sound. Pitch intervals typically last from 5 to 20 msec for men and from 2.5 to 10 msec for women. Since a typical voiced sound lasts for approximately 100 msec, there may be as many as 20 to 40 pitch intervals in a single sound. Although pitch interval encoding can provide significant reductions in bit rates, the pitch is sometimes very difficult to detect. (Not all voiced sounds produce a readily identifiable pitch interval as in Figure 3.23.) If the pitch gets encoded erroneously, strange sounds result.

An interesting aspect of pitch interval encoding is that it provides a means of speeding up speech while maintaining intelligibility. By deleting some percentage of pitch intervals from each sound (phoneme) the rate of sound generation is effectively increased in a manner analogous to more rapid word formation. The pitch of the sounds remains unchanged. In contrast, if the rate of reconstruction is merely increased, all frequencies including the pitch increase proportionately. Moderate speedups produce obvious distortion while greater speedups become unintelligible. Devices designed to simulate faster word formation have demonstrated that we are capable of assimilating spoken information much faster than we normally generate it.

3.3.5 Inactivity Factors

Analyses of telephone conversations have indicated that a party is typically active about 40% of a call duration. Most inactivity occurs as a result of one person listening while the other is talking. Thus a conventional (circuit-switched) full-duplex, connection is significantly underutilized. Time assignment speech interpolation (TASI) described in Chapter 1 is a technique to improve

channel utilization on expensive analog links. Digital speech interpolation (DSI) is a term used to refer to a digital circuit counterpart of TASI. In essence DSI involves sensing speech activity, seizing a channel, digitally encoding and transmitting the utterances, and releasing the channel at the completion of each speech segment.

Digital speech interpolation is obviously applicable to digital speech storage systems where the duration of a pause can be encoded more efficiently than the pause itself. In recorded messages, however, the pauses are normally short since a "half-duplex" conversation is not taking place. DSI techniques are sometimes used to expand the voice channel capacity of an existing digital TDM link. As such the inputs are standard PCM signals that are digitally processed to detect speech activity. The DSI operation is often combined with speech compression algorithms to implement digital circuit multiplication (DCM) equipment. When a 2:1 voice compression algorithm is combined with a 2.5:1 DSI concentration factor an overall 5:1 circuit expansion is achieved. The use of such equipment in a network must be carefully managed to ensure that voiceband data and digital data channels bypass the DCM operations.

3.3.6 Nonuniform Long-Term Spectral Densities

The time domain redundancies described in the preceding sections exhibit certain characteristics in the frequency domain that can be judiciously processed to reduce the encoded bit rate. Frequency domain redundancies are not independent of the redundancies in the time domain. Frequency domain techniques merely offer an alternate approach to analyzing and processing the redundancies.

A totally random or unpredictable signal in the time domain produces a frequency spectrum that is flat across the bandwidth of interest. Thus a signal that produces uncorrelated time domain samples makes maximum use of its bandwidth. On the other hand, a nonuniform spectral density represents inefficient use of the bandwidth and is indicative of redundancy in the waveform.

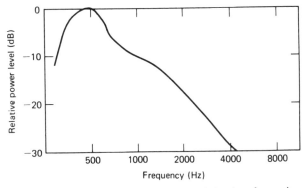

Figure 3.25. Long-term power spectral density of speech.

Figure 3.25 shows the long-term spectral density of speech signals averaged across two populations: men and women [22]. Notice that the upper portions of the 3-kHz bandwidth passed by the telephone network have significantly reduced power levels. The lower power levels at higher frequencies are a direct consequence of the time domain sample-to-sample correlations discussed previously. Large amplitude signals cannot change rapidly because, on the average, they are predominantly made up of lower-frequency components.

A frequency domain approach to more efficient coding involves flattening the spectrum before encoding the signal. The flattening process can be accomplished by passing the signal through a high-pass filter to emphasize the higher frequencies before sampling. The original waveform is recovered by passing the decoded signal through a filter with a complementary, low-pass characteristic. An important aspect of this process is that a high-pass filter exhibits time domain characteristics similar to a differentiator and a low-pass filter has time domain characteristics analogous to an integrator. Thus the spectrum flattening process essentially means the slope of the signal is encoded at the source, and the signal is recovered by integrating at the destination—the basic procedure described previously for sample-to-sample redundancy removal in the time domain.

In studying Figure 3.25 it is natural to think that the remarkably low levels of signal energy at the higher frequencies (2 to 3.4 kHz) means that more bandwidth is being allocated to a voice signal than is really necessary. The error in such a conclusion, however, lies in the distinction between energy content and information content of the voice frequency spectrum. As any beginning computer programmer soon learns, the meaning of a program variable can be retained even though it is shortened by deleting all of the vowels. In speech the vowels require most of the energy and occupy primarily the lower portion of the frequency band. The consonants, on the other hand, contain most of the information but use much less power and generally higher frequencies. Hence merely reproducing a high percentage of the original speech energy is an inadequate goal for a digital speech transmission or storage system.

3.3.7 Short-Term Spectral Densities

The speech spectrums shown in Figure 3.25 represent long-term averages of the spectral densities. Over shorter periods of time the spectral densities vary considerably and exhibit sound specific structures with energy peaks (resonances) at some frequencies and energy valleys at others. The frequencies at which the resonances occur are called formant frequencies, or simply formants. Voiced speech sounds typically contain three to four identifiable formants. These features of the short-term spectral density are illustrated in the spectogram of Figure 3.26. A spectogram is a display of speech spectral energy as a function of time and frequency. The horizontal axis represents time, the vertical axis represents frequency, and the shadings represent energy levels. Thus the darker portions in Figure 3.26 indicate relatively high energy levels (formants) at particular instants in time.

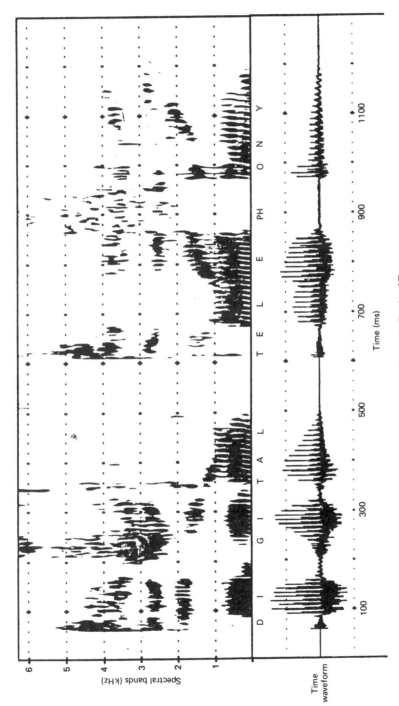

Figure 3.26. Specrogram of phrase: "digital telephony." (Courtesy of Robert L. Davis of Texas Instruments Inc.)

129

Frequency domain voice coders provide improved coding efficiencies by encoding the most important components of the spectrum on a dynamic basis. As the sounds change different portions (formants) of the frequency band are encoded. The period between formant updates is typically 10 to 20 msec. Instead of using periodic spectrum measurements, some higher quality vocoders continuously track gradual changes in the spectral density at a higher rate. Frequency domain vocoders often provide lower bit rates than the time domain coders, but typically produce more unnatural sounding speech.

3.4 DIFFERENTIAL PULSE CODE MODULATION

Differential pulse code modulation (DPCM) is designed specifically to take advantage of the sample-to-sample redundancies in a typical speech waveform. Since the range of sample *differences* is less than the range of individual *samples*, fewer bits are needed to encode difference samples. The sampling rate is often the same as for a comparable PCM system. Thus the bandlimiting filter in the encoder and the smoothing filter in the decoder are basically identical to those used in conventional PCM systems.

A conceptual means of generating the difference samples for a DPCM coder is to store the previous input sample directly in a sample-and-hold circuit and use an analog subtractor to measure the change. The change in the signal is then quantized and encoded for transmission. The DPCM structure shown in Figure 3.27 is more complicated, however, because the previous input value is reconstructed by a feedback loop that integrates the encoded sample differences. In essence, the feedback signal is an estimate of the input signal as obtained by integrating the encoded sample differences. Thus the feedback signal is obtained in the same manner used to reconstruct the waveform in the decoder.

The advantage of the feedback implementation is that quantization errors do not accumulate indefinitely. If the feedback signal drifts from the input signal, as a result of an accumulation of quantization errors, the next encoding of the difference signal automatically compensates for the drift. In a system without

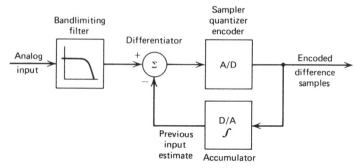

Figure 3.27. Functional block diagram of differential PCM.

feedback the output produced by a decoder at the other end of the connection might accumulate quantization errors without bound.

As in PCM systems, the analog-to-digital conversion process can be uniform or companded. Some DPCM systems also use adaptive techniques (syllabic companding) to adjust the quantization step size in accordance with the average power level of the signal. (See reference [9] for an overview of various techniques.)

Example 3.4. Speech digitization techniques are sometimes measured for quality by use of an 800-Hz sine wave as a representative test signal. Assuming a uniform PCM system is available to encode the sine wave across a given dynamic range, determine how many bits per sample can be saved by using a uniform DPCM system.

Solution. In essence, the solution is obtained by determining how much smaller the dynamic range of the difference signal is in comparison to the dynamic range of the signal amplitude. Assume the maximum amplitude of the sine wave is A, so that

$$x(t) = A \, \sin(2\pi \cdot 800t)$$

The maximum amplitude of the difference signal can be obtained by differentiating and multiplying by the time interval between samples:

$$\frac{dx}{dt} = A \cdot (2\pi) \cdot (800) \cdot \cos(2\pi \cdot 800t)$$

$$|\Delta x(t)|_{\max} = A \cdot (2\pi) \cdot (800) \cdot \left(\frac{1}{8000}\right) = 0.628A$$

The savings in bits per sample can be determined as

$$\log_2\left(\frac{1}{0.628}\right) = 0.67 \text{ bits}$$

Example 3.4 demonstrates that a DPCM system can use $\frac{2}{3}$ bit per sample less than a PCM system with the same quality. Typically DPCM systems provide a full 1 bit reduction in code word size. The larger savings is achieved because, on average, speech waveforms have a lower slope than an 800 Hz tone (see Figure 3.25).

3.4.1 DPCM Implementations

Differential PCM encoders and decoders can be implemented in a variety of ways depending on how the signal processing functions are partitioned between analog and digital circuitry. At one extreme the differencing and integration functions can be implemented with analog circuitry, while at the other extreme

all signal processing can be implemented digitally using conventional PCM samples as input. Figure 3.28 shows block diagrams of three different implementations with differing amounts of digital signal processing.

Figure 3.28a depicts a system using analog differencing and integration. Analog-to-digital (A/D) conversion is performed on the difference signal, and conversion from digital-to-analog (D/A) for the feedback loop is immediately performed on the limited-range difference code. Analog summation and storage in a sample-and-hold (S/H) circuit is used to provide integration.

Figure 3.28b shows a system that performs the integration function digitally. Instead of immediately converting the difference code back to analog for feedback, the difference code is summed and stored in a data register to generate a digital representation of the previous input sample. A full scale D/A converter

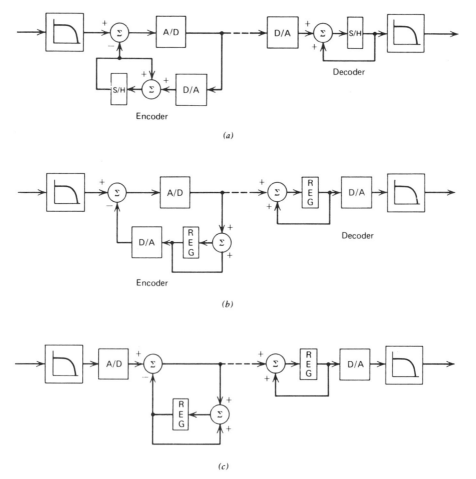

Figure 3.28. DPCM implementations. (a) Analog integration. (b) Digital integration. (c) Digital differencing.

is then used to produce the analog feedback signal for differencing. Notice that the D/A converters in Figure 3.28b must provide full amplitude range conversion whereas the D/A converters in Figure 3.28a convert the more limited difference signals.

Figure 3.28c shows a system where all signal processing is performed by digital logic circuits. The A/D converter produces full amplitude range sample codes which are compared to digitally generated approximations of the previous amplitude code. Notice that the A/D converter in this case must encode the entire dynamic range of the input whereas the A/D converters in the other two versions operate on only the difference signals.

Due to the availability of digital signal processing components, some of which contain internal A/D converters, digital processing (as in Figure 3.28c) is generally the most effective means of implementing a DPCM algorithm. In fact, many DPCM applications involve processing speech signals that have already been digitized into standard 64-kbps PCM formats. Thus the DPCM implementation may require no analog processing. As an aid in processing log-PCM signals, some DSP components provide internal μ-law and A-law conversion functions.

The decoders in all three implementations shown in Figure 3.28 are exactly like the feedback implementations in the corresponding encoder. This reinforces the fact that the feedback loop generates an approximation of the input signal (delayed by one sample). If no channel errors occur, the decoder output (before filtering) is identical to the feedback signal. Thus the closer the feedback signal matches the input, the closer the decoder output matches the input.

3.4.2 Higher Order Prediction

A more general viewpoint of a DPCM encoder considers it a special case of a linear predictor with encoding and transmission of the prediction error. The feedback signal of a DPCM system represents first-order prediction of the next sample value, and the sample difference is a prediction error. Under this

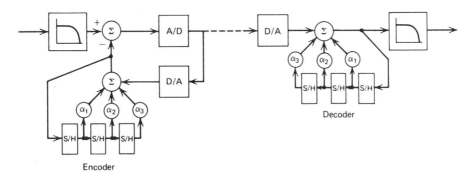

Figure 3.29. Extension of DPCM to third-order prediction.

viewpoint the DPCM concept can be extended to incorporate more than one past sample value into the prediction circuitry. Thus the additional redundancy available from all previous samples can be weighted and summed to produce a better estimate of the next input sample. With a better estimate, the range of the prediction error decreases to allow encoding with fewer bits. For systems with constant predictor coefficients, results have shown that most of the realizable improvement occurs when using only the last three sample values. The basic implementation of linear prediction using the last three sample values is shown in Figure 3.29. For conceptual purposes this implementation shows analog differencing and integration as in Figure 3.28a. The most effective implementations use digital memory, multiplication, and addition in a DSP component in lieu of the analog processing shown—particularly because many applications involve already digitized (PCM) signals.

As mentioned in Section 3.4, analysis of differential PCM systems with first-order predication typically provides a 1 bit per sample reduction in code length relative to PCM systems with equivalent performance. Extended DPCM systems utilizing third-order prediction can provide reductions of $1\frac{1}{2}$ to 2 bits per sample [23]. Thus a standard DPCM system can provide 64-kbps PCM quality at 56 kbps, and third-order linear prediction can provide comparable quality at 48 kbps. However, subjective evaluations often indicate that somewhat higher bit rates are needed to match 64-kbps PCM quality.

3.4.3 Adaptive Differential PCM

Relatively straightforward implementations of DPCM can provide savings of 1 to 2 bits per sample with respect to standard PCM encoding. Even greater savings can be achieved by adding adaptation logic to the basic DPCM algorithm to create what is referred to as adaptive differential PCM (ADPCM). Many forms of ADPCM have been investigated and used in various applications. Two of the most prevalent applications are voice messaging and DCM equipment for increasing the number of voice channels on a T1 line. With respect to the latter application CCITT has established a 32-kbps ADPCM standard (Recommendation G.721) [24]. This algorithm has been extensively tested and characterized to not significantly degrade toll quality voice circuits when inserted into the internal portions of the network. Design considerations of the standard are

1. Multiple tandem encodings and decodings between both PCM and analog interfaces.
2. End-to-end signal quality for voice, voiceband data, and facsimile.
3. Effects of random and bursty channel errors.
4. Performance on analog signals degraded by loss, noise, amplitude distortion, phase distortion, and harmonic distortion.
5. Easy transcoding with μ-law and A-law PCM.

The 32-kbps rate implies a 2:1 savings in channel bandwidth with respect to standard PCM. The only significant impairment introduced by implementations of the ADPCM standard is the corruption of modem signals carrying data rates greater than 4800 bps [24]. Voiceband data at rates of 4800 bps and below are adequately supported.

The ADPCM algorithm is conceptually similar to that shown in Figure 3.29 but more sophisticated in that it uses an eighth-order predictor, adaptive quantization, and adaptive prediction. Furthermore, the algorithm is designed to recognize the difference between voice or data signals and use a fast or slow quantizer adaption mode, respectively.

Subjective evaluation [25] of the algorithm using the mean opinion score (MOS) method of evaluating speech quality is shown in Figure 3.30. The MOS method uses trained listeners to rate the speech quality on a scale of 1 to 5. Figure 3.30 shows the average scores of 32-kbps ADPCM and 64-kbps PCM as a function of the number of analog tandem encodings. The speech quality associated with multiple digital transcodings (to and from PCM) does not degrade beyond the first transcoding as long as there are no digital transmission impairments such as channel errors or slips (Chapter 7).

3.5 DELTA MODULATION

Delta modulation (DM) is another digitization technique that specifically exploits the sample-to-sample redundancy in a speech waveform. In fact, DM can be considered as a special case of DPCM using only 1 bit per sample of the difference signal. The single bit specifies merely the polarity of the difference sample and thereby indicates whether the signal has increased or decreased since the last sample. An approximation to the input waveform is constructed in the feedback path by stepping up one quantization level when the difference is

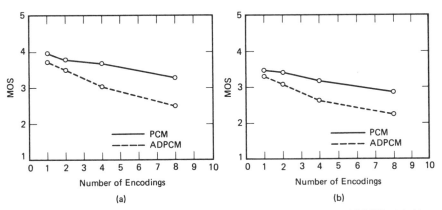

Figure 3.30. Average MOS vs. number of encodings for PCM and ADPCM. (a) Linear microphone. (b) Carbon microphone.

Figure 3.31. Waveform encoding by delta modulation.

positive ("one") and stepping down when the difference is negative ("zero"). In this way the input is encoded as a sequence of "ups" and "downs" in a manner resembling a staircase. Figure 3.31 shows a DM approximation of a typical waveform. Notice that the feedback signal continues to step in one direction until it crosses the input at which time the feedback step reverses direction until the input is crossed again. Thus when tracking the input signal the DM output "bounces" back and forth across the input waveform allowing the input to be accurately reconstructed by a smoothing filter.

Since each encoded sample contains a relatively small amount of information (1 bit), delta modulation systems require a higher sampling rate than PCM or multibit DPCM systems. In fact, the sampling rate is necessarily much higher than the minimum (Nyquist) sampling rate of twice the bandwidth. From another viewpoint, "oversampling" is needed to achieve better prediction from one sample to the next.

The main attraction of DM is its simplicity. Figure 3.32 shows a basic implementation of a DM encoder and decoder. Notice that the A/D conversion function is provided by a simple comparator. A positive difference voltage produces a 1, and a negative difference voltage produces a 0. Correspondingly, the D/A function in the feedback path, and in the decoder, is provided by a two-polarity pulse generator. In the simplest form the integrator can consist of nothing more than a capacitor to accumulate the charge from the pulse generator.

In addition to these obvious implementation simplicities, a delta modulator

Figure 3.32. Delta modulation encoder and decoder.

also allows the use of relatively simple filters for bandlimiting the input and smoothing the output [26]. As discussed in Section 3.1, the spectrum produced by a sampling process consists of replicas of the sampled spectrum centered at multiples of the sampling frequency. The relatively high sampling rate of a delta modulator produces a wider separation of these spectrums, and, hence, foldover distortion is prevented with less stringent roll-off requirements for the input filter.

3.5.1 Slope Overload

The conceptual operation of a delta modulator shown in Figure 3.31 indicates that the encoded waveform is never much more than a step size away from the input signal. Sometimes, however, a delta modulator may not be able to keep up with rapid changes in the input signal and thus fall more than a step size behind. When this happens the delta modulator is said to be experiencing "slope overload." A slope overload condition is shown in Figure 3.33.

Basically, slope overload occurs when the rate of change of the input exceeds the maximum rate of change that can be generated by the feedback loop. Since the maximum rate of change in the feedback loop is merely the step size times the sampling rate, a slope overload condition occurs if

$$\left|\frac{dx(t)}{dt}\right| > qf_s$$

where $x(t)$ = input signal
$\qquad q$ = step size,
$\qquad f_s$ = sampling frequency

(3.13)

The design of a delta modulator necessarily involves a trade-off between two types of distortion: the more or less random quantization noise, sometimes referred to as granular noise, and the slope overload noise. As indicated in Figure 3.33, granular noise is a predominant consideration for slowly changing signals, whereas slope overload is dominant during rapidly changing signals. Obviously, granular noise is small if step sizes are small, but small step sizes increase the likelihood of slope overload. The optimum step size in terms of

Figure 3.33. Slope overload and granular noise of delta modulation system.

minimizing the total of granular and slope overload noise has been considered by Abate [27].

The perceptual effects of slope overload on the quality of a speech signal are significantly different from the perceptual effects produced by granular noise. As indicated in Figure 3.33, the slope overload noise reaches its peaks just before the encoded signal reaches its peaks. Hence, slope overload noise has strong components identical in frequency and approximately in phase with a major component of the input. Distortion that is correlated in this manner to the speech signal is effectively "masked" by the speech energy and therefore is less noticeable than "uncorrelated" distortion. In fact, overload noise is much less objectionable to a listener than random or granular noise at an equivalent power level [28]. Hence, from the point of view of perceived speech quality, the optimum mix of granular and slope overload noise is difficult to determine.

Slope overload is not a limitation of just a delta modulation system, but an inherent problem with any system, such as DPCM in general, that encodes the difference in a signal from one sample to the next. A difference system encodes the slope of the input with a finite number of bits and hence a finite range. If the slope exceeds that range, slope overload occurs. In contrast, a conventional PCM system is not limited by the rate of change of the input, only by the maximum encodable amplitude. Notice that a differential system can encode signals with arbitrarily large amplitudes, as long as the large amplitudes are attained gradually.

3.5.2 Linear Delta Modulation

The simplest delta modulator uses a constant step size for all signal levels and is therefore referred to as a uniform or linear delta modulator (LDM). The design of a linear delta modulator is fundamentally concerned with selecting a fixed step size and a sampling rate to satisfy two criteria. First, the signal-to-granular noise ratio must be some minimum value for the lowest level signal to be encoded. Second, the signal-to-slope overload distortion ratio must be some minimum for the highest level signal to be encoded.

A convenient procedure for satisfying the slope overload criterion is to use Equation 3.13 and design the system so that slope overload is just on the verge of occurring at the highest level of input. Thus the step size for the highest level signal is not optimum in a perceptual sense or even in terms of minimizing the sum of granular noise and overload distortion. Nevertheless, when only granular noise is present, a delta modulator's SQR performance can be appropriately compared to the SQR performance of a PCM system. Comparing the two encoding methods under other conditions, for example when overloads occur in either system, requires listener evaluations.

A significant consideration in the implementation of a delta modulator, or DPCM systems in general, is maintaining equal positive and negative steps in the encoder and decoder. If a step size imbalance exists in the decoder, the reconstructed waveform accumulates a dc offset that may lead to circuit

Filtered output

Figure 3.34. Effect of imbalance in positive and negative step sizes in a delta modulator with a zero level input.

saturations. As discussed in the next section, modest step size imbalances in the decoder are relatively easy to accommodate. On the other hand, step size imbalances in the encoder are more troublesome and therefore require close attention.

When a delta modulator encodes the signal of an idle channel (speech pause), the output data ideally alternates between a 1 and a 0. When alternation does occur, the output filter of the decoder eliminates the relatively high-frequency sampling noise. However, if an imbalance in step sizes exists, as shown in Figure 3.34, two 1's or two 0's occasionally occur together. Since double 0's or double 1's occur relatively infrequently, they produce low-frequency distortion, which is not filtered out, and idle channel noise results. Idle channel noise is a particular problem in delta modulators because of the relatively large quantization intervals. One means of balancing the step sizes in the encoder is to subtract an offset obtained by low-pass filtering (integrating) the feedback signal over a long time period [26].

3.5.3 Adaptive Delta Modulation

Numerous researchers have proposed and studied a large variety of algorithms for adapting the step size of a delta modulator. Basically, all algorithms increase the step size when the onset of slope overload is detected and reduce the step size as the input slope decreases. Some of the algorithms measure the input slope directly and transmit explicit step size information. Other algorithms derive the step size information from the transmitted bit stream for both the encoder and decoder. The rate of adaption is sometimes instantaneous in that adjustments are made on a sample-to-sample basis [21] but more often syllabic in that significant changes in the step size occur only once every 10 msec or so. All of the basic algorithms appear to provide approximately the same voice quality [23], but some algorithms have certain features making them more attractive than others in specific applications. In particular, the algorithms vary in terms of their channel error sensitivity, transmission format, interconnectability with other coders, and amenability to digital signal processing.

This section describes the basic operation of one particular type of adaptive-delta modulation (ADM) generally referred to as continuously variable slope

delta modulation (CVSD). The basic CVSD encoding algorithm was first described by Greefkes and de Jager in 1968 [29]. A subsequent design by Greefkes and Riemens [26, 30] incorporated what is referred to as delta modulation with digitally controlled companding. This particular type of adaptive delta modulation is chosen because it is simple and embodies many generally desirable features. Furthermore, the basic algorithm is available from several manufacturers as a single integrated circuit.

Digitally controlled companding, as the name implies, derives its step size information from the transmitted bit stream. As shown in Figure 3.35, the adaption logic monitors the transmitted data to detect the occurrence of four successive ones or four successive zeros.* A string of ones indicates that the feedback signal is probably not rising as fast as the input, while a string of zeros indicates the feedback is probably not falling as fast as the input. In either case, all ones or all zeros implies that slope overload is occurring and the step size should be increased. Thus the all ones and all zeros signals are "or"ed together to control a pulse generator. During overload the pulse generator is enabled so that the step size voltage, as stored on capacitor C_1, is increased.

The CVSD system shown in Figure 3.35 makes no explicit measurements to determine if the step size is too large and therefore should be reduced. Instead, a resistor is connected to capacitor C_1 to allow the step voltage to decay with time. Thus, in the absence of explicit increases, the step size automatically becomes smaller through exponential decay. Eventually the step size decays to a minimum value or to a point at which overload is detected, causing the step size to increase again.

The resistor–capacitor combination is sometimes referred to as a "leaky integrator," implying that the capacitor integrates the input pulses but the resistor allows the accumulated step size voltage to leak off. Besides providing an automatic way to reduce the step size, the leaky integrator also eliminates the long-term effects of channel errors in the receiving terminal. If "perfect" integration were used, an erroneous increment in the step size would be retained indefinitely. With leaky integration, however, the duration of an erroneous increment (or lack of increment) is limited to a few time constants of the leaky integrator.

Notice that leaky integration is also used to reconstruct the input signal at the output of the decoder and, correspondingly, in the feedback path of the encoder. Thus an offset in the output of the receiving terminal, caused by a channel error, naturally decays in time and is effectively eliminated within a few time constants of the second leaky integrator. Elimination of an offset in the output signal is not so important from a listener's point of view since a fixed offset is inaudible, but is useful for preventing saturation in the decoder. If "perfect" integration were used, a large excess of positive or negative errors would eventually cause the electronics in the decoder to saturate and distort the voice waveform. In addition, leaky integrators provide compensation for small

*The number of successive ones or zeros causing a step size increase is reduced to three on lower-rate delta modulators to improve the response time.

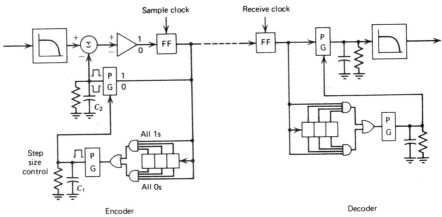

Figure 3.35. Continuously variable slope delta modulation with digitally controlled companding.

imbalances in positive and negative step sizes that would otherwise lead to saturation also.

Figure 3.36 shows the performance of two CVSD codecs as a function of the input level of a 1-kHz test tone. The performance curves were obtained from specification sheets for Motorola's MC3417 and MC3418 CVSD codecs. The 16-kbps codec (MC3417) provides what is referred to as general communications quality. The 37.7-kbps codec (MC3418) provides commercial telephone quality. Notice, however, that the 37.7-kbps performance does not neet the D3 channel bank requirements of 33 dB minimum signal-to-noise ratio (SQR) across 30 dB dynamic range. Recall that the D3 specification arose out of a need for higher quality because of the possibility of multiple conversions. Thus, although a CVSD at 37.7 kbps provides acceptable end-to-end quality, the data

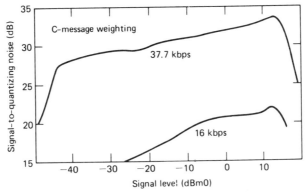

Figure 3.36. Signal-to-quantizing noise ratios of CVSD.

rate must be increased for tandem coding applications. The idle channel noise of the CVSD is also inferior if step size imbalances exist.

The performance curves shown in Figure 3.36 do not completely reflect the output quality when applied to an input speech signal. Since the 1-kHz tone is a steady state input, it does not test the adaptive aspects of the codec. In particular, adaptive delta modulators are generally noted for a certain lack of "crispness" in the speech output. The lack of crispness occurs at the beginning of words and relatively strong syllables that experience temporary slope overloads.

In conclusion, it should be mentioned again that numerous varieties of delta modulation algorithms have been proposed and implemented. A CVSD codec is described here because it represents a very basic form of delta modulation and illustrates basic concepts of DPCM in general.

3.6 ADAPTIVE PREDICTIVE CODING

The differential systems described in the previous sections (DPCM, ADPCM, ADM) operate with lower data rates than PCM systems because they encode a difference signal that has lower average power than the raw input signal. The ratio of the input signal power to the power of the difference signal is referred to as the prediction gain. Simple DPCM systems (first-order predictors) provide about 5 dB of prediction gain. ADPCM provides greater levels of prediction gain depending on the sophistication of the adaption logic and the number of past samples used to predict the next sample. The prediction gain of ADPCM is ultimately limited by the fact that only a few past samples are used to predict the input and the adaption logic only adapts the quantizer—not the prediction weighting coefficients (the αs in Figure 3.29).

Adaptive predictive coding (APC) [31, 32] provides greater levels of prediction gain by adapting the prediction coefficients to individual speech segments and, in most cases, using higher orders of prediction (e.g., up to 12). If the coefficients are determined from past history and used to predict subsequent speech segments (backward estimation) 13 dB of prediction gain is possible [10]. If speech segments are delayed so predictor coefficients can be used on the same speech segments from which they were derived (forward estimation) 20 dB of prediction gain is possible [33].

A block diagram of a basic APC encoder/decoder is shown in Figure 3.37. The input to the encoder and the output from the decoder are assumed to be uniform PCM—most likely representing conversions from and to log PCM. The transmitted data stream is necessarily composed of blocks containing three types of information: (1) the encoded difference signal (residual), (2) a gain factor, and (3) the predictor coefficients. The most significant difference between this coder and a DPCM or ADPCM coder involves the periodic determination and transmission of the predictor coefficients. Notice that the integrated residual signal at point A is identical to the input signal delayed by one sample (except for residual quantization error). Thus the corresponding point in the decoder is the reconstructed PCM output.

Figure 3.37. Adaptive predictive encoder/decoder.

143

Explicit transmission of a gain factor, as opposed to deriving a gain factor from the transmitted residual, is useful in this application because a block structured transmission format is required for transmitting the predictor coefficients anyway. Residual encoding may use any of the waveform algorithms described previously. Arbitrarily accurate reconstruction of the input can be achieved if enough bits are used in encoding the residual. Of course, the purpose of the adaptive prediction is to achieve a low data rate. Single-bit PCM that conveys only the polarity of the residual provides adequate performance for low data rate applications [33]. Notice that because of the feedback path, single-bit encoding of the residual essentially produces a delta modulator with very sophisticated feedback logic that is periodically changed to match the shape and energy level of corresponding (forward estimated) speech segments.

A large variety of algorithms have been investigated for calculating the predictor coefficients—most of which involve extensive computation. If the application is for voice response systems the computations do not have to occur in real time and are therefore sometimes done on a large mainframe computer. Real time encoding restricts the algorithm to one that can be realized with a DSP or special purpose IC designed to implement a specific coefficient determining algorithm. A linear predictive coding (LPC) algorithm as described in Section 3.8.3 is a common algorithm because it provides good performance and is available in custom parts. The U.S. defense department had adopted an APC algorithm using a fourth-order LPC coefficient analysis as a government standard for secure voice transmission at 9.6 kbps [34]. This system transmits blocks of 240 bits containing 180 one-bit samples of the residual, 54 bits for parameters and gain factors, 5 bits for error corrections of critical msb's, and 1 framing bit.

3.7 SUBBAND CODING

A subband coder is one form of coder using a frequency domain analysis of the input signal instead of a time domain analysis as in previously described coders. As shown in Figure 3.38, the coder first divides the input spectrum into separate bands using a bank of bandpass filters. The signal passing through each of the relatively narrow subbands is individually encoded with separate adaptive APCM, PCM, or ADPCM encoders. After each subband is encoded, the individual bit streams are multiplexed for transmission to the decoder where they are demultiplexed, decoded, and combined to reconstruct the input.

Separately encoding each subband is advantageous for several reasons. First, by using separate adaption for each band, the quantization step sizes can be adjusted according to the energy level in each band. Those bands with relatively high energy levels can be encoded with relatively coarse quantization. In this manner the spectrum of the quantization noise is matched to the short-term spectrum of the signal. This property is very desirable in a perceptual sense because it allows the speech signal to mask the quantization noise. (The human

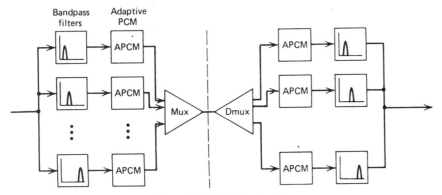

Figure 3.38. Subband coder.

ear perceives speech by measuring the short-term energy level of individual frequency bands. Hence, relatively low noise in a band with no speech energy is perceptually more significant than greater noise in a band with significant speech energy.)

A second advantage of subband coding is that the bit rate (quality) assigned to each band can be optimized according to the perceptual importance of each individual band. In particular, a relatively large number of bits per sample can be used for low frequencies where it is important to preserve the pitch and formant structure of voiced sounds. A higher frequencies, however, fewer bits per sample can be used because noise-like fricatives do not require comparable quality in reproduction.

As reported in reference [10], subband coders provide significant bit rate reductions compared to the more common and simpler coding algorithms: adaptive delta modulation and adaptive differential PCM. Specifically, subband coders at 16 kbps are reported to be perceptually equivalent to ADPCM coders at 22 kbps. A subband coder at 9.6 kbps is reported to be equivalent to an ADM coder at 19.5 kpbs. An extensive description and performance analysis of subband coding is available in reference [35].

A particularly significant example of subband coding is the CCITT recommendation G.722 for wideband speech coding [36, 37]. This standard provides for encoding 7-kHz speech bandwidth with 64 kbps. Thus it provides a means of significantly improving voice fidelity when end-to-end digital channels are available. Applications that can benefit most from the higher fidelity are teleconferencing and speakerphones. Because it is not intended for internal equipmemt of the network, it does not have to process voiceband data signals or support tandem analog conversions. Transcodings to and from uniform PCM are required in support of conference bridge arrangements.

As shown in Figure 3.39 [36] the G.722 algorithm divides the input speech bandwidth from 50 to 7000 Hz into two approximately equal sized subbands. Both subbands are encoded with ADPCM: the lower subband at 48 kbps and

Figure 3.39. G.722 7-kHz audio codec. (From reference [36], p. 10.)

the upper subband at 16 kbps. The upper band does not require as high a data rate because it is not as important perceptually and has significantly less energy. The algorithm permits the lower band to be encoded at 40 or 32 kbps, which allows 8 or 16 kbps for an auxiliary data channel in teleconferencing applications or for the use of 56-kbps channels when 64 kbps is not available. Figure 3.40 shows the MOS performance of the G.722 algorithm for speech and music at three basic rates [38].

Figure 3.40. Average MOS vs. bit rate of G.722 7-kHz audio codec. (From reference [38], p. 15.)

3.8 VOCODERS

For the most part, the encoding/decoding algorithms described previously have been concerned primarily with reproducing the input waveform as accurately as possible. Thus they assume little or no knowledge of the nature of the signal they process and are basically applicable to any signal occurring in a voice channel. Exceptions occur when subband coding and adaptive predictive coding are designed for relatively low bit rates (20 kbps or less). At these bit rates the encoders have been closely tailored to the statistics of a speech signal and cannot provide comparable quality for other signals. Differential systems, such as DPCM and delta modulation, also exhibit a certain amount of speech specific properties by virtue of their high-frequency encoding deficiencies (slope overload).

The digitization procedures described in this section very specifically encode speech signals and speech signals only. For this reason these techniques are referred to collectively as "vocoders," an acronym for voice coders. Since these techniques are designed specifically for voice signals, they are not applicable to portions of the public telephone network in which other analog signals (such as modem signals) must be accommodated. Furthermore, vocoders typically produce unnatural or synthetic sounding speech.

The basic goal of a vocoder is to encode only the perceptually important aspects of speech with fewer bits than the more general waveform encoders. Thus they can be used in limited bandwidth applications where the other techniues cannot.

Some of the main applications for vocoders are recorded (e.g., "wrong number") messages, encrypted voice transmission over narrowband HF radio, digital cellular radio, analog telephone circuits, computer output, and games. A particularly interesting use of one type of vocoder, linear predicative coding, provides multiple voice channels over a single voice frequency leased line. Using a well-conditioned leased line to obtain a 9600-bps circuit, four 2400 bps voice signals can be time division multiplexed into a single leased line [39]. Hence, in this case, digitization actually decreases the bandwidth of a voice signal. Although these systems provide intelligible voice, the overall quality is below telephone standards.

This section describes three of the most basic vocoding techniques: the channel vocoder, the formant vocoder, and the linear predictive coder. Many other forms and variations of vocoders have been proposed and studied. For a discussion of some of the other techniques and an extensive bibliography on the subject (see reference [10]).

A fundamental requirement for maintaining good speech quality involves preserving the short-term power spectrum of the signal. The phase relationship between individual frequency components is perceptually much less important. One of the best examples of the ear's insensitivity to phase is demonstrated when two notes are played simultaneously, or nearly simultaneously, on a piano. The composite sound, as perceived by a listener, is seemingly no different if one note

is struck slightly later than the other. In essence, the ear senses how much energy is present at various frequencies in the speech spectrum but does not sense the phase relationship between individual frequency components.

The effect of a phase shift in one component of a composite signal is shown in Figure 3.41. The first composite waveform is produced when two individual frequency components have identical starting phases. The second composite waveform occurs when the two frequency terms have starting phases shifted by 90° with respect to each other. Notice that the composite waveforms are markedly different even though the difference is imperceptible to the ear. For these reasons the time waveform produced by a vocoder generally bears little resemblance to the original input waveform. Instead, the emphasis of a vocoder is in reproducing the short-term power spectrum of the input.

3.8.1 Channel Vocoder

Channel vocoders were first developed in 1928 by Homer Dudley [40]. Dudley's original implementation compressed speech waveforms into an analog signal with a total bandwidth of about 300 Hz. Based on the original concept, digital channel vocoders have been developed operating in the range of 1 to 2 kbps.

A major part of the encoding process of a channel vocoder involves determining the short-term signal spectrum as a function of time. As indicated in Figure 3.42, a bank of bandpass filters is used to separate the speech energy into subbands that are full wave rectified and filtered to determine relative power levels. The individual power levels are encoded and transmitted to the destination. Notice that this much of a channel vocoder is very similar to the subband coder discussed previously. A subband coder, however, typically uses wider bandpass filters, which necessitate sampling the subband waveforms more often (determining a waveform instead of just a power level). Since a subband coder encodes waveforms, it also includes phase information that is ignored by a channel vocoder.

In addition to measuring the signal spectrum, modern channel vocoders also determine the nature of speech excitation (voice or unvoiced) and the pitch frequency of voiced sounds. The excitation measurements are used to synthesize the speech signal in the decoder by passing an appropriately selected source

$\sin(\omega t)$
$+$
$\sin(2\omega t)$

$\sin(\omega t)$
$+$
$\sin(2\omega t + \frac{\pi}{2})$

Figure 3.41. Effect of phase shift in the superposition of two tones.

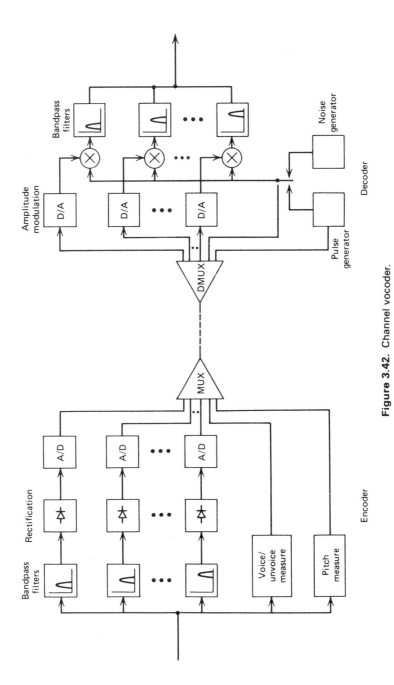

Figure 3.42. Channel vocoder.

signal through a frequency domain model of the vocal tract transfer function. Voiced excitation is simulated by a pulse generator using a repetition rate equal to the measured pitch period. Unvoiced excitation is simulated by a noise generator. Owing to the synthesized nature of the excitation, this form of a vocoder is sometimes referred to as a pitch excited vocoder.

As indicated in Figure 3.42, a decoder implements a vocal tract transfer function as a bank of bandpass filters whose input power levels are determined by respective subband power levels in the encoder. Thus outputs of each bandpass filter in the decoder correspond to outputs of respective bandpass filters in the encoder. Superposing the individual bands recreates, in a spectral sense, the original signal.

Many variations in the basic channel vocoder have been developed involving the nature of the excitation and the means of encoding the power levels. Recent advances in digital technology have introduced the use of digital signal processing to determine the input spectrum by way of Fourier transform algorithms in lieu of the bank of analog filters. All forms of vocoders that measure the power spectral density are sometimes referred to collectively as "spectrum channel vocoders" to distinguish them from time domain vocoders such as a linear predictive coder to be discussed later.

The most difficult aspect of most vocoder realizations involves determining the pitch of voiced sounds. Furthermore, certain sounds are not clearly classifiable as purely voiced or purely unvoiced. Thus a desirable extension of the basic vocoder involves more accurate characterization of the excitation. Without accurate excitation information, vocoder output quality is quite poor and often dependent on both the speaker and the particular sounds being spoken. Some of the more advanced channel vocoders have produced highly intelligible, although somewhat synthetic sounding, speech at 2400 bps [23].

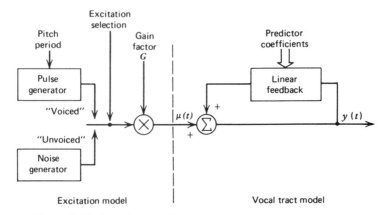

Figure 3.43. Speech generation model of linear predictive coding.

3.8.2 Formant Vocoder

As indicated in the spectogram of Figure 3.26, the short-term spectral density of speech is rarely distributed across the entire voice band (200 to 3400 Hz). Instead, speech energy tends to be concentrated at three or four peaks called formants. A formant vocoder determines the location and amplitude of these spectral peaks and transmits this information instead of the entire spectrum envelope. Thus a formant vocoder produces lower bit rates by encoding only the most significant short-term components in the speech spectrum.

The most important requirement for achieving useful speech from a formant vocoder involves accurately tracking changes in the formants. Once this is accomplished, a formant vocoder can provide intelligible speech at less than 1000 bps [10].

3.8.3 Linear Predictive Coding

A linear predictive coder (LPC) is a popular vocoder that extracts perceptually significant features of speech directly from a time waveform rather than from frequency spectra, as does a channel vocoder and formant vocoder. Fundamentally, an LPC analyzes a speech waveform to produce a time-varying model of the vocal tract excitation and transfer function. A synthesizer in the receiving terminal recreates the speech by passing the specified excitation through a mathematical model of the vocal tract. By periodically updating the parameters of the model and the specification of the excitation, the synthesizer adapts to changes in either. During any one specification interval, however, the vocal tract is assumed to represent a linear time-invariant process. A block diagram of the basic model for speech generation is shown in Figure 3.43. Thus Figure 3.43 is also a model of an LPC decoder/synthesizer.

The equation of the vocal tract model shown in Figure 3.43 is defined as follows:

$$y(n) = \sum_{k=1}^{p} a_k y(n-k) + Gx(n) \tag{3.14}$$

where $y(n) = n$th output sample
$a_k = k$th predictor coefficient
G = gain factor
$x(n)$ = input at sample time n
p = order of the model

Notice that the speech output in Equation 3.14 is represented as the present input value plus a linear combination of the previous p outputs of the vocal tract. The model is adaptive in that the encoder periodically determines a new set of predictor coefficients corresponding to successive speech segments. A basic LPC does not measure and encode difference waveforms or error signals.

Instead, the error signals are minimized in a mean squared sense when the predictor coefficients are determined. The ability to avoid encoding the predictor errors comes from the fact than an LPC uses prediction parameters based on the actual input segments to which the parameters are applied (forward estimation). In contrast, most of the predictive coders mentioned previously base their prediction on past measurements only (backward estimation). Rapid changes in the vocal tract or excitation cause models based on past measurements to be more inaccurate. The negative aspect of forward estimation is the delay it inserts into the encoding process, which, depending on the application, may or may not be a problem.

The information than LPC encoder/analyzer determines and transmits to the decoder/synthesizer consists of

1. Nature of excitation (voiced or unvoiced)
2. Pitch period (for voiced excitation)
3. Gain factor
4. Predictor coefficients (parameters of vocal tract model).

The nature of the excitation is determined, as in other vocoders, by determining whether strong periodic components exist in the waveform. Pitch is determined by measuring periodicity when it exists. In addition to measuring pitch with techniques similar to those used by other vocoders, an LPC encoder/analyzer has particular properties that aid in pitch determination [41].

The predictor coefficients can be determined using one of several different computational procedures. All procedures use actual waveform samples as the desired outputs of the synthesizer. Using these sample values, a set of p linear equations in p unknown coefficients is produced. Thus the coefficients are determined by inverting a p-by-p matrix. Since the order of p may vary from 6 to 12, depending on the speech quality desired, straightforward matrix inversion represents a formidable amount of computation. Depending on specific assumptions made in the model, however, the matrices have special properties that greatly simplify the solution of the equations [41].

Although linear predictive coders process time domain representations of speech, their operation is such that they provide good estimates of the peaks of the speech spectrum. Furthermore, an LPC is capable of effectively tracking gradual changes in the spectrum envelope. The overall result is that LPCs provide more natural sounding speech than the purely frequency domain based vocoders [23]. Most LPC research has concentrated on encoding speech in the range of 1.2 to 2.4 kbps.

The US Department of Defense established a 2400-bps, tenth-order LPC (LPC-10) as a standard for secure communications over dial up telephone lines [42]. This system transmits blocks of 54 bits every 22.5 msec. The 54 bits are allocated as 7 bits for pitch, 5 bits for energy level (gain factor), 41 bits for 10 coefficients (voiced excitation), and one bit for framing.

The performance of the coder with respect to natural speech is provided in Table 3.4. A diagnostic rhyme test (DRT) [43] is a means of determining the intelligibility of a speech system as a percentage of correct word recognitions from a standardized list of pairs of rhyming words (e.g., goat vs. coat or thick vs. tick). The DRT-with-noise test measures the word recognition percentage when background noise appropriate to the application is added to the test words before encoding. The diagnostic acceptability measure (DAM) [44] rates intelligibility and subjective acceptability using procedures that eliminate much of the dependence on personal preferences of the listeners. The scores are normalized from 0 to 100.

TABLE 3.4 Performance Comparison of Natural Speech and LPC-10

Test	Natural Speech (3.6 kHz)	LPC-10
DRT	95%	90%
DRT-with-noise	92%	82%
DAM	65	48

3.8.4 Enhanced Excitation Linear Predictive Coding

The basic LPC algorithm described in the previous section synthesizes speech in the decoder using a very simple model for the excitation that requires only about 10% of the aggregate data rate. The simplicity of the model inevitably produces synthetic sounding speech. To overcome this shortcoming several techniques have been developed to enhance the excitation. Three of the most prevalent enhanced excitation algorithms are multipulse LPC (MPLPC), residual excited LPC (RELP), and code excited LPC (CELP).

Multipulse LPC

As shown in Figure 3.44 where it can be compared with a conventional LPC system an MPLPC [45] is a conceptually simple extension of a conventional LPC. Where an LPC uses the residual to determine the pitch period of voiced signals an MPLPC uses the residual to determine a sequence of pulses to more closely match the residual. In the simplest case the MPLPC uses a fixed number of pulses in a pulse train and determines only the starting phase of the train and the amplitudes and polarities of each pulse within the train. A significant side benefit of an MPLPC is that it does not have to identify whether a speech segment is voiced or not and consequently determine the pitch period of voiced signals. Instead, the multipulse determination in each analysis segment automatically adapts to the nature of the actual excitation. MPLPC at 9600 bps is used in AT&T's 14A Announcement System for recorded messages to telephone users [45].

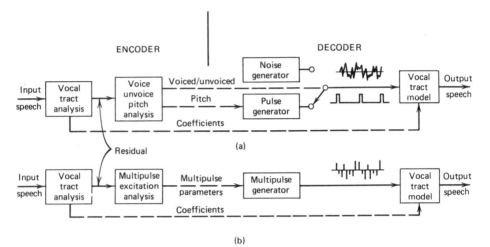

Figure 3.44. Multipulse linear predictive coding. Comparison of LPC (a) and MPLPC (b) encoder/decoder.

MPLPC is also the algorithm chosen for the Skyphone Aeronautical Telephone service using 9.6-kbps channels on the **INMARSAT** maritime satellite [46]. Another application of an MPLPC algorithm is the Pan-European digital mobile radio system [47] that allocates a 13-kbps data rate to voice encoding. The speech codec that has been standardized by CEPT for this application is referred to as regular-pulse excitation LPC (RPE-LPC) combined with long-term prediction (LTP) [48]. This system transmits 260 bit blocks consisting of 72 bits of prediction parameters and 188 bits of excitation encoding. Because the codec does not support voiceband data rates at 1200 bps and above the system has provisions for data bypass of the codec.

Residual Excited LPC
The APC algorithm described in Section 3.6 transmits the encoded difference signal (prediction error or residual) determined in the encoder so the decoder can integrate it to recover the original input. Notice that the decoded residual waveform in the APC decoder of Figure 3.37 acts as the excitation of the prediction loop. If the prediction loop uses an LPC formulation the APC essentially becomes a residual-waveform excited LPC.

The formal term "residual excited LPC" or RELP refers to a structure that is identical to the APC block diagram shown in Figure 3.37 but differs in the manner in which the residual is encoded. A RELP encoder does not encode the residual directly but preprocesses it to allow transmission at a lower data rate. The fundamental premise of the preprocessing is that the residual consists of a fundamental frequency and multiplies (harmonics) of the fundamental. Thus, a

RELP encoder encodes only the fundamental. The decoder reconstructs the residual (in a frequency domain sense) by decoding the fundamental and adding in the harmonics. In the sense that extracting and encoding the fundamental is "residual encoding" and that decoding and adding harmonics is "residual decoding" the diagram in Figure 3.37 serves as a basic diagram of a RELP.

As commonly implemented a RELP encoder [49] extracts the fundamental by low pass filtering the residual and down sampling to reduce the sampling frequency to the Nyquist rate. As an example, if a (DSP-based) low pass filter has a cut off of 800 Hz only every fifth sample of the filtered residual is needed to meet the 1600-Hz Nyquist rate. In this manner, a RELP decoder is excited by samples occurring at a 1600-Hz rate—approximately the same rate as a MPLPC. Thus the residual decoder in a RELP receiver is similar to the pulse generator of an MPLPC synthesizer. In fact, the regular pulse excitation LPC of the European digital mobile system determines its pulse excitation by testing each of four sequences obtained by taking every fourth sample of a filtered residual and selecting the sequence with maximum correlation to the residual sequence. Thus even the encoding algorithms of an MPLPC and a RELP are sometimes similar. In a comparison of three specific coders (subband, MPLPC, and RELP) MPLPC provided the best performance [50].

RELP encoding has been used in various mobile radio and satellite applications. An example of the latter is a 9.6-kbps RELP codec designed to support conversion of analog FM voice channels to digital voice channels [51].

Code Excited LPC

A CELP [52] is yet another example of an enhanced excitation LPC whose basic block diagram can be represented by Figure 3.37. Instead of encoding residue waveforms on a sample-by-sample basis as in an APC, a CELP assumes residuals are not random waveforms with independent samples but rather that each block of residue samples can be represented by one of a manageable number of waveform templates. "Residual encoding" in this case involves selecting one codeword that represents a block of samples in a codebook that best matches the block of residual samples. "Residual decoding" in this context implies using the received codeword as an index into the table of codewords to obtain the residual sequence chosen to represent the original residue sequence. Because a block of residue samples can be considered as a residue vector this form of coding is also referred to as vector quantization (VQ) excited LPC [53].

Maximum coding efficiency is achieved by encoding long sequences (i.e., vectors with many dimensions), which, of course, implies large codebooks. Thus, most of the research effort of this class of codecs involves establishing a large enough set of vectors in the codebook that all residue vectors can be adequately matched and then establishing an efficient procedure for searching the codebook to find the vector with the best match. Because the number of vectors is necessarily large, tree-structured searches are required, which implies the entries in the codebook are grouped into hierarchical families of vectors.

3.9 ENCODER/DECODER SELECTION CONSIDERATIONS

This chapter has described several techniques for digitizing voice signals and indicated that many other variations are possible. This section compares the various algorithms in terms of

1. Cost/complexity
2. Voice quality
3. Data rate
4. Transparency for nonvoice signals
5. Tolerance of transmission errors
6. Effects of tandem encodings
7. Coding formats
8. Signal processing requirements.

Cost/Complexity

Of all the algorithms, delta modulation is by far the simplest and therefore can be implemented with the lowest cost. With respect to PCM, however, the cost advantage is insignificant because of high volume, competitive production of standard PCM codecs. Other algorithms that have been used in sufficient numbers to warrant customized integrated circuit implementations are ADPCM and LPC. With regards to voice response systems the encoding complexity is not an issue because a fixed vocabulary is encoded only once.

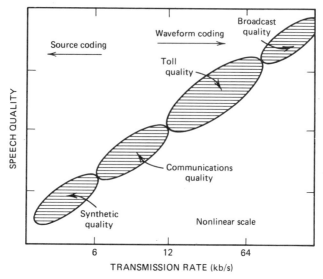

Figure 3.45. General speech quality vs. transmission rate. (From reference [11], p. 36.)

Voice Quality

If it were not for the fact that some applications can tolerate lower levels of quality there would not be much variety in the algorithms. Otherwise, waveform coders such as PCM or DPCM would be the only acceptable algorithms— particularly when transparency is required for nonvoice signals. Generally speaking, the applications and the corresponding quality levels are commonly identified as broadcast quality, toll quality, communications quality, and synthetic quality. These categories and their respective ranges of data rates are qualitatively shown in Figure 3.45 as obtained from reference [11]. Table 3.5, obtained from reference [38], compares six different coders in terms of the quality measures: DRT [43], DAM [44], and MOS [9, 11, 18].

TABLE 3.5 Speech Quality Comparisons of Common Coders

Coder	DRT	DAM	MOS
64-kbps PCM	95	73	4.3
32-kbps ADPCM	94	68	4.1
16-kbps LD-CELP	94[a]	70[a]	4.0
8-kbps CELP	93	68[a]	3.7
4.8-kbps CELP	93	67	3.0
2.4-kbps LPC	90	54	2.5[a]

[a]Estimates.

Notice that the low rate coders do reasonably well on intelligibility (DRT) but fall off significantly in terms of subjective measures. Additional comparisons of low rate coders (APC, RELP, LPC) are available in references [54] and [55].

Data Rate

As indicated in Figure 3.45 speech quality and data rate are almost directly related. Within a particular class of coders the data rate can be reduced by about 30% with increased complexity. The 32-kbps ADPCM G.721 standard, for example, is about equivalent to simple versions of ADPCM at 48 kbps. CELP coders are another example of where increased complexity in the encoder can provide either improved quality or lower data rates.

Transparency for Nonvoice Services

The separation between communications quality and toll quality coders in Figure 3.45 also separates those algorithms that do not perform well on nonvoice signals from those that do. The lower data rates of communications quality coders are achieved by tailoring the algorithms to the specifics of voice signals thereby sacrificing modem and facsimile performance. For a comprehensive theoretical treatment of voiceband data signal digitization see reference [56]. As end-to-end digital channels become more available there will be less need to support the nonvoice applications as voiceband signals.

In addition to user-oriented applications coders installed in the internal portions of a network must provide acceptable quality for network-related signaling tones such as DTMF, MF, and SF. DTMF tones, in particular, must be accurately reconstructed because they are used for end-to-end control by the users.

Tolerance of Transmission Errors

Of the waveform coders, the differential systems (DPCM, ADPCM, DM) are the most tolerant of transmission errors and PCM the least. The threshold of perceptibility of random errors on delta modulation is 10^{-3}. For PCM the threshold is 10^{-4}. Delta modulation is intelligible at random error rates as high as 10% but PCM is unintelligible at 1% error rates. In burst error environments the tolerances of PCM and differential systems are more nearly the same. (If the msb of a PCM codeword is in error it does not matter if the lsb are also in error.)

Those parameters of synthetic quality coders that are critical for voice reconstruction are usually redundantly encoded. Speech storage applications are least concerned with the effects of bit errors while digital mobile applications need to function with error rates as high as 1%.

Effects of Tandem Encoding

Some low bit rate vocoders provide acceptable performance by themselves but do not perform well in combination with other coders [57]. Thus, whenever an encoding algorithm is to be used within the public network its performance when connected in tandem with other coders needs to be evaluated. Examples of tests that need to be performed are (1) multiple conversions to and from analog, (2) multiple conversions to and from standard PCM, (3) intervening conversions to ADPCM, and (4) the effects of delays and echoes on the performance.

Coding Formats

The encoding algorithm with the simplest coding format for transmission is delta modulation because no level of framing in the bit stream is required. Standard PCM word formats are easily supported because T1 transmission links were specifically designed for word-interleaved multiplexing. Algorithms with variable-length block codes are the most cumbersome to deal with because framing patterns must be established within the channel. This extra level of framing can cause particular sensitivity to losses of the internal frame structure due to channel errors or slips (discussed in Chapter 7). Block structured coding algorithms present no special problems to packetized voice transmission or storage systems because the systems themselves support the block structures.

Another consideration of the coding format is the data patterns themselves. Because older versions of T1 lines do not support all zeros transmission codes, algorithms such as ADPCM must preclude the all zeros codeword when used in North America.

Signal Processing

When μ-255 and A-law PCM coding algorithms were chosen particular attention was paid to signal processing functions such as gain adjustments, conferencing, echo canceling, and tone generation and detection. Most transmission applications require at least some of these functions so the ease with which they can be implemented is an important consideration. Most often, the signal processing functions are implemented by first converting to standard PCM.

REFERENCES

1 J. M. Kasson, "The Rolm Computerized Branch Exchange: An Advanced Digital PBX," *Computer Magazine*, June 1979, p 24.

2 G. Frantz and K.-S. Lin, "Speech Technology in Consumer Products," *Speech Technology*, April 1982, pp 25–27.

3 S. Finkelstein, "Customized LPC Vocabulary Lets Cars Talk," *Speech Technology*, September/October 1983, pp 65–69.

4 M. G. Carasso, J. B. H. Peek, and J. P. Sinjou, "The Compact Disc Digital Audio System," *Phillips Technical Review*, Vol 40, No. 6, 1982, pp 151–155.

5 S. E. Levinson and D. B. Roe, "A Perspective on Speech Recognition," *IEEE Communications Magazine*, January 1990, pp 28–34.

6 Y. Sagiska, "Speech Synthesis from Text," *IEEE Communications Magazine*, January 1990, pp 35–41.

6 J. L. Flanagan, *Speech Analysis, Synthesis, and Perception*, 2nd ed. Springer-Verlag, New York, 1972.

8 K. F. Lee, H. W. Hon, and R. Reddy, "An Overview of the SPHINX Speech Recognition System," *IEEE Transactions on Acoustics Speech and Signal Processing*, January 1990, pp 35–45.

9 N. S. Jayant and P. Noll, *Digital Coding of Waveforms: Principles and Applications to Speech and Video*, Prentice-Hall, Englewood Cliffs, New Jersey, 1984.

10 J. Flanagan, M. Schroeder, B. Atal, R. Crochiere, N. Jayant, and J. Tribolet, "Speech Coding," *IEEE Transactions on Communications*, April 1979, pp 710–737.

11 N. Kitawaki and H. Nagabuchi, "Quality Assessment of Speech Coding and Speech Synthesis Systems," *IEEE Communications Magazine*, October 1988, pp 36–44.

12 B. J. McDermott, C. Scagliola, and D. J. Goodman, "Perceptual and Objective Evaluation of Speech Processed by Adaptive Differential PCM," *Bell System Technical Journal*, May–June 1978, pp 1597–1618.

13 "The D3 Channel Bank Compatibility Specification—Issue 3," Technical Advisory No. 32, American Telephone and Telegraph Company, October 1977.

14 M. B. Akgun, "Transmission Objectives for a Subscriber Loop System Digital Voice Codec," *International Symposium—Subscriber Loops and Services*, Ottawa, 1974, pp 7.1.1–7.1.7.

15 H. Nann, H. M. Straube, and C. P. Villars, "A Companded Coder for an Experimental PCM Terminal," *Bell System Technical Journal*, January 1962, pp 173–226.

16 K. E. Fultz and D. B. Penick, "The T1 Carrier System," *Bell System Technical Journal*, September 1965, pp 1405–1451.

17 H. H. Henning and J. W. Pan, "D2 Channel Bank: System Aspects," *Bell System Technical Journal*, October 1972, pp 1641–1657.

18 W. R. Daumer, "Subjective Evaluation of Several Efficient Speech Coders," *IEEE Transactions on Communications*, April 1982, pp 655–662.

19 N. S. Jayant, "Step-Size Transmitting Differential Coders for Mobile Telephony," *Bell System Technical Journal*, November 1975, pp 1557–1581.

20 D. L. Duttweiler and D. G. Messerschmitt, "Nearly Instantaneous Companding and Time Diversity as Applied to Mobile Radio Transmission," *International Communications Conference*, 1975, pp 40-12–40-15.

21 I. M. McNair, Jr., "Digital Multiplexer for Expanded Rural Service," *Bell Labs Record*, March 1972, pp 80–86.

22 H. K. Dunn and S. D. White, "Statistical Measurements on Conversational Speech," *Journal of Acoustic Society of America*, January 1940, pp 278–288.

23 J. W. Bayless, S. J. Campanella, and A. J. Goldberg, "Voice Signals: Bit by Bit," *IEEE Spectrum*, October 1973, pp 28–34.

24 M. Decina and G. Modena, "CCITT Standards on Digital Speech Processing," *IEEE Journal on Selected Areas in Communications*, February 1988, pp 227–234.

25 G. Williams and H. Suyderhoud, "Subjective Performance Evaluation of the 32 kbit/s ADPCM Algorithm," *IEEE Globecom Proceedings*, 1984, pp 23.2.1–8.

26 J. A. Greefkes and K. Riemens, "Code Modulation with Digitally Controlled Companding for Speech Transmission," *Phillips Technical Review*, Vol. 31, 1970, pp 335–353.

27 J. E. Abate, "Linear and Adaptive Delta Modulation," *Proceedings of IEEE*, March 1967, pp 298–308.

28 N. S. Jayant and A. E. Rosenberg, "The Preference of Slope Overload to Granularity in the Delta Modulation of Speech," *Bell System Technical Journal*, December 1971, pp 3117–3125.

29 J. A. Greefkes and F. de Jager, "Continuous Delta Modulation," *Phillips Research Reports*, Vol. 23, 1968, pp 233–246.

30 J. A. Greefkes, "A Digitally Companded Delta Modulation Modem for Speech Transmission," *Proceedings of IEEE International Communications*, 1970, pp 7-33–7-48.

31 B. S. Atal and M. R. Schroeder, "Adaptive Predictive Coding of Speech Signals," *Bell System Technical Journal*, October 1970, pp 1973–1986.

32 P. Noll, "On Predictive Quantizing Schemes," *Bell System Technical Journal*, May–June 1978, pp 1499–1532.

33 M. R. Schroeder, "Linear Predictive Coding of Speech: Review and Current Directions," *IEEE Communications Magazine*, August 1985, pp 54–61.

34 T. Tremain, "The Government Standard Adaptive Predictive Coding Algorithm: APC-04," *Speech Technology*, February/March 1985, pp 52–62.

35 R. E. Crochiere, "An Analysis of 16 kb/s Sub-Band Coder Performance, Dynamic Range, Tandem Connections, and Channel Errors," *Bell System Technical Journal*, October 1978, pp 2927–2951.

36 P. Mermelstein, "G.722 A New CCITT Coding Standard for Digital Transmission of Wideband Audio Signals," *IEEE Communications Magazine*, January 1988, pp 8–15.

37 X. Maitre, "7 kHz Audio Coding within 64 kbit/s," *IEEE Journal on Selected Areas in Communications*, February 1988, pp 283–298.

38 N. S. Jayant, "High-Quality Coding of Telephone Speech and Wideband Audio," *IEEE Communications Magazine*, January 1990, pp 10–20.

39 L. Radin, "Low-Bit-Rate Voice Digitizer Hits the Market at 2,400 b/s," *Speech Technology*, September/October 1983, pp 92–95.

40 H. Dudley, "Remaking Speech," *Journal of the Acoustic Society of America*, October 1939, pp 169–177.

41 L. R. Rabiner and R. W. Schafer, *Digital Processing of Speech Signals*, Prentice-Hall, Englewood Cliffs, New Jersey, 1978.

42 T. E. Tremain, "The Government Standard Linear Predictive Coding Algorithm: LPC-10," *Speech Technology*, April 1982, pp 40–49.

43 W. D. Voiers, "Evaluating Processed Speech Using the Diagnostic Rhyme Test," *Speech Technology*, January/February 1983, pp 30–39.

44 W. D. Voiers, "Diagnostic Acceptability Measure for Speech Communication Systems," *Proceedings International Conference on Acoustics, Speech and Signal Processing*, May 1977, pp 204–207.

45 B. S. Atal and R. Q. Hofacker, "The Telephone Voice of the Future," *Bell Labs Record*, July 1985, pp 4–10.

46 I. Boyd, C. B. Southcott, D. P. Crowe, and P. J. Bolingbroke, "Speech Codec for the Skyphone Aeronautical Telephone Service," *British Telecommunications Engineering*, July 1986, pp 83–91.

47 J. E. Natig, S. Hansen, and J. deBrito, "Speech Processing in the Pan European Digital Mobile Radio System (GSM)-System Overview," *IEEE Globecom Proceedings*, 1989, pp 29B.1.1–5.

48 K. Hellwig, P. Vary, D. Massaloux, J. P. Petit, C. Galand, and M. Russo, "Speech Codec for the European Mobile Radio System," *IEEE Globecom Proceedings*, 1989, 29B.2.1–5.

49 C. K. Un and D. T. Magill, "The Residual-Excited Linear Prediction Vocoder with Transmission Rate Below 9.6 kbot/s," *IEEE Transactions on Communications*, December 1975, pp 1466–1474.

50 K. K. Paliwal and T. Svendsen, "A Study of Three Coders (Sub-band, RELP, and MPE) for Speech with Additive White Noise," *Proceedings IEEE International Conference on Acoustics, Speech and Signal Processing*, 1985, pp 43.5.1–4.

51 A. J. Morris and S. Kay, "Telephony Earth Station," *IEEE Globecom Proceedings*, 1989, pp 48.1.1–4.

52 M. R. Schroeder and B. S. Atal, "Code-Excited Linear Prediction (CELP): High-Quality Speech at Very Low Bit Rates," *IEEE International Conference on Acoustics, Speech and Signal Processing*, 1985, pp 25.1.1–4.

53 J. H. Chen and A. Gersho, "Gain-Adaptive Vector Quantization with Application to Speech Coding," *IEEE Transactions on Communications*, September 1987, pp 918—930.

54 R. D. Hoyle and D. D. Falconer, "A Comparison of Digital Speech Coding Methods for Mobile Radio Systems," *IEEE Journal on Selected Areas in Communications,* June 1987, pp 915–920.

55 N. Kitawaki, H. Nagabuchi, and K. Itoh, "Objective Quality Evaluation for Low-Bit-Rate Speech Coding Systems," *IEEE Journal on Selected Areas in Communications,* February 1988, pp 242–247.

56 J. B. O'Neal, Jr., "Waveform Encoding of Voiceband Data Signals," *Proceedings of the IEEE,* February 1980, pp 232–247.

57 D. J. Goodman, C. Scagliola, R. E. Crochiere, L. R. Rabiner, and J. Goodman, "Objective and Subjective Performance of Tandem Connections of Waveform Coders with an LPC Vocoder," *Bell System Technical Journal,* March 1979, pp 601–628.

PROBLEMS

3.1 Assume a signal consists of three tones: one at 1 kHz, one at 10 kHz, and one at 21 kHz. What tones will be present at the output of a PAM decoder if the sampling rate is 12 kHz and the input is unfiltered? (Assume the output filter cutoff frequency is 6 kHz.)

3.2 Derive an expression for the average quantization noise power that occurs when the decoder output samples are offset from the center of a quantization interval by a distance equal to 25% of the interval. (The output values are at the 75% point instead of the 50% point.) How much degradation in decibels does this offset represent? (Assuming uncorrelated offsets.)

3.3 How much does the signal-to-noise ratio of a uniform PCM encoder improve when 1 bit is added to the codeword?

3.4 A black and white television signal has a bandwidth of about 4.2 MHz. What bit rate is required if this signal is to be digitized with uniform PCM at an SQR of 30 dB? Use a sampling-rate to Nyquist-rate ratio comparable to that used for PCM voice encoding.

3.5 How much dynamic range is provided by a uniform PCM encoder with 12 bits per sample and a minimum SQR of 33 dB?

3.6 What is the signal-to-quantizing noise ratio produced by a segmented μ255 PCM coder when the input signal is a full-range triangular wave? (Assume the repetition frequency is low enough that the bandlimiting filter does not change the waveform significantly.)

3.7 Given a sample value of 420 mV for a μ255 PCM encoder capable of encoding a maximum level of 2 V, determine each of the following:
(a) The compressed μ255 codeword.
(b) The linear representation of the compressed code.

(c) The A-law code obtained by converting from the $\mu255$ code.

(d) The μ-law code obtained by converting back from the A-law code.

3.8 Given the following $\mu255$ code words, determine the code word that represents the linear sum: (0 110 1001), (1 011 0111).

3.9 Generate an encoding table (i.e., list the quantization intervals and corresponding codes) for the magnitude of a piecewise linear code with segment slopes: $1, \frac{1}{2}, \frac{1}{4}, \frac{1}{8}$. Assume four equally spaced intervals are in each segment. Assume all intervals in the first segment are of equal length (as in A-law coding).

3.10 What is the signal-to-noise ratio of a full amplitude sample for the coder of Problem 3.9?

3.11 What is the signal-to-noise ratio of a maximum amplitude sample in the first linear segment of the preceding coder?

3.12 What is the dynamic range implied by Problems 3.10 and 3.11?

3.13 For the encoder in Problem 3.9, how many bits are required for uniform encoding of the same dynamic range and same minimum quantization interval?

3.14 A uniform PCM system is defined to encode signals in the range of -8159 to $+8159$ in quantization intervals of length 2. (The quantization interval at the origin extends from -1 to $+1$.) Signals are encoded in sign-magnitude format with a polarity bit $= 1$ denoting a negative signal.

(a) How many bits are required to encode the full range of signals?

(b) How many unused codes are there?

(c) Determine the quantization noise, noise power, and signal-to-noise ratio (in decibels) of each of the following sample values: 30.2, 123.2, -2336.4, 8080.9.

3.15 Repeat part c in Problem 3.14 for piecewise linear $\mu255$ PCM.

3.16 Given two A-law piecewise linear codewords (00110110 and 00101100) determine their linear representations, add them, and convert back to compressed representation.

3.17 A D3 channel bank uses "robbed digit" signaling wherein the least significant bit of every sixth frame is stolen for signaling. Determine the relative increase in overall quantization noise produced by this process under the following conditions:

(a) The decoder merely treats the signaling bit as a voice bit and decodes the PCM sample accordingly.

(b) The decoder knows which bits are signaling bits and generates an output sample corresponding to the middle of the double-length quantization interval defined by the seven available bits. (This is the actual technique specified for D3 channel bank decoders.)

3.18 If two bits per sample are added o a PCM bit stream, how much can the dynamic range be increased if the quantization intervals are adjusted to improve the SQR by 3 dB?

3.19 How much can the SQR of a digitized voice signal be increased if the bandwidth is reduced by 30% and the dynamic range is unchanged?

3.20 An A-law PCM encoder with a perfect zero-level threshold detects an arbitrarily small input sine wave because the first quantization interval does not straddle the origon. What is the power level of the decoded output signal?

3.21 Determine the sequence of $4 \mu 255$ PCM samples for a 2000 Hz tone at -6 dBm when the first sample occurs at a phase of $45°$.

3.22 Repeat Problem 3.21 for A-law signals.

3.23 Determine the A-law codewords that correspond to the μ-law codewords of the digital milliwatt signal given in Example 3.3. Assume both systems are scaled to the same maximum signal power.

4

DIGITAL TRANSMISSION AND MULTIPLEXING

A fundamental consideration in the design of digital transmission systems is the selection of a finite set of discrete electrical waveforms for encoding the information. In the realm of digital communications theory these discrete waveforms are referred to as signals. The same terminology is used in this chapter with the understanding that signals in the present context refer to internal waveforms (pulses) of a transmission system and not the control information (signaling) used to set up and monitor connections within a network. In communications theory terminology "signal processing" refers to filtering, shaping, and transformations of electrical waveforms—not the interpretation of control signals by the processor of a switching machine.

A second aspect of digital transmission involves defining certain time relationships between the individual transmission signals. The source terminal transmits the individual signals using predefined time relationships so the receiving terminal can recognize each discrete signal as it arrives. Invariably the establishment of a proper time base at the receiver requires transmission capacity above that needed for the digital information itself. Over relatively short distances (as within a switching system or a computer complex) the timing information (clock) is usually distributed separately from the information bearing signals. Over long distances, however, it is more economical to incorporate the timing information into the signal format itself. In either case the timing information requires channel capacity in terms of bandwidth, data rate, or code space.

This chapter discusses the most common digital signaling techniques for wire-line transmission systems. Thus the subject of this chapter is often referred to as line coding. These techniques are generally applicable to any transmission system transmitting digital signals directly in the form of pulses (such as coaxial cable or optical fiber). The fact that these techniques include low-frequency components in their transmitted spectrum leads to their also being called low-pass or baseband transmission systems. In Chapter 6 we discuss bandpass transmission systems, that is, radio systems that require modulation and carrier frequency transmission.

The following discussions concentrate on system and application level considerations of digital transmission. Analytic details of pulse transmission systems are not emphasized since material of this nature is available in all works on digital communications theory. Some fundamentals of pulse transmission systems are presented in Appendix C where many of the equations presented in this chapter are derived.

4.1 PULSE TRANSMISSION

All digital transmission systems are designed around some particular form of pulse response. Even carrier systems must ultimately produce specific pulse shapes at the detection circuitry of the receiver. As a first step, consider the perfectly square pulse shown in Figure 4.1. The frequency spectrum correspon-

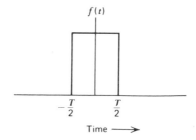

Figure 4.1. Definition of a square pulse.

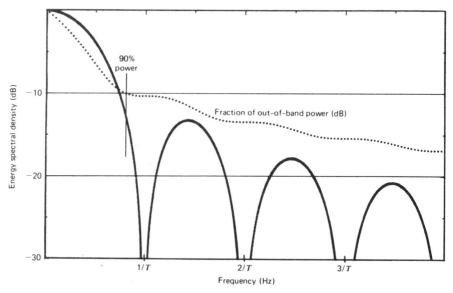

Figure 4.2. Spectrum of square pulse with duration T.

ding to the rectangular pulse is derived in Appendix A and shown in Figure 4.2. It is commonly referred to as a $\sin(x)/x$ response:

$$F(\omega) = (T)\frac{\sin(\omega T/2)}{\omega T/2} \tag{4.1}$$

where $\omega = 2\pi f$ is the radian frequency, and T is the duration of a signal interval.

Notice that Figure 4.2 also provides the percentage of total spectrum power at various bandwidths. As indicated, 90% of the signal energy is contained within the first spectral null at $f = 1/T$. The high percentage of energy within this band indicates that the signal can be confined to a bandwidth of $1/T$ and still pass a good approximation to the ideal waveform. In theory, if only the sample values at the middle of each signal interval are to be preserved, the bandwidth can be confined to $1/2T$. From this fact the maximum signaling rate in a specified bandwidth is determined as

$$R_{max} = 2BW \tag{4.2}$$

where R is the signaling rate $= 1/T$, and BW is the available bandwidth. Equation 4.2 states a fundamental result from communications theory credited to Harry Nyquist: the maximum signaling rate achievable through a low-pass bandwidth with no intersymbol interference is equal to twice the bandwidth. This rate R_{max} is sometimes referred to as the "Nyquist rate."

Although discrete, square-shaped pulses are easiest to visualize, preservation of the square shape requires wide bandwidths and is therefore undesirable. A more typical shape for a single pulse is shown in Figure 4.3. The ringing on both sides of the main part of the pulse is a necessary accompaniment to a channel with a limited bandwidth. Normally, a digital transmission link is excited with square pulses (or modulated equivalents thereof), but bandlimiting filters and the transmission medium itself combine to produce a response like the one shown. Figure 4.3 shows pulse output in negative time so the center of the pulse occurs at $t = 0$. Actually, the duration of the preringing is limited to the delay of the filters and equalizers.

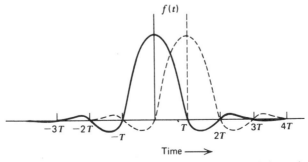

Figure 4.3. Typical pulse response of a bandlimited channel.

An important feature of the pulse response shown in Figure 4.3 is that, despite the ringing, a pulse can be transmitted once every T seconds and be detected at the receiver without interference from adjacent pulses. Obviously, the sample time must coincide with the zero crossings of the adjacent pulses. Pulse responses like the one shown in Figure 4.3 can be achieved in channel bandwidths approaching the minimum (Nyquist) bandwidth equal to one-half of the signaling rate. Appendix C describes pulse transmission design in more detail.

Intersymbol Interference

As the signaling rate of a digital transmission link approaches the maximum rate for a given bandwidth, both the channel design and the sample times become more critical. Small perturbations in the channel response or the sample times produce nonzero overlap at the sample times called intersymbol interference. The main causes of intersymbol interference are

1. Timing inaccuracies
2. Insufficient bandwidth
3. Amplitude distortion
4. Phase distortion.

Timing Inaccuracies

Timing inaccuracies occurring in either the transmitter or the receiver produce intersymbol interference. In the transmitter, timing inaccuracies cause intersymbol interference if the rate of transmission does not conform to the ringing frequency designed into the channel. Timing inaccuracies of this type are insignificant unless extremely sharp filter cutoffs are used while signaling at the Nyquist rate.

Since timing in the receiver is derived from noisy and possibly distorted receive signals, inaccurate sample timing is more likely than inaccurate transmitter timing. Sensitivity to timing errors is small if the transmission rate is well below the Nyquist rate (e.g., if the transmission rate is equal to or less than the channel bandwidth, as opposed to being equal to the theoretical maximum rate of twice the bandwidth; see Appendix C).

Insufficient Bandwidth

The ringing frequency shown in Figure 4.3 is exactly equal to the theoretical minimum bandwidth of the channel. If the bandwidth is reduced further, the ringing frequency is reduced and intersymbol interference necessarily results.

Some systems purposely signal at a rate exceeding the Nyquist rate, but do so with prescribed amounts of intersymbol interference accounted for in the receiver. These systems are commonly referred to as partial response systems— so called because the channel does not fully respond to an input during the time of a single pulse. The most common forms of partial response systems are discussed in a later section.

Amplitude Distortion

Digital transmission systems invariably require filters to bandlimit transmit spectrums and to reject noise and interference in receivers. Overall, the filters are designed to produce a specific pulse response. When a transmission medium with predetermined characteristics is used, these characteristics can be included in the overall filter design. However, the frequency response of the channel cannot always be predicted adequately. A departure from the desired frequency response is referred to as amplitude distortion and causes pulse distortions (reduced peak amplitudes and improper ringing frequencies) in the time domain. Compensation for irregularities in the frequency response of the channel is referred to as *amplitude equalization.*

Phase Distortion

When viewed in the frequency domain, a pulse is represented as the super-position of frequency components with specific amplitude and phase relationships. If the relative amplitudes of the frequency components are altered, amplitude distortion results as discussed. If the phase relationships of the components are altered, phase distortion occurs. Basically, phase distortion results when the frequency components of a signal experience differing amounts of delay in the transmission link. Compensation of phase distortion is referred to as *phase equalization.* For a good tutorial on equalization, including a description of an automatic equalizer for data transmission, see reference [1].

4.2 ASYNCHRONOUS VERSUS SYNCHRONOUS TRANSMISSION

There are two basic modes of digital transmission involving two fundamentally different techniques for establishing a time base (sample clock) in the receiving terminal of a digital transmission link. The first of these techniques is asynchronous transmission, which involves separate transmissions of groups of bits or characters. Within an individual group a specific predefined time interval is used for each discrete signal. However, the transmission times of the groups are unrelated to each other. Thus the sample clock in the receiving terminal is reestablished for reception of each group. With the second technique, called synchronous transmission, digital signals are sent continuously at a constant rate. Hence the receiving terminal must establish and maintain a sample clock that is synchronized to the incoming data for an indefinite period of time.

4.2.1 Asynchronous Transmission

Between transmissions an asynchronous line is in an inactive or idle state. The beginning of each transmission group is signified by a start bit. The middle of the start bit is determined and succeeding information bits are sampled at a nominal rate beginning at the middle of the second bit interval. Following the in-

formation symbols, one or more stop bits are transmitted to allow the line to return to the inactive state.* Figure 4.4 shows an asynchronous mode of operation commonly used for low-speed data communication.

As shown in Figure 4.4, the detection of each information bit is accomplished by ideally sampling the input waveform at the middle of each signal interval. In practice, sample times depart from the ideal depending on how much the start bit is corrupted by noise and distortion. Since the sample time for each information bit is derived from a single start bit, asynchronous systems do not perform well in high-noise environments. Of course, more than one start bit could be used to improve the accuracy of the starting phase of the sample clock, but this would complicate the receiver and add more overhead for transmission of timing information.

Timing errors also arise if the nominal rate of the sample clock in the receiver is different from the nominal rate of transmission at the source. Even though the start bit might define the proper starting phase for the sample clock, an offset in the clock frequency of the receiver causes each successive sample time to drift farther from the center of the respective signal intervals. Since the very use of the term "asynchronous" implies a free-running clock in the receiver, a certain amount of drift is inevitable in all asynchronous systems. The maximum length of each symbol group or character is determined by the limits of the initial phase inaccuracies and the maximum expected frequency difference between the transmitter and receiver clocks.

The main attraction of asynchronous transmission is the ease with which it determines the sample times in the receiver. In addition, asynchronous transmission automatically provides character framing and is inherently flexible in the range of average data rates that can be accommodated. For high rates, one character after another is transmitted. Lower data rates are automatically accommodated by increasing the idle time between characters. In contrast, a

Ideal sample times for each 8-bit character

Figure 4.4. Asynchronous transmission.

*Originally, stop bits were inserted to allow electromechanical equipment enough time to reset before the next character arrived. With electronic equipment the only purpose of stop bits is to allow a start bit to always be a transition to a space (logic "0").

synchronous receiver must track changes in a transmitter rate before it can sample the incoming signals properly. Normally the receive clock of a synchronous system can be adjusted only quite slowly and only over a narrow range. Hence an asynchronous system is more naturally suited to applications where the data rate varies.

Synchronous transmission systems can support variable information rates, but the task of adjusting the information rate falls upon higher level processes (data link protocols) that insert null codes into the bit stream. The null codes are used as filler when a source has nothing to send. This form of transmission is sometimes referred to as "isochronous." An isochronous mode of operation is required whenever a synchronous line carries data from an asynchronous source.

The major drawback of asynchronous transmission is its poor performance in terms of error rates on noisy lines. Thus asynchronous transmission evolved for applications where implementation costs dominated performance considerations. Asynchronous transmission is used in voiceband data sets (modems) for transmission rates up to 1200 bps. For digital telephony, something similar to asynchronous transmission is used on some two-wire digital subscriber loops [2]. These systems provide a full duplex (four-wire) circuit by transmitting bursts of data alternately in each direction on a single pair of wires. Thus these systems are sometimes referred to as "ping-pong" transmission systems; they are not truly asynchronous since each transmission in each direction occurs at prescribed times allowing timing information in one burst to carry over into the next burst.

4.2.2 Synchronous Transmission

T1 lines, and all other interoffice digital transmission links used in the telephone networks use synchronous transmission exclusively. Thus the line coding format for these systems must incorporate special considerations to ensure that each regenerative repeater or receiver can synchronize a local sample clock to the incoming signaling rate. Generally speaking, the synchronization requirements imply that a certain minimum density of signal transitions is required to provide continuous indication of signaling boundaries. Often, purely random data cannot be relied upon to provide a sufficient number of transitions. In these cases certain provisions must be made to insert artificial transitions into the transmitted waveforms. Although these extra transitions imply a certain amount of transmission overhead, the loss in capacity can be relatively small. Following are descriptions of five techniques for ensuring the existence of signal transitions for timing recovery:

1. Source code restriction
2. Dedicated timing bits
3. Bit insertion
4. Data scrambling
5. Forced bit errors.

A sixth technique—inserting the transitions into the signal waveforms themselves—is discussed in Section 4.3 under "line coding."

Source Code Restriction

One means of ensuring a sufficient number of signal transitions is to restrict the code set or data patterns of the source so that long transition-free data sequences do not occur. Historically, PCM channel banks in the North American telephone network precluded all-zeros code words since the original line code of Tl lines produced no transitions for zeros. In the case of 8 bit PCM code words, the exclusion of a single code word represents a loss in transmission capacity of only one part in 256.*

Relying on source coding to ensure sufficient transitions in the line code has one significant drawback: the transmission link cannot be used for new applications where the source does not exclude the unwanted data patterns. For this reason, the total capacity of a T1 line could not be used for random data until a new line code (B8ZS) was introduced.

Dedicated Timing Bits

As an alternative to excluding transition-free data patterns, the line itself can periodically insert transition bearing bits into the data stream. These bits are inserted at regular intervals, independently of the source data, to ensure the existence of a minimum number of signal transitions. Thus some fraction of the channel capacity is dedicated to timing bits.

As an example, the original Dataphone® Digital Service (DDS) offering for data communications over T1 lines provided a maximum capacity of 56 kbps for each channel. Thus only 7 of the 8 bits in each time slot of each channel were available for the user. Among other functions, the unused bit in each time slot provides an assurance that all 8 bits of a time slot are not zero. Several fiber systems described in Chapter 8 also use explicit timing bits as inherent aspects of their line codes. The density of timing pulses in these systems ranges from 1 in 5 bits to 1 in 20 bits. Notice that insertion of a dedicated timing bit is essentially the same procedure used to establish timing for asynchronous transmission. In a synchronous receiver, however, a sample clock is obtained by averaging the timing information over a large number of timing transitions, not just one.

Bit Insertion

In the preceding DDS example 1 bit of every 8 bits in a time slot is dedicated to ensuring sufficient timing information in the bit stream. Another possibility for precluding unwanted line patterns is to use "bit insertion" only when necessary. As an example, the source data over a T1 line could be monitored for all zeros in the first 7 bits of a time slot. Whenever the zeros occur, a one could be inserted into the data stream as the eighth bit of the time slot. At the other end of the line

*From an information theory point of view the loss in voice information is even lower since the probability of occurrence of the all-zeros code word is much less than 1 in 256.

the one following seven zeros is removed automatically. Each insertion delays the source data by the time of 1 bit, but otherwise the full capacity of the channel is available.

This procedure is directly analogous to the "zero bit insertion" algorithm used in IBM's Synchronous Data Link Control (SDLC) for data communications. In this protocol [and other similar bit-oriented protocols such as CCITT's High Level Data Link Control (HDLC) or American National Standards Institute's Advanced Data Communications Control Procedure (ADDCP)] a specific data pattern called a "flag" is used to indicate the end of a data block. The transmitter must be precluded from inadvertently sending the flag as part of the user data. The means of preventing inadvertent flags is to insert a 0 following a string of five 1's in the user data. Since a flag contains a string of six 1's, zero bit insertion effectively precludes unintended flag transmissions. The receiving node of an SDLC (or HDLC) data link removes any 0 following five 1's. The receipt of six 1's, however, can only be part of a legitimate flag (01111110). If HDLC data are inverted for transmission it never produces all 0's (except during the idle state).

Although a bit insertion algorithms allows for more efficient use of a channel than dedicated timing bits, the insertion procedure has a number of drawbacks. First, this process causes the source data to be delayed every time an insertion is made. Hence a continuous, real-time transmission application (such as voice) requires smoothing the arrival rate with data buffers at the destination. Second, the bit insertion process causes any character structure in the user's data to be become unrelated to the time slot structure of a time division multiplexed transmission link. Thus if user data consists of 8 bit characters (like PCM voice samples), character boundaries cannot be maintained with respect to 8 bit time slots in a T1 transmission link.

Data Scrambling

Many digital transmission systems use data scramblers to randomize the data patterns on their transmission links. Although these data scramblers are similar to those used for encryption, the fundamental purpose of these scramblers is to prevent the transmission of repetitive data patterns, and not to encrypt the traffic. Repetitive data patterns generate line spectra that can be significantly more degrading from an interference point of view than continuously distributed spectra produced by random data patterns. Voiceband data models, for example, are allowed to operate at higher power levels if they include scramblers to randomize the data traffic. Also, digital radio systems are required by the FCC to not transmit line spectra, which essentially means that repetitive data patterns must be excluded.

Even when not required, data scramblers are useful in transforming data sequences with low transition densities into sequences with strong timing components. Scrambling is not used on lower rate T-carrier systems (T1 and T2), but was used on the 274 Mbps T4M coaxial transmission system [3] and many fiber systems.

Data scramblers (with equal input and output bit rates) do not prevent long strings of zeros in an absolute sense. They merely ensure that relatively short repetition patterns are transformed to randomized traffic with a minimum density of transitions. If purely random input data are scrambled, the output data are also purely random and therefore have a certain statistical probability of producing arbitrarily long strings of zeros. The probability of a random occurrence, however, is acceptable when compared to the probability of nonrandom sequences representing speech pauses or idle data terminals. To determine which seemingly random data sequence produces all zeros at the output of a scrambler, apply all zeros to the corresponding descrambler.

The T4M coaxial transmission system used a data scrambler as the basic means of producing adequate timing information. This system could tolerate much longer strings of zeros because the timing recovery circuits in the regenerative repeaters used phase-locked-loops that maintain timing over relatively long periods of time. In contrast, the original T1 systems recovered timing with tuned circuits that resonated at the desired clock frequency (1.544 mbps) when excited by a received pulse. Because the tuned circuits have lower effective Qs than a phase-locked-loop, the oscillations drift from the proper frequency and die out more rapidly. Hence the original T1 receivers could not tolerate as long a string of zeros as could T4M receivers. Phase-locked-loop clock recovery circuits are used on all later generation wire line and fiber transmission systems so much longer strings of 0's can be tolerated.

Forced Bit Errors

A fifth method of maintaining sufficient timing information in the line signals involves having the transmission terminal at the source force an occasional bit error in order to interrupt a long, transition-free data pattern. If the transition-free sequences are long enough and uncommon enough, the intentional bit errors might be less frequent than random channel errors on the digital transmission link. Thus the intentional errors may not represent a significant degradation over and above that already existing. Nevertheless, forced bit errors are not generally recommended as part of a line coding procedure but are mentioned in the interest of completeness.

As mentioned previously, North American channel banks force a bit error in the second least significant bit of an all-zeros transmission code to ensure sufficient signal transitions. An important aspect of this procedure is that it is performed by the source where the significance of the bit error is known. If the transmission link itself inserted the bit errors, the effects would not be as controllable—particularly when a variety of traffic types are being serviced.

A more subtle problem with forced transmission errors arises if the digital transmission link is used for automatic repeat request (ARQ) data transmission. An ARQ data communication link is designed to provide error-free transmission, despite random channel errors, by inserting redundancy into the data stream and checking the received data for error-free reception. If errors are detected, a retransmission is requested. When the errors are not random, but

forced by the transmission link, the ARQ system will become frustrated if it ever encounters the restricted sequence, no matter how unlikely it is.* Once again, if forced errors are used they should be incorporated into the source as part of the source code restriction process and not as a function of the transmission link.

4.3 LINE CODING

In the preceding section various techniques for establishing timing information are described in general terms. The choice of any particular technique is dependent on the specific line code in use. This section describes the most common line codes used for digital transmission and indicates what additional steps, if any, are needed to maintain synchronization between the transmitter and receiver. Some line coding techniques provide adequate timing information themselves and do not require any of the procedures discussed previously. In addition to synchronization requirements, other considerations for choosing a line code are the spectrum of the line code and the available bandwidth (particularly at low frequencies), noise and interference levels, synchronization acquisition times, performance monitoring, and implementation costs.

4.3.1 Level Encoding

Conceptually, the simplest form of line coding uses a different signal level to encode each discrete symbol transmitted. Within a computer system the most common form of coding is an on–off code using a 3 V level for a 1 and near 0 V for a 0. Over a transmission link, however, it is more efficient in terms of power to encode binary data with an equivalent difference in levels but symmetrically balanced about 0 V. For example, the average power required for equally likely +3 and 0 V encodings is 4.5 W (assuming 1 Ω resistance). With +1.5 and −1.5 V encodings, the same error distance is achieved with half the power requirements (2.25 W). Communications engineers commonly refer to the unbalanced code as a unipolar code and the balanced code as a polar code. A representative sequence of binary data and its corresponding balanced and unbalanced level encoding is shown in Figure 4.5. Notice that the level of each signal is maintained for the duration of a signal interval. For this reason the balanced (polar) encoding is also referred to as a nonreturn-to-zero (NRZ) code.†

As indicated in Figure 4.5, an NRZ signal contains no transitions for long strings of ones or zeros. Hence one of the procedures described previously to ensure timing transitions must be employed if NRZ encoding is used on a synchronous transmission link.

*If the transmission link uses a scrambler, the unlikely sequence will not be repeated.

†Some communications theorists refer to a balanced two-level code as a "bipolar code." The North American telephone industry, however, uses the term bipolar to refer to a three-level code described in the next section.

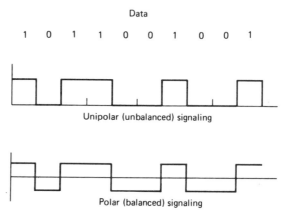

Figure 4.5. Unipolar and polar (NRZ) line codes.

An NRZ line code is a pulse transmission system wherein the pulse (before filtering) lasts for the duration of a signaling interval T. Hence the frequency spectrum (assuming random data) of an NRZ code is the $\sin(x)/x$ spectrum of Equation 4.1 and shown in Figure 4.2. As indicated, the frequency spectrum is significantly nonzero at zero frequency (dc). Most wire-line transmission links, however, do not pass dc signals by virtue of their being ac coupled with transformers or capacitors to eliminate dc ground loops. Furthermore, some systems purposely remove dc components from the signal to allow line powering of repeaters or to facilitate single sideband transmission. The elimination of the low-frequency components in the waveform causes long strings of ones or zeros to decay gradually in amplitude. Hence not only would a receiver lose timing information during these strings, but it would also lose its amplitude reference for optimally discriminating between a 1 level and a 0 level. The effect of low-frequency cutoff, called dc wander, is shown in Figure 4.6 for a typical transmission sequence. Notice that following the long string of ones, the output of the link is such that 1 to 0 errors are more likely than 0 to 1 errors. Similarly, long strings of zero increase the likelihood of a 0 to 1 error. This problem arises not only for long strings of ones or zeros, but whenever there is an imbalance in the number of ones and zeros. Hence periodic timing pulses are not sufficient to remove dc wander.

The existence of low frequencies in a random data signal is the basic reason why modems are needed for data communications over the analog telephone network. (Analog telephone circuits do not pass direct current.) It is also the reason that NRZ coding is not often used for long-distance transmission. Dc wander is not unique to data transmission systems. It is a phenomenon that must be reconciled in television receivers, radar receivers, or radiation detectors.

One technique of offsetting dc wander is referred to as dc or baseline restoration [4]. As illustrated in Figure 4.7, dc restoration basically involves passing received pulses through a capacitor, detecting them, and then removing

Figure 4.6. Dc wander of NRZ signal.

Figure 4.7. Dc restoration for unipolar pulses.

the charge on the capacitor before the next pulse arrives. Charge on the capacitor is removed by driving the voltage to a specific threshold (0 V in Figure 4.7) and then removing the driving voltage when the threshold is reached. Since all charge on the capacitor is removed after each pulse, the baseline or decision reference level is constant at the beginning of a signal interval. An obvious disadvantage of this technique is that the signal input must have zero amplitude or be disabled during the reset time.

A generally more useful technique for overcoming baseline wander is to use decision feedback, also called quantized feedback equalization [5, 6]. In contrast to dc restoration, which drives the capacitor voltage to a constant, predetermined level, quantized feedback compensates for dc wander by locally generating the unreceived low-pass response and adding it to the received signal. To accomplish this, the original data stream is reconstructed. As shown in Figure 4.8, the reconstructed data stream is passed through a low-pass filter that generates a pulse equal to the tail or droop characteristic of the channel. The feedback signal adds to the received signal to eliminate the droop (intersymbol interference). Using a frequency domain analysis, the feedback response is complimentary to the channel response [5]. Quantized feedback is used in ISDN basic rate line interfaces [7].

4.3.2 Bipolar Coding

The dc restoration techniques mentioned in the preceding section simplify pulse detection by creating a low-pass pulse response in the receiver. There are numerous line codes that are specifically designed to not contain dc energy and thereby be unaffected by dc removal. Bipolar coding solves the dc wander

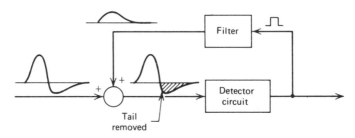

Figure 4.8. Decision feedback restoration.

problem by using three levels to encode binary data. Specifically, a logic 0 is encoded with zero voltage while a logic 1 is alternately encoded with positive and negative voltages. Hence the average voltage level is maintained at zero—to eliminate dc components in the signal spectrum. Since bipolar coding uses alternate polarity pulses for encoding logic 1's, it is also referred to as "alternate mark inversion" (AMI).*

Bipolar coding is the basic line coding procedure used by T1 lines in the telephone network. Rather than use full period pulses, however, T1 lines use a 50% duty cycle pulse to encode each logic 1. Fifty percent duty cycle pulses (Figure 4.9) were selected to minimize intersymbol interference and simplify timing recovery in the regenerative repeaters of a T1 line [8]. The power spectrum of a bipolar code is obtained from [9] as

$$S(\omega) = \frac{2p(1-p)}{T} |G(\omega)|^2 \frac{1 - \cos \omega T}{1 - 2(2p-1)\cos \omega T + (2p-1)^2} \tag{4.3}$$

where p is the probability of a 1, and $G(\omega)$ is the spectrum of an individual pulse.

$$G(\omega) = \left(\frac{T}{2}\right) \frac{\sin(\omega T/4)}{\omega T/4} \quad \text{for 50\% duty cycle pulses.}$$

Equation 4.3 is plotted in Figure 4.10 for various values of p. For purely random data, $p = \frac{1}{2}$. Recall, however, that source coding for $\mu255$ PCM channel

Figure 4.9. Bipolar coding.

*A mark is a term arising from telegraphy to refer to the active or 1 state of a level encoded transmission line.

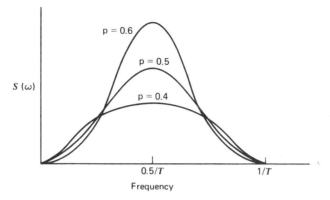

Figure 4.10. Spectral density of bipolar coding.

banks produces more 1's than 0's in the interest of establishing a strong clock signal. Hence the appropriate value of p for a T1 voice line is normally somewhat larger than 0.5 and depends on the amplitude of the voice signal. Low-level signals that remain in the first encoding segment produce a value of p approximately equal to 0.65. On the other hand, full-scale sine waves produce a value for p that is somewhat below 0.5 since most of the samples occur near maximum amplitude.

Because a bipolar code uses alternating polarities for encoding 1's, strings of "1"s have strong timing components. However, a string of 0's contains no timing information and therefore must be precluded by the source. The specifications for T1 line repeaters state that the repeaters will maintain timing as long as no string of greater than 15 zeros is allowed to occur [10]. A string of 15 zeros can only occur if a 0 framing bit falls between a 10000000 code in time slot 24 and a 00000001 code in time slot 1.

Code Space Redundancy

In essence, bipolar coding uses a ternary code space, but only two of the levels during any particular signal interval. Hence bipolar coding eliminates dc wander with an inefficient and redundant use of the code space. The redundancy in the waveform also provides other benefits. The most important additional benefit is the opportunity to monitor the quality of the line with no knowledge of the nature of the traffic being transmitted. Since pulses on the line are supposed to alternate in polarity, the detection of two successive pulses of one polarity implies an error. This error condition is known as a "bipolar violation." No single error can occur without a bipolar violation also occurring. Hence the bipolar code inherently provides a form of line code parity. The terminals of T1 lines are designed to monitor the frequency of occurrence of bipolar violations, and if the frequency of occurrence exceeds some threshold, an alarm is set.

In T-carrier systems, bipolar violations are used merely to detect channel errors. By adding some rather sophisticated detection circuitry, the same

redundancy can be used for correcting errors in addition to detecting them. Whenever a bipolar violation is detected, an error has occurred in one of the bits between and including the pulses indicating the violation. Either a pulse should be a 0, or an intervening 0 should have been a pulse of the opposite polarity. By examining the actual sample values more closely, a decision can be made as to where the error was most likely to have occurred. The bit with a sample value closest to its decision threshold is the most likely bit in error. This technique belongs to a general class of decision algorithms for redundant signals called maximum likelihood or Viterbi decoders [11]. Notice that this method of error correction requires storage of pulse amplitudes. If decision values only are stored, error correction cannot be achieved (only error detection).

An additional application of the unused code space in bipolar coding is to purposely insert bipolar violations to signify special situations such as time division multiplex framing marks, alarm conditions, or special codes to increase the timing content of the line signals. Since bipolar violations are not normally part of the source data, these special situations are easily recognized. Of course, the ability to monitor the quality of the line is compromised when bipolar violations occur for reasons other than channel errors.

4.3.3 Binary *N* Zero Substitution

A major limitation of bipolar (AMI) coding is its dependence on a minimum density of 1's in the source code to maintain timing at the regenerative repeaters. Even when strings of zeros greater than 14 are precluded by the source, a low density of pulses on the line increases timing jitter and therefore produces higher error rates. Binary *N* zero substitution (BNZS) [12] augments a basic bipolar code by replacing all strings of *N* zeros with a special *N* length code containing several pulses that purposely produce bipolar violations. Thus the density of pulses is increased while the original data are obtained by recognizing the bipolar violation codes and replacing them at the receiving terminal with *N* zeros.

As an example, a three zero substitution algorithm (B3ZS) is described. This particular substitution algorithm is specified for the standard DS-3 signal interface in North America [13]. It was also used in the LD-4 coaxial transmission system in Canada [14].

In the B3ZS format, each string of three zeros in the source data is encoded with either 00V or B0V. A 00V line code consists of 2 bit intervals with no pulse (00) followed by a pulse representing a bipolar violation (V). A B0V line code consists of a single pulse in keeping with the bipolar alternation (B), followed by no pulse (0), and ending with a pulse with a violation (V). With either substitution, the bipolar violation occurs in the last bit position of the three zeros replaced by the special code. Thus the position of the substitution is easily identified.

The decision to substitute with 00V or B0V is made so that the number of B pulses (unviolated pulses) between violations (V) is odd. Hence if an odd number

TABLE 4.1 B3ZS Substitution Rules

Polarity of the Preceding Pulse	Number of Bipolar Pulses (ones) Since Last Substitution	
	Odd	Even
−	00−	+0+
+	00+	−0−

of ones has been transmitted since the last substitution, 00V is chosen to replace three zeros. If the intervening number of ones is even, B0V is chosen. In this manner all purposeful violations contain an odd number of intervening bipolar pulses. Also, bipolar violations alternate in polarity so that dc wander is prevented. An even number of bipolar pulses between violations occurs only as result of a channel error. Furthermore, every purposeful violation is immediately preceded by a 0. Hence considerable systematic redundancy remains in the line code to facilitate performance monitoring. Table 4.1 summarizes the substitution algorithm.

Example 4.1. Determine the B3ZS line code for the following data sequence: 10100011000000010001. Use + to indicate a positive pulse, − to indicate a negative pulse, and 0 to indicate no pulse.

Solution. There are two possible sequences depending on whether an odd or even number of pulses has been transmitted following the previous violation.

Substitutions

	101	$\widehat{000}$	11	$\widehat{000}$	$\widehat{000}$	001	$\widehat{000}$	1
Case 1 (odd):	+0−	00−	+−	+0+	−0−	00+	00+	−
Case 2 (even):	+0−	+0+	−+	−0−	+0+	00−	00−	+

Violations

Example 4.1 indicates that the process of breaking up strings of zeros by substituting with bipolar violations greatly increases the minimum density of pulses in the line code. In fact, the minimum density is 33% while the average density is just over 60%. Hence the B3ZS format provides a continuously strong timing component. Notice that all BNZS coding algorithms guarantee continuous timing information with no restrictions on source data. Hence BNZS coding supports any application in a completely transparent manner.

TABLE 4.2 B6ZS Substitution Rules

Polarity of Pulse Immediately Preceding Siz Zeros to Be Substituted	Substitution
−	0− +0+ −
+	0+ −0− +
Example	
	1 0 0 0 0 0 0 1 0 1 1 0 0 0 0 0 0 0 0 0 0 0 0 0 0 000 1
+	−0− +0+ − +0− +0+ −0− +0+ −0− +000 −
−	+0+ −0− + −0+ −0− +0+ −0− +0+ −000 +

Another BNZS coding algorithm is the B6ZS algorithm for T2 transmission lines [15]. The B6ZS algorithm is defined in Table 4.2. This algorithm produces bipolar violations in the second and fifth bit positions of the substituted sequence.

CCITT recommends another BNZS coding format referred to as high-density bipolar (HDB) coding [16]. As implemented in the CEPT primary digital signal, HDB replaces strings of four zeros with sequences containing, a bipolar violation in the last bit position. Since this coding format precludes strings of zeros greater than three, it is referred to as HDB3 coding. The encoding algorithm is basically the same as the B3ZS algorithm described earlier. Table 4.3 presents the basic algorithm. Notice that substitutions produce violations only in the fourth bit position, and successive substitutions produce violations with alternating polarities.

A fundamental feature of end-to-end digital connectivity as provided by ISDN is 64-kbps transparent channels referred to as clear channel capability (CCC) [17]. Two aspects of a bipolar/AMI line code as used on T1 lines preclude CCC: robbed signaling in the lsb of every sixth frame and the need to avoid all zeros code words on the channel. Bit robbing for signaling is avoided with common channel signaling (also an inherent requirement for ISDN deployment). Two means of augmenting T1 lines to allow transparent channels have been developed. The first method is zero byte time slot interchange

TABLE 4.3 HDB3 Substitution Rules

Polarity of Preceding Pulse	Number of Bipolar Pulses (ones) Since Last Substitution	
	Odd	Even
−	000 −	+00+
+	000 +	−00−

TABLE 4.4 B8ZS Substitution Rules

Polarity of Preceding Pulse	Substitution
−	000 − + 0 + −
+	000 + − 0 − +

(ZBTSI) described in reference [18]. ZBTSI is an attractive method for upgrading existing T1 facilities for CCC but is not a good long-term solution.

The second method, which is most desirable in the long run, involves the use of a B8ZS line code for T1 lines. As such, the B8ZS line code became the new line code standard for T1 lines with the introduction of D5 channel banks. (D5 channel banks can also operate with bipolar/AMI line codes for backward compatibility.) As indicated in Table 4.4 the B8ZS algorithm is similar to the B6ZS algorithm in that each individual substitution is DC balanced. Notice that there are two bipolar violations, in positions 4 and 7, in every substitution. The purposeful introduction of bipolar violations requires replacement of any equipment that monitors violations for performance monitoring.

4.3.4 Pair Selected Ternary

The BZNS substitution algorithms described in the preceding section are examples of selecting codes in the ternary code space to increase the timing content of a binary signal. Pair selected ternary (PST) is another example [19].

The PST coding process begins by pairing the binary input data to produce sequences of 2 bit code words. These code words are then translated into two ternary digits for transmission. Since there are nine two-digit ternary codes but only four two-digit binary codes, there is considerable flexibility available in selecting the codes. The most useful of the possible coding formats is shown in Table 4.5. This particular format not only ensures a strong timing component but also it prevents dc wander by switching modes to maintain a balance between positive and negative pulses. The codes are selected from one column until a single pulse is transmitted. At that time the encoder switches modes and

TABLE 4.5 Pair Selected Ternary Encoding

Binary Input	+ Mode	− Mode
00	− +	− +
01	0 +	0 −
10	+ 0	− 0
11	+ −	+ −

selects codes from the other column until another single pulse (of the opposite polarity) is transmitted.

Example 4.2. Encode the following binary data stream into a PST line code: 01001110101100.

Solution. There are two possible solutions depending on whether the encoder is in the $+$ or $-$ mode at the beginning of the sequence.

	01	00	11	10	10	11	00
Case 1 ($+$ mode):	$0+$	$-+$	$+-$	-0	$+0$	$+-$	$-+$
Case 2 ($-$ mode):	$0-$	$-+$	$+-$	$+0$	-0	$+-$	$-+$

One potential drawback of the PST coding algorithm is that the binary data stream must be framed into pairs. Hence a PST decoder must recognize and maintain pair boundaries. Recognition of the boundaries is not difficult if random data is being transmitted since a pairwise misframe eventually produces unallowed codes (00, $++$, $--$). Furthermore, time division multiplex formats typically provide character and pairwise framing automatically.

The power spectrum of a PST line code with equal probabilities for 1's and 0's is obtained from reference [19] and plotted in Figure 4.11. Also shown is the B6ZS power spectrum [15] and the conventional bipolar power spectrum.

An important point to notice in Figure 4.11 is that bipolar coding and its extensions require equal bandwidths. Their only significant difference is that B6ZS and PST have higher energy levels as a result of greater pulse densities. The higher energy levels have the undesirable effect of increasing crosstalk interference in multipair cables. However, the degradation from the increased crosstalk is somewhat offset by improved accuracy of the recovered sample clock.

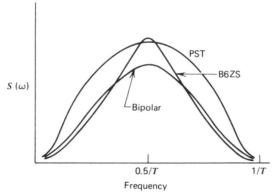

Figure 4.11. Spectrum of bipolar, B3ZS, and PST line codes for equally likely 1's and 0's.

4.3.5 Ternary Coding

Since bipolar and PST coding use a ternary code space to transmit binary data, they do not achieve as high an information rate as is possible with more efficient use of the code space. For example, an eight-element ternary code is capable of representing $3^8 = 6561$ different codes. In contrast, eight bits of binary data produce only $2^8 = 256$ different codes. The previously described line codes do not take advantage of the higher information content of ternary codes—they select codes for their timing content and spectral properties.

One ternary encoding procedure involves mapping successive groups of 4 bits into three ternary (4B3T) digits. Since binary words of 4 bits require only 16 of the 27 possible three-digit ternary code words, considerable flexibility exists in selecting the ternary codes. Table 4.6 presents one possible encoding procedure. Ternary words in the middle column are balanced in their dc content. Code words from the first and third columns are selected alternately to maintain dc balance. If more positive pulses than negative pulses have been transmitted, column 1 is selected. When the disparity between positive and negative pulses changes, column 3 is chosen. Notice that the all-zeros code word is not used. Hence a strong timing content is maintained. Because of the higher information efficiency, however, the ability to monitor performance is sacrificed, and framing is required on three-digit boundaries. 4B3T coding is used on the T148 span line developed by ITT Telecommunications [20]. This system provides T-carrier

TABLE 4.6 Encoding Table for 4B3T Line Code

Binary Word	Ternary Word (Accumulated Disparity)		
	−	0	+
0000	− − −		+ + +
0001	− − 0		+ + 0
0010	− 0 −		+ 0 +
0011		0 − −	0 + +
0100	− − +		+ + −
0101	− + −		+ − +
0110	+ − −		− + +
0111	− 0 0		+ 0 0
1000	0 − 0		0 + 0
1001	0 0 −		0 0 +
1010		0 + −	
1011		0 − +	
1100		+ 0 −	
1101		− 0 +	
1110		+ − 0	
1111		− + 0	

transmission for two DS1 signals (48 channels) using a bandwidth that is only 50% greater than a T1 bandwidth (carrying 24 channels). A generalized discussion of 4B3T coding and other ternary coding techniques is contained in reference [21].

4.3.6 Digital Biphase

Bipolar coding and its extensions BNZS and PST use extra encoding levels for flexibility in achieving desirable features such as timing transitions, no dc wander, and performance monitorability. These features are obtained by increasing the code space and not by increasing the bandwidth. (The first spectral null of all codes discussed so far, including an NRZ code, is located at the signaling rate $1/T$.)

Many varieties of line codes achieve strong timing and no dc wander by increasing the bandwidth of the signal while using only two levels for binary data. One of the most common of these codes providing both a strong timing component and no dc wander is the digital biphase code, also referred to as "diphase" or a "Manchester" code.

A digital biphase code uses one cycle of a square wave at a particular phase to encode a one and one cycle of an opposite phase to encode a zero. An example of a digital biphase coding sequence is shown in Figure 4.12. Notice that a transition exists at the center of every signaling interval. Hence strong timing components are present in the spectrum. Furthermore, logic 0 signals and logic 1 signals both contain equal amounts of positive and negative polarities. Thus dc wander is nonexistent. A digital biphase code, however, does not contain redundancy for performance monitoring. If in-service performance monitoring is desired, either parity bits must be inserted into the data stream or pulse quality must be monitored. (A later section of this chapter discusses performance monitoring in more detail.)

The frequency spectrum of a digital biphase signal is derived in Appendix C and plotted in Figure 4.13, where it can be compared to the spectrum of an NRZ signal. Notice that a digital biphase signal has its first spectral null at $2/T$. Hence the extra timing transitions and elimination of dc wander come at the expense of a higher frequency signal. In comparison to three-level bipolar codes, however, the digital biphase code has a lower error rate for equal signal-to-noise ratios (see Appendix C).

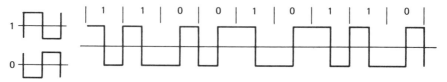

Figure 4.12. Digital biphase (Manchester) line code.

Figure 4.13. Power spectral density of digital biphase.

Examination of the frequency spectra in Figure 4.13 shows that the diphase spectrum is similar to an NRZ spectrum but translated so it is centered about $1/T$ instead of direct current. Hence digital biphase actually represents digital modulation of a square wave carrier with one cycle per signal interval. Logic ones cause the square wave to be multiplied by $+1$ while logic zeros produce multiplication by -1. Diphase is primarily used on shorter links where terminal costs are more significant than bandwidth utilization. The "Ethernet," IEEE 802.3 local area data network uses digital biphase (Manchester) coding.

4.3.7 Differential Encoding

One limitation of NRZ and digital biphase signals, as presented up to this point, is that the signal for a 1 is exactly the negative of a signal for a 0. On many transmission media, it may be impossible to determine an absolute polarity or an absolute phase reference. Hence the decoder may decode all 1's as 0's and vice versa. A common remedy for this ambiguity is to use differential encoding that encodes a 1 as a change of state and encodes a 0 as no change in state. In this manner no absolute reference is necessary to decode the signal. The decoder merely detects the state of each signal interval and compares it to the state of the previous interval. If a change occurred, a 1 is decoded. Otherwise, a 0 is determined.

Differential encoding and decoding do not change the encoded spectrum of purely random data (equally likely and uncorrelated ones and zeros), but do double the error rate. If the detector makes an error in estimating the state of one interval, it also makes an error in the next interval. An example of a differentially encoded NRZ code and a differentially encoded diphase signal is shown in Figure 4.14. All signals of differentially encoded diphase retain a transition at the middle of an interval, but only the 0's have a transition at the beginning of an interval.

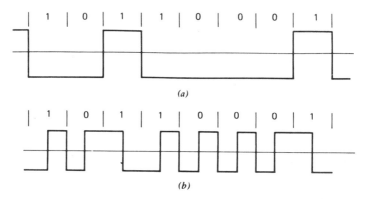

Figure 4.14. Differentially encoded NRZ and diphase signals. (a) Differentially encoded NRZ. (b) Differentially encoded diphase.

4.3.8 Coded Mark Inversion

A variety of line codes have evolved similar to the digital biphase code described previously. One of these is referred to as coded mark inversion (CMI) in CCITT recommendations [16]. CMI encodes ones (marks) as an NRZ level opposite to the level of the previous one. Zeros are encoded as a half-cycle square wave of one particular phase. Figure 4.15 shows a sample encoding for CMI. There is no dc energy in the signal and an abundance of signal transitions exist as in diphase. Furthermore, there is no ambiguity between 1's and 0's. Elimination of the ambiguity actually leads to a major drawback of CMI coding: its error performance is 3 dB worse than diphase.* CMI is the specified interface code for the fourth-level CCITT multiplex signal at 139.264 Mbps.

4.3.9 Multilevel Signaling

In the line codes discussed so far, two-level (binary) signaling has been assumed. In applications where the bandwidth is limited but higher data rates are desired,

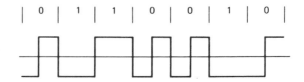

Figure 4.15. Coded mark inversion.

*The error performance of CMI is 3 dB worse than diphase when bit-by-bit detection is used. The inefficiency arises because for one-half of an interval a "1" looks like a "0." Because CMI has redundancy, some of the inefficiency can be recovered with maximum likelihood (Viterbi) detection.

the number of levels can be increased while maintaining the same signaling rate. The data rate R achieved by a multilevel system is

$$R = \log_2(L)\left(\frac{1}{T}\right) \tag{4.4}$$

where L = number of levels that can be freely chosen during each interval
$\quad\ T$ = signaling interval

The signaling rate $1/T$ is often referred to as the symbol rate and is measured in bauds. Within the data communications industry it is common practice to use "baud" as being synonymous with bit rate. Strictly speaking, however, the bit rate is only equal to the baud rate if binary signaling (1 bit per signal interval) is used. Figure 4.16 shows an example of an eight-level transmission format that achieves 3 bits per signal interval (i.e., 3 bits per baud).

Multilevel transmission systems achieve greater data rates within a given bandwidth but require much greater signal-to-noise ratios for a given error rate. One aspect of wireline transmission that favors multilevel line coding is the lower baud rate for a given data rate, which in turn reduces the crosstalk. Thus in crosstalk-limited systems the signal-to-noise penalty of a multilevel line code is not as significant. The T1G T-carrier system developed by AT&T [22] uses a four-level line code at the T1C baud rate (3.152 MHz) to double the capacity of a T1C system (from 48 channels to 96).

Another example of multilevel transmission of particular significance is the ISDN basic rate digital subscriber line (DSL), which uses four-level transmission at a signaling rate of 80 baud to achieve 160 kbps. The primary factors that led to selecting a multilevel line code in this application are (1) near-end crosstalk that cannot be eliminated by pair isolation as in T1 systems and (2) high levels of intersymbol interference caused by bridged tap reflections. Both of these factors are easier to control when lower frequency signals are used [23]. Additional aspects of this application are described in the ISDN section of Chapter 9.

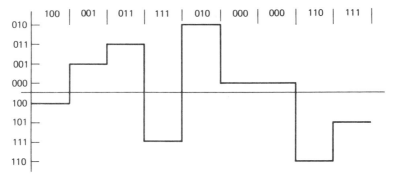

Figure 4.16. Multilevel transmission with 3 bits per signal interval.

4.3.10 Partial Response Signaling

Conventional bandlimiting filters of a digital transmission system are designed to restrict the signal bandwidth as much as possible without spreading individual symbols so much that they interfere with sample values of adjacent intervals. One class of signaling techniques, variously referred to as duobinary [24], correlative level encoding [25], or partial response signaling [26], purposely introduces a prescribed amount of intersymbol interference that is accounted for in the detection circuitry of the receivers. By overfiltering the encoded signal, the bandwidth is reduced for a given signaling rate, but the overlapping pulses produce multiple levels that complicate the detection process and increase the signal power requirements for a given error rate.

Figure 4.17 shows the pulse response of a typical partial response system. If the channel is excited by a pulse of duration T, channel filters (defined in Appendix C) limit the spectrum such that the main part of the pulse extends across three signal intervals and contributes equally to two sample times. The reason for the term partial response is now apparent: the output only responds to one-half the amplitude of the input.

If the input pulse of Figure 4.17 is followed by another pulse of the same amplitude, the output will reach full amplitude by virtue of the overlap between pulses. However, if the next input pulse has a negative polarity, the overlap produces zero amplitude at the sample time. Thus, as shown in Figure 4.18 a partial response system with two-level inputs $(+1, -1)$ produces an output with three levels $(+1, 0, -1)$. In similar fashion, Figure 4.19 shows a system with four

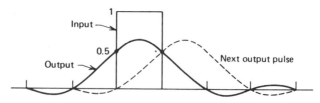

Figure 4.17. Output pulse of partial response channel.

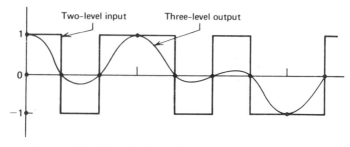

Figure 4.18. Three-level partial response inputs and outputs.

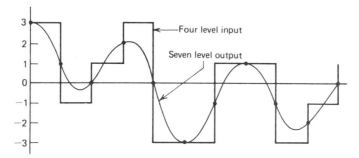

Figure 4.19. Seven level partial response inputs and outputs.

input levels $(+3, +1, -1, -3)$ and seven output levels $(+3, +2, +1, 0, -1, -2, -3)$.

Detection of a partial response signal is complicated by the additional levels produced by the channel filters and the fact that sample values are dependent on two adjacent pulse amplitudes. One method of detecting a PRS signal involves subtracting the overlap of a previously detected pulse from the incoming signal to generate a difference signal representing the channel response to the new pulse. Only the overlap (intersymbol interference) at the sample times needs to be subtracted. The difference samples are nominally equal to one-half the amplitude of the unknown input pulse. This technique doubles the error rate in the same manner that differentially encoded error rates are doubled.

Another method of detecting partial response systems involves a technique called "precoding" at the source. Precoding transforms the input data in such a manner that the output level at the detector directly indicates the original data without comparison to the previous sample value. In a binary system, for example, a 1 is encoded with a pulse of the same polarity as the previous pulse. Hence logic 1's show up at the the detector as either a $+1$ or a -1 sample. Transmission of a 0 is encoded with a pulse of opposite polarity to the previous pulse. Hence logic 0's always show up at the detector as a zero-level signal. Similar precoding techniques exist for multilevel systems [27].

The partial response technique just described is actually a special case of a more general class of signaling techniques referred to as "correlative level encoding" [25]. To describe the properties of generalized partial response or correlative level systems, it is convenient to introduce a delay operator D to denote delay equal to one signal interval T. Physically, D can be thought of as a delay line of length T. Two units of delay is implemented with two delay lines in series and is denoted as D^2.

Using this notation, the partial response system described is referred to as $1 + D$ PRS: the output represents the superposition of the input with a delayed version of the same input. Other forms of overlap are possible. These systems do not necessarily produce the overlapping pulses by overfiltering an input. An alternative approach is to overlap and add the pulses directly in the encoding

process (hence the term: correlative level encoding). An interesting special case of correlative level encoding is the $1 - D$ system shown in Figure 4.20 and extended in Figure 4.21 to show the effect of overlapping pulses.

The $1 - D$ encoder uses a single cycle of a square wave across two signal intervals to encode each bit. Since neither of the two individual signals $(+-,$ or $-+)$ produce dc energy, the encoded signal has no dc wander. Furthermore, the positive and negative levels of the composite signal alternate in a manner reminiscent of bipolar coding. In fact, if differential encoding is used (i.e., if a zero is encoded with the same phase as the previous interval and a one as the opposite phase) this form of encoding is identical to bipolar coding (assuming the NRZ levels are replaced by 50% RZ pulses). Thus $1 - D$ correlative level encoding is used to shape the bandwidth rather than to limit it. The spectra of unfiltered $1 + D$, $1 - D$, and $1 - D^2$ signals are obtained from reference [26] and plotted in Figure 4.22. The T1D T-carrier system of AT&T uses precoded $1 + D$ level encoding (also called duobinary).

The spectrum of the $1 - D^2$ signal is particularly interesting. It has no dc component and an upper limit equal to $1/2T$: the same upper limit as a maximally filtered NRZ code with no intersymbol interference. The $1 - D^2$ partial response systems have been used by GTE in a modified T-carrier system

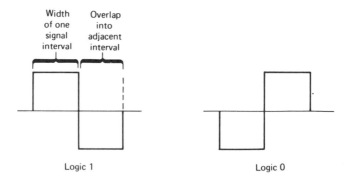

Width Overlap
of one into
signal adjacent
interval interval

Logic 1 Logic 0

Figure 4.20. Individual signal element for $1 - D$ correlative level encoding.

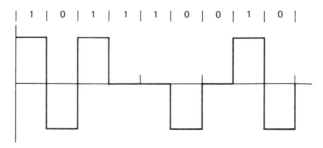

| 1 | 0 | 1 | 1 | 1 | 0 | 0 | 1 | 0 |

Figure 4.21. Representative waveform of $1 - D$ correlative level encoding.

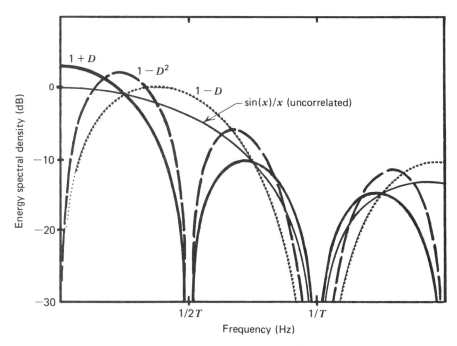

Figure 4.22. Spectra of unfiltered $1 + D$, $1 - D$, and $1 - D^2$ correlative encoded signals.

providing 48 voice channels [28] and for digital transmission over analog microwave radios using very low baseband frequencies for pilots and service channels (where no message spectrum exists) [29]. The 1A-RDS (data under voice) system of AT&T also used $1 - D^2$ coding with four-level inputs to get 1.544 Mbps into 500 kHz of bandwidth [30].

Notice that the $1 - D^2$ spectrum is essentially the product of the $1 + D$ and the $1 - D$ spectra. Indeed, a $1 - D^2$ system can be implemented by concatenating a $1 - D$ encoder with a $1 + D$ channel response: $(1 - D)(1 + D) = 1 - D^2$. Thus correlative level system polynomials are very useful for simultaneously representing correlations and spectrum shaping. Since all correlative level systems use more levels than necessary for encoding the data, they are inferior to uncorrelated or full response systems in terms of error performance.

A summary of digital transmission systems identifying various line codes is provided in Table 4.7. Other line codes used specifically in fiber systems are described in Chapter 8.

4.4 ERROR PERFORMANCE

The preceding sections of this chapter emphasized the timing and power spectrum requirements of various transmission codes. Another fundamental consideration in choosing a line code is the error rate performance in the

TABLE 4.7 Digital Transmission Systems

Designation	Country or Administration	Bit Rate (Mbps)	Line Code	Media	Repeater Spacing
T1	AT&T	1.544	AMI/B8ZS	Twisted pair	6000 ft
CEPT1	CCITT	2.048	HDB3(B4ZS)	Twisted pair	2000 m
T1C	AT&T	3.152	Bipolar	Twisted pair	6000 ft
T148	ITT	2.37, ternary	4B3T	Twisted pair	6000 ft
9148A	GTE	3.152	$1 - D^2$, duobinary	Twisted pair	6000 ft
T1D	AT&T	3.152	$1 + D$, duobinary	Twisted pair	6000 ft
T1G	AT&T	6.443	4-level	Twisted pair	6000 ft
T2	AT&T	6.312	B6ZS	Low capacitance twisted pair	14800 ft
LD-4	Canada	274.176	B3ZS	Coax	1900 m
T4M	AT&T	274.176	Polar	Coax binary (NRZ)	5700 ft

presence of noise. Except on relatively short lines, where noise may be insignificant, error performance requirements can impact system cost significantly. If a certain minimum error rate is specified, those coding schemes providing the desired error rate at lower signal-to-noise ratios allow regenerative repeaters to be spaced farther apart, thereby reducing installation and maintenance.* Repeater spacing is an important economic factor in wire line transmission, optical fiber transmission, and on point-to-point radio links.

The error rate results and comparisons presented in this section are based on white Gaussian noise. This is the most common form of noise and the best understood. In some applications, particularly within an older, electromechanical telephone network, impulse noise may actually be more prevalent. Thus these analysis do not provide a complete error rate analysis for some applications. The intent of this section is to present relative error performance comparisons of various line coding techniques. For this purpose a white noise analysis is most appropriate, since if impulses are large enough they cause errors independent of the coding scheme in use.

For the most part, the following sections present only the results of the error analyses in the form of graphs of error rate as a function of signal-to-noise ratios. Appendix C derives the basic equations used to produce the results.

*On T1 lines the repeater spacing was predetermined by the locations where loading coils needed to be removed. Thus the error rate performance essentially determined the maximum data rate.

4.4.1 Signal Detection

Invariably, the detection circuitry of a digital receiver processes incoming signal waveforms to measure each possible discrete signal. In most cases the measures are nothing more than samples of a filtered receive signal. Depending on the signal shape and the level of performance desired, the receiver uses more sophisticated processing. In any case, the end measurement of a binary signal nominally produces one voltage level for a 0 and another voltage level for a 1. A decision of which signal was transmitted is made by comparing the measurement (at the appropriate time) to a threshold located halfway between these nominal voltages. Naturally, the error probability depends on the nominal distance between the voltages and the amount of fluctuation in the measurements caused by noise.

Since signal measurements are normally linear in nature, the error distance between 1's and 0's is proportional to the received signal voltage. Equivalently, the amount of noise power required to produce an error is a direct function of the signal-to-noise ratio. The optimum detector for a particular signal set maximizes the signal-to-noise ratio at the input to the decision circuit.

4.4.2 Noise Power

White Gaussian noise is characterized as having a uniform frequency spectrum across an arbitrarily large bandwidth and an amplitude distribution that varies according to a normal (Gaussian) probability distribution. A parameter N_0 conventionally represents the power spectral density of white noise. N_0 is the amount of power measured in a bandwidth of 1 Hz. Hence the rms power of white noise coming out of a filter with a bandwidth BW is $N_0 \cdot BW$.*

To determine the amount of noise power present at a decision circuit it is necessary to determine N_0 and the effective bandwidth of the detection circuitry. When the detection circuitry consists of nothing more than a filter, the effective bandwidth is usually very close to the 3 dB bandwidth of the filter. With other more sophisticated detectors, the effective bandwidth can be more difficult to determine. As derived in Appendix C, the effective bandwidth is usually referred to as the effective noise bandwidth (NBW) of the receiver. Hence the noise power at the decision circuit is $N_0 \cdot NBW$.

4.4.3 Error Probabilities

An error in detection occurs whenever noise causes the signal measurement to cross the threshold between the two nominal output levels of the detection

*The power spectral density of white noise is also specified as $\frac{1}{2}N_0$ for a two-sided spectral density. As a practical matter, there is no difference in the specifications, since a real filter has the mathematical equivalent of identical positive and negative frequency bands. Thus the measured power coming through a filter with a one-sided (positive frequency) bandwidth BW is N_0BW in either case.

circuitry. Appendix C derives this probability for white Gaussian noise as

$$\text{prob(error)} = \frac{1}{\sqrt{2\pi}\,\sigma} \int_{V}^{\infty} e^{-t^2/2\sigma^2}\, dt \tag{4.5}$$

where V is the nominal distance (voltage) to a decision threshold and σ^2 is the noise power at the detector, $\sigma^2 = N_0 \cdot NBW$.

Equation 4.5 is nothing more than the area under the probability density function of a normal distribution. As shown in Figure 4.23, the equation represents the error probability as the probability of exceeding V/σ standard deviations in a normal distribution with zero mean and unit variance $N(1, 0)$.

The error rate is completely determined by the ratio of V to σ. Since V is the noise-free sample voltage and σ^2 is the rms noise power, V^2/σ^2 is a signal power-to-noise power ratio at the detector. This ratio is sometimes referred to as a post detection SNR, since it is measured after the detection circuitry. It is usually more important to express error rates in terms of a signal-to-noise ratio at the input to the receiver. Figure 4.24 depicts a basic channel and detection model and the relation between a predetection SNR and a postdetection SNR. For reasons discussed in Appendix C, the most appropriate predetection SNR for comparing line codes and digital modulation formats is an energy-per-bit to noise density ratio E_b/N_0. The relationship between E_b/N_0 and the signal power

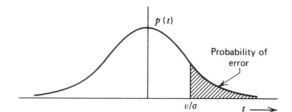

Figure 4.23. Probability of error for binary signaling.

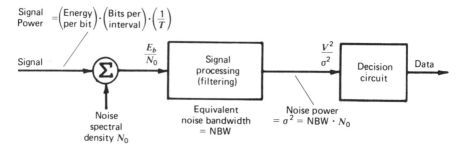

Figure 4.24. Signal detection model.

to noise power ratio is

$$\text{SNR} = \frac{\text{signal power}}{\text{noise power}}$$

$$= \frac{d \cdot E_s(1/T)}{N_0 \cdot NBW}$$

$$= \frac{d \cdot E_b \log_2 L \cdot (1/T)}{N_0 \cdot NBW} \tag{4.6}$$

where d is the pulse density, E_s is the energy per symbol, E_b is the energy per bit, $\log_2 L$ is the number of bits per symbol (i.e., L is the number of levels), $1/T$ is the signaling rate, and NBW is the effective noise bandwidth of the receiver.

In determining the signal power in Equation 4.6 notice the dependence on the pulse density d. In an NRZ line code the pulse density is 1, but in many of the other line codes the pulse density is dependent on the data and any substitution algorithms that may be in use. In these codes, increasing the pulse density increases the SNR but does not reduce the error rate. The error rate is determined by the energy-per-bit-to-noise density ratio. In fact, increasing the pulse density might worsen the error rate if interference between cable pairs is a significant consideration. (Interference is directly proportional to the signal power.)

Antipodal Signaling

The optimum signaling format for binary signaling maximizes the error distance for a given received signal power and simultaneously minimizes the noise bandwidth. This condition arises only when two signal levels are allowed, and only when one signal is the exact negative of the other. Since the signal for a 1 is the exact opposite (the antipode) of the signal for a 0, optimum signaling is often referred to as "antipodal signaling." Since no other bit-by-bit signaling scheme is any better, antipodal performance is often used as a basis for comparisons. Of the line codes described previously, only balanced two-level (NRZ) encoding and digital biphase can provide antipodal performance. Figure 4.25 shows the optimum error performance provided by antipodal signaling as a function of E_b/N_0 and the signal-to-noise ratio SNR.

Error Rate of Level Encoded Signals

As already mentioned, a balanced, two-level line code is capable of providing optimum error rate performance. If an unsymmetric level code is used, such as unipolar, the same basic detector is used as in symmetric level encoding. The only difference is that the decision threshold must be moved from zero to half the amplitude of the "on" signal. To maintain the same error distance, the transmit power is increased by a factor of 2 (assuming 50% "on" pulses). Hence a unipolar code carries a performance penalty of 3 dB with respect to antipodal performance.

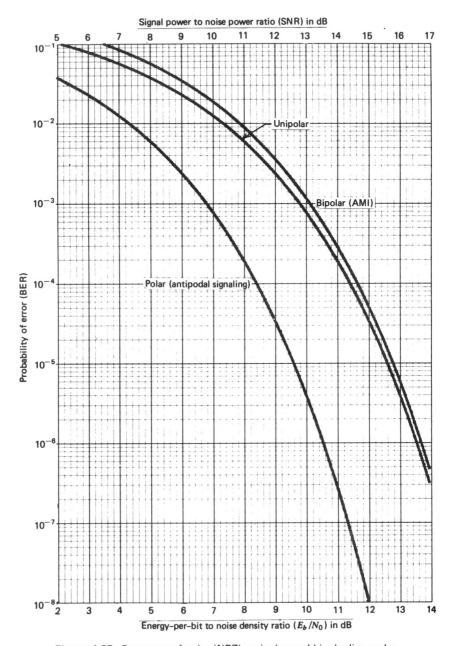

Figure 4.25. Error rates of polar (NRZ), unipolar, and bipolar line codes.

Figure 4.25 shows the ideal performance of a unipolar (on–off) code where it is compared to antipodal performance. Notice that for all signal-to-noise ratios the error rate of the on—off system is exactly equal to the error rate of an antipodal system with 3 dB less SNR.

Bipolar Signaling

With respect to error performance, bipolar signaling is basically identical to a unipolar code. During any particular signal interval, the receiver must decide between one of two possible levels: zero or a pulse with the appropriate polarity. Hence the decision threshold pertinent to a particular signal interval lies halfway between zero and the amplitude level of the allowed pulse. Narrowing the pulse to 50% of the signal interval does not change the theoretical error performance (with respect to average pulse energy). Thus the error rate curve for on–off keying in Figure 4.25 can be used to determine theoretical bipolar error rates.

One consideration in a bipolar line code contributes to a slightly higher error rate than in on–off keying. This increase occurs because both positive and negative noise can cause an erroneous threshold crossing when a zero-level signal is transmitted. In contrast, a unipolar code is affected only by positive noise when a lower-level signal is transmitted, and by negative noise when the upper-level signal is transmitted. If the bipolar detector treats the erroneous pulse as a 1 (despite a bipolar violation), the error probability when 0's are transmitted is doubled. Thus the overall error probability is increased by 50% if zeros and ones are equally likely.

Because of the steepness of the curve, an increase in the error rate of 50% does not represent much of a performance penalty. For example, if the error rate is increased from 1×10^{-6} to 1.5×10^{-6}, the source power of a line code needs to increase by only 0.2 dB to get back to a 1×10^{-6} error rate. At higher error rates, a larger penalty occurs because the curve is not as steep. These effects are demonstrated in Figure 4.25, where the ideal performance of a bipolar code (with 50% zeros) can be compared to the performance of a unipolar code.

The fact that bipolar coding incurs a 3.2 dB penalty (at 10^{-6} error rate) with respect to digital biphase is indicative that timing and dc wander problems are solved by increasing the number of signal levels. In contrast, digital biphase incurs a bandwidth penalty. Not all of the 3.2 dB penalty for bipolar coding, and its extensions BNZS and PST, can be attributed to removing direct current and adding a consistent timing pattern. Bipolar coding contains considerable redundancy for performance monitoring or possible error correction.*

Example 4.3. Assume each section of a T1 transmission system on 22-gauge cable is near-end crosstalk limited and operating with a 10^{-6} error rate. What design changes are needed to reduce the error rate to 10^{-8}?

*In terms of logic-level decisions, a bipolar code cannot provide error correction. If a pulse is removed to eliminate a bipolar violation, half of the time the error is corrected but half of the time another error is made. Error correction is possible only with a Viterbi-like detector described previously. In this case a bipolar code provides better performance than a unipolar code.

Solution. Since the system is crosstalk limited, the error rate cannot be improved by increasing the power out of the regenerative repeaters. The solution is to space the repeaters closer together (ignore the impractical aspects of this solution). From Figure 4.25 it can be determined that the signal power must be increased by 1.6 dB to improve the error rate from 10^{-6} to 10^{-8}. In Figure 4.10 it is seen that the main lobe of the bipolar spectrum extends up to 1.544 MHz. However, most of the energy in the spectrum lies below 1 MHz. Using Figure 1.9, we determine that the attenuation of 22-gauge cable is approximately 20 dB/mile at 1 MHz. Hence $1.6/20 = 0.08$ miles or 422 ft is the reduction in repeater spacing required.

Example 4.3 demonstrates a number of important aspects of digital transmission systems. First, arbitrarily good transmission quality can be obtained with only a small penalty in transmit power or repeater spacing.* Thus, as mentioned in Chapter 2, a digital transmission and switching network can be readily designed to impart no degradation to voice quality—except where analog-to-digital and digital-to-analog conversions take place.

Second, the dramatic improvement in error rate for a relatively small increase in the signal-to-noise ratio implies extreme sensitivity in the opposite direction also. A slight increase in noise power or signal attenuation would cause a large increase in the error rate. Hence the nominal design of a digital link often provides considerably better performance than normally necessary.

Third, the solution to Example 4.3 applies to any digital transmission format using the same band of frequencies, and hence, the same attenuation per mile. Since all line codes have approximately the same steepness at a 10^{-6} error rate, the relative change in SNR is the same for all systems. Thus a T1 system with suboptimum detection would exhibit the same performance improvement if the repeaters were moved 422 ft closer together. (T1 lines are designed to provide less than a 10^{-6} error rate to begin with. See reference [31] for a thorough report of error rate surveys on T1 lines.)

Multilevel Error Rates

The multilevel transmission system shown in Figure 4.16 does not require bandwidth in excess of a binary system using the same signaling rate—yet it achieves three times the data rate. The penalty for multilevel transmission comes in the form of greatly increased signal power for a given error rate. For example, the average power of the eight-level system in Figure 4.16 is 8.7 dB above the average power a symmetric two-level system with the same error distance. To make matters worse, some systems are peak power limited, in which case the eight-level system has a 12.4 dB disadvantage with respect to a two-level system. The error rates of multilevel systems are derived in Appendix C and plotted in

*Some systems, such as the radio systems discussed in Chapter 6, require a greater decrease in repeater spacing to achieve the same improvement in performance.

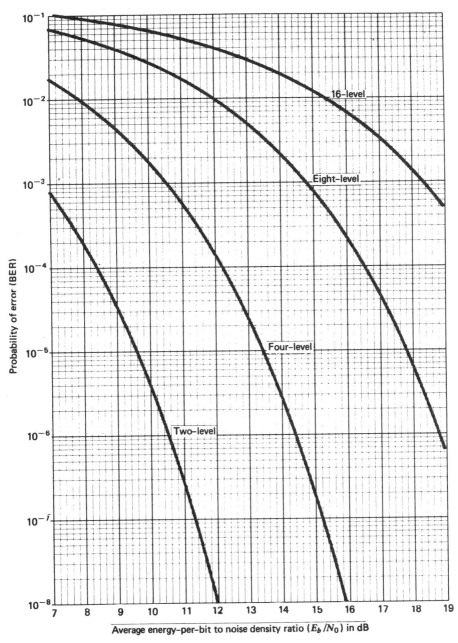

Figure 4.26. Error rates of balanced multilevel signals (all systems providing an identical data rate).

Figure 4.26 as a function of E_b/N_0. For those cases where the peak power is of interest, the relationship between the peak and average power of a multilevel system is derived in Appendix C as

$$\text{peak-to-average (dB)} = 10 \log_{10} \frac{(L-1)^2}{(2/L)\sum_{i=1}^{L/2}(2i-1)^2} \qquad (4.7)$$

where L is the number of equally spaced levels centered about zero (for example, $\pm 1, \pm 3, \pm 5, \ldots, \pm L - 1$).

The error rates of $1 + D$ partial response systems are also derived in Appendix C and plotted in Figure 4.27. These error rates are derived under the assumption that bit-by-bit detection is used. Since partial response systems exhibit redundancy (correlation) in adjacent samples, better performance can be achieved with Viterbi decoders [11].

4.5 PERFORMANCE MONITORING

Two basic techniques exist for directly monitoring the quality of a digital transmission link: redundancy checks and pulse quality measurements. Both techniques are designed to provide an indication of the bit error rate (BER) of the channel.

4.5.1 Redundancy Checks

Redundancy can be incorporated into a digital signal using one of two common methods. First, the line code itself may contain redundancy as in bipolar coding. In a random, independent error environment the frequency of bipolar violations is very nearly equal to the channel BER (except for extremely high BERs). Second, logic-level redundancy can be inserted into the data stream in the form of overhead bits. For example, parity bits are inserted into DS3 and DS4 signals for the express purpose of monitoring the channel error rate. (The frame structures of these and other higher-level multiplex signals are provided in Chapter 7). Cyclic redundancy check (CRC) codes have also been incorporated into a number of transmission systems as a means of monitoring bit error rates and validating framing acquisition. Two prevalent examples of CRC use are (1) extended superframe (ESF) on T1 lines [32] introduced with the D4 channel banks and (2) optical fiber transmission using the SONET standard described in Chapter 8.

The use of logic level redundancy (either parity bits or CRC codes) is generally better than the use of line code redundancy (such as bipolar violations) because the latter is dependent on the technology of the transmission link itself. Logic level redundancy, on the other hand, is unaffected by a change from cable pairs to fiber or radio. As described later in this chapter a major impetus for the

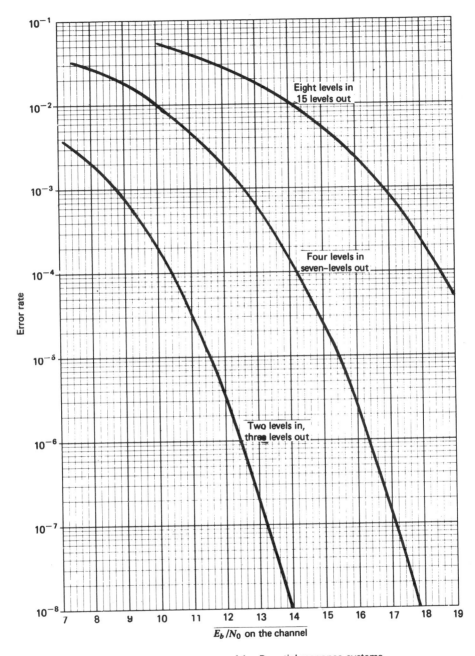

Figure 4.27. Error rates of $1 + D$ partial response systems.

The figure shows error rate (vertical axis, 10^{-1} to 10^{-8}) versus $\overline{E_b/N_0}$ on the channel (horizontal axis, 7 to 19). Three curves are labeled:

- Eight levels in 15 levels out
- Four levels in seven-levels out
- Two levels in, three levels out

introduction of ESF into T1 transmission systems was to get performance monitorability independent of bipolar line codes. DS1 signals are often multiplexed into higher level digital signals and carried on a variety of transmission systems such as fiber and radio. The CRC codes in the ESF framing format provide the means for end-to-end performance monitoring independent of whatever transmission systems carry the data stream.

In contrast to line code redundancies, parity bits and CRC codes do not provide a one-to-one indication of the BER. The following equation relates the parity error rate (PER) to the channel bit error rate (BER):

$$\text{PER} = \sum_{i=1}^{N} \binom{N}{i} \rho^i (1 - \rho)^{N-i} \quad (i \text{ odd}) \tag{4.8}$$

where N = the length of a parity field (number of bits over which parity is generated)
ρ = the bit error rate (BER) assuming random, independent errors

The relationship between the PER and the BER is plotted in Figure 4.28 for DS3 and DS4 signals. Notice that at low error rates the PER is essentially N times the BER (N_ρ). At high error rates, however, this relationship changes because any odd number of errors in a frame is indistinguishable from a single error. When the BER is high enough that more than one error in a parity field is likely, the PER is useless as an absolute estimate of the error rate. In these cases the PER indicates only that the BER is above a threshold approximately equal to $1/N$. Because a DS4 frame format contains a higher density of parity bits, DS4 signals can be measured for higher error rates than can DS3 signals.

Determination of a CRC error rate is simplified because it is virtually impossible for multiple errors to not produce a CRC error. Thus the probability of a CRC error (CRCER) is 1 minus the probability that no errors occur:

$$\text{CRCER} = 1 - (1 - \rho)^N \tag{4.9}$$

where N = the length of the CRC field (including CRC bits)
ρ = the bit error rate (BER) assuming random, independent errors

Again, at low error rates Equation 4.9 reverts to N times the BER ($N\rho$). Equation 4.9 is plotted in Figure 4.28 for ESF frames where the CRC field is 4614 bits. (The CRC field does not include F bits in Table 4.8 except the C bits themselves.)

To have confidence in error rate measurements, the sample sequence must be long enough to allow an average of about 10 errors in the sample size (e.g., the sample size must be 10/BER). Hence when trying to measure low BERs (e.g., 10^{-6} or 10^{-7}), the measurement time may be too long to respond to changing channel conditions such as radio channel fading.

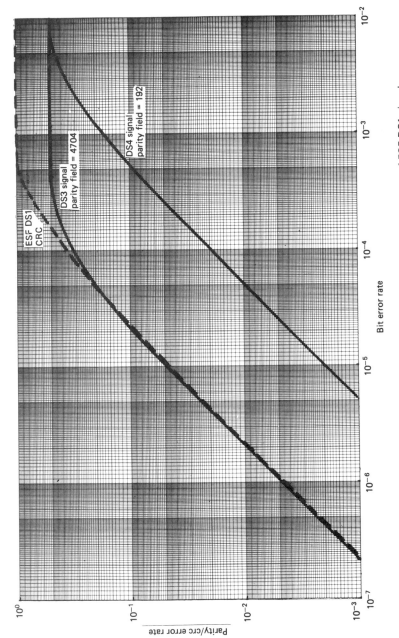

Figure 4.28. Parity error rates of DS3 and DS4 signals and CRC error rate of ESF DS1 signal vs. bit error rate.

4.5.2 Signal Quality Measurements

The second basic technique for monitoring digital transmission quality is to process the digital signal directly and measure certain properties related to the error rate. A simple approach involves merely measuring the received signal power—a common technique in analog systems. In a fixed noise environment this approach is adequate. However, on transmission links where the noise level can vary or where signal distortions can arise, the quality of the pulses themselves must be measured.

Figure 4.29 demonstrates the operation of a "pseudo"error detector designed to detect received pulses with abnormal amplitudes at the sample times. In the example shown binary data is detected by use of a single threshold located midway between the normal pulse amplitudes. Two additional thresholds are included to detect the presence of pulses with abnormal amplitudes. Sample values falling into the central decision region are not necessarily data errors, but a high pseudoerror rate is a good indication that the channel is not performing properly.

In a random (Gaussian) noise environment the occurrence rate of pseudo-errors is directly related to the actual error rate. Figure 4.30 shows a Gaussian noise distribution and decision thresholds chosen to produce pseudo errors at 100 times a 10^{-6} error rate. Hence an atrractive feature of this error rate

Figure 4.29. Pseudoerror detection.

Figure 4.30. Pseudoerror decision thresholds for error multiplication of 100.

measurement is that it can measure very low error rates using comparatively short test intervals. Note, however, that the error multiplication factor is dependent on the error rate.

A major disadvantage of pseudoerror detection is that pseudoerror rates are not accurately related to actual error rates when the noise is non-Gaussian. Channel distortions, in particular, cause high pseudoerror rates when the actual error rates are small. Even though abnormal pulse amplitudes are present, errors do not occur unless noise adds to the distortion. On the other hand, impulse noise might produce high actual error rates but relatively low pseudo error rates.

The correspondence between measured and deduced error rates is improved when the decision threshold is narrowed. Unfortunately, narrower pseudoerror decision regions are more difficult to implement and provide lower "error" rate multiplication factors.

4.5.3 Framing Channel Errors

As described in Section 4.6.2, identification of individual channels of a TDM data stream requires overhead bits with a unique data pattern for frame identification. After a receiving terminal has "locked" onto the framing pattern it ignores occasional bit errors in the framing pattern but declares an out-of-frame (OOF) condition when the errors are consistent. Thus framing bit errors can be used to determine low error rates but not high error rates that cause false OOF indications and subsequent loss of framing. Error rates that are high enough to cause false OOF indications are usually so high that the link is effectively out of service.

4.5.4 Performance Objectives

The most basic measure of quality of a digital transmission link is its bit error rate (BER). If the errors occur in a truly random and independent manner (i.e., they conform to a simple Poisson probability model) the average BER completely specifies the performance. Unfortunately, error rate observations [31] show that errors often occur in bursts and because bursts affect different services in different ways specifying the quality of a transmission link requires more than just the average BER. For example, data traffic is generally transmitted in blocks that are retransmitted no matter how many errors occur in a block. Thus, a burst has little more effect than a single bit error. The rate of retransmissions of data blocks is more a function of the frequency of the bursts than of the long-term average BER. Voice traffic, on the other hand, is increasingly degraded by all bit errors whether or not they occur in bursts.

Channel errors that occur in bursts can also be much more deleterious to certain technical operations of a network than are distributed errors at a similar average rate. A loss of framing, for example, occurs much more often in a burst

error environment.* Certain timing operations (pulse stuffing) described in Chapter 7 are also similarly affected by bursts. Losses of framing and time synchronization both produce outages (e.g., continuous errors) in the associated traffic channels or even inadvertent disconnects if the conditions last too long.

The most common performance parameters used to characterize the error bursts are as follows:

1. Error free second (EFS): Because data blocks transmitted at 64 kbps require much less than a second, the percentage of EFS essentially specifies the percentage of time that the channel is available for data transmission applications. (When the block transmissions are much shorter than a second a shorter measurement interval is more appropriate.)

2. Errored second (ES): A second with at least one error.

3. Severely errored second (SES): A second with a BER greater than $1(10)^{-3}$.

4. Degraded minute (DM): A minute with a BER greater than $1(10)^{-6}$.

The last three parameters are defined in CCITT recommendation G.821 for performance objectives of a digital network. The objectives for a 27,500-km reference connection (implying error contributions from many sources) are ES < 8%, SES < 0.2%, and DM < 10%.

4.6 TIME DIVISION MULTIPLEXING

Although frequency division multiplexing of digital transmission signals is possible and is used in some special situations, time division multiplexing is by far the most common and economical means of subdividing the capacity of a digital transmission facility. One application where FDM techniques have been used for digital signals is on multidrop data communications lines where the sources and destinations of the data are distributed along the line. Most telephone network applications, however, involve clusters of channels in the form of trunk groups between switching offices. Even for local distribution, where it may be tempting to add and drop channels on a distributed basis, clustered structures prevail because of maintenance and reliability considerations.

There are two basic modes of operation for time division multiplexing: those that repeatedly assign a portion of the transmission capacity to each source, and those that assign capacity only as it is needed. The first form of operation is referred to as *synchronous* time division multiplexing when necessary to distinguish it from the "as needed" mode of operation. Otherwise, time division multiplexing is understood to imply the synchronous variety. The "as needed"

*Errors in a burst are not necessarily contiguous. They merely occur in a short interval at a rate that is too high to be reasonably explained as chance independent events.

form of time division multiplexing is variously referred to as "asynchronous time division multiplexing" (ATDM), asynchronous transfer mode (ATM), "statist-ical time division multiplexing" (Stat-Mux), or packet switching. Circuit-switched telephone networks use synchronous TDM whereas data networks typically use ATDM techniques. Discussion of these latter techniques is deferred to Chapter 9.

Another aspect of time division multiplexing that occurs in digital networks is the method of multiplexing lower level TDM signals into higher level signals (for example, multiplexing 4 DS1s into a DS2 or 28 DS1s into a DS3). Here again the term "asynchronous" arises to describe one particular method of multiplexing multiple, independent tributary signals into a higher rate signal. In this context "asynchronous" is totally unrelated to either asynchronous trans-mission as described at the beginning of this chapter or to packet like multiplexing mentioned in the previous paragraph. The rest of this chapter is concerned only with (synchronous) TDM as used in first level (DS1 or CEPT1) signals. Network level (asynchronous) TDM is discussed in Chapter 7 as a fundamental aspect of network synchronization.

4.6.1 Bit Interleaving versus Word Interleaving

Two different structures of (synchronous) time division multiplex frames are shown in Figure 4.31. In the first instance each channel is assigned a time slot corresponding to a single bit—hence the term: bit interleaving. In the second instance each channel is assigned a longer time slot corresponding to some larger number of bits, referred to as a word—hence the term: word interleaving.

The decision as to which structure to use in a particular application is primarily dependent on the nature of the sources. In T-carrier systems each channel produces a complete 8 bit word at a time. Hence word interleaving is desirable so all bits can be transmitted as generated.

In contrast, higher-level digital TDM multiplexers typically use bit interleav-ing of the lower-level multiplex signals since the lower-level signals represent continuous, one bit at a time, data streams. The specific formats of the higher-

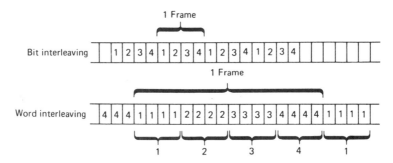

Figure 4.31. Bit interleaving and word interleaving of four-channel TDM multiplexers.

level multiplex signals are described in Chapter 7 when synchronization of bit streams is considered. One notable exception to higher-level bit interleaving is the word interleaved structure of fiber-based SONET multiplexing described in Chapter 8.

4.6.2 Framing

To identify individual time slots within a TDM frame, a receiving terminal uses a counter synchronized to the frame format of the transmitter. Just as for synchronization of sample clocks, a certain amount of transmission overhead is required to establish and maintain frame synchronization. In fact, most of the techniques used for frame synchronization are directly analogous to clock synchronization techniques discussed previously. Specifically, the basic means of establishing frame synchronization are

1. Added digit framing
2. Added channel framing
3. Unique line signal framing.

The main considerations in choosing a framing procedure are time required to establish framing, effects of channel errors in maintaining frame synchronization, relationships between the line clock and sample clocks derived from the line clock, transmission overhead, and complexity of the framing circuitry.

The severity of a loss of framing and the time required to reestablish synchronization depends on the nature of the traffic. Since a loss of framing implies a loss of data on all channels, the mean time between misframes must be as long as possible. For voice traffic infrequent misframes can be tolerated if frame synchronization is reestablished rapidly enough to minimize the length of the "glitch" in the output speech. For data traffic the duration of reframe times is not as critical as the frequency of occurrence since most data communications protocols initiate recovery procedures and retransmit a message no matter how many data bits are lost.

A critical requirement for reframe time in the telephone network comes from the possibility that various in-channel control signals may be lost and interpreted as disconnects. Thus the maximum reframe time on a particular digital transmission link is often determined from analog network signaling conventions. A loss of framing is also used as a performance monitor for the transmission link and usually sets alarm conditions, which in turn cause automatic switching to spare terminals or transmission facilities. Generally speaking, terminals serving 600 or more channels require automatic protection switching [33].

Out-of-frame conditions occur in two ways. First, the locally derived sample clock may lose synchronization with the line clock and produce a slip in the

counter sequence. Normally, the timing information in the line clock is sufficiently strong to prevent misframes of this type. An exception occurs on radio links when deep fades reduce the signal power to the point that clock synchronization is impossible. Receiving terminals can also lose clock synchronization through no fault of their own or of the transmission link. If the source clock at the transmitting terminal has too much jitter or generates abrupt phase shifts receiving terminals may be unable to track the clock signal. Both phenomena are related to network synchronization practices as discussed in Chapter 7.

A second source of misframes results from channel errors creating false out-of-frame framing patterns. Thus considerable redundancy in the framing pattern is required to minimize the probability of false misframes. A loss of framing is determined when the occurrence of framing pattern violations exceeds some short-term density threshold.

In all of the framing techniques discussed in the following paragraphs, special framing bits or codes are inserted into the information stream. These insertions do not have to occur with each frame. Instead, they can be sent only once for a predetermined number of information frames. In this manner the transmission overhead for framing can be reduced—which is particularly important in bit-interleaved systems. An individual information frame contained between framing indicators is sometimes referred to as a "subframe."

Added Digit Framing

One common technique of framing a digital TDM information stream is to periodically insert a framing bit with an identifiable data sequence. Usually the framing bit is added once for every frame and alternates in value. This particular format, in fact, is the procedure used to establish framing in the original D1 channel banks. When the T1 line carries only voice traffic, this framing format is particularly useful since no information bits can sustain an alternating 1, 0 pattern. (An alternating pattern represents a 4-kHz signal component, which is rejected by the bandlimiting filter in the PCM coder.)

Framing is established in a receiving D1 channel bank by monitoring first one bit position within a 193 bit frame and then another, until the alternating pattern is located. With this framing strategy, the expected framing time from a random starting point with random data is derived in Appendix A as

$$\text{frame time} = \left(\begin{array}{l}\text{average number}\\\text{of bit}\\\text{positions before}\\\text{framing bit is found}\end{array}\right)\left(\begin{array}{l}\text{average number of bits}\\\text{to determine that an}\\\text{information position is}\\\text{not a framing position}\end{array}\right)$$

$$= \left(\frac{N}{2}\right)\cdot(2N+1)$$

$$= N^2 + \frac{N}{2} \text{ bit times} \tag{4.10}$$

where N is the number of bits in a frame including the framing bit.

For D1 channel banks, $N = 193$ so that the framing time is 37,249 bits or 24.125 msec. Also of interest is the maximum framing time. Unfortunately, there is no absolute maximum framing time for a T1 system with random data. It is very unlikely, however, that the framing time would ever exceed the average search time for all bit positions, or 48.25 msec. This latter measure of framing time is referred to as the "maximum average frame time." It is the average time required to establish framing, but with the assumption that all bit positions must be tested for the framing sequence before the actual framing bit is found. Obviously, the maximum average frame time is twice the average value from a random starting point defined in Equation 4.10.

The framing time can be reduced by using more sophisticated frame search strategies. One approach examines one bit at a time, as before, but during a reframe the search begins a few clock positions in front of the present position under the assumption that short lapses of clock synchronization cause small counter offsets. A second approach [34, 35] uses a parallel search by monitoring all bit positions simultaneously for the framing pattern. With this framing procedure, framing is established when the last of the $N - 1$ information bit positions finally produces a framing pattern violation. The probability that all information bit positions produce a framing violation in n or less frames is derived in Appendix A as follows:

$$\text{prob(frame time} < n) = [1 - (\tfrac{1}{2})^n]^{N-1} \tag{4.11}$$

where N is the number of bits in a frame and the probability of ones $= \tfrac{1}{2}$.

Using Equation 4.11, we determine the median framing time by setting prob(frame time $< n) = \tfrac{1}{2}$. Hence

$$n = -\log_2[1 - (\tfrac{1}{2})^{1/(N-1)}] \tag{4.12}$$

Setting $N = 193$ for the frame length of D1 channel bank produces the result that $n = 8.1$ frames or approximately 1 msec.

An even more sophisticated framing strategy involves continually monitoring all bit positions, even while the system is synchronized. Then, when a misframe is detected, a new frame position is immediately established or the frame search time is significantly reduced. Furthermore, the continuous search for framing patterns provides additional information for declaring an out-of-frame condition. (There is little point is discarding the present frame position unless another bit position exhibits a more consistent framing pattern.)

The framing pattern of second generation channel banks (D3, D4) from AT&T was changed from the alternating 1, 0 pattern to establish a longer sequence for identification of signaling frames. As mentioned previously, these channel banks provide 8 bits of voice for all time slots except in every sixth frame, which uses the least significant PCM bit for signaling. The signaling channels thus derived are divided into an A and a B subchannel—implying each

subchannel sends a bit in every twelfth frame. Hence a 12 bit framing sequence is needed to identify the signaling bits and the two signaling subchannels. The 12-frame superframe (SF) structure and associated framing sequence are shown in Figure 4.32.

When describing or processing the framing sequence shown in Figure 4.32, it is convenient to divide the framing bits into two separate sequences. During the odd frames the framing bit alternates, while during the even frames the framing bit sequence is 000111000111000. Figure 4.32 indicates that the A signaling frame can be identified by a 0 to 1 transition in the even-numbered frame sequence. Correspondingly, a 1 to 0 transition in the even-numbered frame sequence signifies a B signaling frame. Frame acquisition begins by finding the alternating bit sequence (with 385 intervening bits). Then, the 000111 framing pattern is located. Another framing sequence, extended superframe (ESF), for DS1 signals is described in Section 4.6.3.

Added-Channel Framing

Added-channel framing is basically identical to added-digit framing except that framing digits are added in a group such that an extra channel is established. Hence the transmission rate of individual channels is integrally related to the line rate of the entire multiplex.

The fact that the frame boundaries are identified by whole code words adds considerable performance and flexibility to the framing process. First, framing

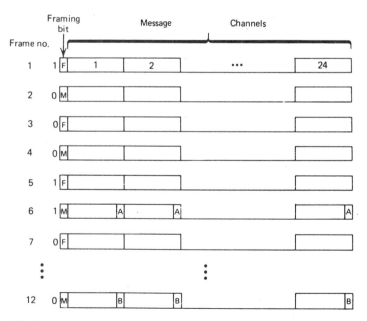

Figure 4.32. Twelve-frame superframe structure of DS1 signal with robbed digit signaling. F, Frame alignment signal = 101010. M, Multiframe alignment signal = 001110.

can be established more rapidly since random 8 bit code words are very unlikely to appear as framing codes. (See the problem set at the end of the chapter.) Second, the larger code space simplifies identification of auxiliary functions such as superframe boundaries, parity bits, or equipment status. In most systems the added channel contains more than framing bits.

The first-level digital multiplex signal of CCITT (CEPT1) is an example of a system using added-channel framing. The CEPT1 multiplex standard establishes 32 channels per frame with one channel providing framing and one more channel dedicated to signaling. Thus 30 of 32 channels are available for message channels. Figure 4.33 shows the frame structure of the primary multiplex signal CEPT1.

The average frame acquisition time of a multibit frame code is derived in Appendix A as

$$\text{frame time} = \frac{N^2}{2(2^L - 1)} + \frac{N}{2} \tag{4.13}$$
$$\text{(in bits)}$$

where N is the length of a frame including the frame code, L is the length of the frame code, and it is assumed that 1's and 0's are equally likely.

From Figure 4.33 it can be seen that for the CEPT1 signal of CCITT, $N = 512$ and $L = 7$. Thus the average frame time from a random starting point is determined from Equation 4.13 as 0.5 msec. Again, the "maximum average" frame time is twice the average from a random starting point, or 1 msec. Notice that these frame times are much faster than added digit frame times.

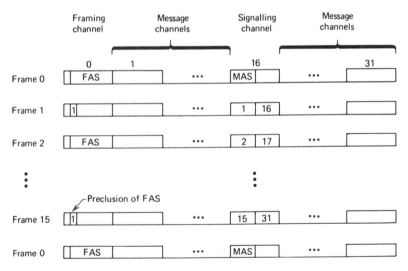

Figure 4.33. Framing sequence of primary digital signal of CCITT using channel-associated signaling. FAS, 0011011; MAS, 0000.

Unique Line Code Framing

Bipolar coding manages to shape the spectrum (remove the DC component) of the line code by adding extra signal levels to provide more flexibility in selecting signals. The same technique can be used to establish frame synchronization. In fact, with bipolar coding and added-digit framing, bipolar violations can be used to identify the framing boundaries uniquely and rapidly. A particularly significant example of using line code violations for framing is the ISDN S/T basic rate interface described in Chapter 9.

Even without added-digit framing, unique line codes can be used to carry information while simultaneously conveying frame positioning. If the number of signal levels is doubled for the framing bit only, the extra levels provide sufficient code space for the information but uniquely identify that bit position as a frame boundary. This procedure does not necessarily increase the error rate, since for any particular bit (information or framing), half of the levels can be disallowed.

The main advantage of using unique line codes for framing is that the information bit positions cannot generate framing patterns. Hence framing is established as soon as a frame bit occurs, and misframes are detected almost immediately.

The main disadvantages of unique line codes are the added signal processing requirements (if new levels are established just for framing) and the dependence on the transmission terminals to locate framing. With the other framing techniques described in this section, the framing patterns are represented in the logic levels of the bit stream. Hence the transmission equipment can be changed independently of the multiplex equipment. For additional information on framing techniques and performance analyses see references [36] and [37].

4.6.3 DS1 Extended Superframe

When a T1 line is used as an interoffice transmission link the operating telephone companies have access to both ends of the line for maintenance and performance monitoring. Furthermore, most installations include protection switching facilities that can also be used to routinely test a line while the traffic normally assigned to that line is diverted to a spare line. A significantly different situation arises when a T1 line is used by a customer as access to the public switched network or as a leased line in a private network. The explosive use of T1 lines in such applications created the need for enhanced maintenance features for T1 customers. The extended superframe (ESF) as provided in the D5 channel bank provides the desired features.

Foremost among the customer needs is nonintrusive performance monitoring of an end-to-end T1 circuit. Monitoring bipolar violations at the customer service unit (CSU) provides nonintrusive monitoring of the receive signal but provides no information regarding the quality of the signal at the other end of the line (which terminates at the provider's facilities). Furthermore, leased T1 circuits generally involve intermediate transmission, multiplexing, and cross-

connect equipment. Because the interfaces of each of these equipments remove bipolar violations the customer has no access to performance statistics internal to the provider's facilities. Determination of end-to-end error rate performance in this environment requires taking at least a portion of the T1 circuit out of service (for which there is usually no spare).

ESF allows customers to achieve end-to-end in service performance monitoring by redefining the 8-kbps F bit of each DS1 frame to consist of a 2-kbps framing channel, a 4-kbps data link channel, and a 2-kbps CRC channel representing a check sum over all information bits. Because the CRC channel is carried intact from one customer location to another channel errors occurring in any intermediate facility produce CRC errors at the far end.

The data link supports numerous services, one of which is the ability to interrogate the far end, or any intermediate equipment, for performance statistics. Thus the data link provides both the customer and the service provider with an extremely useful tool for isolating faulty span lines or equipment.

As indicated in Table 4.8 obtained from reference [32], the three subchannels are established by "extending" the D4 superframe format to encompass 24 DS1 frames. Because the actual framing bits occur only once every four DS1 frames there are 771 bits intervening between bits carrying the Frame Pattern Sequence (FPS) = 001011.

The 6 CRC bits (C_1 to C_6) of each extended superframe represent a CRC check of all 4608 information bits in the previous superframe.* Besides providing end-to-end performance monitoring, the CRC virtually precludes the chances of false framing on a data bit position. Even though static user data can easily simulate the FPS, it is extremely unlikely that user data can spuriously generate valid CRC codes in successive superframes. (Six bits of random data match a 6-bit code with a probability of 1 in 64.)

The performance parameters measured and reported by the 4-kbps data link (DL) are framing bit errors, CRC errors, out-of-frame (OOF) events, line code (bipolar) violations, and controlled slip events (described in Chapter 7). Individual events are reported as well as event summaries. The four performance summaries reported are

1. Errored seconds (ESs) (ES = at least one CRC event).
2. Bursty second (BSs) (BS = 2–319 ESs).
3. Severely errored seconds (SESs) (SES = >319 ESs or OOF).
4. Failed seconds state (FS) (10 consecutive SES).

ESF CSUs typically determine the above parameters on 15-min intervals and store them for up to 24 hr for polling by a controller [38]. The SES report conforms to CCITT recommendation G.821.

*Calculation of the CRC actually includes F bits that are set to 1 for purposes of CRC calculation only. Thus, channel errors in the F bits do not create CRC errors (unless they occur in the CRC bits themselves).

TABLE 4.8 Extended Superframe Framing Channel Format[a]

ESF Frame Number	ESF Bit Number	F-Bit Assignment		
		FPS	FDL	FRC
1	0	—	m	—
2	193	—	—	CB1
3	386	—	m	—
4	579	0	—	—
5	772	—	m	—
6	965	—	—	CB2
7	1158	—	m	—
8	1351	0	—	—
9	1544	—	m	—
10	1737	—	—	CB3
11	1930	—	m	—
12	2123	1	—	—
13	2316	—	m	—
14	2509	—	—	CB4
15	2702	—	m	—
16	2895	0	—	—
17	3088	—	m	—
18	3281	—	—	CB5
19	3474	—	m	—
20	3667	1	—	—
21	3860	—	m	—
22	4053	—	—	CB6
23	4246	—	m	—
24	4439	1	—	—

[a]FPS, framing pattern sequence (...001011...); FDL, 4 kbps facility data link (message bits m); CRC, CRC-6 cyclic redundancy check (check bits CB1-CB6).

In addition to supporting remote interrogation of performance statistics the data link carries alarm information, loopback commands, and protection switching commands.

4.7 TIME DIVISION MULTIPLEX LOOPS

In Chapter 2 it is mentioned that time division multiplexing is not as amenable to applications with distributed sources and sinks of traffic as is frequency division multiplexing. In this section a particular form of a TDM network is described that is quite useful in interconnecting distributed nodes. The basic structure of interest is referred to as a "time division multiplex loop" and is shown in Figure 4.34.

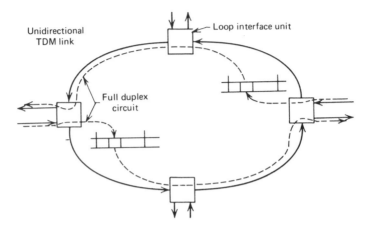

Figure 4.34. Time division multiplex loop.

Basically, a time division multiplex loop is configured as a series of unidirectional (two-wire) links arranged to form a closed circuit or loop. Each node of the network is implemented with two fundamental operational features. First, each node acts as a regenerative repeater merely to recover the incoming bit stream and retransmit it. Second, the network nodes recognize the TDM frame structure and communicate on the loop by removing and inserting data into specific time slots assigned to each node. As indicated in Figure 4.34, a full-duplex connection can be established between any two nodes by assigning a single time slot or channel for a connection. One node inserts information into the assigned time slot that propagates around the loop to the second node (all intervening nodes merely repeat the data in the particular time slot). The destination node removes data as the assigned time slot passes by and inserts return data in the process. The return data propagates around the loop to the original node where it is removed and replaced by new data, and so forth.

Since other time slots are not involved with the particular connection shown, they are free to be used for other connections involving arbitrary pairs of nodes. Hence a TDM loop with C time slots per frame can support C simultaneous full-duplex connections.

If, as channels become available, they are reassigned to different pairs of nodes, the transmission facilities can be highly utilized with high concentration factors and provide low blocking probabilities between all nodes. Thus a fundamental attraction of a loop network is that the transmission capacity can be assigned dynamically to meet changing traffic patterns. In contrast, if a star network with a centralized switching node is used to interconnect the nodes with four-wire links, many of the links to particular nodes would be underutilized since they can not be shared as in a loop configuration.

Another feature of the loop-connected network is the ease with which it can

be reconfigured to accommodate new nodes in the network. A new access node is merely inserted into the nearest link of the network and the new node has complete connectivity to all other nodes by way of the TDM channels. In contrast, a star structured network requires transmission to the central node and expansion of the centralized switching facilities.

The ability to reassign channels to arbitrary pairs of nodes in a TDM loop implies that the loop is much more than a multiplexer. It is, in fact, a distributed transmission and switching system. The switching capabilities come about almost as a by-product of TDM transmission. TDM loops represent the epitome of integrated transmission and switching.

TDM loops have been used within computer complexes to provide high capacity and high interconnectivity between processors, memories, and peripherals [39]. The loop structure in this application is sometimes more attractive than more conventional bus structures since all transmission is unidirectional and therefore avoids timing problems on bidirectional buses that limit their physical length. Furthermore, as more nodes are added to a bus, the electrical loading increases—causing a limitation on the number of nodes that can be connected to a bus. Loops, on the other hand, have no inherent limits of transmission length or numbers of nodes.

The loop structure of Figure 4.34 is topologically identical to the token passing ring developed by IBM and standardized by the IEEE as a 802.5 local area network. However, a token passing ring operates differently than a TDM loop in that there is only one channel. When a node on a ring becomes active it uses the entire capacity of the outgoing link until it is through sending its message. A node on a loop, on the other hand, uses only specific time slots in the TDM structure allowing other nodes to be simultaneously "connected" using other time slots. In essence, a TDM loop is a distributed circuit switch and a ring is distributed packet switch.

Although TDM loops have not been used within the public telephone network in the past, they are currently being considered as a new exchange-area structure [40, 41] that takes particular advantage of high bandwidth fiber transmission systems. As shown in Figure 4.35 the combination of "free" excess capacity on the fiber with drop-and-insert access at switching nodes allows direct functional connectivity between any two offices. In this case, large groups of channels are semipermanently assigned between all pairs of offices. Thus a logical mesh is established with a topological loop. The flexibility of this architecture allows easy reconfiguration of the network in response to changes in traffic patterns or system failures.

One obvious limitation of a loop is its vulnerability to failures of any link or node. The effect of a node failure can be minimized by having bypass capabilities included in each node. When bypassed, a node becomes merely a regenerative repeater as on T-carrier transmission links. Link failures can be circumvented by providing alternate facilities. Figure 4.36 shows one particular structure using a second, reverse direction loop to provide backup capabilities in the case of failures. When fully operational, the network can use the reverse loop as a

Figure 4.35. Functional mesh, fiber loop, and add/drop multiplexers (ADM).

Figure 4.36. Use of reverse loop to circumvent link failures in TDM loops.

separate, independent network for traffic as needed. Whenever a failure occurs, the nodes adjacent to the break establish a new loop by connecting the forward path to the reverse path at both places. Hence all nodes continue to have full connectivity to any node on the new loop.

A particular example of the use of the dual reverse loop for both protection and distributed queued access to the channels is the distributed queued dual bus (DQDB) [42] systems developed by QPSX in Australia and standardized by the IEEE as a 802.6 metropolitan area network.

REFERENCES

1 F. deJager and M. Christiaens, "A Fast Automatic Equalizer for Data Links," *Philips Technical Review*, Vol. 36, 1977.

2 R. Komiya, K. Yoshida, and N. Tamaki, "The Loop Coverage Comparison between TCM and Echo Canceller under Various Noise Considerations," *IEEE Transactions on Communications*, November 1986, pp 1058–1067.

3 F. D. Waldhauer, "A 2-Level, 274 Mb/s Regenerative Repeater for T4M," *IEEE International Communications Conference*, 1975, pp 48-13–48-17.

4 N. Karlovac and T. V. Blalock, "An Investigation of the Count Rate Performance of Baseline Restorers," *Nuclear Sciences Symposium*, Washington, D.C., 1974.

5 F. D. Waldhauer, "Quantized Feedback in an Experimental 280-Mb/s Digital repeater for coaxial transmission," *IEEE Transactions on Communications*, January 1974, pp 1–5.

6 J. M. Cioffi, W. L. Abbott, H. K. Thapar, C. M. Melas, and K. D. Fisher, "Adaptive Equalization in Magnetic-Disk Storage Channels," *IEEE Communications Magazine*, February 1990, pp 15–29.

7 S. Sugimoto, K. Hayashi, and F. Mano, "Design of 2B1Q Transceiver for ISDN Subscriber Loops," *IEEE International Conference on Communications*, June 1989, pp 8.1.1–8.1.5.

8 M. R. Aaron, "PCM Transmission in the Exchange Plant," *Bell System Technical Journal*, January 1962, pp 99–141.

9 Members of Technical Staff, Bell Telephone Laboratories, *Transmission Systems for Communications*, Western Electric Co., 1971, p 668.

10 "1.544 Mbps Digital Service," *Bell System Technical Reference Publication*, No. 41451, May 1977.

11 G. D. Forney, "The Viterbi Algorithm," *Proceedings of IEEE*, March 1973, pp 268–278.

12 V. I. Johannes, A. G. Kaim, and T. Walzman, "Bipolar Pulse Transmission with Zero Extraction," *IEEE Transactions on Communications*, April 1969, pp 303–310.

13 "The D3 Channel Bank Compatibility Specification—Issue 3," *Technical Advisory No. 32*, American Telephone and Telegraph Company, October 1977.

14 B. Johnston and W. Johnston, "LD-4 A Digital Pioneer in Action," *Telesis*, Vol. 5, No. 3, June 1977, pp 66–72.

15 J. H. Davis, "T2: A 6.3 Mb/s Digital Repeatered Line," *IEEE International Communications Conference*, 1969, pp 34-9–34-16.

16 Recommendation G.703, *CCITT Orange Book*, Vol. 3, No. 2.

17 R. M. Wienski, "Evolution to ISDN within the Bell Operating Companies," *IEEE Communications Magazine*, January 1984, pp 33–41.

18 American National Standard: Digital Hierarchy—Formats Specifications, ANSI T1.107-1988, Editorial Revised April 1989.

19 J. M. Sipress, "A New Class of Selected Ternary Pulse Transmission Plans for Digital Transmission Lines," *IEEE Transactions on Communication Technology*, September 1965, pp 366–372.

20 E. E. Schnegelberger and P. T. Griffiths, "48 PCM Channels on T1 Facilities," *National Electronics Conference*, 1975, pp 201–205.

21 P. A. Franaszek, "Sequence-State Coding for Digital Transmission," *Bell System Technical Journal*, December 1967, pp 143–157.

22 J. O. Azaret, J. F. Graczyk, W. M. Hauser, and J. V. Mahala, "New Improved T-Carrier Doubles Capacity, Cuts Costs," *Bell Labs Record*, July 1985, pp 26–31.

23 J. W. Lechleider, "Line Codes for Digital Subscriber Lines," *IEEE Communications Magazine*, September 1989, pp 25–31.

24 A. Lender, "The Duobinary Technique for High Speed Data Transmission," *IEEE Transactions on Communication Electronics*, May 1963, pp 214–218.

25 Adam Lender, "Correlative Level Coding for Binary Data Transmission," *IEEE Spectrum*, February 1960, pp 104–110.

26 P. Kabal and S. Pasupathy, "Partial Response Signaling," *IEEE Transactions on Communications*, September 1975, pp 921–934.

27 A. M. Gerrish and R. D. Howson, "Multilevel Partial Response Signaling," *International Communications Conference*, 1967, p 186.

28 D. W. Jurling and A. L. Pachynski, "Duobinary PCM System Doubles Capacity of T1 Facilities," *International Communications Conference*, 1977, pp 32.2-297–32.2-301.

29 T. Seaver, "An Efficient 96 PCM Channel Modem for 2 GHz FM Radio," *National Telecommunications Conference*, 1978, pp 38.4.1–38.4.5.

30 "Special Issue: 1ARDS," *Bell System Technical Journal*, March 1984.

31 M. B. Brilliant, "Observations of Errors and Error Rates on T1 Digital Repeatered Lines," *Bell System Technical Journal*, March 1978, pp 711–746.

32 Carrier-to-Customer Installation—DS1 Metallic Interface, ANSI T1.403-1989, American National Standards Institute.

33 Members of Technical Staff, Bell Telephone Laboratories, *Engineering and Operations in the Bell System*, Western Electric Co., 1978.

34 D. E. Dodds, L. R. Button, and S.-M. Pan, "Robust Frame Synchronization for Noisy PCM Systems," *IEEE Transactions on Communications*, May 1985, pp 465–469.

35 D. E. Dodds, S.-M. Pan, and A. G. Wacker, "Statistical Distribution of PCM Framing Times," *IEEE Transactions on Communications*, November 1988, pp 1236–1241.

36 R. A. Scholtz, "Frame Synchronization Techniques," *IEEE Transactions on Communications*, August 1980, pp 1204–1212.

37 D. T. R. Munhoz, J. R. B. de Marca, and D. S. Arantes, "On Frame Synchronization of PCM Systems," *IEEE Transactions on Communications*, August 1980, pp 1213–1218.

38 K. Stauffer and A. Brajkovic, "DS-1 Extended Superframe Format and Related Performance Issues," *IEEE Communications Magazine*, April 1989, pp 19–23.

39 *C-System General Description*, Radio Company Report No. 523-0561697-20173R, Collins Radio Company, Dallas, Texas, May 1, 1970.

40 D. J. Kolar and T. H. Wu, "A Study of Survivability versus cost for several Fiber network Architectures," *IEEE International Conference on Communications*, 1988, pp 2.6.1–2.6.6.

41 T.-H. Wu, D. J. Kolar, and R. H. Cardwell, "High-Speed Self-Healing Ring

Architectures for Future Interoffice Networks," *IEEE Globecom Proceedings*, 1989, pp 23.1.1–23.1.7.

42 "New Proposal Extends the Reach of Metro Area Nets," *Data Communications*, February 1988, pp 139–145.

PROBLEMS

4.1 **Using the symbols** $+$, 0, $-$ to represent a positive pulse, no pulse, and a negative pulse, respectively, determine the following line code sequences of the binary data sequence:

$$0110\ 1000\ 0100\ 0110\ 0000\ 0010$$

(a) Bipolar with most recent pulse being positive.
(b) Bipolar with most recent pulse being negative.
(c) Pair selected ternary beginning in " $-$ " mode.
(d) B3ZS with a $+0+$ substitution having just been made.
(e) B6ZS with the most recent pulse being positive.

4.2 Assume that two identical cable systems are used for digital transmission with equal pulse amplitudes. One system uses conventional bipolar signaling and the other uses pair selected ternary. Compare the crosstalk levels of the two systems. (Assume that ones and zeros are equally probable.)

4.3 A digital transmission system operating at an error rate of 10^{-6} is to have its data rate increased by 50%. Assuming that the same transmitted power is to be used in the second system, what is the new error rate?

4.4 What is the average reframe time of a D3 channel bank if we assume a random starting point? What is the "maximum average" reframe time of a D3 channel bank? (Assume 1's and 0's in the message traffic are equally likely.)

4.5 Repeat Problem 4.4 for the primary TDM multiplex signal specified by CCITT.

4.6 A TDM system operating at 2 Mbps is to have a "maximum average" reframe time of 20 msec. What is the maximum possible frame length if framing is established with a bit-by-bit frame search? (Assume that 1's and 0's in message channels are equally likely.)

4.7 A T1 transmission system using a D1 frame format is to have an average reframe time (from a random starting point) of 10 msec. How large a block of bit positions must be examined in parallel to achieve the desired result?

4.8 What is the expected framing time for a T1 line (D3 frame format) if the

framing strategy is bit-by-bit and the data stream has 60% ones and 40% zeros?

4.9 What is the average pulse density of 4B3T coding? (Assume ones and zeros are equally likely.)

4.10 A TDM transmission link using 4B3T coding can transmit 32 64-kbps voice channels using the same symbol rate as a T1 line (1544 kbps). Assuming a fixed Gaussian noise environment, how much must the average transmit power of the 4B3T system be increased to provide the equivalent error rate of a bipolar code?

4.11 Assume that crosstalk interference in a multipair cable system produces an effect equivalent to Gaussian noise at an equal power level. Using an error rate of 10^{-6} as a design objective, determine the effective degradation of the crosstalk on binary (polar) NRZ coding under each of the following conditions. (The effective degradation is determined as the increased transmit power, in decibels, required to achieve the desired error rate.)

 (a) The crosstalk level is 16 dB below the average signal level, but the crosstalk is to be overcome on only one pair (i.e., all other pairs stay at a power level for 10^{-6} BER with no crosstalk)

 (b) The crosstalk level is 16 dB below the average signal level, but the effects of the crosstalk are to be overcome on all pairs. (Hint: Use signal-power-to-noise-power ratios, not E_b/N_0.)

4.12 Repeat Problem 4.11 for bipolar coding.

4.13 How much does -18 dB of crosstalk degrade the error rate performance of a polar binary NRZ signal for a BER of 10^{-7}? (Assume that all transmitters are at equal power levels.) If crosstalk increases 15 dB per decade increase in frequency, what is the relative crosstalk level of a four-level NRZ code carrying the same data rate? What is the overall performance penalty of the four-level system compared to the two-level system?

4.14 How many distinct code words of length four can be constructed from ternary symbols? How many of these code words contain an equal number of positive and negative pulses? How many code words can be used to ensure a minimum of one timing pulse occurs in each code word?

4.15 Can a set of ternary code words of length eight be constructed to encode binary data using exactly four pulses per 8-bit word and containing equal numbers of positive and negative pulses?

4.16 Given the following sequence of signal levels determine the sequence of output signal levels for each of the following correlative encodings.
 Input sequence: $+1, -3, +1, -1, +3, +3, -3$

 (a) $1 + D$ encoder:
 (b) $1 - D$ encoder:
 (c) $1 - D^2$ encoder:

5

DIGITAL SWITCHING

Of the three basic elements in a communications network (terminals, transmission media, and switches), switches are the most invisible to the users yet represent the most important element in terms of available service offerings. As mentioned in Chapter 1, a major milestone was established in 1965 when stored program control switching was first introduced into the U.S. public telephone network. Stored program control provides the means for implementing many innovative user services and for greatly simplifying switch administration and maintenance.

The use of computers to control the switching functions of a central office led to the designation "electronic" switching (e.g., electronic switching system: ESS or electronic automatic exchange: EAX). However, the switching matrices of these first generation electronic switches are actually electromechanical in nature. The first use of electronic switching matrices occurred in France in 1971 when digital switching was applied to an end-office environment. Ironically, these first digital switches did not use stored program control. Electronic switching matrices were first introduced into the U.S. public network in 1976 with AT&T's No. 4 ESS digital toll switch.

In the late 1970s digital class 5 switches began to be installed in the United States—mostly in smaller end offices as replacements for step-by-step switches. At that time most major metropolitan switching offices already had stored program control from No. 1 ESS or No. 1 EAX machines. Because the digital toll and end office switching machines were initially installed in predominantly analog transmission environments their digital matrix provided no direct benefit to network customers. The motivation for the digital machines was reduced costs for the operating companies: lower maintenance, reduced floor space, simplified expansion, and lower manufacturing costs [1].

By the mid 1980s the interoffice transmission environment had changed to be almost exclusively digital. Thus analog-to-digital conversion costs became associated with analog equipment further sealing the fate of analog toll or tandem switching technology. At this time frame, end office transmission environments also began the swing to a digital environment. Interoffice trunks

were already digital, digital loop carrier systems became cost effective in metropolitan applications, remotely controlled switching modules with digital fiber interconnect became common for service to outlying communities, digital cross-connect systems (DCSs) were being deployed, and the feeder portion of the subscriber loop plant began to use fiber.

The combination of cost penalties in a digital transmission environment and higher maintenance costs leads to the older electromechanical switches being replaced by digital machines when expansion or consolidation of an office occurs. The ability to offer ISDN services is a lesser motivation for changing to digital switches because low cost fiber transmission in conjunction with DCS systems allows provisioning of these services from other offices.

This chapter describes the basic operation and implementation of digital time division switching as applied to PBXs, end offices, and toll switches. Before digital switching is discussed, certain basic switching concepts and terminology are introduced.

5.1 SWITCHING FUNCTIONS

Obviously, the basic function of any switch is to set up and release connections between transmission channels on an "as-needed basis." The structure and operation of a switch vary significantly, however, depending on particular applications. Three main switching categories for voice circuits are local (line-to-line) switching, transit (tandem) switching, and call distribution.

The most common switching function involves direct connections between subscriber loops at an end office or between station loops at a PBX. These connections inherently require setting up a path through the switch from the originating loop to a specific terminating loop. Each loop must be accessible to every other loop. This level of switching is sometimes referred to as line switching.

Transit connections require setting up a path from a specific incoming (originating) line to an outgoing line or trunk group. Normally, more than one outgoing circuit is acceptable. For example, a connection to an interoffice trunk group can use any one of the channels in the group. Hence transit switching structures can be simplified because alternatives exist as to which outgoing line is selected. Furthermore, it is not even necessary that every outgoing line be accessible from every incoming line. Transmit switching functions are required by all switching machines in the telephone network. Some machines such as remote concentrators and toll or tandem switches service only transit traffic (e.g., do not provide local connections). These concepts are illustrated in Figure 5.1.

Call distributors are often implemented with the same basic equipment as PBXs. The mode of operation (software) is significantly different, however, in that incoming calls can be routed to any available attendant. Normally, the software of an automatic call distributor (ACD) is designed to evenly distribute the arriving calls among the attendants. Although it is not an inherent

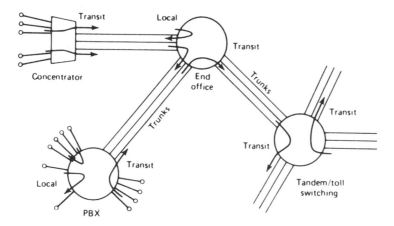

'**Figure 5.1.** Local and transit traffic switching examples.

requirement that every incoming line (trunk) be connectable to every attendant, call distributors are normally designed to provide accessibility to all attendants. Furthermore, it is often desirable that nonblocking operations be provided. (No matter what switch paths are in use, a new request can be serviced by any available attendant.)

5.2 SPACE DIVISION SWITCHING

Conceptually, the simplest switching structure is a rectangular array of crosspoints as shown in Figure 5.2. This switching matrix can be used to connect any one of N inlets to any one of M outlets. If the inlets and outlets are connected to two-wire circuits, only one crosspoint per connection is required.*

Rectangular crosspoint arrays are designed to provide intergroup (transit) connections only, that is, from an inlet group to an outlet group. Applications for this type of an operation occur in the following:

1. Remote concentrators.
2. Call distributors.
3. The portion of a PBX or end-office switch that provides transit switching.
4. Single stages in multiple stage switches.

In most of the foregoing applications, it is not necessary that the inlets be connectable to every outlet. In situations involving large groups of outlets,

*In actual fact, two (and sometimes three) switching contacts are associated with each crosspoint of a two-wire switch. Since these contacts are part of a single unit and operate in unison, they are considered a single crosspoint.

N Inlets

M Outlets

Figure 5.2. Rectangular crosspoint array.

considerable savings in total crosspoints can be achieved if each inlet can access only a limited number of outlets. When such a situation occurs "limited availability" is said to exist. By overlapping the available outlet groups for various inlet groups, a technique called "grading" is established. An example of a graded switching matrix is shown in Figure 5.3. Notice that if outlet connections are judiciously chosen, the adverse effect of limited availability is minimized. For example, if inlets 1 and 8 in Figure 5.3 request a connection to the outlet group, outlets 1 and 3 should be chosen instead of outlets 1 and 4 to avoid future blocking for inlet 2.

Graded switching structures were often used for access to large trunk groups in electromechanical switches where crosspoints were expensive and individual switching modules were limited in size. Gradings were also used in individual switching stages of large multiple-stage switches where more than one path to any particular outlet exists.

Intragroup switching, as for loop-to-loop switching, requires each loop to be connectable to every other loop. Thus full availability from all inlets to all outlets of the switching matrix is required. Figure 5.4 shows two matrix structures that can be used to fully interconnect two-wire lines. The dashed lines indicate that corresponding inlets and outlets of two-wire switching matrices are actually connected together to provide bidirectional transmission on two-wire

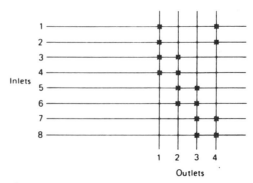

Inlets

Outlets

Figure 5.3. Graded rectangular switching matrix.

circuits. For purposes of describing switching matrices, however, it is convenient to consider the inlets and outlets of two-wire switching matrices as being distinct.

Both structures in Figure 5.4 allow any connection to be established by selecting a single crosspoint. However, the square matrix, which is also called a two-sided matrix, allows any particular connection to be established in two ways. For example, if input link i is to be connected to input link j, the selected crosspoint can be at the intersection of inlet i and outlet j—or at the intersection of inlet j and outlet i. For simplicity these crosspoints are referred to as (i,j) and (j,i), respectively. In a typical implementation crosspoint (i,j) is used when input i requests service, and crosspoint (j,i) is used when input j requests service.

In the triangular matrix of Figure 5.4, the redundant crosspoints are eliminated. The crosspoint reduction does not come without complications, however. Before setting up a connection between switch input i and switch input j, the switch control element must determine which is larger: i or j. If i is larger, crosspoint (i,j) is selected. If i is smaller, crosspoint (j,i) must be selected. With computer controlled switching, the line number comparison does not represent a significant imposition. In the older, electromechanically controlled switches, however, the added complexity of the switch control is more significant.

Switching machines for four-wire circuits require separate connections for the go and return branches of a circuit. Thus two separate connections must be established for each service request. Figure 5.5 depicts a square matrix structure used to provide both connections. The structure is identical to the square matrix shown in Figure 5.4 for two-wire switching. The difference, however, is that corresponding inlets and outlets are not connected to a common two-wire input. All of the inlets of the four-wire switch are connected to the wire pair carrying the incoming direction of transmission, and all of the outlets are connected to the outgoing pairs. When setting up a connection between four-wire circuits i and j, the matrix in Figure 5.5 must select both crosspoints (i,j) and (j,i). In actual operation these two crosspoints may be selected in unison and implemented as a common module.

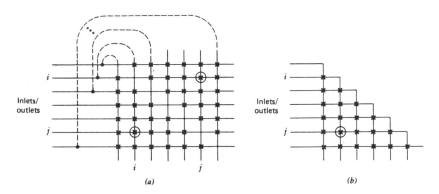

Figure 5.4. Two-wire switching matrices. (a) Square matrix. (b) Triangular (folded).

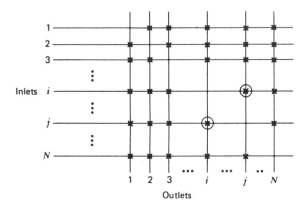

Figure 5.5. Four-wire switching matrix.

5.2.1 Multiple Stage Switching

In the switching structures described to this point, an inlet is connected directly to an outlet through a single crosspoint. (Four-wire switches use two crosspoints per connection, but only one for an inlet to outlet connection.) For this reason, these switching structures are referred to as "single stage" switches. Single stage switches have the property that each individual crosspoint can only be used to interconnect one particular inlet/outlet pair. Since the number of inlet/outlet pairs is equal to $N(N-1)/2$ for a triangular array, or $N(N-1)$ for a square array, the number of crosspoints required for a large switch is prohibitive. Furthermore, the large number of crosspoints on each inlet and outlet line imply a large amount of capacitive loading on the message paths. Another fundamental deficiency of single stage switches is that one specific crosspoint is needed for each specific connection. If that crosspoint fails, the associated connection cannot be established. (An exception is the square, two-wire switch that has a redundant crosspoint for each potential connection. Before the redundant crosspoint could be used as an alternate path, however, the inlet-oriented selection algorithm would have to be modified to admit outlet-oriented selection.)

Analysis of a large single stage switch reveals that the crosspoints are very inefficiently utilized. Only one crosspoint in each row or column of a square switch is ever in use, even if all lines are active. To increase the utilization efficiency of the crosspoints and thereby reduce the total number, it is necessary that any particular crosspoint be usable for more than one potential connection. If crosspoints are to be shared, however, it is also necessary that more than one path be available for any potential connection so that blocking does not occur. The alternate paths serve to eliminate or reduce blocking and also to provide protection against failures. The sharing of crosspoints for potential paths through the switch is accomplished by multiple stage switching. A block

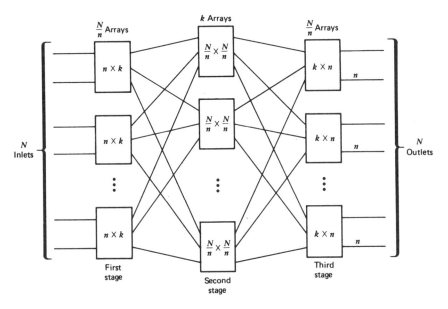

Figure 5.6. Three-stage switching matrix.

diagram of one particular form of a multiple stage switch is shown in Figure 5.6.

The switch of Figure 5.6 is a three-stage switch in which the inlets and outlets are partitioned into subgroups of n inlets and n outlets each. The inlets of each subgroup are serviced by a rectangular array of crosspoints. The inlet arrays (first stage) are $n \times k$ arrays where each one of the k outputs is connected to one of the k center stage arrays. The interstage connections are often called junctors. The third stage consists of $k \times n$ rectangular arrays that provide connections from each center stage array to the groups of n outlets. All center stage arrays are N/n by N/n arrays that provide connections from any first-stage array to any third-stage array. Notice that if all arrays provide full availability, there are k possible paths through the switch for any particular connection between inlets and outlets. Each of the k paths utilizes a separate center stage array. Thus the multiple stage structure provides alternate paths through the switch to circumvent failures. Furthermore, since each switching link is connected to a limited number of crosspoints, capacitive loading is minimized.

The total number of crosspoints N_x required by a three-stage switch, as shown in Figure 5.6 is

$$N_x = 2Nk + k \left(\frac{N}{n}\right)^2 \tag{5.1}$$

where $N =$ the number of inlets/outlets
$\quad n =$ the size of each inlet/outlet group
$\quad k =$ the number of center stage arrays

As is demonstrated shortly, the number of crosspoints defined in Equation 5.1 can be significantly lower than the number of crosspoints required for single stage switches. First, however, we must determine how many center stage arrays are needed to provide satisfactory service.

Nonblocking Switches

One attractive feature of a single stage switch is that it is strictly nonblocking. If the called party is idle, the desired connection can always be established by selecting the particular crosspoint dedicated to the particular input/output pair. When crosspoints are shared, however, the possibility of blocking arises. In 1953 Charles Clos [2] of Bell Laboratories published an analysis of three-stage switching networks showing how many center stage arrays are required to provide a strictly nonblocking operation. His result demonstrated that if each individual array is nonblocking, and if the number of center stages k is equal to $2n - 1$, the switch is strictly nonblocking.

The condition for a nonblocking operation can be derived by first observing that a connection through the three-stage switch requires locating a center stage array with an idle link from the appropriate first stage and an idle link to the appropriate third stage. Since the individual arrays themselves are nonblocking, the desired path can be set up any time a center stage with the appropriate idle links can be located. A key point in the derivation is to observe that since each first-stage array has n inlets, only $n - 1$ of these inlets can be busy when the inlet corresponding to the desired connection is idle. If k is greater than $n - 1$, it follows that, at most, $n - 1$ links to center stage arrays can be busy. Similarly, at most $n - 1$ links to the appropriate third-stage array can be busy if the outlet of the desired connection is idle.

The worst case situation for blocking occurs (as shown in Figure 5.7) if all $n - 1$ busy links from the first-stage array lead to one set of center stage arrays, and if all $n - 1$ busy links to the desired third-stage array come from a separate set of center stage arrays. Thus these two sets of center stage arrays are unavailable for the desired connection. However, if one more center stage array exists, the appropriate input and output links must be idle, and that center stage can be used to set up the connection. Hence if $k = (n - 1) + (n - 1) + 1 = 2n - 1$ the switch is strictly nonblocking. Substituting this value of k into Equation 5.1 reveals that for a strictly nonblocking operation of a three stage switch:

$$N_x = 2N(2n - 1) + (2n - 1)\left(\frac{N}{n}\right)^2 \qquad (5.2)$$

As expressed in Equation 5.2, the number of crosspoints in a nonblocking three-stage switch is dependent on how the inlets and outlets are partitioned into subgroups of size n. Differentiating Equation 5.2 with respect to n and setting the resulting expression equal to 0 to determine the minimum reveals

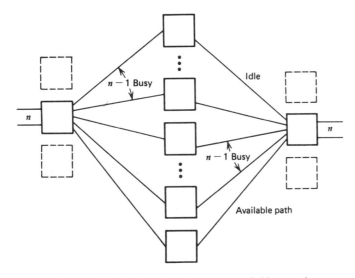

Figure 5.7. Nonblocking three-stage switching matrix.

that (for large N) the optimum value of n is $(N/2)^{1/2}$. Substituting this value of n into Equation 5.2 then provides an expression for the minimum number of crosspoints of a nonblocking three-stage switch.

$$N_x(\text{min}) = 4N(\sqrt{2N} - 1) \qquad (5.3)$$

where N = total number of inlets/outlets.

Table 5.1 provides a tabulation of $N_x(\text{min})$ for various sized nonblocking three-stage switches and compares the values to the number of crosspoints in a single stage square matrix. Both switching structures inherently provide four-wire capabilities, a requirement for digital switches because voice digitization implies four-wire circuits.

TABLE 5.1 Crosspoint Requirements of Nonblocking Switches

Number of Lines	Number of Corosspoints for Three-Stage Switch	Number of Crosspoints for Single-Stage Switch
128	7,680	16,256
512	63,488	261,632
2,048	516,096	4.2 million
8,192	4.2 million	67 million
32,768	33 million	1 billion
131,072	268 million	17 billion

As indicated in Table 5.1, a three-stage switching matrix provides significant reductions in crosspoints, particularly for large switches. However, the number of crosspoints for large three-stage switches is still quite prohibitive. Large switches typically use more than three stages to provide greater reductions in crosspoints. For example, the No. 1 ESS uses an eight-stage switching matrix that can service up to 65,000 lines. However, the most significant reductions in crosspoint numbers are achieved not so much from additional stages but by allowing the switch to introduce acceptably low probabilities of blocking.

5.2.2 Blocking Probabilities: Lee Graphs

Strictly nonblocking switches are rarely needed in most voice telephone networks. Both the switching systems and the number of circuits in interoffice trunk groups are sized to service most requests as they occur, but economics dictates that network implementations have limited capacities that are occasionally exceeded during peak traffic situations. Equipment for the public telephone network is designed to provide a certain maximum probability of blocking for the busiest hour of the day. The value of this blocking probability is one aspect of the telephone company's grade of service. (Other aspects of grades of service are: availability, transmission quality, and delay in setting up a call.)

A typical residential telephone is busy 5–10% of the time during the busy hour. Business telephones are often busy for a larger percentage of their busy hour (which may not coincide with a residential busy hour). In either case, network blocking occurrences on the order of 1%* during the busy hour do not represent a significant reduction in the ability to communicate since the called party is much more likely to have been busy anyway. Under these circumstances, end-office switches and, to a lesser degree, PBXs can be designed with significant reductions in crosspoints by allowing acceptable blocking probabilities.

There are a variety of techniques that can be used to evaluate the blocking probability of a switching matrix. These techniques vary according to complexity, accuracy, and applicability to different network structures. One of the most versatile and conceptually straightforward approaches of calculating blocking probabilities involves the use of probability graphs as proposed by C. Y. Lee [3]. Although this technique requires several simplifying approximations, it can provide reasonably accurate results, particularly when comparisons of alternate structures are more important than absolute numbers. The greatest value of this approach lies in the ease of formulation and the fact that the formulas directly relate to the underlying network structures Thus the formulations help provide insight into the network structures and how these structures might be modified to change the performance.

*Transmission and switching equipment in the public network is normally designed for even lower blocking probabilities to provide for growth in the traffic volume.

In the following analyses we are determining the blocking probabilities of various switching structures using utilization percentages or "loadings" of individual links. The notation p will be used, in general, to represent the fraction of time that a particular link is in use (i.e., p is the probability that a link is busy). In addition to a utilization percentage or loading, p is also sometimes referred to as an occupancy. The probability that a link is idle is denoted by $q = 1 - p$.

When any one of n parallel links can be used to complete a connection, the composite blocking probability B is the probability that all links are busy:*

$$B = p^n \qquad (5.4)$$

When a series of n links are all needed to complete a connection, the blocking probability is most easily determined as 1 minus the probability that they are all available:*

$$B = 1 - q^n \qquad (5.5)$$

A probability graph of a three-stage network is shown in Figure 5.8. This graph relates the fact that any particular connection can be established with k different paths: one through each center stage array. The probability that any particular interstage link is busy is denoted by p'. The probability of blocking for a three-stage network can be determined as

$\begin{aligned} B &= \text{probability that all paths are busy} \\ &= (\text{probability that an arbitrary path is busy})^k \\ &= (\text{probability that at least one link in a path is busy})^k \\ &= (1 - q'^2)^k \end{aligned}$ $\qquad (5.6)$

where k is the number of center stage arrays, and $q' = 1 - p'$ is the probability that an interstage link is idle.

If the probability p that an inlet is busy is known, the probability p' that an

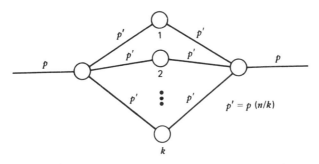

Figure 5.8. Probability graph of three-stage network.

*These formulas assume each link is busy or idle independently of other links.

interstage link is busy can be determined as

$$p' = p/\beta \qquad (p < \beta) \tag{5.7}$$

where $\beta = k/n$. Equation 5.7 presents the fact that when some number of inlets (or outlets) are busy, the same number of first-stage outputs (or third-stage inputs) are also busy. However, there are $\beta = k/n$ times as many interstage links as there are inlets or outlets. Hence the percentage of interstage links that are busy is reduced by the factor β.

The factor β is defined as though k is greater than n, which implies that the first stage of the switch is providing space expansion (i.e., switching some number of input links to a larger number of output links). Actually, β may be less than 1, implying that the first stage is concentrating the incoming traffic. First-stage concentration is usually employed in end-office or large PBX switches where the inlets are lightly used (5–10%). In tandem or toll offices, however, the incoming trunks are heavily utilized, and expansion is usually needed to provide adequately low-blocking probabilities.

Substituting Equation 5.7 into Equation 5.6 provides a complete expression for the blocking probability of a three-stage switch in terms of the inlet utilization p.

$$B = [1 - (1 - p/\beta)^2]^k \tag{5.8}$$

Table 5.2 tabulates numbers of crosspoints obtained from Equation 5.1 for the same switch sizes presented in Table 5.1. The number of center arrays was chosen in each case to provide a blocking probability on the order of 0.002. The inlet utilization in each example was assumed to be 10%. Notice that the designs with small but finite blocking probabilities are significantly more cost effective than nonblocking designs.

The switch designs in Table 5.2 assume that the inlets are only 10% busy, as

TABLE 5.2 Three-Stage Switch Designs for Blocking Probabilities of 0.002 and Inlet Utilizations of 0.1

Switch Size N	n	k	β	Number of Crosspoints	Number of Crosspoints in Nonblocking Design	
128	8	5	0.625	2,560	7,680	$(k = 15)$
512	16	7	0.438	14,336	63,488	$(k = 31)$
2,048	32	10	0.313	81,920	516,096	$(k = 63)$
8,192	64	15	0.234	491,520	4.2 million	$(k = 127)$
32,768	128	24	0.188	3.1 million	33 million	$(k = 255)$
131,072	256	41	0.160	21.5 million	268 million	$(k = 511)$

TABLE 5.3 Three-Stage Switch Designs for Blocking Probabilities of 0.002 and Inlet Utilizations of 0.7

Switch Size N	n	k	β	Number of Crosspoints	Number of Crosspoints in Nonblocking Design	
128	8	14	1.75	7,168	7,680	($k = 15$)
512	16	22	1.38	45,056	63,488	($k = 31$)
2,048	32	37	1.16	303,104	516,096	($k = 63$)
8,192	64	64	1.0	2.1 million	4.2 million	($k = 127$)
32,768	128	116	0.91	15.2 million	33 million	($k = 255$)
131,072	256	215	0.84	113 million	268 million	($k = 511$)

might be the case for an end-office switch or a PBX. The dramatic savings in crosspoints for large switches by introducing significant concentration factors ($1/\beta$) into the middle stage. When the inlet utilization is higher (as typically occurs in tandem switches), high concentration factors are not acceptable, and the crosspoint requirements therefore increase. Table 5.3 lists corresponding crosspoint requirements and implementation parameters for inlet loadings of 70%.

The results presented in Tables 5.2 and 5.3 indicate that very large switches still require prohibitively large numbers of crosspoints, even when blocking is allowed. As mentioned previously, very large switches use more than three stages to provide further reductions in crosspoints. Figure 5.9 shows a block diagram of a five-stage switch obtained by replacing every center stage array in Figure 5.6 with a three-stage array. This particular structure is not optimum in terms of providing a given level of performance with the fewest crosspoints, but it is a useful design because of its modularity. (Furthermore, it is a lot easier to analyze than some other five-stage structures.)

If the middle three stages of a five-stage switch as shown in Figure 5.9 are strictly nonblocking ($k_2 = 2n_2 - 1$), the design provides a savings of 8704 crosspoints in each center stage array of the 32,768 line, three-stage switch designs presented earlier. Hence a little over 1 million crosspoints are saved in the 32,768 line tandem switch design of Table 5.3. Since the middle stages do not introduce blocking, the performance of this five-stage switch is identical to the performance of the three-stage design. Naturally, a more cost effective design could be obtained by allowing small amounts of blocking in the middle stages. The probability graph of the five-stage switch is shown in Figure 5.10. From this graph, the blocking probability is determined as follows:

$$B = \{1 - (q_1)^2[1 - (1 - q_2^2)^{k_2}]\}^{k_1} \tag{5.9}$$

where $q_1 = 1 - p_1$ and $q_2 = 1 - p_2$.

Even greater crosspoint reductions can, of course, be achieved by using more stages to replace the rather large first- and third-stage arrays. For example, the

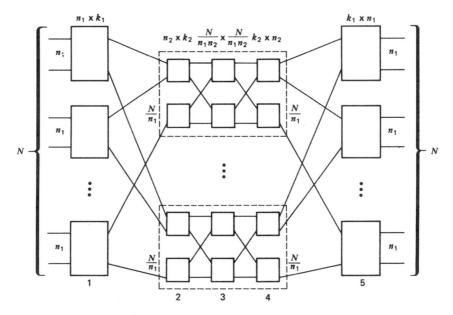

Figure 5.9. Five-stage switching network.

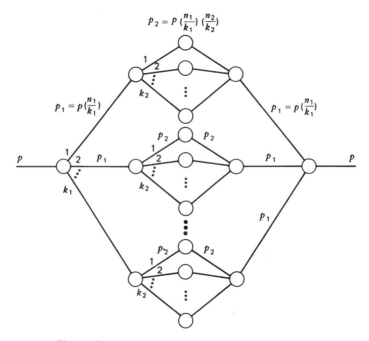

Figure 5.10. Probability graph of five-stage network.

total number of crosspoints in the 32,000 line switch can be reduced to less than 3 million. The 130,000 line switch is not practical with electromechanical switching matrices, but is well within the capabilities of a digital time division switch.

5.2.3 Blocking Probabilities: Jacobaeus

The formulations of blocking probability obtained from probability graphs entail several simplifying assumptions. One of these assumptions involves expressing the composite blocking probability of the alternate paths as the product of the blocking probabilities of each individual path. This step assumes that the individual probabilities are independent. In actual fact, the probabilities are not independent, particularly when significant amounts of expansion are present. Consider a switching matrix with $k = 2n - 1$. Equation 5.8 produces a finite blocking probability even though the switch is known to be strictly nonblocking. The inaccuracy results because when $2n - 2$ paths are busy, the remaining path is assumed to be busy with a probability of $1 - (q')^2$. In fact, the remaining path is necessarily idle.

In general, when space expansion exists, the assumption of independent individual probabilities produces higher than actual blocking probabilities. The inaccuracy results because as more and more paths in a switch are found to be busy, the remaining paths are less and less likely to be in use (only a subset of n of the interstage links can ever be busy at any one time).

A more accurate but not exact analysis of multistage switching matrices was presented in 1950 by C. Jacobaeus [4]. Although the analysis is conceptually straightforward, it does involve a considerable amount of manipulation that is not presented here. The resulting equation for a three-stage switch is obtained from reference [5] as

$$B = \frac{(n!)^2}{k!(2n - k)!} \, p^k(2 - p)^{2n - k} \qquad (5.10)$$

where n = number of inlets (outlets) per first (third) stage array
k = number of second-stage arrays
p = inlet utilization

In the interest of comparing the two methods, Equations 5.6 and 5.10 have been evaluated for three-stage switches with varying amounts of space expansion. The results were obtained for inlet utilizations of 0.7 and are presented in Table 5.4.

Table 5.4 reveals that the two analyses are in close agreement for near-unity expansion factors. In fact, if β is exactly equal to 1, the two formulations produce identical results. As expected, the Lee graph analysis (Equation 5.8) produces overly pessimistic values for the blocking probability when β is greater than 1.

TABLE 5.4 Comparison of Blocking Probability Analyses ($p = 0.7$)[a]

Number of Center Stages k	Space Expansion β	Lee Equation 5.8	Jacobaeus Equation 5.10
14	0.875	.548	.598
16	1.0	.221	.221
20	1.25	.014	.007
24	1.50	$3.2(10)^{-4}$	$2.7(10)^{-5}$
28	1.75	$3.7(10)^{-6}$	$7.7(10)^{-9}$
31[b]	1.94	$8.5(10)^{-8}$	$.1(10)^{-12}$

[a]Switch size $N=512$; inlet group size $n = 16$; inlet utilization $p=0.7$.
[b]Nonblocking.

As another comparison between the two approaches, Table 5.5 is included to demonstrate the use of Equations 5.8 and 5.10 for switches with significant amounts of concentration made possible by a relatively low inlet utilization of .1.

Table 5.5 reveals that a Lee graph analysis (Equation 5.8) consistently underestimates the blocking probability when concentration exists. Actually, the Jacobaeus analysis presented in Equation 5.10 also underestimates the blocking probability if large concentration factors and high blocking probabilities are used. When necessary, more accurate techniques can be used for systems with high concentration and high blocking. However, switches with high blocking probabilities normally have no practical interest so they are not considered here.

Users of older PBXs sometimes experienced high blocking probabilities, but blocking in these cases usually arose from too few tie lines to other corporate

TABLE 5.5 Comparison of Blocking Probability Analyses ($p = 0.1$)[a]

Number of Center States k	Space Expansion β	Lee Equation 5.8	Jacobaeus Equation 5.10
6	0.375	.0097	.027
8	0.5	$2.8(10)^{-4}$	$8.6(10)^{-4}$
10	0.625	$4.9(10)^{-6}$	$1.5(10)^{-5}$
12	0.75	$5.7(10)^{-8}$	$1.4(10)^{-7}$
14	0.875	$4(10)^{-10}$	$7.8(10)^{-10}$
16	1.0	$2.9(10)^{-12}$	$2.9(10)^{-12}$

[a]Switch size $N = 512$; inlet group size $n = 16$; inlet utilization $p = 0.1$.

locations or too few trunk circuits to the public network. The subject of blocking in trunk groups is treated in Chapter 10.

Up to this point, the blocking probability analyses have assumed that a specific inlet is to be connected to a specific outlet. Also, it has been assumed that the requests for service on the individual lines are independent. These assumptions are generally valid for switching one subscriber line to another in an end-office switch, or for connecting one station to another in a PBX. When connecting to or from trunks, however, these assumptions are not valid.

When connecting to a trunk circuit, any one circuit in a trunk group is acceptable. Thus the blocking probability to a specific circuit is only as important as its significance in the overall blocking to the trunk group. The blocking probability to any particular circuit in a trunk group can be relatively large and still achieve a low composite blocking probability to the trunk group as a whole. If the blocking probabilities to the individual trunks are independent, the composite blocking probability is the product of the individual probabilities. However the paths to the individual trunk circuits normally involve some common links (e.g., the junctors from a first-stage array to all second-stage arrays). For this reason the individual blocking probabilities are normally dependent, which must be considered in an accurate blocking probability analysis.

As an extreme example, consider a case where all trunks in a trunk group are assigned to a single outlet array in a three-stage switch. Since the paths from any particular inlet to all trunks in the group are identical, the ability to select any idle trunk is useless. In actual practice the individual circuits of a trunk group should be assigned to separate outlet arrays.

Another aspect of trunk groups that must be considered when designing a switch or analyzing the blocking probabilities involves the interdependence of activity on the individual circuits within a trunk group. In contrast to individual subscriber lines or PBX stations, individual circuits in a trunk group are not independent in terms of their probabilities of being busy or idle. If some number of circuits in a trunk group are tested and found to be busy, the probability that the remaining circuits are busy is increased. The nature of these dependencies is discussed more fully in Chapter 10. At this point it is only necessary to point out that these dependencies cause increased blocking probabilities if the individual trunks are competing for common paths in the switch. Again, the effect of these dependencies is minimized by assigning the individual trunks to separate inlet arrays so that independent paths are involved in connections to and from the trunk group. This process is sometimes referred to as decorrelating the trunk circuits.

One last aspect of the blocking probability as a grade of service parameter that must be mentioned involves variations in the loading of the network by individual users. In the design examples for an end office presented earlier, it was tacitly assumed that all subscribers are busy 10% of the time during a busy hour. In actual fact, some subscribers are active much more than 10% of the time, and other subscribers are active less than 10% of the time. In terms of traffic theory,

some subscribers present more than 0.1 erlangs* of traffic to the network, but others present less.

When a switch is partitioned into subgroups (as all large switches must be) and the traffic is concentrated by first-stage switching arrays, a few overactive subscribers in one subgroup can significantly degrade service for the other subscribers in the subgroup. It does not matter that the subscribers in some other subgroup may be experiencing lower than average blocking. Their essentially nonblocking service is no compensation for those subscribers experiencing a relatively poor grade of service.

Operating companies have traditionally solved the problem of overactive subscribers by specifically assigning the most active lines (businesses) to separate inlet groups of the switch. Sometimes this procedure requires making traffic measurements to determine which lines are most active and reassigning these lines to different parts of the switch. These procedures fall into the general category of line administration. If the subgroups are large enough, or if the designs provide adequate margin for overactive users, this aspect of line administration can be minimized. One feature of a digital switch that can be utilized in this regard is the ability to design economical switches with very low nominal blocking probabilities so that wide variations in traffic intensities can be accommodated.

5.2.4 Folded Four-Wire Switches

Multiple stage switches can be used for either two-wire or four-wire switching operations. Figure 5.11 depicts a four-wire connection through a four-stage switch. Notice that two paths must be established for the complete connection. A two-wire connection requires only one path since each outlet is externally connected to its corresponding inlet.

The two paths shown in Figure 5.11 demonstrate a particularly useful relationship: one path is a mirror image of the other path. If the switch diagram is folded about the vertical center line, the paths coincide. Hence this method of setting up connections is sometimes referred to as a folded operation. When all connections in the switch are set up with a folded relationship, several benefits result.

First of all, only one pathfinding operation is needed since the reverse path is automatically available as a mirror image of the forward path. In essence, every crosspoint on one side is paired with another crosspoint in a corresponding array on the opposite side of the switch. Whenever one crosspoint of a pair is used in a connection, the other crosspoint in the pair is also used. For example, the third inlet array in the first stage uses crosspoint 6, 4 to connect its sixth inlet to its fourth outlet (leading to the fourth array of the second stage). The corresponding crosspoint in the third outlet array of the last stage connects its

*An erlang is a measure of traffic intensity specifying the proportion of time that a device is busy. A circuit is said to carry 0.1 erlangs if it is busy 10% of the time.

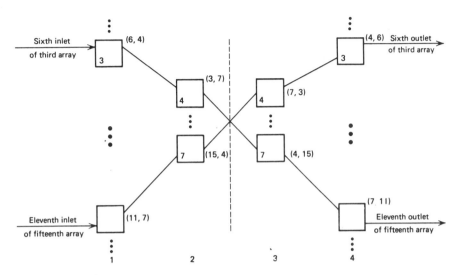

Figure 5.11. Four-wire connection through four-stage switch.

fourth inlet (coming from the fourth array in the fourth stage) to its sixth outlet. In general, crosspoint i, j in one array is paired with crosspoint j, i in the corresponding array on the opposite side of a switch. Since the availability of one crosspoint in a pair ensures the availability of the other, the reverse path is automatically specified and available.

A second advantage of the folded four-wire operation results because the amount of information specifying the status of the switch can be cut in half. Only the status of each pair of crosspoints or associated junctor is needed to find an available path through the switch.

A third benefit of the folded structure occurs because the blocking probability is one-half of the probability of finding two paths independently. It might seem that pairing the crosspoints in the described manner would restrict the paths available for a particular connection. On the contrary, the crosspoint pairings merely guarantee that a reverse path is automatically available for any selected path in the forward direction.

A fourth potential advantage of the folded operation can be obtained by sharing the crosspoint selection logic for corresponding arrays in the switch. For example, the same control signal that selects crosspoint 6, 4 in the first stage can be used to select crosspoint 4, 6 in the last stage. In an actual implementation, the paired crosspoints may even be in a common module.

The folded operation in the preceding paragraphs referred to a switch with an even number of switching stages. An even number was chosen because the concept is easiest to demonstrate when no center stage is present. The basic approach can be extended to odd numbers of switching stages if the center stage contains an even number of arrays and is folded about a horizontal line at the

center of the stage. In this manner, crosspoint i, j in the top center-stage array is paired with crosspoint j, i in the bottom center-stage array, and so on.

5.2.5 Pathfinding

Determining a path through a single stage switch is virtually automatic since the necessary crosspoint is uniquely specified by the inlet/outlet pair to be connected. In contrast, availability of more than one path in a multiple stage switch complicates the path selection process in several ways. First, the control element of a switch must keep track of (or be able to sense) which potential paths for a particular connection are available. With computer controlled switching, the information as to which switching paths are available is retained in memory and referred to as a state store. A programmed pathfinding routine is then used to process the state store information and select an available path. Whenever a new connection is established or an old one released, the state store is updated with the appropriate information.

Pathfinding Times
Pathfinding operations require the use of common equipment and must therefore be analyzed to determine the rate at which connect requests can be processed. The time required to find an available path is directly dependent on how many potential paths are tested before an idle one is found. Some systems can test a number of paths in parallel and thereby shorten the processing time. Since the expected number of potential paths that must be tested to find an idle path is a function of link utilizations, pathfinding times unfortunately increase when the common control equipment is busiest.

Assume that the probability of a complete path through the switch being busy is denoted by p. If each of k possible paths through the switch has an equal and independent probability of being busy, the expected number of paths N_p that must be tested before an idle path is found is determined in Appendix A as follows:

$$N_p = \frac{1 - p^k}{1 - p} \tag{5.11}$$

Example 5.1. What is the expected number of potential paths that must be tested to find an idle path in the three-stage, 8192 line switch defined in Table 5.2?

Solution. As indicated in the table, a space expansion factor of 0.234 is used to provide a blocking probability of 0.002. Hence the utilization of each interstage link is $0.1/0.234 = 0.427$. The blocking probability of each path through the switch is merely the probability that one of two links in series is busy. Hence the probability $p = 1 - (1 - 0.427)^2 = 0.672$, and the expected number of paths to

be tested is

$$N_p = \frac{1 - (0.672)^{15}}{1 - 0.672}$$

$$= 3.04$$

Example 5.1 demonstrates that, on average, only 3 of the 15 potential paths need to be tested before an idle path is found. However, when the switch is experiencing greater than normal traffic loads, the average number of paths tested increases. For example, if the input line utilization increases from 10% to 15%, the blocking probability increases from 0.002 to 0.126, and the expected number of paths to be tested in the pathfinding operation increases from 3 to 4.9.

Although this simple pathfinding example is, by itself, not particularly important it demonstrates a very important aspect of the loading of common equipment in a switching system (or a network as a whole): linear increases in the offered traffic produce geometric increases in the utilization of the network resources. If a system is designed and analyzed under nominal traffic conditions and the results are extrapolated to determine total capacity, greatly optimistic estimates usually occur. In the pathfinding example the processing requirements increase just when the call processor (stored program control) is already loaded with greater than average traffic volumes. For further discussion of the effects of traffic loads on the common equipment of a switching system see Schwartz [7].

5.2.6 Switch Matrix Control

When an available path through a common control switching network is determined, the control element of the switch transfers the necessary information to the network to select the appropriate crosspoints. Crosspoint selection within a matrix is accomplished in one of two ways. The control may be associated with the output lines and therefore specify which inputs are to be connected to the associated outputs, or the control information may be associated with each input, and subsequently specify to which outputs the respective inputs are to be connected. The first approach is referred to as "output-associated control" while the second is accordingly referred to as "input-associated control." These two control implementations are presented in Figure 5.12.

Input-associated control is inherently required in step-by-step switches where the information (dial pulses) arrives on the input link and is used to directly select the output links to each successive stage. In common control systems, however, the address information of both the originating line and the terminating line is simultaneously available. Hence the connection can be established by beginning at the desired outlet and proceeding backward through the switch while selecting inputs to each stage.

The implementation of both types of digital crosspoint arrays using standard

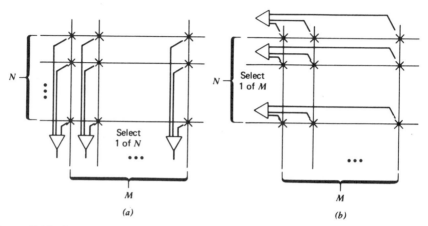

Figure 5.12. Switch matrix control. (a) Output associated control. (b) Input associated control.

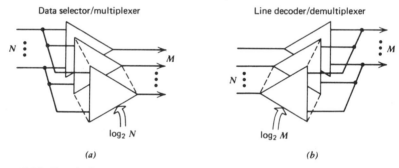

Figure 5.13. Standard component implementations of digital crosspoint array. (a) Output-associated control. (b) Input-associated control.

components is shown in Figure 5.13. Output-associated control uses a conventaional data selector/multiplexer for each matrix output. The number of bits required to control each data selector is $\log_2 N$ where N is the number of inlets. Thus the total number of bits required to completely specify a connection configuration is $M \log_2 N$.

Input-associated control can be implemented using conventional line decoders/demultiplexers. The outputs are commoned using a "wired or" logic function. Thus the output gates of each decoder circuit must be open collector or tristate devices if TTL logic is used. The total number of bits required to specify a connection configuration in this case is $N \log_2 M$.

A significant drawback of input-associated control arises from the need to

disable unused inputs to prevent crossconnects when another input selects the same output. With output-associated control, unused outputs can remain connected to an input without preventing that input from being selected by another output. For this reason and for generally greater speeds of operation, digital switching networks typically use output-associated control. Notice, however, that the total amount of information needed to specify a connection configuration with input-associated control is less than that with output control if the number of inputs N is much smaller than the number of outputs M. ($N \log_2 M < M \log_2 N$.) Furthermore, input-associated control is more flexible in terms of "wired-or" expansion.

5.3 TIME DIVISION SWITCHING

As evidenced by multiple stage switching, sharing of individual crosspoints for more than one potential connection provides significant savings in implementation costs of space division switches. In the cases demonstrated, the crosspoints of multistage space switches are shared from one connection to the next, but a crosspoint assigned to a particular connection is dedicated to that connection for its duration. Time division switching involves the sharing of crosspoints for shorter periods of time so that individual crosspoints and their associated interstage links are continually reassigned to existing connections. When the crosspoints are shared in this manner, much greater savings in crosspoints can be achieved. In essence, the savings are accomplished by time division multiplexing the crosspoints and interstage links in the same manner that transmission links are time division multiplexed to share interoffice wire pairs.

Time division switching is equally applicable to either analog or digital signals. Analog time division switching is attractive when interfacing to analog transmission facilities, since the signals are only sampled and not digitally encoded. However, large analog time division switches experience the same limitations as do analog time division transmission links: the PAM samples are particularly vulnerable to noise, distortion, and crosstalk. In digital switches, the voice signals are regenerated every time they pass through a logic gate. Thus at some point in switching sizes, the cost of digitizing PAM samples is mandated by the need to maintain end-to-end signal quality.

Although numerous advances in analog integrated circuit technology appropriate for switching matrices have been made analog technology still lags behind digital technology in terms of cost and performance. Thus a digital matrix is always less expensive than an analog matrix of the same size. The only situation in which an analog matrix is cost effective is in an analog transmission environment—and then only in small line sizes. Due to the low cost of PCM codecs analog switching is only used in very small PBXs, key systems, or small hybrid key/PBX systems.

5.3.1 Analog Time Division Switching

Although analog time division switching has become obsolete it is a good starting point to establish the basic principle of time division switching. Figure 5.14 depicts a particularly simple analog time division switching structure. A single switching bus supports a multiple number of connections by interleaving PAM samples from receive line interfaces to transmit line interfaces. The operation is depicted as though the receive interfaces are separate from the respective transmit interfaces. When connecting two-wire analog lines, the two interfaces are necessarily implemented in a common module. Furthermore, in some PAM-PBX systems, analog samples were simultaneously transferred in both directions between the interfaces [6].

Included in Figure 5.14 are two cyclic control stores. The first control store controls gating of inputs onto the bus one sample at a time. The second control store operates in synchronism with the first and selects the appropriate output line for each input sample. A complete set of pulses, one from each active input line, is referred to as a frame. The frame rate is equal to the sample rate of each line. For voice systems the sampling rate ranges from 8 to 12 kHz. The higher sampling rates are used to simplify the band-limiting filter and reconstructive filters in the line interfaces.

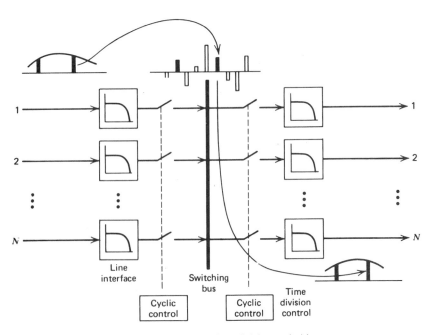

Figure 5.14. Analog time division switching.

5.3.2 Digital Time Division Switching

The analog switching matrix described in the preceding section is essentially a space division switching matrix. By continually changing the connections for short periods of time in a cyclic manner, the configuration of the space division switch is replicated once for each time slot. This mode of operation is referred to as time multiplexed switching. While this mode of operation can be quite useful for both analog and digital signals, digital time division multiplexed signals usually require switching between time slots, as well as between physical lines. This second mode of switching represents a second dimension of switching and is referred to as time switching.

In the following discussion of digital time division switching it is assumed, unless otherwise stated, that the switching network is interfaced directly to digital time division multiplex links. This assumption is generally justified since, even when operating in an analog environment, the most cost-effective switch designs multiplex groups of digital signals into TDM formats before any switching operations take place. Thus most of the following discussion is concerned with the internal structures of time division switching networks, and possibly not with the structure of an entire switching complex.

The basic requirement of a time division switching network is shown in Figure 5.15. As an example connection, channel 3 of the first TDM link is connected to channel 17 of the last TDM link. The indicated connection implies that information arriving in time slot 3 of the first link is transferred to time slot 17 of the last link. Since the voice digitization process inherently implies a four-wire operation, the return connection is required and realized by transferring information from time slot 17 of the last input link to time slot 3 of the first link. Thus each connection requires two transfers of information: each involving translations in both time and space.

A variety of switching structures are possible to accomplish the transfers indicated in Figure 5.15. All of these structures inherently require at least two stages: a space division switching stage and a time division switching stage. As discussed later, larger switches use multiple stages of both types. Before discussing switching in both dimensions, however, we discuss the characteristics and capabilities of time switching alone.

Figure 5.15. Time and space division switching.

A Digital Memory Switch

Primarily owing to the low cost of digital memory, time switching implementations provide digital switching functions more economically than space division implementations. Basically, a time switch operates by writing data into and reading data out of a single memory. In the process the information in selected time slots is interchanged as shown in Figure 5.16. When digital signals can be multiplexed into a single TDM format, very economical switches can be implemented with time switching alone. However, practical limitations of memory speed limit the size of a time switch so that some amount of space division switching is necessary in large switches. As demonstrated in later sections, the most economical multistage designs usually perform as much switching as possible in the time stages.

The basic functional operation of a memory switch is shown in Figure 5.17. Individual digital message circuits are multiplexed and demultiplexed in a fixed manner to establish a single TDM link for each direction of travel. The multiplexing and demultiplexing functions can be considered as part of the

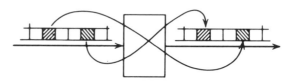

Figure 5.16. Time slot interchange operation.

Figure 5.17. MUX/TSI/DEMUX memory switch.

switch itself, or they may be implemented in remote transmission terminals. In either case, a byte-interleaved form of multiplexing is required. The asynchronous transmission hierarchy (DS2, DS3, DS4) uses bit-interleaving and therefore requires back-to-back demultiplexing and multiplexing operations before switching can be accomplished. In contrast, the SONET transmission format described in Chapter 8 can provide byte-interleaving specifically so it can be more directly interfaced to a digital switching system.

The exchange of information between two different time slots is accomplished by a time slot interchange (TSI) circuit. In the TSI of Figure 5.17 data words in incoming time slots are written into sequential locations of the data store memory. Data words for outgoing time slots, however, are read from addresses obtained from a control store. As indicated in the associated control store, a full-duplex connection between TDM channel i and TDM channel j implies that data store address i is read during outgoing time slot j and vice versa. The data store memory is accessed twice during each link time slot. First, some control circuitry (not shown) selects the time slot number as a write address. Second, the content of the control store for that particular time slot is selected as a read address.

Since a write and a read is required for each channel entering (and leaving) the TSI memory, the maximum number of channels c that can be supported by the simple memory switch is

$$c = \frac{125}{2t_c} \tag{5.12}$$

where 125 is the frame time in microseconds for 8 kHz sampled voice, and t_c is the memory cycle time in microseconds.

As a specific example, consider the use of a 125 nsec memory. Equation 5.12 indicates that the memory switch can support 500 full-duplex channels (250 connections) in a strictly nonblocking mode of operation. The complexity of the switch (assuming digitization occurs elsewhere) is quite modest: the TSI memory stores one frame of data organized as 500 words by 8 bits each. The control store also requires 500 words, but each word has a length equal to $\log_2(c)$ (which is 9 in the example). Thus the memory functions can be supplied by 500×8 and 500×9 bit random access memories (RAMs). The addition of a time slot counter and some gating logic to select addresses and enable new information to be written into the control store can be accomplished with a handful of conventional integrated circuits (ICs).

This switch should be contrasted to a space division design (Table 5.1) that requires 63,488 crosspoints for a nonblocking three-stage switch. Although modern integrated circuit technology might be capable of placing that many digital crosspoints in a few integrated circuits, they could never be reached because of pin limitations. As mentioned in Chapter 2, one of the main advantages of digital signals is the ease with which they can be time division

multiplexed. This advantage arises for communication between integrated circuits as well as for communication between switching offices.

Time Stages in General

Time switching stages inherently require some form of delay element to provide the desired time slot interchanges. Delays are most easily implemented using random access memories that are written into as data arrive and read from when data are to be transferred out. If one memory location is allocated for each time slot in the TDM frame format, the information from each TDM channel can be stored for up to one full frame time without being overwritten.

There are two basic ways in which the time stage memories can be controlled: written sequentially and read randomly, or written randomly and read sequentially. Figure 5.18 depicts both modes of operation and indicates how the memories are accessed to translate information from time slot 3 to time slot 17. Notice that both modes of operation use a cyclic control store that is accessed in synchronism with the time slot counter.

The first mode of operation in Figure 5.18 implies that specific memory locations are dedicated to respective channels of the incoming TDM link. Data for each incoming time slot are stored in sequential locations within the memory by incrementing a modulo-c counter with every time slot. As indicated, the data received during time slot 3 are automatically stored in the third location within the memory. On output, information retrieved from the control store specifies which address is to be accessed for that particular time slot. As indicated, the seventeenth word of the control store contains the number 3, implying that the contents of data store address 3 is transferred to the output link during outgoing time slot 17.

The second mode of operation depicted in Figure 5.18 is exactly the opposite of the first one. Incoming data are written into the memory locations as specified by the control store, but outgoing data are retrieved sequentially under control

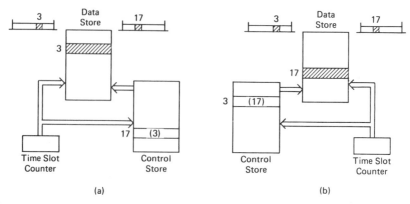

Figure 5.18. Time switch modules. (a) Sequential writes/random reads. (b) Random writes/sequential reads.

of an outgoing time slot counter. As indicated in the example, information received during time slot 3 is written directly into data store address 17 where it is automatically retrieved during outgoing TDM channel number 17. Notice that the two modes of time stage operation depicted in **Figure 5.18** are forms of output-associated control and input-associated control, respectively. In a multiple stage design example presented later, it is convenient to use one mode of operation in one time stage and the other mode of operation in another time stage.

5.4 TWO-DIMENSIONAL SWITCHING

Larger digital switches require switching operations in both a space dimension and a time dimension. There are a large variety of network configurations that can be used to accomplish these requirements. To begin with, consider a simple switching structure as shown in **Figure 5.19**. This switch consists of only two stages: a time stage T followed by a space stage S. Thus this structure is referred to a time–space TS switch.

The basic function of the time stage is to delay information in arriving time slots until the desired output time slot occurs. At that time the delayed information is transferred through the space stage to the appropriate output link. In the example shown the information in incoming time slot 3 of link 1 is delayed until outgoing time slot 17 occurs. The return path requires that information arriving in time slot 17 of link N be delayed for time slot 3 of the next outgoing frame. Notice that a time stage may have to provide delays ranging from one time slot to a full frame.

Associated with the space stage is a control store that contains the information needed to specify the space stage configuration for each individual time slot of a frame. This control information is accessed cyclically in the same manner as the control information in the analog time division switch. For example, during each outgoing time slot 3, control information is accessed that

Figure 5.19. Time–space (TS) switching matrix.

specifies interstage link number 1 is connected to output link N. During other time slots, the space switch is completely reconfigured to support other connections.

As indicated, a convenient means of representing a control store is a parallel end-around-shift register. The width of the shift register is equal to the number of bits required to specify the entire space switch configuration during a single time slot. The length of the shift register conforms to the number of time slots in a frame. Naturally, some means of changing the information in the control store is needed so that new connections can be established. In actual practice, the control stores may be implemented as random access memories with counters used to generate addresses in a cyclic fashion (as in the time stage control stores shown previously).

Implementation Complexity of Time Division Switches

In previous sections, alternative space division switching structures were compared in terms of the total number of crosspoints required to provide a given grade of service. In the case of solid state electronic switching matrices, in general, and time division switching, in particular, the number of crosspoints alone is a less meaningful measure of implementation cost. Switching structures that utilize integrated circuits with relatively large numbers of internal cross-points are generally more cost effective than other structures that may have fewer crosspoints but more packages. Hence a more relevant design parameter for solid state switches would be the total number of integrated circuit packages. If alternate designs are implemented from a common set of integrated circuits, the number of packages may closely reflect the number of crosspoints.

Another useful cost parameter is the total number of IC pin-outs required in a particular implementation. Although this parameter is obviously related closely to the total number of packages, it is generally more useful since it more accurately reflects package cost and circuit board area requirements. Pin-out measurements also provide a direct indication of implementation reliability, since external interconnections are generally less reliable than internal con-nections of an IC.

Medium scale integrated (MSI) circuits typically provide the equivalent of one crosspoint (AND gate) for $1\frac{1}{2}$ external pins to access the crosspoint. Thus when MSI technology is used, the total number of crosspoints is a useful indication of the total number of pins. We therefore continue to use crosspoints as an implementation cost measurement with the understanding that medium scale integration would be used in all comparative designs. To do so we must be sure that all systems operate at about the same speed, since higher speeds require lower levels of integration.

In addition to the number of crosspoints in space division stages, a digital time division switch uses significant amounts of memory that must be included in an estimate of the overall cost.* The memory count includes the time stage memory arrays and the control stores for both the time stages and the space stages. In the following analyses we assume that 100 bits of memory corresponds

to $1\frac{1}{2}$ IC interconnections. (A 1024 bit random access memory typically requires 14 pins.) With this assumption, we can relate memory costs to crosspoint costs by a 100 bits per crosspoint factor. Hence the following analyses of implementation complexity for digital time division switching matrices include the total numer of crosspoints and the total number of bits of memory divided by 100.

The implementation complexity is expressed as follows:

$$\text{complexity} = N_X + \frac{N_B}{100} \tag{5.13}$$

where N_X = number of space stage crosspoints
N_B = number of bits of memory

Example 5.2. Determine the implementation complexity of the TS switch shown in Figure 5.19 where the number of TDM input lines $N = 80$. Assume each input line contains a single DS1 signal (24 channels). Furthermore, assume a one-stage matrix is used for the space stage.

Solution. The number of crosspoints in the space stage is determined as

$$N_X = 80^2 = 6400$$

(The crosspoints on the main diagonal are included since two channels within a single TDM input may have to be connected to each other.) The total number of memory bits for the space stage control store is determined as

$$\begin{aligned} N_{BX} &= (\text{number of links})(\text{number of control words}) \\ &\quad (\text{number of bits/control word}) \\ &= (80)(24)(7) \\ &= 13440 \end{aligned}$$

The number of memory bits in the time stage is determined as the sum of the time slot interchange and the control store bits:

$$\begin{aligned} N_{BT} &= (\text{number of links})(\text{number of channels}) \\ &\quad (\text{number of bits/channel}) + (\text{number of links}) \\ &\quad (\text{number of control words})(\text{number of bits/control word}) \\ &= (80)(24)(8) + (80)(24)(5) \\ &= 24{,}960 \end{aligned}$$

*It is worth noting that digital memories are inherently implemented with at least two crosspoints per bit. In this case the crosspoints are gates used to provide write and read access to the bits. These crosspoints, however, are much less expensive than message crosspoints that are accessed from external circuits.

Thus the implementation complexity is determined as

$$\text{complexity} = N_X + (N_{BX} + N_{BT})/100 = 6784 \text{ equivalent crosspoints}$$

The implementation complexity determined in Example 5.2 is obviously dominated by the number of crosspoints in the space stage. A significantly lower complexity (and generally lower cost) can be achieved if groups of input links are combined into higher-level multiplex signals before being switched. The cost of the front end multiplexers is relatively small if the individual DS1 signals have already been synchronized for switching. In this manner, the complexity of the space stage is decreased appreciably while the overall complexity of the time stage increases only slightly. (See the problem set at the end of this chapter.) The implementation costs are reduced proportionately—up to the point that higher speeds dictate the use of a more expensive technology.

A significant limitation of the TS structure of Figure 5.19 occurs when a connection has to be made to a specific channel of an outlet as opposed to any channel of an outlet. A connection to a specific channel is blocked whenever the desired time slot of the inlet TSM is already in use. For example, if time slot 17 of the first inlet is in use for some connection from the first inlet to some link other than link N, the connection from channel 3 of inlet 1 to channel 17 of inlet N cannot be made. Because of this limitation the TS structure is useful only if the outlets represent trunk groups, which implies any channel of an outlet is suitable.* Applications that require connections to specific channels require additional stages for adequate blocking probability performance.

Multiple Stage Time and Space Switching

As discussed in the preceding section, an effective means of reducing the cost of a time division switch is to multiplex as many channels together as practical and perform as much switching in the time stages as possible. Time stage switching is generally less expensive than space stage switching—primarily because digital memory is much cheaper than digital crosspoints (AND gates). To repeat, the crosspoints themselves are not so expensive, it is the cost of accessing and selecting them from external pins that makes their use relatively costly.

Naturally, there are practical limits as to how many channels can be multiplexed into a common TDM link for time stage switching. When these practical limits are reached, further reductions in the implementation complexity can be achieved only by using multiple stages. Obviously, some cost savings result when a single space stage matrix of a TS or ST switch can be replaced by multiple stages.

A generally more effective approach involves separating the space stages by a time stage, or, conversely, separating two time stages by a space stage. The next two sections describe these two basic structures. The first structure,

*A full duplex connection requires the reverse connection to be established, which adds restrictions to which outlet channels can be used. See the problem set at the end of the chapter.

consisting of a time stage between two space stages, is referred to as a space–time–space (STS) switch. The second structure is referred to as a time–space–time (TST) switch.

5.4.1 STS Switching

A functional block diagram of an STS switch is shown in Figure 5.20. Each of the space switches is assumed to be a single stage (nonblocking) switch. For large switches, it may be desirable to implement the space switches with multiple stages. Establishing a path through an STS switch requires finding a time switch array with an available write access during the incoming time slot and an available read access during the desired outgoing time slot. When each individual stage (S, T, S) is nonblocking, the operation is functionally equivalent to the operation of a three-stage space switch. Hence a probability graph in Figure 5.21 of an STS switch is identical to the probability graph of Figure 5.8 for three-stage space switches. Correspondingly, the blocking probability of an STS switch is

$$B = (1 - q'^2)^k \tag{5.14}$$

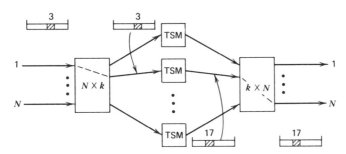

Figure 5.20. Space–time–space (STS) switching structure.

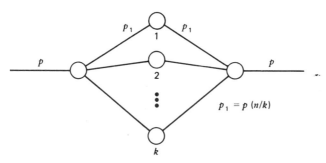

Figure 5.21. Probability graph of STS switch with nonblocking stages.

where $q' = 1 - p' = 1 - p/\beta$ $(\beta = k/N)$
 k = number of center stage time switch arrays

Assuming the space switches are single stage arrays and that each TDM link has c message channels, we may determine the implementation complexity of an STS switch as*

complexity = (number of space stage crosspoints)

$\quad\quad\quad$ + [(number of space stage control bits)

$\quad\quad\quad$ + (number of time stage memory bits)

$\quad\quad\quad$ + (number of time stage control bits)]/100

$$= 2kN + \frac{[2 \cdot k \cdot c \cdot \log_2 N + k \cdot c \cdot 8 + k \cdot c \cdot \log_2 c]}{100} \quad\quad\quad (5.15)$$

Example 5.3. Determine the implementation complexity of a 2048 channel STS switch implemented for 16 TDM links with 128 channels on each link. The desired maximum blocking probability is 0.002 for channel occupancies of 0.1.

Solution. The minimum number of center stage time switches to provide the desired grade of service can be determined from Equation 5.14 as $k = 7$. Using this value of k, the number of crosspoints is determined as $(2)(7)(16) = 224$. The number of bits of memory can be determined as $(2)(7)(128)(4) + (7)(128)(8) + (7)(128)(7) = 20{,}608$. Hence the composite implementation complexity is 430 equivalent crosspoints.

The value of implementation complexity obtained in Example 5.3 should be compared to the number of crosspoints obtained for an equivalent sized three-stage switch listed in Table 5.2. The space switch design requires 41,000 crosspoints while the STS design requires only 430 equivalent crosspoints. The dramatic savings comes about as a result of the voice signals having already been digitized and multiplexed (presumably for transmission purposes). When the STS switch is inserted into an analog environment, the dominant cost of the switch occurs in the line interface. Modern digital switches are not unlike modern digital computers. The cost of a central processing unit has become relatively small compared to the costs of peripheral devices.

5.4.2 TST Switching

A second form of multiple-stage time and space switching is shown in Figure 5.22. This switch is usually referred to as a time–space–time (TST) switch. Information arriving in a TDM channel of an incoming link is delayed in the

*This derivation assumes output-associated control is used in the first stage and input-associated control is used in the third stage. A slightly different result occurs if the space stages are controlled in different manners.

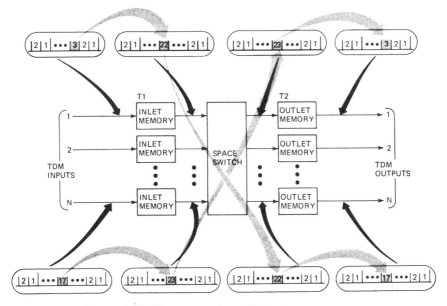

Figure 5.22. Time–space–time (TST) switching structure.

inlet time stage until an appropriate path through the space stage is available. At that time the information is transferred through the space stage to the appropriate outlet time stage where it is held until the desired outgoing time slot occurs. Assuming the time stages provide full availability (i.e., all incoming channels can be connected to all outgoing channels), any space stage time slot can be used to establish a connection. In a functional sense the space stage is replicated once for every internal time slot. This concept is reinforced by the TST probability graph of Figure 5.23.

An important feature to notice about a TST switch is that the space stage operates in a time divided fashion, independently of the external TDM links. In

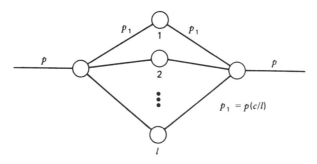

Figure 5.23. Probability graph of TST switch with nonblocking stages.

fact, the number of space stage time slots l does not have to coincide with the number of external TDM time slots c.

If the space stage is nonblocking, blocking in a TST switch occurs only if there is no internal space stage time slot during which the link from the inlet time stage and the link to the outlet time stage are both idle. Obviously, the blocking probability is minimized if the number of space stage time slots l is made to be large. In fact, as a direct analogy of three-stage space switches, the TST switch is strictly nonblocking if $l = 2c - 1$. The general expression of blocking probability for a TST switch with nonblocking individual stages (T, S, T) is

$$B = [1 - q_1^2]^l \qquad (5.16)$$

where $q_1 = 1 - p_1 = 1 - p/\alpha$
$\quad \alpha =$ time expansion (l/c)
$\quad l =$ number of space stage time slots

The implementation complexity of a TST switch can be derived as follows:

$$\text{complexity} = N^2 + \frac{[N \cdot l \cdot \log_2 N + 2 \cdot N \cdot c \cdot 8 + 2 \cdot N \cdot l \log_2 c]}{100} \qquad (5.17)$$

Example 5.4. Determine the implementation complexity of a 2048 channel TST switch with 16 TDM links and 128 channels per link. Assume the desired maximum blocking probability is 0.002 for incoming channel occupancies of 0.1.

Solution. Using Equation 5.16, we can determine the number of internal time slots required for the desired grade of service as 25. Hence time concentration of $1/\alpha = 5.12$ is possible because of the light loading on the input channels. The implementation complexity can now be determined from Equation 5.17 as 656.

The results obtained in Examples 5.3 and 5.4 indicate that the TST architecture is more complex than the STS architecture. Notice, however, that the TST switch operates with time concentration whereas the STS switch operates with space concentration. As the utilization of the input links increase, less and less concentration is acceptable. If the input channel loading is high enough, time expansion in the TST switch and space expansion in the STS switch are required to maintain low blocking probabilities. Since time expansion can be achieved at less cost than space expansion, a TST switch becomes more cost effective than an STS switch for high channel utilizations. The implementation complexities of these two systems are compared in Figure 5.24 as a function of the input utilization.

As can be seen in Figure 5.24, TST switches have a distinct implementation advantage over STS switches when large amounts of traffic are present. For small switches, the implementation complexities favor STS architectures. The choice of a particular architecture may be more dependent on such other factors

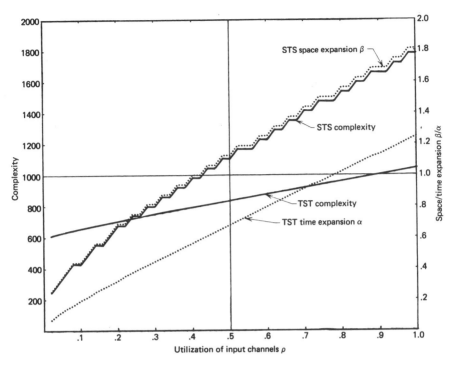

Figure 5.24. Complexity comparison of STS and TST switching structures for a blocking probability of 0.002.

as modularity, testability, expandability. One consideration that generally favors an STS structure is its relatively simpler control requirements [8]. For very large switches with heavy traffic loads, the implementation advantage of a TST switch is dominant. Evidence of this fact is provided by the No. 4 ESS, a TST structure that services in excess of 100,000 lines.

TSSST Switches

When the space stage of a TST switch is large enough to justify additional control complexity, multiple space stages can be used to reduce the total crosspoint count. Figure 5.25 depicts a TST architecture with a three-stage space switch. Because the three middle stages are all space stages, this structure is sometimes referred to as a TSSST switch. The EWSD switch of Siemens uses a TSSST structure [9].

The implementation complexity of a TSSST switch can be determined as follows:*

*The assumed control orientations by stages are output, output, output, input, input.

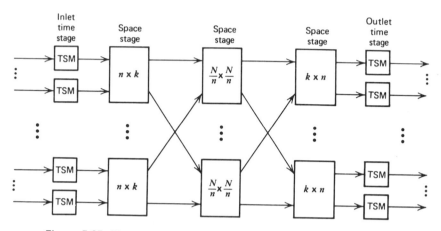

Figure 5.25. Time–space–space–space–time (TSSST) switching structure.

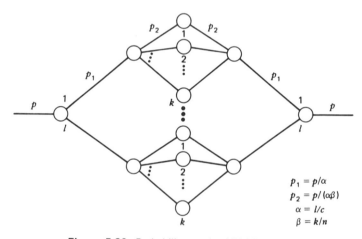

Figure 5.26. Probability graph of TSSST switch.

$$\text{complexity} = N_X + \frac{(N_{BX} + N_{BT} + N_{BTC})}{100} \qquad (5.18)$$

where N_X = number of crosspoints = $2 \cdot N \cdot k + k \cdot (N/n)^2$

N_{BX} = number of space stage control store bits = $2 \cdot k \cdot (N/n) \cdot l \cdot \log_2(n) + k \cdot (N/n) \cdot l \cdot \log_2(N/n)$

N_{BT} = number of bits in time stages = $2 \cdot N \cdot c \cdot 8$

N_{BTC} = number of time stage control store bits = $2 \cdot N \cdot l \cdot \log_2(c)$

The probability graph of a TSSST switch is shown in Figure 5.26. Notice that

this diagram is functionally identical to the probability graph of a five-stage space switch shown in Figure 5.10. Using the probability graph of Figure 5.26, we can determine the blocking probability of a TSSST switch as

$$B = \{1 - (q_1)^2[1 - (1 - q_2^2)^k]\}^l \tag{5.19}$$

where $q_1 = 1 - p_1 = 1 - p/\alpha$
$q_2 = 1 - p_2 = 1 - p/\alpha\beta$

Example 5.5. Determine the implementation complexity of a 131,072 channel TSSST switch designed to provide a maximum blocking probability of 0.002 under channel occupancies of 0.7. Assume the switch services 1024 TDM input links with 128 channels on each link. Also assume that unity time expansion is used on the space stages.

Solution. The space switch can be designed in a variety of ways depending on how many links are assigned to each array in the first (and third) space stages. A value of 32 is chosen as a convenient binary number near the theoretical optimum presented earlier $(N/2)^{1/2}$. With this value for n, the only unknown in Equation 5.19 is the number of center stage arrays k. Thus k is determined to be 24 for a blocking probability of 0.0015. The complexity is determined from Equation 5.18 as

$$N_X = (2)(1024)(24) + (24)(32)^2 = 73,728$$

$$N_{BX} = 2(24)(32)(128)(5) + (24)(32)(128)(5) = 1,474,560$$

$$N_{BT} = 2(1024)(128)(8) = 2,097,152$$

$$N_{BTC} = 2(1024)(128)(7) = 1,835,008$$

$$73,728 + 3,571,712/100 = 127,795 \text{ equivalent crosspoints}$$

5.4.3 No. 4 ESS Toll Switch

As shown in Figure 5.27 the basic structure of a No. 4 ESS matrix is time–space–time with four stages in the space switch (i.e., a TSSSST) [10, 11]. The inputs to the matrix are 120-channel TDM links formed by multiplexing a 5 DS1 signals together. Before these inputs are interfaced to the matrix they are passed through a decorrelator to distribute the channels of incoming TDM trunk groups across multiple TDM links into the matrix. Decorrelation is used because the No. 4 ESS is a toll switch in which the incoming TDM links represent trunk groups from other switching machines. In contrast to TDM links of an end office switch formed by multiplexing independent subscriber channels, the channels of a TDM trunk group are not independently busy. In fact, the act of concentrating multiple, independent source onto a trunk group

Figure 5.27. No. 4 ESS matrix architecture.

causes high levels of correlation between the activity on the individual channels. If the channels of a trunk group were not decorrelated they would experience much higher blocking probabilities because they would all be vying for the same paths through the matrix. Decorrelation shuffles the trunk groups so the alternate paths available for any particular connection are more likely to be statistically independent. Notice that, besides shuffling channels, the decorrelator provides space expansion (7 to 8) and time expansion (120 to 128).

The maximum size of the No. 4 ESS uses 128 decorrelators with seven 120-channel inputs each. Thus the maximum channel capacity is $(128)(7)(120) = 107{,}520$ channels. The space stage is a 1024 by 1024 matrix with four possible paths provided during each of the 128 space stage time slots. The probability graph of the 4 ESS switch is shown in Figure 5.28 from which the following blocking probability is derived:

$$B = [1 - (1 - p_1)(1 - p_2)(1 - p_1)]^{128} \tag{5.20}$$

where $p_1 = (7/8)(120/128)p$, p is the external channel utilization, and p_2 is the blocking probability of the four-stage space switch derived in Appendix A as

$$p_2 = 2p_1^2 q_1^6 + 16p_1^3 q_1^5 + 50p_1^4 q_1^4 + 52p_1^5 q_1^3 + 28p_1^6 q_1^2 + 8p_1^7 q_1 + p_1^8 \tag{5.21}$$

Example 5.6. Determine the blocking probability and number of paths that have to be searched in a No. 4 ESS with inlet channel occupancies of 0.7 and 0.9.

Solution. The occupancy of the interstage links (p_1) is determined to be 0.574 and 0.738, respectively. Using Equation 5.21 the corresponding values of p_2 are

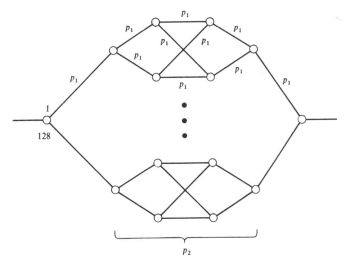

Figure 5.28. Probability graph of No. 4 ESS matrix.

determined as 0.737 and 0.934. Equation 5.20 can now be used to determine the blocking probabilities as 0.002 and 0.56 for inlet occupancies of 0.7 and 0.9, respectively.

The probability that all four space paths are busy in a particular time slot is $1 - q_1 q_2 q_1$ (the base term in Equation 5.20), which evaluates to 0.952 for $p = 0.7$ and 0.9955 for $p = 0.9$. Equation 5.11 now determines the average number of time slots (four paths per time slot) that must be tested to find an idle path as 21 and 98, respectively.

Example 5.6 demonstrates that low blocking probabilities can be achieved even when individual paths are highly utilized if there are enough alternatives—a feature that is most practical with time division switching. This example also demonstrates the sensitivity of the call processing time as a function of load in that path search times more than quadrupled when the node becomes overloaded with traffic. Reference [11] reports that the path search time is only 12% of the total processing load at $p = 0.7$ but this example implies that it would become more than 50% of the load at $p = 0.9$ if other aspects of call processing were linear functions of the traffic volume.

5.4.4 System 75 Digital PBX

As originally released, the System 75 PBX of AT&T is a mid-range PBX that can support 400 stations, 200 data instruments, and 200 trunks [12]. The switching matrix consists of two byte-wide TDM buses as shown in Figure 5.29. Because the buses operate at 2.048 MHz there are 256 time slots per bus per frame. Thus the total number of time slots on the buses is 512, which support 256 full-duplex connections.* All ports of the switch have access to all time slots on both buses. The dual bus architecture allows the use of slower electronics and provides redundancy in the case of failure.

As long as there is an idle time slot on either bus any source can be connected to any destination. Thus the blocking probability is zero from the standpoint of matching loss. Blocking can occur, however, as a result of congestion: if more connections are requested than can be supported by the number of time slots. Blocking analyses of this type require a different form of mathematics referred to as congestion theory. As presented in Chapter 10, congestion theory deals with the probability that the offered traffic load exceeds some value. Matching loss, on the other hand, is concerned with average or busy hour traffic volumes.

As an example, congestion analyses in Chapter 10 show that if 800 stations are 37.5% busy, on average, the probability that a station requests service when 400 stations are already busy is 10^{-5}. Thus the system is virtually nonblocking for voice applications. Because data connections are often full-period connections the blocking probability for voice connections may become more

*Because some of the time slots are used for overhead and continuous distribution of various tones the maximum traffic capacity is 7200 CCS [12], which relates to 200 connections.

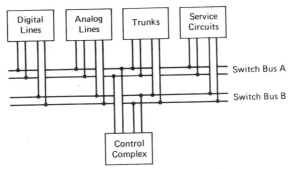

Figure 5.29. System 75 matrix architecture.

significant in applications involving intensive data switching. See the problem set in Chapter 10 for some examples.

5.5 DIGITAL CROSS-CONNECT SYSTEMS

A digital cross-connect system (DCS) is basically a digital switching matrix with an operations interface for setting up relatively static connections between input and output signals or channels. Instead of establishing connections in response to signaling information pertaining to call-by-call connection requests, the (cross)-connections are established in response to network configuration needs. The most basic function of a DCS is to act as an electronic patch panel in lieu of a manual cross-connect facility. Manual cross-connect frames are typically installed in switching offices as demarcation points between transmission facilities and switching machines and in wire centers as demarcation points between feeder cables and distribution facilities. In both cases, the major purpose of the cross-connect frame is to allow rearrangement of transmission circuits and access for testing the circuits in both directions.

Figure 5.30 shows a typical manual cross-connect environment while Figure 5.31 shows the same function implemented with a DCS. Manual cross connects are implemented with bridging clips and punch down wires. Cross connections within the DCS are established by entering commands at the management console (either locally or remotely). As indicated in the two figures, a major feature of a DCS is its elimination of back-to-back multiplexing functions when cross-connecting individual channels within TDM transmission links. Additional advantages of a DCS system with respect to manual cross-connect systems are as follows:

1. Automatic record keeping. Because the cross-connects are under processor control, circuit connection reports are readily available through the management interface. In contrast, records in manual systems are inherently error prone and often out of date.

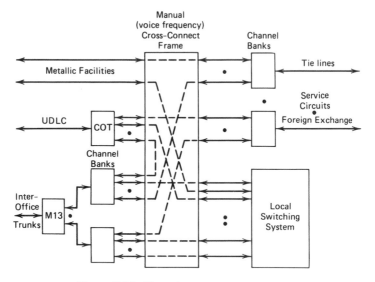

Figure 5.30. Manual cross-connect system.

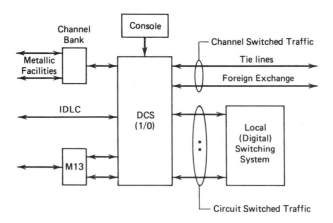

Figure 5.31. Electronic digital cross-connect system.

2. Remote and rapid provisioning. Provisioning is the basic process of providing (or discontinuing) service to a particular customer. The basic operations involved are outside plant cross-connections, inside plant cross-connections, configuration changes in switching system data base, and customer record updates in business (billing) systems. Obviously, the more these processes are automated the faster and more accurately they can be performed.

3. Automated test access. Testing analog circuits at a manual cross-connect frame involves physically breaking the connection (by removing bridging clips) and attaching the test equipment to the side of the circuit to be tested. All manual operations are eliminated with an electronic patch panel by entering commands at the management console to connect the desired test tones and (DSP) measurement channels to the circuit under test.

Figure 5.31 depicts two types of network traffic: circuit-switched traffic and channel-switched traffic. Circuit-switched traffic represents locally switched traffic (DS0 circuits typically) and channel switched-traffic refers to leased line equivalents of digital channels. Channel-switched traffic might terminate at another public network switching office as in a foreign exchange (FX) circuit or at a distant customer premise as a tie line of a private network. In the latter case, more than one DS0 channel might be concatenated together to form a single higher-rate channel referred to as a fractional T1 circuit or $M \times 64$ kbps channel. Channel-switched services typically account for over one-half of the transmission bandwidth between U.S. metropolitan offices [13]. The process of separating channel-switched services from circuit-switched services is often referred to as "grooming." Figure 5.30 also shows that universal digital loop carrier (UDLC) becomes integrated digital 100p carrier (IDLC) in a digital environment (Figure 5.31).

5.5.1 Consolidation and Segregation

Two basically different DCS grooming functions are depicted in Figures 5.32 and 5.33: consolidation and segregation. When multiple access lines carrying traffic destined to a common distant node are partially filled, the per channel costs of transport to the remote node can be reduced by consolidating the traffic. Conversely, when different types of traffic originate at a single location, it is desirable to allow a single facility (e.g., a T1 line) to carry all types of traffic and segregate it at the DCS. Examples of such traffic are circuit-switched channels, DDS channels, tie lines, multidrop data circuits, FX circuits, or other special services that might be available only at another office.

Figure 5.32. Consolidation. **Figure 5.33.** Distribution.

5.5.2 DCS Hierarchy

The cross-connect system depicted in Figure 5.31 terminates DS1 signals and interchanges DS0 signals, which leads to the designation DCS 1/0. Similarly, a digital cross-connect system that terminates DS3 signals and rearranges DS1 signals within the DS3s is referred to as a DCS 3/1. If a DCS, such as a DCS 3/0, provides rearrangement of lower level signals, such as DS0s, it does not necessarily mean that it also provides cross-connect services for intermediate level signals, such as DS1s or DS2s. Cross-connection of DS1s requires transparent mapping of the entire 1.544 Mbps, which includes the framing channel—a capability that may not be provided when 64-kbps DS0 channels are the primary focus. When a DCS provides multiple levels of cross-connect the intermediate levels are sometimes designated—as in DCS 3/1/0 for terminating DS3s and rearranging DS1s and DS0s. In most cases, it is not necessary to provide cross-connect services at all levels of the digital signal hierarchy for all of the terminations. Figure 5.34 depicts a DCS hierarchy that provides rearrangement of lower level signals on only subsets of higher level signals. The higher level (e.g., DS3) signals that are not cross-connected to a lower level DCS may be unchannelized high speed (44.736 Mbps) signals or transit DS3s that terminate on a DCS 3/1 or 3/0 at a distant location. The primary purpose of the DCS 3/3 is to provide network restoration or protection switching of the DS3 signals and to possibly provide time-of-day rearrangement of trunk group assignments.

5.5.3 Integrated Cross-Connect Equipment

Figure 5.31 depicts distinct multiplexing equipment and two distinct and colocated switching systems: the DCS and the local digital switch. Although DCS functions within the public network are traditionally implemented with separate equipment, private networks often utilize equipment that provides integrated multiplexing and cross-connect functions. This equipment has

Figure 5.34. DCS hierarchy.

Figure 5.35. Transposition of data in concatenated time slots.

evolved from CPE based T1 multiplexers because TSI circuits are insignificant hardware additions to T1 multiplexing hardware. Such equipment are variously referred to as intelligent multiplexers [14], nodal processors, or networking T1 multiplexers.

Integration of cross-connect functions with higher-level digital signals (such as a DS3) has not occurred because the manner in which the higher-level signals are multiplexed (described in Chapter 7) is not amenable for direct termination on a switching system. A newer form of (synchronous) multiplexing, as specified by the SONET standard described in Chapter 8, allows integration of higher-level digital multiplexers with cross-connect systems for both public and private equipment.

Although cross-connect functions are basically nothing more than "pegged" or "nailed-up" circuit switched connections, the two functions are traditionally implemented separately because of the following differences in the application requirements:

1. A DCS needs to be strictly nonblocking at the DS0 level, which is generally uneconomical in large public network switches.

2. Transparent cross-connection of a 1.544-Mbps signal requires transport of the framing bit that is not possible in typical digital circuit switches because DS1 interface equipment assumes the DS1 signals are channelized with framing in the 193rd bit.

3. DCS functionality does not involve processing of signaling bits so fully functional circuit switch interfaces have excess costs when used as interfaces to cross-connect systems.

4. Cross-connecting multiple, concatenated DS0 channels (fractional T1 channels) require maintaining order in the concatenated data stream. Although the order can always be maintained by careful mapping of the individual 64-kbps connections it is a function that is typically not provided in DS0 circuit switching software. Figure 5.35 shows how the order of bytes in the concatenated data stream can be transposed by ill-chosen 64-kbps connections.

5.6 DIGITAL SWITCHING IN AN ANALOG ENVIRONMENT

When digital end-office switches are installed in an analog environment the analog interfaces are necessarily unchanged. Although the digital switch may interface with digital subscriber carrier or digital fiber feeder systems, these systems merely extend the analog interface point closer to the subscriber. This section describes the basic considerations of using digital switching in such an analog environment. Chapter 9 describes digital end-office switching with digital subscriber loops in relation to the integrated services digital network.

5.6.1 Zero Loss Switching

As already mentioned, a well-designed digital transmission and switching system adds no appreciable degradation to the end-to-end quality of digitized voice. In particular, the derived analog signal coming out of a decoder can be adjusted to the same level as that presented to the encoder at the far end. Curiously, zero loss transmission presents some significant problems when digital switching is used in the analog transmission environment of a class 5 central office.

Analog end-office switches are two-wire switches designed to interconnect bidirectional two-wire customer loops. The voice digitization process, however, inherently requires separation of the go and return signal paths involved in a connection. Thus, a class 5 digital end office must be a four-wire switch. When inserted into a two-wire analog environment, hybrids are required to separate the two directions of transmission. As shown in Figure 5.36, hybrids at each end of the internal digital subnetwork produce a four-wire circuit with the potential for echos and singing. (Amplifiers are shown in conjunction with the encoders to offset forward path loss inherent in the hybrids.) Instability arises as a result of impedance mismatches at the hybrids causing unwanted coupling from the receive portion to the transmit portion of the four-wire connection. Impedance

Figure 5.36. Four-wire digital switch with two-wire analog interfaces.

mismatches occur because of the variability in the lengths and wire sizes of the subscriber loops. In particular, loaded and unloaded wire pairs have markedly different impedance characteristics.

The instability problems are compounded by certain amounts of artificial delay that are required in a digital time division switch. Although the delay through a digital switch (several hundred microseconds, typically) is basically unnoticeable to a user, it represents the equivalent of as much as 30 to 40 miles of wire. This increased delay has the effect of lowering the oscillation frequencies that might otherwise be outside the voiceband and effectively removed by encoder/decoder filters.

As mentioned in Chapter 1, hybrids are used extensively at the interface to the toll network where the transmission medium is basically four-wire. In these instances, the instability of the four-wire circuit is controlled by designing a prescribed amount of net attenuation (by way of net loss) into the transmission path of shorter toll network circuits. On the longer circuits, the echos and singing are eliminated by echo suppressors or echo cancellers. Class 5 digital switches could prevent instability in the same manner: by designing a certain amount of loss into the encoding/decoding process. While the necessary amount of signal attenuation (approximately 2 to 3 dB in each path [15]) could be tolerated on local connections, the added loss to some toll connections would be unacceptably large.

From the foregoing discussion it can be seen that the use of selectable attenuations is one solution to the instability problems. The necessary loss is inserted into the talking path for local connections but not used on long-distance calls, which already have loss designed into them. A second solution involves matching the impedances at the hybrids more closely. Although nearly complete elimination of the unwanted coupling is prohibitively expensive, adequate isolation of the transmission paths can be accomplished with just two different matching networks: one for loaded loops and one for unloaded loops [16]. Impedance matching has also been simplified by the use of advanced interface electronics that contain trainable and automatic impedance matching circuitry. The impedance matching circuits essentially represent short-delay echo cancellers. The use of conventional bulk echo cancellers has also been considered [17].

Notice that an all-digital network (with four-wire telephones) avoids the instability problems because there are no two-wire analog lines. The voice is digitized at the telephone and uses a separate path from the receive signal all the way to the destination telephone. Thus an end-to-end four-wire circuit completely eliminates echos and singing—allowing all connections to operate on a zero loss basis.

5.6.2 BORSCHT

In Chapter 1 the basic functional requirements of the subscriber loop interface are described. These requirements are repeated here with two additional

requirements for a digital switch interface: coding and hybrid. The complete list of interface requirements is unaffectionately known as BORSCHT [18]:

B: Battery feed
O: Overvoltage protection
R: Ringing
S: Supervision
C: Coding
H: Hybrid
T: Test

As mentioned in Chapter 1, the high-voltage, low-resistance, and current requirements of many of these functions are particularly burdensome to integrated circuit implementations. First-generation digital end-office switches reduced the termination costs by using analog switching (concentrators) to common codecs. The DMS-100 [19] of Northern Telecom and the No. 5 ESS of AT&T [20] use analog concentration at the periphery. Integrated circuit manufacturers have worked diligently to implement the BORSCHT functions in what is called a subscriber loop interface circuit (SLIC). Thus, per-line SLICs allow implementation of per-line BORSCHT functions. The GTD-5 EAX switch is such an example [21].

5.6.3 Conferencing

In an analog network conference calls are established by merely adding individual signals together using a conference bridge. If two people talk at once, their speech is superposed. Furthermore, an active talker can hear if another conferee begins talking. Naturally, the same technique can be used in a digital switch if the signals are first converted to analog, added, and then converted back to digital.

As described in Chapter 3, $\mu255$ and the A-law code recommended by CCITT were designed with the specific property of being easily digitally linearizable (EDL). With this property, the addition function can be performed digitally by first converting all codes to linear formats, adding them, and then converting back to compressed formats, To the user, the operation is identical to the customary analog summation. For a conference involving N conferees, N separate summations must be performed—one for each conferee and containing all signals but his own. For a description of the conferencing algorithm in the System 75 PBX of AT&T see reference [12]. For more general descriptions of conferencing implementations in digital switches see reference [22].

Another conferencing technique involves monitoring the activity of all conferees and switching the digital signal of the loudest talker to all others. Although this technique is functionally different from a customary analog conference bridge it is advantageous for large conferences because the idle

channel noise of the inactive talkers does not get added into the output of the conference bridge. High quality conference circuits also include echo cancellers so higher signal powers can be provided.

5.6.4 Transmultiplexing

A transmultiplexer (transmux) is designed to directly convert TDM signals in the digital hierarchy to FDM signals in the analog hierarchy and vice versa. A common example is shown in Figure 5.31 where five DS1 signals with 24 channels each are translated to two supergroups with 60 channels each. The primary need for transmultiplexing occurs at the interface between digital toll switches and long haul analog transmission systems such as TD-2, TH-3 and AR-6A analog radios.

Functionally, the transmultiplexer in Figure 5.37 is equivalent to a back-to-back interconnection of a digital demultiplexer (digital channel bank) and an analog FDM multiplexer. A transmux is much less expensive than back-to-back demultiplexing/multiplexing because it avoids the costly signaling interfaces provided with voice channel interfaces in conventional multiplex equipment.

Transmultiplexers have been developed using two basic techniques. The first technique merely implements back-to-back demultiplexers and multiplexers, but without the costly signaling interfaces. Examples of these systems are AT&T's LT-1 and GTE's 4691A/B FDM/PCM converter [23].

The second approach involves digitizing the FDM signals directly and using digital signal processing to perform traditional analog domain operations such as filtering, mixing, and amplification. Thus, digital equivalents of single sideband signals are processed to produce digitized baseband (PCM) signals. In the reverse direction (TDM to FDM) the digital TDM signals are digitally processed to produce digitized representations of single sideband multiplex groups. Then the FDM signal is produced by a conventional digital-to-analog converter. Examples of the latter type of transmux can be obtained in reference [24].

Figure 5.37. Transmultiplexer for converting five DS1 signals into two supergroups.

REFERENCES

1 M. R. Aaron, "Digital Communications—The Silent (R)evolution?" *IEEE Communications Magazine*, January 1979, pp 16–26.

2 C. Clos, "A Study of Non-Blocking Switching Networks," *Bell System Technical Journal*, March 1953, pp 406–424.

3 C. Y. Lee, "Analysis of Switching Networks," *Bell System Technical Journal*, November 1955, pp 1287–1315.

4 C. Jacobaeus, "A Study of Congestion in Link Systems," *Ericsson Techniques*, No. 48, Stockholm, 1950, pp 1–70.

5 A. A. Collins and R. D. Pedersen, *Telecommunications, A Time For Innovation*, Merle Collins Foundation Dallas, Texas, 1973.

6 T. E. Browne, D. J. Wadsworth, and R. K. York, "New Time Division Switching Units for No. 101 ESS," *Bell System Technical Journal*, February 1969, pp 443–476.

7 M. Schwartz, *Telecommunications Network Protocols, Modeling and Analysis*, Addison Wesley, Reading MA, 1987.

8 S. G. Pitroda, "Telephones Go Digital," *IEEE Spectrum*, October 1979, pp 51–60.

9 N. Skaperda, "EWSD: Where It Is," *Siemens Telecom Report*, Vol 12, No. 2–3, 1989, pp 56–59.

10 J. H. Huttenhoff, J. Janik, G. D. Johnson, W. R. Schleicher, M. F. Slana, and F. H. Tendick, "No. 4 ESS: Peripheral Systems," *Bell System Technical Journal*, September 1977, pp 1029–1042.

11 A. E. Ritchie and L. S. Tuomenoksa, "No. 4 ESS: System Objectives and Organization," *Bell System Technical Journal*, 1977, pp 1017–1027.

12 L. A. Baxter, P. R. Berkowitz, C. A. Buzzard, J. J. Horenkamp and F. E. Wyatt, "System 75: Communications and Control Architecture," *AT&T Technical Journal*, January 1985, pp 153–173.

13 R. K. Berman, R. W. Lawrence and, P. C. Whitehead, "Providing and Managing Channel Switched Services in an Intelligent ISDN Environment," *IEEE Globecom*, 1987, pp 4.6.1–5.

14 J. L. Melsa and H. R. Scull, "The Application of Intelligent T1 Multiplexers in Hybrid and ISDN Networks," *IEEE Globecom*, 1989 pp 15.3.1–6.

15 James L. Neigh, "Transmission Planning for an Evolving Local Switched Digital Network," *IEEE Transactions on Communications*, July 1979, pp 1019–1024.

16 R. Bunker, F. Scida, and R. McCabe, "Zero Loss Considerations in Digital Class 5 Office," *IEEE Transactions on Communications*, July 1979, pp 1013–1018.

17 D. L. Duttweiler, "A Twelve Channel Echo Cancellor," *IEEE Transactions on Communications*, May 1978, pp 647–653.

18 F. D. Reese, "Memo to Management—You Must Appraise How New Technology Fits Customers," *Telephone Engineering and Management*, October 1, 1975, pp 116–121.

19 J. B. Terry, D. R. Younge, and R. T. Matsunaga, "A Subscriber Line Interface for the DMS-100 Digital Switch," *IEEE National Telecommunications Conference*, 1979, pp 28.3.1–28.3.6.

20 D. L. Carney, J. I. Cochrane, L. J. Gitten, E. M. Prell, and R. Staehler, "The 5 ESS Switching System: Architectural Overview," *AT&T Technical Journal*, August 1985, pp 1339–1356.

21 S. E. Pucini and R. W. Wolff, "Architecture of GTD-5 EAX Digital Family," *IEEE International Conference on Communications*, 1980, pp 18.2.1–18.2.8.

22 R. J. D'Ortenzio, "Conferencing Fundamentals for Digital PABX Equipments," *IEEE International Conference on Communications*, 1979, pp 2.5-29–2.5-36.

23 D. W. Surling and G. McGibbon, "The 4691B FDM/PCM Converter," *International Conference on Communications*, 1980.

24 C. M. Puckette and T. G. Marshall, Eds., "Special Issue on Transmultiplexers," *IEEE Transactions on Communications*, July 1982.

PROBLEMS

5.1 How many four-wire voice circuit connections can be provided by a bidirectional PAM switching bus if the minimum achievable pulse width is 250 nsec?

5.2 The TS switch of Figure 5-19 uses DS1 signals on each TDM link. What is the implementation complexity if groups of 5 DS1 inputs are first multiplexed to form 16 input links with 120 channels on each link? What is the implementation complexity of the multiplexers (i.e., how many crosspoints and how many bits of memory are in the multiplexers)?

5.3 Determine the Lee graph and Jacobaeus blocking probabilities of the first switch in Table 5.5 ($k = 6$) if two inlets out of 16 become connected to 1 erlang subscribers. (Possibly these lines could be dial-up ports to a computer.) (Hint: two inlets are permanently busy but the remaining inlets continue to be 0.1 erlang ports.)

5.4 Repeat Problem 5.3 but instead of two inlets being busy assume that two of the output links of the first stage module have failed.

5.5 How many crosspoints are needed in a 1024 line, three-stage space switch if the input loading is 6 CCS per line and the maximum acceptable blocking probability (using a Lee graph analysis) is .005?

(a) $N = 16, N/n = 64$
(b) $N = 32, N/n = 32$
(c) $N = 64, N/n = 16$

5.6 What is the (Lee graph) blocking probability of the TS switch in Figure 5-19 for channel loadings of 0.2 erlangs?

(a) Assume each TDM input is a 24-channel interoffice trunk group.
(b) Assume the TDM inputs are derived from 24-channel channel banks with each analog interface connected to a dedicated 0.2 erlang line.

5.7 Design an STS switch for 128 primary TDM signals of the CCITT hierarchy (30 voice channels per input). Blocking should be less than 0.002 and the loading is 0.2 erlangs per channel. How many time slot interchange modules are needed? What is the complexity of the switch?

5.8 Repeat Problem 5.7 for a TST design.

5.9 Determine the number of crosspoints N_X and the total number of memory bits N_B required for a TST switch defined as follows:

Number of lines $= 32$
Single stage space switch
Number of channels per frame $= 30$
Time expansion $= 2$

5.10 What is the blocking probability of the switch in Problem 5.9 if the channel loading is 0.9 erlangs per channel?

5.11 How many bits of memory are needed in a time slot interchange circuit for a 60-channel signal with nine bits per time slot?

6

DIGITAL MODULATION AND RADIO SYSTEMS

Chapter 4 discusses various means of encoding digital information for transmission over wireline (or fiber) transmission links. In the context of this chapter the data encoding described in Chapter 4 is referred to as baseband coding. To impress the same information onto a carrier for radio transmission a baseband signal is commonly used to modulate the carrier. Modulation essentially shifts the dc-centered spectrum of a baseband signal to the carrier frequency.

In contrast to Chapter 4, which emphasizes binary transmission, this chapter emphasizes multilevel digital modulation to achieve high data rates within the rigidly defined bandwidth of a radio channel. As described later in this chapter, most multilevel modulation techniques can be implemented by directly modulating a carrier with one of the multilevel baseband signals described in Chapter 4. Although this chapter is primarily concerned with radio systems the same multilevel modulation techniques are also used to achieve high data rates in voiceband modem equipment.

In the interest of ensuring a minimum utilization efficiency of the radio frequency spectrum, the FCC in the United States established certain data rates that must be achieved by digital microwave radios [1]. Basically, these minimum rates (shown in Table 6.1) specify approximately the same number of voice circuits (at 64 kbps) as were available in the analog FM radios in use at the time (1974). Although individual voice channels in a FDM signal occupy 4 kHz of bandwidth, frequency modulation typically expands the signal bandwidth by a factor of about 4, depending on the amount of FM deviation applied. Thus the digital systems were required to compete with an equivalent voice channel bandwidth of about 16 kHz. If the digital radios were forced to compete with analog single sideband radios [2, 3] introduced in the early 1980s they could not have done so without using lower bit rate voice encoding.

Over and above the minimum bit rates established by the FCC, competition among digital microwave manufacturers and the economics of maximizing the number of voice circuits per radio have stimulated the development of advanced digital modulation techniques, achieving even greater transmission rates. The

279

TABLE 6.1 Information Densities Required by FCC for Common Carrier Microwave Channels Using 64 kbps per Voice Circuit

Frequency Band (gHz)	Channel BW (MHz)	Required Bit Rate (Mbps)	Information Density (bps/Hz)
2	3.5	6.144	1.8
4	20	73.7	3.7
6	30	73.7	2.5
11	40	73.7	1.8

first part of this chapter describes these basic modulation techniques and the transmission efficiencies they provide. The latter sections describe basic system design considerations for point-to-point microwave relay systems.

Information Density

A useful parameter for characterizing the bandwidth efficiency of a digital modulation system is the information density, defined as

$$\delta = \frac{R}{BW} \tag{6.1}$$

where R = data rate in bits per second
BW = bandwidth of the digital signal in hertz

The units of information density are sometimes referred to loosely as bits per hertz. However, as defined in Equation 6.1, the units should be bits per second per hertz or simply bits per cycle. Since bits per second per hertz conveys the nature of information density more completely, it is the preferred unit.

The bandwidth factor in Equation 6.1 can be defined in a variety of ways. In theoretical studies the bandwidth of a signal is usually determined as the width of the ideal filter used to strictly bandlimit the signal (i.e., the Nyquist bandwidth). In practical systems, where the spectrum can never be strictly bandlimited, the bandwidth is more difficult to define. In radio systems a channel is usually specified with a certain minimum signal attenuation at the band edges. In this case, the information density is easily determined as the bit rate of a signal meeting these requirements divided by the alloted bandwidth.

When no particular emission specifications are provided, however, the bandwidth required by a digital signal must be defined in another more or less arbitrary manner. Using a 3-dB bandwidth is usually inappropriate since the signal spectrum may drop rather slowly and thereby allow significant amounts of energy to spill over into adjacent channels. Using a greater attenuation level to define the band edges would be more appropriate, but is still somewhat arbitrary. A generally more relevant criterion for practical systems defines the bandwidth of a signal as the channel spacing required to achieve a specified

maximum level of interference into identical adjacent channels. Yet another definition of bandwidth, commonly used in Europe, is the 99% power bandwidth.

The theoretical maximum information density for binary signaling is 2 bps/Hz for a two-level line code or 1 bps/Hz for a modulated double-sideband signal. If a four-level line code is used to achieve 2 bits per signal interval, the theoretical information density is 4 bps/Hz for the line code, or 2 bps/Hz for a double-sideband carrier signal. The information density of amplitude modulated signals can be doubled by using single-sideband transmission to effectively achieve the same efficiency as the line codes (baseband signals).

As a practical example, a representative high-speed data rate on a dial-up analog telephone line is 9600 bps (usually with 4 bits per signal interval and a 2400-Hz signaling rate). Since the usable bandwidth of a telephone channel is approximately 3 kHz, dial-up lines provide an information density of 3.2 bps/Hz. Correspondingly, a typical maximum data rate on leased analog lines is 19,200 bps, implying an information density of 6.4 bps/Hz.

As a starting point for digital microwave radios, Table 6.1 includes the minimum bit rates and corresponding information densities required in common carrier bands of the United States. Competition and compatibility with the digital hierarchy have led to realizations of even greater transmission efficiencies.

6.1 DIGITAL MODULATION

Digital transmission on a radio link differs from wire line transmission in two important regards. First, the digital information must modulate a carrier in some manner to produce the radio frequency (RF) signal. In many cases the modulation process can be viewed as a special form of amplitude modulation by a nonreturn to zero (NRZ) line code signal. Thus the line code represents a baseband signal that amplitude modulates the carrier in the transmitter and is reproduced by demodulation in the receiver. Representing the modulation process in this manner has the advantage that the RF spectra can be determined by merely translating the line code (baseband) spectra to the selected carrier frequency.

Second, a radio link differs from wireline transmission due to the necessity of strictly bandlimiting the transmitted signals to prevent interference into other channels. Although wireline transmission links automatically filter the line signal to some extent, explicit filter requirements sometimes occur only in the receivers to reject as much noise as possible.* Since radio links are *bandlimited* in the transmitter and *noisefiltered* in the receiver, the end-to-end filter function must be partitioned between the two ends. Figure 6.1 shows a block diagram of

*Adherence to EMI compatibility specifications defined by the FCC (Part 15) often requires some amount of transmit filtering in on premises wireline transmission systems.

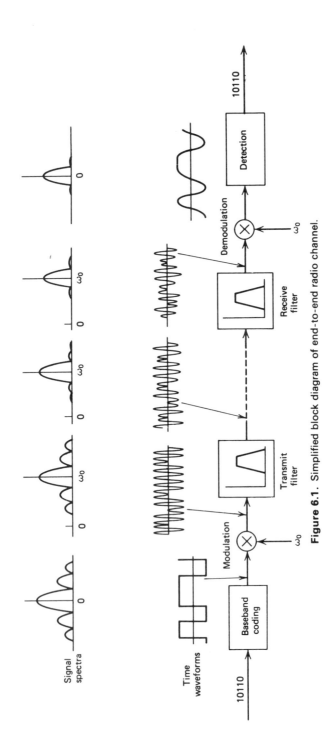

Figure 6.1. Simplified block diagram of end-to-end radio channel.

282

a radio link showing representative baseband and RF waveforms along with corresponding frequency spectrums. For modulation, Figure 6.1 shows multiplication of the carrier by the baseband waveform.

In the actual design of a digital radio, the modulation process must be designed in conjunction with the filter functions. For ease of description, however, modulation techniques are considered first. Later on, the filtering requirements of each type of modulation are discussed. Figure 6.1 also shows the baseband encoding process to be independent of the modulation process. For most of the modulation techniques described in this chapter this view is appropriate. In the case of trellis coded modulation (TCM), described later, encoding is an inherent aspect of the modulation process.

6.1.1 Amplitude Modulation

Historically, the simplest form of modulation to generate and detect is amplitude modulation (AM). A conceptual illustration of amplitude modulation is shown in Figure 6.2. The mathematical definition is

$$x(t) = [1 + a \cdot m_n(t)] \cos \omega_c t \tag{6.2}$$

where a = modulation index $(0 < a \leqslant 1)$
$m_n(t)$ = an n-level, symmetric NRZ baseband signal normalized to a maximum amplitude of 1
$\omega_c = 2\pi f_c$, the radian carrier frequency

Amplitude modulation is an example of a class of special modulation techniques referred to as "linear modulation." Linear modulation implies that the spectrum of the modulated signal is obtained by translating the baseband spectrum to the selected carrier frequency band. As shown in Figure 6.2, amplitude modulation by a two-level digital baseband signal essentially translates the $\sin(x)/x$ baseband spectrum up to the carrier frequency f_c. Other linear modulation techniques are double-sideband modulation, single-sideband modulation, and vestigial-sideband modulation.

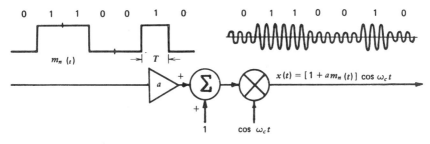

Figure 6.2. Digital binary amplitude modulation.

Figure 6.3. On/off keying.

Inspection of Equation 6.2 or Figure 6.2 indicates that if 100% modulation is used ($a = 1$), no carrier is produced for a logic 0. For obvious reasons, this form of amplitude modulation is often referred to as on/off keying, or "amplitude shift keying" (ASK). As shown in Figure 6.3, on/off keying can be obtained by direct multiplication of a carrier with a two-level (unipolar) line code described in Chapter 4.

Amplitude modulated signals are usually demodulated with a simple envelope detector. The cost effectiveness of this detector is the basic reason that commercial analog broadcasting uses amplitude modulation. Unfortunately, the error performance of digital amplitude modulation in general, and envelope detection in particular, is inferior to other forms of digital modulation and detection. For this reason, amplitude modulation is used only where the cost of the receiver is a significant consideration. Digital microwave links use other forms of modulation and demodulation to minimize the error rate for a given signal-to-noise ratio.

Conventional amplitude modulation provides suboptimum error performance for two basic reasons. First, if $a < 1$, a discrete (informationless) spectral line occurs at the carrier frequency. Although the existence of this spectral line simplifies carrier recovery, it increases the transmitter power without aiding discrimination between information signals.

With 100% modulation (on/off keying) no line spectra are produced, but the system is still inefficient in its use of transmitted power. As discussed in Chapter 4 for two-level line coding, the maximum use of transmitted power is achieved when one signal is the negative of the other. Thus, a second deficiency of amplitude modulation arises because a 0 signal is not the exact negative of a 1 signal. To achieve optimum performance, a symmetric two-level baseband signal should directly modulate (multiply) the carrier. As shown in Figure 6.4, this form of modulation produces two identical signals except for a 180° phase reversal. Hence this modulation is sometimes referred to as phase reversal keying (PRK), or more often, two-level phase shift keying (2-PSK). Notice that PRK cannot be detected by envelope detection. Instead, a PRK signal must be detected by comparing it to a coherent carrier reference.

Coherent detection involves comparing the incoming signal to a local carrier

Figure 6.4. Phase reversal keying.

synchronized in phase to that used at the transmitter. With conventional amplitude modulation there is no information in the phase. With PRK signaling, however, all of the information is in the phase. The use of a coherent reference for PRK signals allows optimum (antipodal) error rate performance.

The basic equation defining a PRK signal is

$$x(t) = m_2(t) \cdot \cos \omega_c t \qquad (6.3)$$

where $m_2(t)$ is the binary digital baseband signal [$m_2(t) = +1$ for a one and $m_2(t) = -1$ for a zero]. The demodulation equation is

$$
\begin{aligned}
y(t) &= x(t) \cdot [2 \cos \omega_c t] \\
&= [m_2(t) \cdot \cos \omega_c t] \cdot [2 \cos \omega_c t] \qquad (6.4) \\
&= m_2(t) + m_2(t) \cdot \cos 2\omega_c t
\end{aligned}
$$

where the double frequency term is removed with a low-pass filter.

As indicated in Figure 6.4 and Equation 6.4, the coherent demodulation process produces a symmetric two-level signal that can be processed further to detect the data. Since the demodulated baseband signal is, in essence, equivalent to a line code, it must include timing considerations as discussed in Chapter 4. In particular, there must be enough signal transitions in the baseband signal to allow recovery of a sample clock. However, the baseband signal need not preclude dc wander since dc levels are translated to the carrier frequency and passed adequately by the double-sideband system.*

*A single-sideband system does not pass dc energy. Thus if single-sideband modulation is used, the baseband signal must exclude dc energy from its spectrum. Some double-sideband systems might also require the elimination of baseband dc energy so that a carrier tone can be inserted into the center of the passband without affecting the signal.

6.1.2 Frequency Shift Keying

In addition to an inefficient use of signal power, conventional amplitude modulated signals (not PRK) have one other undesirable characteristic. By definition, an amplitude modulated signal uses multiple signal levels, which implies AM is quite vulnerable to signal saturations that narrow the distance between amplitude levels and produces spectrum spreading. A common source of saturation in a radio system occurs in the output power amplifier of the transmitter. In most cases output amplifiers are operated at less than maximum power to eliminate saturation and other nonlinearities—so they can accomodate amplitude modulated signals [4].

Angle modulated systems: frequency modulation (FM) or phase modulation (PM) use constant amplitude signals not adversely affected by signal saturations. Hence FM and PM can be transmitted at higher power levels than AM systems. The ability to use saturating power amplifiers is one of the reasons why FM was originally chosen for analog microwave radios. This section discusses digital frequency modulation, commonly referred to as frequency shift keying (FSK). The next section discusses digital phase modulation, commonly referred to as phase shift keying (PSK). Both systems provide a constant amplitude signal. Systems using constant amplitude carriers are also referred to as constant envelope systems.

The general expression for an *n*-ary FSK signal is

$$x(t) = \cos\left[\left(\omega_c + \frac{m_n(t)\cdot\Delta\omega}{2}\right)t\right] \tag{6.5}$$

where ω_c is the radian center frequency, m_n is a symmetric *n*-level NRZ digital baseband signal, and $\Delta\omega$ is the radian difference frequency between signals.

A typical binary FSK signal is shown in Figure 6.5 along with a simple (but low performance) means of implementing the modulator and demodulator. The modulator is a voltage-controlled oscillator (VCO) that is biased to produce the

Figure 6.5. Binary frequency shift keying.

center frequency when no modulation is applied. The amplitude a of the symmetric two-level baseband signal produces a frequency deviation of $+\Delta\omega/2$ for a 1 and $-\Delta\omega/2$ for a 0.

The demodulator is implemented as a phase-locked-loop: a VCO, a phase detector, and a loop filter. The phase detector measures the difference in phase between the FSK signal and the VCO output. A positive voltage is produced when the receive signal leads the VCO, and a negative voltage is produced otherwise. After being filtered to minimize the effects of noise, the phase detector output drives the VCO in such a way as to reduce the phase difference. Ideally, the input control voltage of the demodulator VCO will be identical to the input of the modulator VCO. The loop filter, however, necessarily slows the demodulator response to minimize the effects of noise.

Frequency shift keying generally provides poorer error performance than phase shift keying, particularly for multilevel signaling in a confined bandwidth. For this reason, FSK has not been used in digital microwave applications where high performance and high information densities are desired. The most prevalent usage of FSK is in low rate, low cost, asynchronous models for data transmission over the analog telephone network.

Minimum Shift Keying

One particular form of frequency shift keying that has received considerable attention is minimum shift keying (MSK) [5]. Basically, MSK is binary FSK with the two signaling frequencies selected so that exactly 180° difference in phase shift exists between the two frequencies in one signal interval. In this manner MSK produces a maximum phase differential at the end of an interval using a *minimum* difference in signaling frequencies. Furthermore, an MSK signal maintains continuous phase at signaling transitions. For this reason MSK belongs to a class of FSK signals referred to as continuous phase frequency shift keying (CPFSK). Figure 6.6 depicts a representative MSK waveform. Notice that there is exactly one-half cycle difference between a 1 signal and a 0 signal.

A mathematical expression for MSK signaling can be derived from Equation 6.5 where $m_n(t) = m_2(t)$ is a symmetric binary NRZ signal, and $\Delta\omega = \pi/T$. The result for any particular signal interval is

$$x(t) = \cos\left(\omega_c t + \frac{\pi t}{2T} + \phi_0\right) \qquad \text{(logic 1)}$$

$$= \cos\left(\omega_c t - \frac{\pi t}{2T} + \phi_0\right) \qquad \text{(logic 0)} \qquad (6.6)$$

where $1/T$ is the signaling rate, and ϕ_0 is the phase at the beginning of the signal interval ($\pm\pi$ in Figure 6.6).

The main attraction of minimum shift keying is its comparatively compact spectrum. Furthermore, with an appropriate means of detection, MSK can

$f_0 = 1.5/T$ $f_1 = 2/T$ $1/T$ = Signaling rate

Figure 6.6. Minimum shift keying.

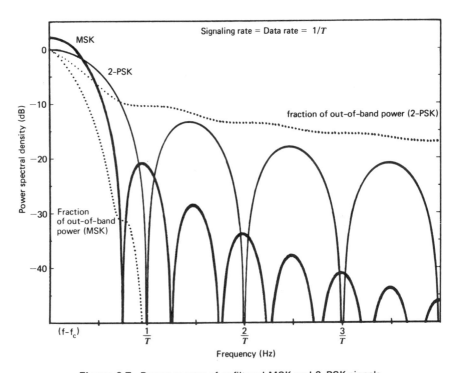

Figure 6.7. Power spectra of unfiltered MSK and 2-PSK signals.

provide optimum error performance in terms of the energy-per-bit to noise-density ratio (E_b/N_0). The expression for the power spectral density of the MSK signal defined in Equation 6.6 is

$$S(\omega) = 32\pi^2 T \left(\frac{\cos z}{\pi^2 - 4z^2} \right)^2 \tag{6.7}$$

where $z = |\omega - \omega_c| T$. The frequency spectrum of an MSK signal is plotted in Figure 6.7 where it is compared to the frequency spectrum of PRK (2-PSK)

signaling with the same data rate. Notice that the MSK spectrum is more compact and has its first spectral null at $3/4T$ instead of $1/T$ for PRK.*

Minimum shift keying is actually just one example of a general class of continuous phase modulation (CPM) techniques that maintain a constant amplitude, narrow power spectrum, and good error performance. For a good overview and bibliography of this class of modulation schemes see reference [6]. CPM schemes have not been used in point-to-point microwave applications because they do not provide high information densities. Many versions of CPM have been used in satellite applications where nonlinear transponder amplifiers preclude the use of modulation techniques with multiple amplitude levels [7].

6.1.3 Phase Shift Keying

The second category of angle modulated, constant envelope signals is referred to as phase shift keying (PSK). Actually, we already discussed one form of PSK when phase reversal keying (PRK) was introduced. PRK is more commonly referred to as 2-PSK, indicating that each signal interval uses one of two phases that are 180° apart to encode binary data. Multiple-phase phase shift keying is also possible. Four-PSK (also called QPSK) and 8-PSK are the most common examples of multiple phase PSK.

Phase shift keying is the most popular modulation technique for intermediate information density high-performance applications. The popularity is primarily due to its constant envelope, insensitivity to level variations, and good error performance. Both 2-PSK and 4-PSK theoretically provide optimum error performance in terms of a signal-to-noise ratio (E_b/N_0).

A general expression for n-ary phase shift keying is provided in Equation 6.8. This expression assumes 100% modulation is employed. That is, the phase shift from one interval to the next can range anywhere from $-180°$ to $+180°$. It is possible to devise PSK systems with lower modulation indices that allow only transitions to neighboring phases.

$$x(t) = \cos\left(\omega_c t + \frac{m_n(t) \cdot \Delta\phi}{2}\right) \qquad (6.8)$$

where $\Delta\phi = 2\pi/n$ is the separation between adjacent signal phases, and $m_n(t)$ is a symmetric n-level NRZ baseband signal with levels ± 1, ± 3,

Examples of typical 2-PSK and 4-PSK waveforms are shown in Figure 6.8. The signaling rate for the 4-PSK system is shown to be exactly one-half the 2-PSK signaling rate so that equal data rates are provided. The same figure also shows corresponding phasor diagrams of the signaling phases of a cosine wave as defined in Equation 6.8. Other phase orientations are possible. The particular

*MSK is actually more closely related to 4-PSK and therefore is compared to it in a later section of this chapter.

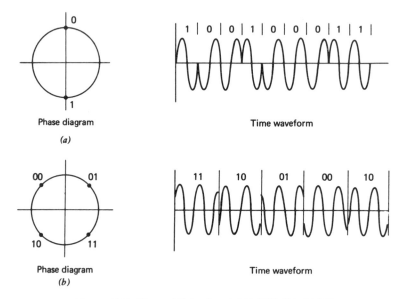

Figure 6.8. Phase shift keying. (a) 2-PSK. (b) 4-PSK.

phases shown, however, are convenient for later discussions relating 4-PSK systems to other types of digital modulation.

Quadrature Signal Representations

Despite the somewhat exotic sounding name, quadrature signal representations are a very convenient and powerful means of describing PSK signals, and many other digitally modulated signals. Quadrature signal representations involve expressing an arbitrary phase sinusoidal waveform as a linear combination of a cosine wave and a sine wave with zero starting phases. The derivation of this representation is provided by the trigonometric identity:

$$\cos(\omega_c t + \phi) = \cos \phi \cdot \cos \omega_c t - \sin \phi \cdot \sin \omega_c t \qquad (6.9)$$

Notice that $\cos \phi$ and $\sin \phi$ are constants over a signaling interval, and hence represent coefficients for expressing $\cos(\omega_c t + \phi)$ as a linear combination of the signals $\cos \omega_c t$ and $\sin \omega_c t$. Since $\cos \omega_c t$ and $\sin \omega_c t$ are 90° out of phase with respect to each other, they are orthogonal in a phasor diagram and hence are said to be "in quadrature."

In essence, $\cos \omega_c t$ and $\sin \omega_c t$ represent basis vectors in a two-dimensional phasor diagram. The cosine signal is usually referred to as the in-phase or I signal, and the sine signal is referred to as the out-of-phase or Q signal. Table 6.2 provides an example of quadrature signal representations for the 4-PSK signals presented in Figure 6.8. Table 6.3 provides a corresponding specification for an

TABLE 6.2 Quadrature Signal Coefficients for 4-PSK Modulation

Data Values	Quadrature Coefficients		Composite Signal
	$\cos \omega_c t$	$\sin \omega_c t$	
01	0.707	−0.707	$\cos(\omega_c t + \pi/4)$
00	−0.707	−0.707	$\cos(\omega_c t + 3\pi/4)$
10	−0.707	0.707	$\cos(\omega_c t − 3\pi/4)$
11	0.707	0.707	$\cos(\omega_c t − \pi/4)$

TABLE 6.3 Quadrature Signal Coefficients for 8-PSK Modulation

Data Values	Quadrature Coefficients		Composite Signal
	$\cos \omega_c t$	$\sin \omega_c t$	
011	0.924	−0.383	$\cos(\omega_c t + \pi/8)$
010	0.383	−0.924	$\cos(\omega_c t + 3\pi/8)$
000	−0.383	−0.924	$\cos(\omega_c t + 5\pi/8)$
001	−0.924	−0.383	$\cos(\omega_c t + 7\pi/8)$
101	−0.924	0.383	$\cos(\omega_c t − 7\pi/8)$
100	−0.383	0.924	$\cos(\omega_c t − 5\pi/8)$
110	0.383	0.924	$\cos(\omega_c t − 3\pi/8)$
111	0.924	0.383	$\cos(\omega_c t − \pi/8)$

8-PSK system using signal phases as provided in the phasor diagram of Figure 6.9. The phasor diagram assumes counterclockwise rotation, and hence the sine function lags the cosine function by 90°.

Most of the rest of this chapter relies heavily on quadrature signal representations to describe various modulation concepts, analyses, and implementations.

Modulator Implementations

A variety of techniques are possible for implementing PSK modulators. As mentioned when discussing PRK modulation, a 2-PSK modulator can be implemented by merely inverting the carrier (multiplying by −1) for a logic 0 and by not inverting for a logic 1. Some of the basic techniques used for generating multiple phase PSK signals are the following:

1. Synthesis of the desired waveforms using digital signal processing at suitably low carrier frequencies (as in voiceband modems).

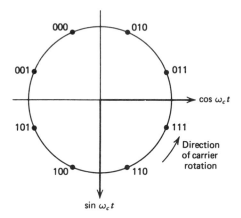

Figure 6.9. Phasor diagram of 8-PSK signal.

2. Generating multiple phases of a single carrier and selecting between the phases depending on the data values.
3. Using controlled delays selected through a switching arrangement to provide the desired phase shifts. Delays are often used to generate the separate signals in method 2.
4. Generating the PSK signals as a linear combination of quadrature signals.

None of the foregoing techniques represents a direct implementation of multiphase PSK modulation as defined in Equation 6.8. To implement Equation 6.8 directly, a device is needed that produces carrier phase shifts in direct proportion to the levels in the baseband signal $m_n(t)$. Except for the special case

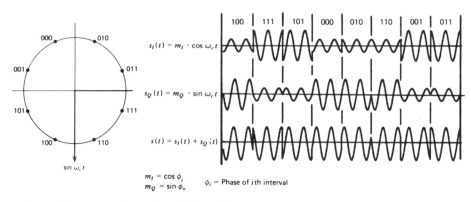

Figure 6.10. Generation of 8-PSK signals by superposition of quadrature amplitude modulated signals.

of 2-PSK where a multiplier can be used to produce $\pm 180°$ phase shifts, such devices do not exist.

Direct modulation by baseband signals can produce PSK signals if a quadrature signal implementation is used as indicated in method 4. Actually, two multilevel baseband signals need to be established: one for the inphase (I) signal and one for the out-of-phase (Q) signal. These baseband signals are referred to as $m_I(t)$ and $m_Q(t)$ for the I and Q signals, respectively. The levels chosen for the two baseband signals correspond to the coefficients needed to represent a PSK signal as a linear combination of the I and Q signals. For example, Figure 6.10 shows how an 8-PSK signal, defined in Table 6.3, can be generated by adding two amplitude modulated quadrature signals. A block diagram of the corresponding implementation is provided in Figure 6.11. This particular form of modulator is chosen, not so much as a recommendation for actual implementation, but because it demonstrates important modulation concepts and is useful in analyzing the spectrum requirements of PSK signaling.

Demodulator Implementation

Owing to the constant envelope,* all PSK systems must be detected with the aid of a local reference. For 2-PSK systems, the ideal reference is coherent with one of the two possible phases. When this reference is multiplied (mixed) with an in-phase signal, a maximum positive output is obtained. When multiplied by the opposite phase, a maximum negative output is obtained. In this manner, 2-PSK

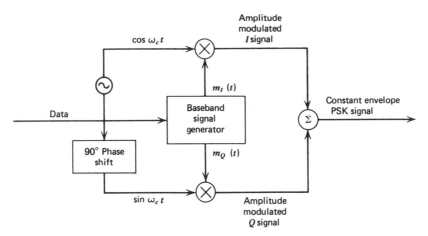

Figure 6.11. Generalized PSK modulator structure.

*Discussions to this point have not considered the effects of filtering on constant envelope signals. A heavily filtered PSK signal does not have a constant envelope. However, as long as filtering occurs after channel nonlinearities (e.g., power amplifiers), the most harmful effect of a nonlinearity, spectrum spreading, is avoided.

provides antipodal performance when a local coherent reference is established. The demodulation process for 2-PSK is presented in Figure 6.4 and Equation 6.4, where the mixing and filtering process effectively recovers the baseband signal $m_2(t)$.

The combination of a mixer and a low-pass filter shown in Figure 6.4 is generally referred to as phase detector. The phase detection property is represented mathematically as

$$y_I(t) = \text{low pass } \{\cos(\omega_c t + \phi) \cdot 2 \cdot \cos \omega_c t\}$$

$$= \text{low pass } \{\cos \phi + \cos \phi \cdot \cos 2\omega_c t - \sin \phi \cdot \sin 2\omega_c t\}$$

$$= \cos \phi \tag{6.10}$$

where low pass $\{\cdot\}$ is a low-pass filter function designed to remove twice carrier terms.

When detecting 2-PSK modulation, a single phase detector indicates whether the received phase is closer to $0°$ or to $180°$. The desired information is directly available as the polarity of the phase detector output $\cos \phi$. In multiphase systems, however, the information provided by a single phase detector is inadequate for two reasons. First, a measure of $\cos \phi$ provides no information as to whether ϕ is positive or negative. Second, the output of the phase detector is proportional to the signal amplitude as well as to $\cos \phi$. Hence the magnitude of the phase detector output is meaningless unless referenced to the signal amplitude.

Both of the aforementioned problems are overcome if a second mixer and filter measures the phase with respect to a different reference. As expected, the best performance is obtained when the second reference is orthogonal to the first. If $y_I(t)$ is the output of the first phase detector defined in Equation 6.10, then the second phase detector output $y_Q(t)$ becomes

$$y_Q(t) = \text{low pass } \{\cos(\omega_c t + \phi) \cdot 2 \cdot \sin \omega_c t\}$$

$$= -\sin \phi \tag{6.11}$$

The second phase detector not only resolves the positive/negative phase ambiguity, but it also eliminates the need to establish an amplitude reference. All decisions can be based on the polarity of a phase detector output and not on the magnitude. As a first example, consider detection of the 4-PSK signals defined in Table 6.2. Notice that the first data bit in a pair is 0 when the phase angle is positive ($\pi/4$ or $3\pi/4$), and a 1 otherwise. Hence the first data bit is completely specified by the polarity of $\sin \phi$: the output of the second phase detector $y_Q(t)$. Similarly, the second data bit is a 1 when the phase is $\pm \pi/4$, indicating that the polarity of $y_I(t)$ provides all information necessary to detect the second bit. The basic implementation of a 4-PSK (QPSK) demodulator/detector is shown in Figure 6.12. Also shown is a 4-PSK modulator to emphasize the relationships

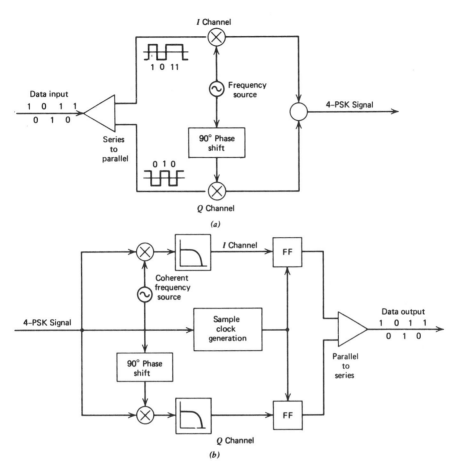

Figure 6.12. Four-PSK modulator/demodulator structure. (a) Modulator. (b) Demodulator/detector.

between the modulator and demodulator. A 4-PSK system is presented specifically because it is a popular system, and because it is a useful foundation for describing other modulation techniques.

An important concept to notice in the 4-PSK system shown in Figure 6.12 is that there are, in essence, two separate data streams. The modulator divides the incoming bit stream so that bits are sent alternately to the in-phase modulator I and the out-of-phase modulator Q. These same bits appear at the output of the respective phase detectors in the demodulator where they are interleaved back into a serial bit stream. In this manner two essentially independent binary PSK channels are established using the orthogonal carriers $\cos \omega_c t$ and $\sin \omega_c t$. The channels are usually referred to as the I and Q channels, respectively. This technique of establishing two channels within an existing bandwidth is some-

times referred to as "quadrature multiplexing."

As long as the carriers in the modulator and the references in the demodulator are truly orthogonal (coherence is maintained for both channels in the receiver), the I and Q channels do not interfere with each other. Any amount of misalignment in these relationships causes crosstalk between the two quadrature channels. Crosstalk also arises if there is unequalized phase distortion in the transmission channel.

At first thought it might seem that quadrature multiplexing increases the capacity of a given bandwidth by a factor of 2. It must be remembered, however, that the binary PSK signal on each quadrature channel is a double-sideband signal. Hence the bandwidth of the channel, *without quadrature multiplexing*, is only 50% utilized in comparison to a single-sideband system. When quadrature channels are used, a single-sideband operation is no longer possible since the sideband separation process destroys the orthogonality of the two signals. In essence, quadrature multiplexing only recovers the loss of capacity incurred by the double-sideband spectrum. In fact, some single-sideband modulators [8] possess a remarkable resemblance to the QPSK modulator shown in Figure 6.12.

Demodulation and detection of higher-level PSK systems are complicated by the fact that the use of only two references does not provide a simple means of detecting all data bits. There are two basic ways in which simple decisions (positive versus negative) can be established to detect all data. One method is to establish more references in the receiver and measure the phase of the received signal with respect to the additional references. The second method is to use only two references and related phase detectors and generate all additional measurements as linear combinations of the first two.

As an example of the first method, consider the 8-PSK system defined in Table 6.3 and shown in Figure 6.9. Optimum detection for this system can be obtained if two additional phase detector outputs are provided with respect to references at $+\pi/4$ and $-\pi/4$. The two new references are designated A and B as shown in Figure 6.13.

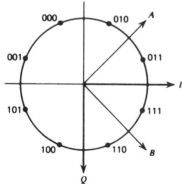

Figure 6.13 Receiver references for 8-PSK detection.

The outputs of four phase detectors corresponding to the four references are determined as follows:

$$y_Q = \text{low pass}\{\cos(\omega_c t + \phi)\cdot 2\cdot \sin \omega_c t\} = -\sin \phi$$

$$y_B = \text{low pass}\left\{\cos(\omega_c t + \phi)\cdot 2\cdot \sin\left(\omega_c t + \frac{\pi}{4}\right)\right\}$$

$$= 0.707 \cos \phi - 0.707 \sin \phi$$

$$y_I = \text{low pass}\{\cos(\omega_c t + \phi)\cdot 2\cdot \cos \omega_c t\} = \cos \phi$$

$$y_A = \text{low pass}\left\{\cos(\omega_c t + \phi)\cdot 2\cdot \cos\left(\omega_c t + \frac{\pi}{4}\right)\right\}$$

$$= 0.707 \cos \phi + 0.707 \sin \phi \qquad (6.12)$$

As an aid in determining the appropriate decision logic, these equations have been evaluated for each of the eight possible signal phases and listed in Table 6.4. Examination of Table 6.4 indicates that the first data bit is 1 whenever y_Q is positive. Similarly, the second data bit is 1 whenever y_I is positive. The third data bit is a 1 whenever y_A, y_I, and y_B are all positive, or when they are all negative. Hence the third data bit is determined as a logical combination of phase detector outputs. In summary,

$$D_1 = Q$$

$$D_2 = I \qquad (6.13)$$

$$D_3 = AIB + \overline{AIB}$$

where D_i is the ith data bit and Q, I, A, B are logic variables representing positive outputs from y_Q, y_I, y_A, and y_B, respectively.

Another method of detecting 8-PSK signals, which avoids two extra references and phase detectors, is revealed in the phase detector Equation 6.12.

TABLE 6.4 Eight-PSK Phase Detector Outputs

Data	Phase	y_Q	y_B	y_I	y_A
011	$\pi/8$	−0.383	0.383	0.924	0.924
010	$3\pi/8$	−0.924	−0.383	0.383	0.924
000	$5\pi/8$	−0.924	−0.924	−0.383	0.383
001	$7\pi/8$	−0.383	−0.924	−0.924	−0.383
101	$-7\pi/8$	0.383	−0.383	−0.924	−0.924
100	$-5\pi/8$	0.924	0.383	−0.383	−0.924
110	$-3\pi/8$	0.924	0.924	0.383	−0.383
111	$-\pi/8$	0.383	0.924	0.924	0.383

The extra measurements y_A and y_Q can be determined as

$$y_A = 0.707y_I - 0.707y_Q$$
$$y_B = 0.707y_I + 0.707y_Q \tag{6.14}$$

Hence the y_A and y_B measurements can be obtained as linear combinations of the quadrature channel phase measurements y_I and y_Q, and no additional phase detectors are required. The resulting 8-PSK demodulator/detector is shown in Figure 6.14. Notice that implementation of the linear combinations in Equation 6.14 can ignore the magnitude of the 0.707 multipliers since only the sign of the result is needed.

The linear combinations in Equation 6.14 essentially represent a rotation of the quadrature channel basis vectors through an angle of $\pi/4$ radians. By changing the angle of rotation, the linear combinations needed for other phase measurements are easily determined. Hence all phase detector based demodulators can be implemented with two phase detectors and as many linear combinations as needed. The general equation for the linear combinations produced by a rotation of α radians is

$$y_A = y_Q \cdot \cos\alpha - y_I \cdot \sin\alpha$$
$$y_B = y_Q \cdot \sin\alpha + y_I \cdot \cos\alpha \tag{6.15}$$

Reference and Clock Recovery
The demodulators discussed in the preceding paragraphs all require a local, phase coherent, carrier reference for their operation. Furthermore, multiple phase systems require at least one more reference in quadrature with the first

Figure 6.14. Eight-PSK demodulator/detector. Using only two references.

one. Recovery of any reference is complicated by the fact that phase shift keying is a double-sideband *suppressed* carrier modulation system. In other words, there is no discrete spectral line at the carrier frequency as there is with some other types of modulation. In fact, the FCC in the United States has ruled specifically that no spectral lines are allowed in the transmitted signal. The absence of a spectral line at the carrier frequency is overcome by using one of several nonlinear processing techniques [9]. After one coherent reference is established, a quadrature reference is obtained by delaying or differentiating the first.

Clock recovery is obtained in several ways. Once carrier recovery is accomplished and demodulation occurs, the clock can be obtained by locking onto transitions in the baseband signal. Baseband clock recovery occurs as though the signal has never been modulated and demodulated. In contrast, the clock can sometimes be recovered directly from the modulated signal. If a PSK signal is heavily filtered, the envelope is not constant and, in fact, contains amplitude modulation at the signaling frequency. Hence the clock can be recovered, independently of signal demodulation, by envelope detecting a heavily filtered PSK signal [10].

Clock transitions are often assured by purposely shifting the carrier reference halfway between signal points in every interval. Using 4-PSK, for example, the references would be shifted by 45°. During any one interval there are still only four possible signal states separated by 90°. However, the signal states alternate betweeen being aligned with the I and Q axes and being aligned at 45° as shown in Figure 6.8. Because a phase shift is ensured with every symbol interval both the (filtered) carrier envelope and the recovered baseband signal have amplitude transitions. Without the reference shifts, proper clock recovery requires a minimum density of data transitions as described in Chapter 4.

Differential Detection

As an alternative to recovering a coherent reference, some systems merely compare the phase in the present interval to the phase in the previous interval. The signal received in the previous interval is delayed for one signal interval and is used as a reference to demodulate the signal in the next interval. Assuming that the data have been encoded in terms of phase shift, instead of absolute phase positions, we can decode the data properly. Hence this technique, referred to as "differential detection," inherently requires differential encoding.

In general, PSK systems require differential encoding since the receivers have no means of determining whether a recovered reference is a sine reference or a cosine reference. Furthermore, the polarity of the recovered reference is ambiguous. Thus error probabilities for PSK systems are doubled automatically because of the differential encoding process. Differential detection, on the other hand, implies an even greater loss of performance since a noisy reference is used in the demodulation process. Typically, differential detection imposes a penalty of 1 to 2 dB in signal-to-noise ratio [8].

PSK Spectra

By far the easiest way to determine the spectrum of a PSK signal is to analyze the baseband waveforms applied to the quadrature channels. Owing to the orthogonality of the two channels, the signals are uncorrelated, and the composite spectrum is merely the sum of the individual (identical) spectra.

In either 2-PSK or 4-PSK systems the baseband signal is a symmetric two-level NRZ waveform. The corresponding spectrum is the common $\sin(x)/x$ spectrum shown in Figure 4.2. High-level systems (8-PSK or greater) use symmetric multilevel NRZ baseband signals similar to that shown in Figure 4.16. The multiphase PSK baseband signal is somewhat different since unevenly spaced levels as defined in Table 6.3 are used.

As mentioned in Chapter 4, a multilevel NRZ signal has the same spectrum as a two-level signal. Hence all conventional PSK systems produce a spectrum that follows the $\sin(x)/x$ response defined in Equation 4.1, but translated to the carrier frequency.* Figure 6.15 shows the PSK spectrum for two-, four-, and eight-phase systems designed to provide the same data rate. Hence the higher-level systems signal at lower rates and have proportionately narrower spectra.

PSK Error Performance

The error performance of any digital modulation system is fundamentally related to the distance between points in a signal space diagram. For example, a 2-PSK system, as represented in the phase diagram of Figure 6.8, is capable of optimum error performance since the two signal points have maximum separation for a given power level (radius of the circle). In other words, one 2-

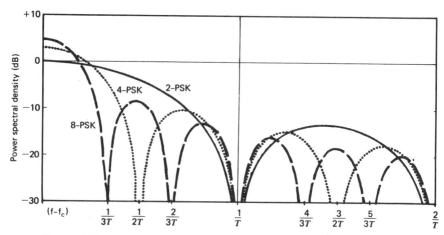

Figure 6.15. Spectrums of unfiltered PSK signals carrying equal data rates.

*The baseband levels must be uncorrelated to each other to produce a $\sin(x)/x$ spectrum. If phase transitions from one interval to the next are restricted in some manner, the baseband levels are correlated, and a different spectrum results.

PSK signal is the exact negative of the other. Hence 2-PSK modulation provides antipodal error performance as defined in Chapter 4.

The error performance of a multiphase PSK system is easily compared to a 2-PSK system by determining the relative decrease in the error distance (voltage output of a properly referenced phase detector). In addition to the error distance, however, the relative values of the noise bandwidths must also be considered. (Recall that the noise bandwidth effectively determines the variance of the noise samples.)

The general expression for the distance between adjacent points in a multiphase PSK system is

$$d = 2 \sin(\pi/N) \qquad (6.16)$$

where N is the number of phases.

A general expression of the bit error probability (or bit error rate) of an N-phase PSK system is derived in Appendix C as

$$P_E = \frac{1}{\log_2 N} \, erf(z) \qquad (6.17)$$

where

$$z = \sin(\pi/N)(\log_2 N)^{1/2} \left(\frac{E_b}{N_0}\right)^{1/2}$$

Equation 6.17 reveals that, with respect to E_b/N_0, 4-PSK provides the same error performance as does 2-PSK. Thus, as mentioned earlier, both systems provide optimum performance, but 4-PSK utilizes half as much bandwidth. The 4-PSK system has an error distance that is 3 dB smaller than the error distance of 2-PSK. However, the shorter error distance is offset by a 3 dB decrease in the noise bandwidth (indicating a 3 dB reduction in noise power at the detector). For the 4-PSK system to have the same noise power at the detector it would have to experience a 3 dB greater noise spectral density. Hence conventional signal-power-to-noise-power ratios (SNRs) can be misleading parameters for comparing digital modulation systems. However, as mentioned in Chapter 4, error rate performances in terms of SNRs are desired when determining the effects of interference or when specifying error rates with respect to measurable quantities. Figure 6.16 displays the error performance of 2-, 4-, 8-, 16-, and 32-PSK as a function of E_b/N_0. Appendix C provides the relationships needed for error rates in terms of SNR.

6.1.4 Quadrature Amplitude Modulation

As described previously, a convenient means of representing phase shift keying with four or more phases involves the use of quadrature signals. In the case of 4-

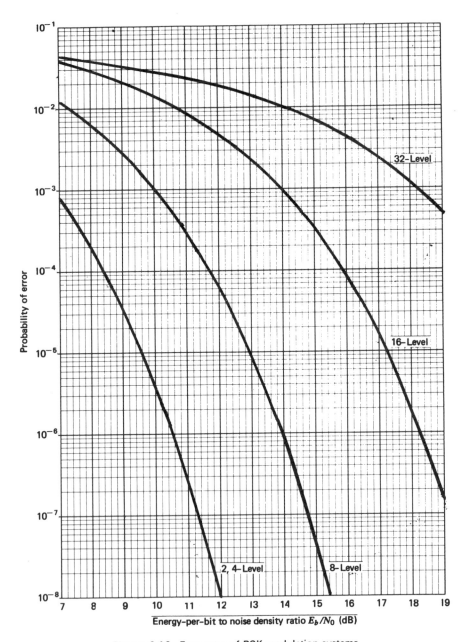

Figure 6.16. Error rates of PSK modulation systems.

PSK, the quadrature signals represent two separate channels by virtue of the independence of the baseband signals for each quadrature channel. In higher-level PSK systems, the level of a baseband signal for the I channel is not independent of the baseband level for the Q channel (see Table 6.3 or Figure 6.10). After the baseband signals have been established, however, the quadrature channel modulation and demodulation processes are independent for all PSK systems.

Quadrature amplitude modulation (QAM) can be viewed as an extension of multiphase PSK modulation wherein the two baseband signals are generated independently of each other. Thus two completely independent (quadrature) channels are established including the baseband coding and detection processes. In the special case of two levels (± 1) on each channel, the system is identical to 4-PSK and is usually referred to as such. Higher-level QAM systems, however, are distinctly different from the higher-level PSK systems. Figure 6.17 shows a signal constellation of a 16-QAM system obtained from four levels on each quadrature channel. The dots represent composite signal points while the hatch marks on the axes represent amplitude levels in each quadrature channel. Figure 6.18 shows a basic QAM modulator and demodulator structure along with a representative waveform for 16-QAM.

Notice that, in contrast to PSK signals, the QAM signal shown in Figure 6.19 does not have a constant envelope. A constant envelope is maintained with PSK modulation by restricting the combination of levels on the quadrature channels. A QAM system does not restrict the combinations since the levels on each channel are selected independently. Thus QAM systems cannot be used with saturating amplifiers.

The spectrum of a QAM system is determined by the spectrum of the baseband signals applied to the quadrature channels. Since these signals have the same basic structure as the baseband PSK signals, QAM spectrum shapes are identical to PSK spectrum shapes with equal numbers of signal points. Specifically, 16-QAM has a spectrum shape that is identical to 16-PSK, and 64-QAM has a spectrum shape identical to 64-PSK.

Even though the spectrum shapes are identical, the error performances of the

Figure 6.17. Signal constellation of 16-QAM modulation.

Figure 6.18. QAM modulator/demodulator.

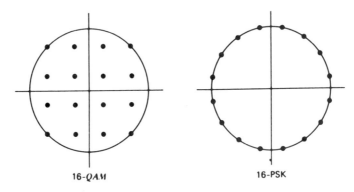

Figure 6.19. Comparison of 16-QAM and 16-PSK signal sets.

two systems are quite different. With large numbers of signal points, QAM systems always outperform PSK systems. The basic reason is that the distance between signal points in a PSK system is smaller than the distance between points in a comparable QAM system. Figure 6.19 compares the signal points of a 16-QAM system with the signal set of a 16-PSK system using the same peak power.

The general expression for the distance between adjacent signal points in a unit peak amplitude QAM system with L levels on each axis is

$$d = \frac{\sqrt{2}}{(L-1)} \tag{6.18}$$

Equations 6.16 and 6.18 reveal that an *n*-ary QAM system has an advantage over an *n*-ary PSK system with the same peak power level. In terms of average power levels, the QAM system has an even greater advantage. The following equation, derived in Appendix C, provides a general expression for the peak-to-average ratio of a QAM system:

$$\frac{\text{peak power}}{\text{average power}} = \frac{L(L-1)^2}{2\sum_{i=1}^{L/2}(2i-1)^2} \tag{6.19}$$

where the QAM system has L^2 total points.

Example 6.1. Determine the error performance of a 16-PSK system relative to a 16-QAM system with the same peak power level. Also determine the relative performance with respect to identical average powers.

Solution. Since the two systems provide identical numbers of signal points, they signal at the same rate and require the same bandwidth for a given data rate. Thus the relative error performance is completely determined by the relative distances between signal points. (When different rates are used, the effect of different noise bandwidths in the receivers must also be considered.) Evaluating Equations 6.16 and 6.18 indicates that 16-QAM has a 1.64 dB advantage over 16-PSK for a given peak power. Equation 6.19 indicates that 16-QAM has a peak-to-average power ratio of 2.55 dB. Since PSK systems have unity peak-to-average ratios, the advantage of a 16-QAM system over a 16-PSK is 4.19 dB for equal average powers.

The results of Example 6.1 show that 16-QAM is a significantly better modulation format than 16-PSK for applications like voiceband models that have no technology limits.* In digital microwave applications the performance advantage of 16-QAM is diminished by available amplifier technology but, nevertheless, it is at 16 signal points that PSK modulation defers to QAM modulation for performance reasons. Some first generation digital radios use 8-PSK modulation [11] and some use 16-QAM [12] with comparable performance. Second generation radios use 64-QAM [13] followed by third generation radios with 256-QAM [14, 15].

Reliable operation of high-density modulation formats like 256-QAM requires extremely linear amplifiers to tolerate the wide range of signal

*Voiceband modems are not restricted by technology but by signal power limits to prevent interference with other signals in the network. It is ironic that these restrictions arise primarily from FDM analog radio systems in which a high powered signal in one channel can create interference (crosstalk) into other channels. In a predominantly digital network the main signal limitation would be the saturation point of PCM encoders. It is also ironic that increasing the signal power of a voiceband modem to fall just short of the PCM saturation point would not improve performance in a mostly digital network. Performance of high speed modems in this case is primarily determined by multiplicative noise created by PCM companders [16].

amplitudes and extremely accurate adaptive equalizers to remove small percentages of intersymbol interference that arise when a high-amplitude pulse is transmitted adjacent to a low-amplitude pulse. Amplifier and equalizer performance requirements are basically a function of the dynamic range of the signal amplitudes*:

$$\text{Dynamic range} = 10 \log_1 (P_{max}/P_{min}) \qquad (6.20)$$

The dynamic range of a square QAM modulation format can be improved by modifying the signal set to eliminate corner points which produce high peak powers. One such technique referred to as 256-stepped square QAM (256-SSQAM) [17] is shown in Figure 6.20 where it can be compared with the signal set of conventional 256-QAM.

Example 6.2. Determine the dynamic range of both the conventional 256-QAM signal set and the 256-SSQAM signal set shown in Figure 6.20.

Solution. Using integral units along each quadrature axis of 1, 3, 5, 7, 9, 11, 13, and 15 for 256 QAM the dynamic range (DR) is

$$DR(256\text{-}QAM) = 10 \log_{10}(15^2 + 15^2)/(1^2 + 1^2)$$
$$= 23.52 \, dB$$

The highest amplitude signal point of 256-SSQAM is the $(5, 17)$ point. Thus the dynamic range of the 256-SSQAM signal is:

$$DR(256\text{-}SSQAM) = 10 \log_{10}(5^2 + 17^2)/(1^2 + 1^2)$$
$$= 21.96 \, dB$$

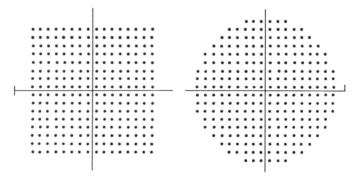

Figure 6.20. 256 QAM and stepped square 256 QAM.

*If the quadrature carriers are perfectly recovered, intersymbol interference would only be a function of the dynamic range along each axis. To make an allowance for quadrature carrier phase error, which causes intersymbol interference between I and Q channels, the composite signal power is used.

The result of Example 6.2 shows that the peak power of 256-SSQAM is 1.56 dB lower than the peak power of conventional 256-QAM for the same signal point separation. Thus in a peak-power limited application 256-SSQAM can operate with 1.56 dB more signal separation than can conventional 256-QAM. When other, less significant factors are considered, reference [18] reports a 140 Mbps 256-SSQAM radio at 6 GHz has a 2-dB performance advantage.

The theoretical error rate equation for QAM modulation, derived in Appendix C, is identical to the multilevel baseband error rate equation with the same number of levels. This equation is used to produce error rate curves for 4-, 16-, 64-, and 256-QAM modulation in Figure 6.21. Table 6.5 compares the most common forms of digital modulation used in point-to-point microwave radio systems.

Offset Keying

Because of the interdependence of the baseband signals in high-level PSK modulation, the signal transitions on the quadrature channels necessarily coincide. However, since QAM systems independently modulate the quadrature channels, they are not constrained to align the signal intervals on the two channels. When the signal intervals overlap each other by 50% (Figure 6.22), the mode of operation is referred to as offset keying. Offset keying is commonly used on 4-QAM systems (more commonly referred to as 4-PSK or QPSK systems). The main advantage of offset keying lies in the ability of the reference recovery circuitry to become synchronized to the incoming carrier at lower signal-to-noise ratios than conventional (aligned) QAM and QPSK systems [19, 20].

TABLE 6.5 Comparison of Various Digital Modulation Techniques Based on Equal Data Rates

System Designation	Information Density (bps/Hz)	Signal-to-Noise Ratios for BER = 10^{-6} (dB)		Peak to Average Ratio (dB)[a]
		E_b/N_o on the Channel	SNR at Decision CKT	
2-PSK	1	10.6	13.6	0.0
4-PSK, 4-QAM	2	10.6	13.6	0.0
QPR	2[b]	12.6	17.6	2.0
8-PSK	3	14.0	18.8	0.0
16-QAM	4	14.5	20.5	2.55
16-QPR	4[b]	16.5	24.5	4.55
16-PSK	4	18.3	24.3	0.0
32-QAM	5	17.4	24.4	2.3
64-QAM	6	18.8	26.6	3.68

[a]Ratio of maximum steady-state signal power to average signal power with random data. Measured on a channel with square root filter partitioning.
[b]In a strict sense, the signal bandwidth of a partial response system is no narrower than the theoretical (Nyquist) bandwidth of a corresponding full response system. As a practical manner, however, the partial response systems require about 17% less bandwidth [21].

Figure 6.21. Error rates of QAM modulation systems.

QAM Representation of Minimum Shift Keying

The preceding discussions of QAM systems have assumed the use of baseband signals with NRZ level encoding. A more general view of quadrature amplitude modulation allows arbitrary pulse shapes to be used in generating the baseband

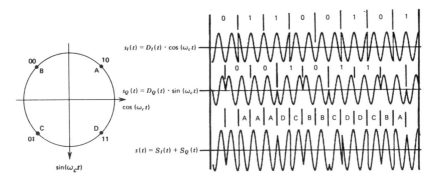

Figure 6.22. Offset keyed 4-PSK signaling.

signals. One particularly interesting pulse shape is a half-sinusoid pulse:

$$A(t) = \cos\frac{\pi t}{T} \qquad \left(-\frac{T}{2} \leqslant t \leqslant \frac{T}{2}\right) \tag{6.21}$$

where T is the duration of a signal interval.

If half-sinusoidal pulse shaping of duration T is used on both channels of an offset keyed QAM system, the quadrature channel signals $p(t)$ and $q(t)$ can be expressed as follows:

$$p(t) = a_i \cos\left(\frac{\pi t}{2T}\right) \cos \omega_c t \qquad \left(-\frac{T}{2} \leqslant t \leqslant \frac{T}{2}\right)$$

$$q(t) = a_q \sin\left(\frac{\pi t}{2T}\right) \sin \omega_c t \qquad (0 \leqslant t \leqslant T) \tag{6.22}$$

where a_i and a_q are data values (± 1) for the I and Q channels, respectively, $\cos(\pi t/2T)$ is shaping on the I channel, and $\sin(\pi t/2T)$ is shaping on the Q channel. (The sine function is used to account for offset keying.) Adding the two quadrature signals together and applying some fundamental trigonometric identities produces the result

$$x(t) = p(t) + q(t)$$

$$= a_i \cos\left(\omega_c t + \frac{\pi t}{2T}\right) \qquad (a_i = a_q)$$

$$= a_i \cos\left(\omega_c t - \frac{\pi t}{2T}\right) \qquad (a_i \neq a_q) \tag{6.23}$$

Equation 6.23 is essentially identical to Equation 6.6 defining minimum shift keying. Hence, except for a logic level transformation of data values, offset QAM

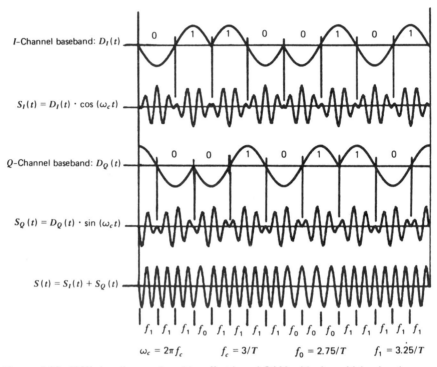

Figure 6.23. MSK signaling produced by offset keyed QAM with sinusoidal pulse shapes.

with sinusoidal pulse shaping is identical to MSK [5, 22]. The relationship is further demonstrated in Figure 6.23, which shows how an MSK signal is generated by offset keyed quadrature channel modulation with sinusoidal pulse shaping.

These foregoing results show that MSK is very closely related to offset keyed 4-PSK. The only difference is in the use of half-sinusoidal baseband pulse shapes for MSK, and square pulse shapes for 4-PSK. Because of this close relationship it is interesting to compare the amplitude spectra of the two systems in Figure 6.24. As indicated, the MSK spectrum has its first spectral null at a 50% higher frequency that 4-PSK's first null. Other than this, the MSK spectrum is more compact than the 4-PSK spectrum and is significantly lower in amplitude at frequencies outside the main lobe of the spectrum. For this reason MSK is an attractive modulation technique where constant envelopes and little or no filtering are desired. One such application is on digital satellite links with frequency division subchannels [7]. In applications requiring transmitter filters, MSK has no particular advantage over 4-PSK and usually requires a more complicated modulator.

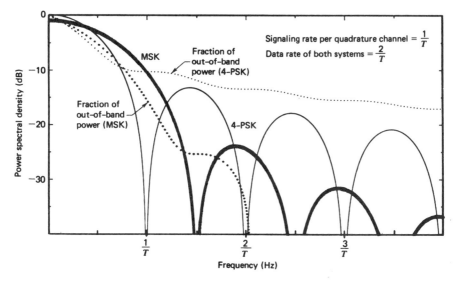

Figure 6.24. Power spectra of MSK and 4-PSK signals.

6.1.5 Partial Response QAM

Another popular modulation technique is quadrature partial response signaling (QPRS). As commonly implemented [23], a QPRS modulator is nothing more than a QAM modulator followed by a narrow bandpass filter that "overfilters" the quadrature signals and produces controlled intersymbol interference in each channel. The most common application of QPRS involves two levels on each channel before filtering and three levels afterward (see Chapter 4).

This system is essentially a 4-PSK system with partial response filtering to increase the information density. As shown in Figure 6.25, the effect of partial response filtering is to produce nine signal points from the original four. In a similar manner, a 16-QAM partial response system, with four levels on each channel before filtering, has seven levels afterward, and 49 signal points in all [24].

Although the modulators of a QPRS system can be conventional QAM modulators, the demodulator/detectors must be modified to account for the extra levels in the waveform. After the signal is demodulated, the detection processes for each channel are independent and identical to the baseband PRS detection procedures described in Chapter 4.

Figure 6.26 compares a QPRS system to a QPSK (4-PSK) system in terms of equal peak power out of the modulator. The average transmit powers are different, however, because the partial response system "overfilters" the signal to to reduce the transmitted bandwidth. In practice the information density is

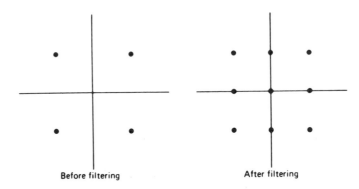

Figure 6.25. QPRS signal constellations.

Figure 6.26. Comparison of QPSK and QPRS systems with equal data rates.

increased by about 17% [21].

As discussed in Chapter 4, partial response filtering cuts the distance between signal points in half, indicating a 6-dB reduction in error performance. However, the noise bandwidth of a PRS receive filter is lower than the noise bandwidth of the corresponding full response system so that some of the error distance degradation is recovered. For the filter systems shown in Figure 6.26 the net performance loss of QPRS system is 4 dB. Notice however, that with respect to power on the channel, the performance loss is only 2 dB.

6.1.6 Trellis Coded Modulation

The 2-dB performance disadvantage of PRS mentioned in the previous section assumes that data are detected by making hard decisions one bit at a time. As

described in Chapter 4, however, a PRS signal has redundancy that overlaps two bit intervals. If the detection circuitry processes the signal spanning two intervals before making hard decisions the performance penalty can be recovered. Processing a redundant signal in this way is an example of a maximum likelihood or Viterbi decoding algorithm [25] for redundant signals. If the redundancy is extended to more levels and more intervals and optimally processed, even greater coding gains are possible. As typically implemented these extensions use extra signal points with restricted sequences (transitions between points). When the allowed sequences are represented with state transition diagrams they form a trellis. Hence the term trellis coded modulation (TCM) [26, 27].

As an example, consider the expansion of a 4-PSK signal set to an 8-PSK signal set as shown in Figure 6.27. Although there are eight signal points in the 8-PSK constellation only four points can be freely chosen in any one interval. Which four points can be chosen is dependent of the signal points chosen in previous intervals. A TCM decoder/demodulator then determines which of the allowed signal sequences most closely matches a received sequence before making hard data decisions.

Figure 6.27 shows that if data decisions are made one interval at a time, without processing the redundancy, the error performance of the expanded signal set is 5.33 dB worse than the original signal set. (A consequence of the noise power margin of adjacent signal points being reduced from 0.5 to 0.146.) Also shown in Figure 6.27 are the noise margins of more widely separated signal points. TCM recovers the 5.33-dB penalty, and more, by ensuring all allowed sequences include some of the larger distances. Furthermore, noise in only one interval is not likely to produce a decision error. When four-state coded 8-PSK uses the transition trellis shown in Figure 6.28, the performance approaches a 3-dB improvement with respect to uncoded 4-PSK [26].

To understand Figure 6.28 it is important to realize that it represents state transitions in the encoding process—not signal waveforms. The use of the states in this manner is just a convenient means of representing past history. (All past histories are encapsulated into just four states.) Transitions from one state to another are labeled with the 8-PSK signal point chosen for encoding two bits of data. Notice that only four of the 8-PSK signal points can be freely chosen and

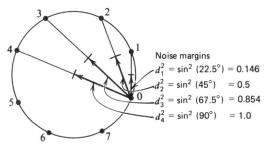

Figure 6.27. Eight-PSK signal points and noise margins.

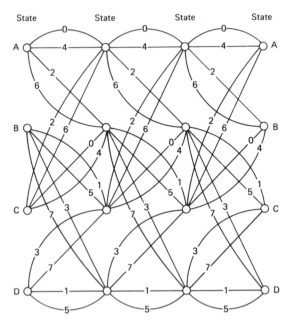

Figure 6.28. Four-state trellis for 8-PSK Modulation.

that these points have the same separation as uncoded 4-PSK. (The fact that there are four states and two bits being encoded in each interval is an unfortunate coincidence.) The significant aspects of the decoding process are summarized as follows:

1. An allowed transition between two states can occur with either of two signals, which implies sequence information does not help discriminate between those two particular signals. However, the two signals in question are chosen to have maximal separation (noise margin = 1) so redundancy is unnecessary.

2. Transitions that originate or terminate in any particular state are encoded with signals having noise power margins of 0.5 (or 1).

3. All sequences that begin and end in common states are at least three intervals long with minimum noise power margins of 0.5, 0.146, and 0.5. Thus the total noise margin between any two minimum length sequences is 1.146.

It may be necessary to discriminate between sequences that are longer than three but in all cases the end points involve transitions with noise power margins of 0.5. Thus the noise margin is always greater than 1, which is 3 dB greater than the noise margin of uncoded 4-PSK. Determination of the exact error rate of a TCM system is much more involved than when interval-by-interval detection is

used. Reference [26] shows that the bit error rate of coded 8-PSK asymptotically approaches 3-dB improvement with respect to 4-PSK as the noise margin would indicate.

Example 6.3. A receiver for a trellis coded modulation system as shown in Figure 6.28 detects a "0, 2, 0" sequence on a symbol-by-symbol basis. Since this sequence represents a coding violation, determine the sequence with the closest distance to the given received sequence. Assume the starting state is state "A."

Solution. By tracing the trellis from state "A" the following list represents all of the allowable sequences that differ from the received sequence by only one symbol.

	Starting State	Sequence	Ending State	Distance
1	A	0, 2, 1	C	d_1
2	A	0, 2, 5	C	d_3
3	A	0, 2, 3	D	d_3
4	A	0, 2, 7	D	d_1
5	A	0, 0, 0	A	d_2
6	A	0, 4, 0	A	d_2

The first and fourth sequences each has distance d_1 separation from the received sequence. However, they terminate in different states, which implies that additional information for decoding the original sequence is contained in subsequent intervals. In fact, the very next interval will contain only even-numbered symbols if the proper state is C or only odd-numbered symbols if the proper state is D. Thus, only one of the candidate sequences (0, 2, 1 or 0, 2, 7) will have distance d_1 after the next interval. Furthermore, states C and D have no common next state so additional information is forthcoming in even later symbol intervals.

Example 6.3 shows that signal decoding requires measuring sequence distances across multiple symbol intervals. As long as candidate sequences terminate in different states additional information is available for decoding the data. Although Example 6.3 implies bit-by-bit decoding, optimum detection actually involves analog summation (or the DSP equivalent) of the phase errors in every interval for all candidate sequences and choosing the sequence with the smallest overall distance.

Coded 8-PSK TCM has been used in satellite circuits [28] where non-linearities dictate the use of a constant envelope signal. TCM with higher density QAM signal sets is used in high-speed voiceband modems such as the 14.4 kbps modem of CCITT recommendation V.33 [29]. TCM can also be used in conjunction with MSK modulation [30]. Because of the high-speed signal

processing requirements TCM has not been used in point-to-point microwave radios (as of 1990) but such a system is under development by NTT in Japan [31].

6.2 FILTER PARTITIONING

The transmitting and receiving equipment of a digital radio system typically contains several filters that limit the signal spectrum to some degree or another. Since the end-to-end frequency response of the channel must conform to certain baseband pulse response objectives, the desired composite filter function must be partitioned among the individual filters. Normally, a number of the filters can be designed to provide their respective functions without significantly impacting the channel pulse response. For example, a mixing process produces a sum and a difference of the input frequencies. Only the sum is wanted when mixing upward, or only the difference is wanted when mixing downward. Usually the undesired terms can be eliminated by a filter that does not significantly affect the pulse shape of the output signal. The following discussion assumes that only two filters significantly influence the baseband pulse response: one in the transmitter and one in the receiver.

6.2.1 Adjacent Channel Interference

One basic purpose of the radio channel receive filter is to minimize the amount of noise present at the detector. A second purpose of this filter is to reject energy in adjacent radio channels. Energy from an adjacent channel that does not get

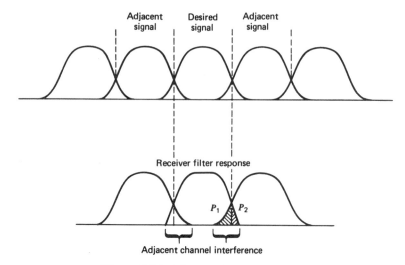

Figure 6.29. Adjacent channel interference.

rejected is referred to as "adjacent channel interference." In this discussion, we assume that the spectrum shapes in the adjacent channels are identical to the desired spectra. This situation is shown in Figure 6.29, which depicts a number of frequency division multiplexed digital channels.

As indicated in Figure 6.29, adjacent channel interference occurs as a result of two phenomena. First, the receive filter passes unwanted power P_1 because the adjacent signal is not completely truncated to prevent overlap into the desired channel. The second source of interference P_2 occurs because the receive filter does not provide infinite attenuation of power properly belonging in the adjacent channel. P_1 is minimized by narrowing the transmit filter, while P_2 is minimized by narrowing the receive filter. Since channel pulse response considerations constrain the composite filter function to some minimum width, one component of the interference cannot be reduced without increasing the other. Hence the total filter function must be partitioned in some manner to minimize the sum of P_1 and P_2.

6.2.2 Optimum Partitioning

The optimum filter design for any particular application may depend on a number of factors including legislated emission specifications, available technology for power amplifiers, the availability of cross-polarization for adjacent channel isolation, and the relative effects of noise versus adjacent channel interference. In the absence of external constraints, a classical result attributed to Sunde [32] and also presented in references [33] and [34] determines optimum partitioning as one matching the output of the transmit filter to the square root of the desired channel response. Hence if the desired output spectrum $Y(\omega)$ is obtained as follows:

$$Y(\omega) = H_{Rx}(\omega) \cdot H_{Tx}(\omega) \cdot X(\omega) \tag{6.24}$$

where $X(\omega) = $ the channel input spectrum
 $H_{Tx}(\omega) = $ the transmit filter response
 $H_{Rx}(\omega) = $ the receive filter response

then, the optimum partitioning is obtained as

$$|H_{Rx}(\omega)| = |Y(\omega)|^{1/2}$$

$$|H_{Tx}(\omega)| = \frac{Y(\omega)^{1/2}}{X(\omega)} \tag{6.25}$$

Equation 6.25 defines the filter amplitude responses to minimize the adjacent channel interference under the condition that the adjacent channels contain identical signals with identical power levels. In addition, if the transmitted spectrum $H_{Tx}(\omega) \cdot X(\omega)$ is the complex conjugate of the receive filter response

$H_{\text{Rx}}(\omega)$, the receive filter is matched to the channel spectrum, and the best possible error performance is obtained with respect to signal power on the channel.

Partitioning as defined by Equation 6.25 is shown in Figure 6.30 for a pulse input and a raised cosine output (see Appendix C). Even though only baseband signal spectra are shown, the concept is extended easily to passband filter functions.

Although Equation 6.25 provides a sound theoretical basis for determining the optimum filter partitioning, practical considerations of a particular application may require deviation from the "optimum." One problem that arises in digital microwave radios is related to the peak at the band edges of the transmit filter in Figure 6.30. With a passive filter this peak can be obtained only by inserting loss into the midband response. If the transmit power can be increased as compensation, no ill effects result. However, the output level of most digital microwave radios is limited by the technology of microwave-frequency power amplifiers (a few watts typically). Since these radios are device-power limited, midband insertion loss cannot be overcome and therefore substracts directly from the received signal level.

In device-power limited applications the optimum transmit filter is one having a flat response in the passband. To achieve the desired channel response at the detector, the receive filter must then be peaked at the band edges. Channel insertion losses at this point do not degrade performance since both the signal and the noise is attenuated equally. However, the peaks have the undesirable effect of increasing the receiver noise bandwidth and the P_1 component of adjacent channel interference. Although these increases represent a poorer performance than that obtained by theoretically optimum partitioning, the degradation is not as great as the insertion loss needed to achieve the "theoretical optimum." As one further note, it should be mentioned that not all theoretically optimum transmit filters have a peak at the band edges. In particular, partial response systems do not require band edge peaking (see Appendix C).

Another aspect of the optimum design to be kept in mind is that optimum error performance is achieved with respect to signal power on the channel, and

Figure 6.30. Theoretical optimum filtering for square pulse excitation and raised-cosine output.

not with respect to unfiltered transmitter power. It is this feature that allows any amount of attenuation to be inserted into the channel at the transmitter and not degrade the optimum performance. If performance is measured with respect to unfiltered transmit power, the best transmit filter may be different from that defined in Equation 6.25. Not only would less midband attenuation be desirable, as discussed, but it might also be desirable to widen the transmit filter response. This decreases the truncation loss in the transmit filter so the receive filter can be narrowed to decrease its noise bandwidth.

As an extreme example of how widening the transmit filter can improve error rate performance, consider removing the transmit filter and incorporating it into the receiver. For a given output at the power amplifier, the signal at the detector is unchanged. The receiver noise bandwidth, however, is reduced because the composite filter is narrower than the original receive filter. Hence a higher signal-to-noise ratio is present at the detector.

The penalty incurred for removing the transmit filter, of course, is a greatly increased P_1 components of adjacent channel interference. If adjacent channels do not exist or if they are adequately isolated by cross-polarization (i.e., if the system is noise limited), the performance can be improved by moving some of the transmitter truncation losses to the receiver until adjacent channel interference matches the noise. Keep in mind, however, that if the system is adjacent channel interference limited, the optimum partitioning is indeed defined by Equation 6.25.

6.3 EMISSION SPECIFICATIONS

One unavoidable consideration when determining the filter functions of a transmitter and receiver is the out-of-band emission specifications established by the FCC. In many cases the emission specifications dictate a narrower transmit filter than the theoretical optimum. Thus the partitioning aspect of the filter designs may be predetermined.

It is somewhat ironic that the FCC emission specifications were intended to control adjacent channel interference but, in some cases, actually cause the interference to increase by forcing the use of a wider than optimum receive filter. The FCC specifications, however, were intended to protect adjacent analog radio channels from out-of-band digital emissions. They were not selected with adjacent digital channels in mind.

In January of 1975, when the FCC established the out-of-band emission specifications, two separate specifications were established: for radios operating below 15 GHz and for radios operating above 15 GHz. The emission limitations for operation below 15 GHz are more restrictive than those for operation above 15 GHz because the lower frequencies were heavily used already for analog FDM-FM radios, which are more sensitive to interference. The higher frequencies have not been used extensively because of vulnerability to rain attenuation. These frequencies are generally used for short-distance, special

purpose applications and are not as congested as the lower frequencies. Initially, the most popular bands for digital microwave were 11 and 6 GHz.

The emission limitations for operation below 15 GHz are

$$A = 35 + 10\,log_{10}\,B + 0.8(P - 50) \qquad (P > 50) \tag{6.26}$$

where A = the power in a 4-kHz band relative to the mean output power (in decibels)

B = authorized bandwidth in MHz

P = percent removed from carrier frequency ($P = 50$ is the band edge)

In addition, the attenuation A must be at least 50 dB everywhere outside the band but does not need to exceed 80 dB at any point. Notice that the emission limitations are not specified in terms of absolute power levels, but only relative to the transmitted power. Thus these emission limitations do not constrain the output power level of the radio. Microwave output powers are often constrained by microwave amplifier technology. (A limit of 10 W of output power does exist when technology can reach it.)

Figure 6.31 displays the emission mask for 11-GHz radios using a 40-MHz bandwidth. Also shown is the power spectrum of a 90-Mbps 8-PSK radio using a signaling rate of 30 MHz. To be compatible with the emission specification, the 8-PSK power spectrum is shown in terms of power per 4-kHz band. The difference between the signal spectrum and the FCC mask represents the minimum attenuation required of the transmit filter. The filtered 8-PSK signal thereby provides an information density of $90/40 = 2.25$ bps/Hz. This is the basic modulation and filtering format used by Collins/Rockwell in their 90-Mbps digital radio, the MDR-11 [16].

Figure 6.31. FCC emission mask at 11 GHz and a 90 Mbps 8-PSK spectrum.

The mission limitations for microwave bands above 15 GHz is defined as

$$A = 11 + 10 \log_{10} B + 0.4(P - 50) \qquad (P > 50) \qquad (6.27)$$

where A = attenuation in a 1 MHz band below mean output power (in decibels)
B = authorized bandwidth in MHz
P = percent removed from the carrier frequency

The attenuation A must be at least 11 dB, but does not have to exceed 56 dB.

In conjunction with the emission limitations, the FCC has stipulated that no discrete lines exist in the transmitted spectrum [35]. Thus no carrier components can exist and no repetitive data patterns are allowed to occur. The repetitive data patterns are effectively eliminated by using a scrambler in the transmitter and a descrambler in the receiver. As mentioned already, PSK and QAM are forms of double-sideband *suppressed* carrier modulation so that the carrier terms are eliminated automatically as long as modulation is continuous.

6.4 RADIO SYSTEM DESIGN

The foremost design requirement of a point-to-point radio system for public telephony is operational dependability, usually referred to as availability. Availability is expressed as the percentage of time that a system provides a specified minimum quality of service. With analog systems, minimum performance is determined by the noise power in the received signal. The performance of a digital system is determined by the bit error rate. Typical objectives for bit error rates range from 10^{-3} for voice traffic to 10^{-6} or 10^{-7} for data traffic. Recall that a bit error rate of 10^{-4} corresponds to the threshold of perceptibility for errors in a PCM voice signal. Typical design objectives for microwave radio systems specify availabilities on the order of 99.98% [36, 37]. Hence the maximum acceptable accumulation of outage, due to all causes, is on the order of 2 hr per year.

Radio system availability is dependent on equipment reliability and path propagation characteristics. The high-availability objectives of a typical radio system mandate redundant equipment and often require redundant paths. The need for redundant paths is determined by the frequency of atmospheric induced outages (rain attenuation or multipath interference). Rain is a dominant consideration at higher carrier frequencies (above 11 GHz), and multipath must be considered at all frequencies. Multipath fading is dependent on prevailing climate and terrain.

Redundant radio equipment typically operates in either a backup or a hot-standby mode with automatic protection switching. The transmission path is backed up with spare channels (frequency diversity) or spare paths (space diversity receivers). In extreme cases a backup route may even be utilized.

6.4.1 Fade Margins

The main technique used to circumvent atmospheric induced outages is to provide an extra strong signal at the receiver during normal path conditions. The difference between the normal received power and the power required for minimum acceptable performance is referred to as the fade margin. Greater fade margins imply less frequent occurrences of minimum performance levels. Radios operating in higher frequency bands generally require greater fade margins because they are more susceptible to rain attenuation. A 50-dB fade margin is typical for a digital radio at 11 GHz, while a 40-dB fade margin is typical for lower microwave frequencies.

The amount of fade margin actually required for a particular route depends on the probability of multipath induced fades and heavy rainfall occurrences. Thus drier climates permit lower fade margins—thereby allowing greater repeater spacings. In some mountain-based microwave links in the western United States, microwave hops can be 100 miles long. By comparison, the average hop in other parts of the country is less than 30 miles long.

When large fade margins are provided, the received signal power during unfaded conditions is so strong that bit errors are virtually nonexistent. Nevertheless, problems with extra strong signals do exist. Namely, automatic gain control in a receiver must operate over a wide dynamic range. If the maximum signal level into the demodulation and detection circuitry is not controlled, saturation is likely to degrade performance—especially in high-density modulation formats such as 64- or or 256-QAM, where information is encoded into the signal amplitudes.

To minimize dynamic range requirements in a receiver and reduce interference between systems adaptive transmitter power control (ATPC) is sometimes used [38]. ATPC uses a feedback data link from a receiving station to control the output power of a transmitting station. Thus when excess power is unnecessary it is not used.

6.4.2 System Gain

One of the most important parameters used to characterize digital microwave system performance is the system gain A_s. System gain is defined to be the difference, in decibels, of the transmitter output power and the minimum receive power for the specified error rate:

$$A_s = 10 \log_{10}\left(\frac{P_T}{P_R}\right) \tag{6.28}$$

where P_T = transmitter output power
 P_R = receive power for specified error rate

The minimum acceptable receive power is sometimes referred to as the

threshold power and is primarily dependent on the receiver noise level, the signal-to-noise ratio required by the modulation format, and various system degradations such as excess noise bandwidth, signal distortions, intersymbol interference, carrier recovery offsets, timing jitter, and coupling and filter losses that either attenuate the signal or increase the noise level.

Noise power in a receiver is usually dominated by thermal noise generated in the front-end receiver amplifiers. In this case, the noise power can be determined as follows:

$$P_N = (F)(N_0)(B)$$
$$= F(kT_0)B \tag{6.29}$$

where F = the receiver noise figure
N_0 = the power spectral density of the noise
B = the receiver bandwidth*
k = $1.38(10)^{-23}$ is Boltzmann's constant
T_0 = the effective receiver temperatures in degrees Kelvin

Equation 6.29 essentially states that the receiver noise power is determined by the spectral noise density of the receiver input resistance and the additional noise introduced by the amplification process (noise figure F). Normally, a reference temperature of 290°K is assumed so that the thermal noise density (kT_0) is $4(10)^{-21}$ W/Hz.

The noise figure of any device is defined as the ratio of the input signal-to-noise ratio to the output signal-to-noise ratio:

$$F = \frac{(S/N)_{in}}{(S/N)_{out}} \tag{6.30}$$

In effect, the noise figure specifies the increase in noise power relative to the increase in signal power. Since all physical devices introduce noise, the noise figure of any system is always greater than one, and is usually expressed in decibels. If a system has no gain or attenuation, the noise figure is exactly equal to the ratio of output noise to input noise. Noise figures of low-noise microwave amplifiers typically range from 2 to 5 (3 to 7 dB). Radio receiver noise figures are typically 6 to 10 without a low-noise amplifier.

Combining Equations 6.28 and 6.29 and incorporating a term D for the degradation from ideal performance produces the following general expression

*Typically, B is assumed to be the minimum theoretical bandwidth for the particular modulation format in use. Excess bandwidth required by practical implementations is then incorporated into a system degradation factor.

for system gain:

$$A_s = 10 \log_{10} \left(\frac{P_T}{(SNR)(F)(kT_0)(B)} \right) - D \qquad (6.31)$$

where SNR is the theoretical signal-power-to-noise-power ratio required for the maximum acceptable error rate, and D includes all degradations from ideal performance. Notice that the SNR term in Equation 6.31 refers to signal-power-to-noise-power ratios and not E_b/N_0. The relationship between SNR and E_b/N_0 is provided in Appendix C.

The system gain, in conjunction with antenna gains and path losses, determines the fade margin:

$$\text{fade margin} = A_s + G_T + G_R + 20 \log_{10} \lambda - A_f - A_0 \qquad (6.32)$$

where G_T = the transmitter antenna gain (dB)
G_R = the receive antenna gain (dB)
λ = the transmitted wavelength
A_f = the antenna feeder and branching loss (dB)
$A_0 = 20 \log_{10} (4\pi d)$ is the free space attenuation (the distance d must be in the same units as λ)

The directivity, and consequently the gain of, an antenna is directly proportional to the size of its aperture and inversely proportional to the square of the transmitted wavelength. In determining the receive power, however, it is actually only the area of the antenna that is important and not the directivity or gain. Thus radio system designers conveniently consider transmit and receive antenna gains as contributing to signal power, but include a wavelength normalization $(20 \log_{10} \lambda)$ to relate the gain of the receiving antenna back to the size of its aperture.

In addition to providing increased antenna gain, greater directivities also reduce multipath problems. The longer secondary paths, which arise from greater emanation angles, have lower power levels when the directivity is increased. Unfortunately, antenna gains are limited by practical considerations in several regards: economically sized tower can support only limited sized antennas, mechanical alignment is difficult, and directional stability of both the antenna and the path is limited.

The feeder and branching losses A_f included in Equation 6.32 arise because single antenna systems typically carry several channels for separate radios. Furthermore, reliability considerations usually dictate that spare transmitters and receivers be available for protection switching. The process of combining signals for transmission or distributing them after reception inherently introduces various amounts of attenuation or splitting of signal power.

Example 6.4. Determine the system gain of a 10-Mbps, 2-GHz digital micro-

wave repeater using 4-PSK modulation and an output power of 2.5 W. Assume the excess bandwidth of the receiver is 30% and that other departures from ideal performance amount to 3 dB degradation. Assume a noise figure of 7 dB for the receiver, and the desired maximum error rate is 10^{-6}. Also determine the fade margin assuming antenna gains of 30 dB each and a path length of 50 km. The branching and coupling losses are 5 dB.

Solution. From Figure 6.16, the required value of E_b/N_0 for 4-PSK modulation can be determined as 10.7 dB. Using Equation 3.42 (in Appendix C), it can be determined that the signal-power-to-noise-power ratio at the detector is 3 dB higher than E_b/N_0. Thus, the required SNR is 13.7 dB.

Since 4-PSK modulation provides 2 bits per baud, the signaling rate is 5 MHz, which is the theoretical minimum (Nyquist) bandwidth. Equation 6.31 can now be used to determine the system gain:

$$A_s = 10\log_{10}\frac{2.5}{4(10)^{-21}(5)10^6} - 13.7 - 7 - 3 - 10\log 1.3$$

$$= 116\,dB$$

At a carrier frequency of 2 GHz, the wavelength is $3(10)^8/2(10)^9 = 0.15$ m. Thus the fade margin can be determined from Equation 6.31:

$$\text{fade margin} = 116 + 60 + 20\log_{10}(0.15) - 5 - 20\log_{10}[4\pi5(10)^4]$$

$$= 38.5\,dB$$

6.4.3 Frequency Diversity

As mentioned previously, neither the transmitting and receiving equipment nor the path is normally reliable enough to provide an acceptable level of system availability. Frequency diversity is one means of providing backup facilities to overcome both types of outages. A deep multipath induced fade occurs when a signal from a secondary path arrives out of phase with respect to the primary signal. Since the phase shift produced by a path is proportional to frequency, when one carrier fades it is unlikely that another carrier fades simultaneously. Frequency diversity involves the use of a spare transmitter and receiver operating in a normally unused channel. Since separate hardware is used, frequency diversity also provides protection against hardware failures.

The simplest means of implementing frequency diversity is to use one-for-one (1 : 1) protection switching as indicated in Figure 6.32. One-for-one protection switching implies that one spare channel is provided for each assigned message channel. When high-spectrum efficiency is required, it is generally necessary to have only one spare for a group of N channels (1 : N protection switching). In fact, the FCC in the United States has stipulated that, in some frequency bands, a system must be implementable in one-for-N configurations. The main impact

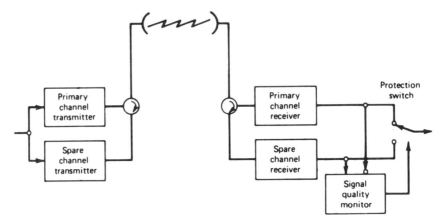

Figure 6.32. Protection switching with one-for-one frequency diversity.

of a $1:N$ protection system is the complexity of the switching unit and the need to switch back to the assigned channel in the first available hop so that a single spare channel can be reused repeatedly on a long route.

Frequency diversity normally does nothing to alleviate rain outages since all channels in a particular frequency band are simultaneously affected. When rain is a particular problem, it can be overcome only by using higher transmit powers or shorter hops.

6.4.4 Space Diversity

Space diversity is implemented, as shown in Figure 6.33, by vertically separating two receive antennas on a single tower. The resulting difference in the two paths is normally sufficient to provide independent fading at the two antennas. Space diversity is the most expensive means of improving the availability, particularly if separate receivers for multiple channels are used for each antenna. The cost

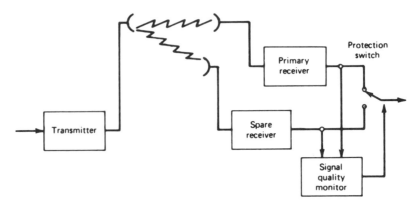

Figure 6.33. Protection switching with space diversity receivers.

can be minimized, however, by combining the two received signals in a phase coherent manner for input to a common receiver [36]. This technique provides less hardware backup than when completely separate receivers are used for each antenna.

6.4.5 Angle Diversity

Because multipath fading is produced by multiple incident rays arriving at slightly different angles, protection from fading can be achieved by discriminating on the angle of arrival. Angle diversity utilizes two side-by-side receiving antennae with slightly different angular elevations to provide the discrimination. Although angle diversity is generally not as effective as space diversity, it can improve the availability of digital signals [39, 40] in applications where the tower requirements of space diversity are impractical or disallowed.

6.4.6 Adaptive Equalization

Since multipath effects are frequency dependent, not all frequencies in a particular channel simultaneously experience the same amount of fading. Thus multipath fading can produce not only a general attenuation of the received signal, but also the equivalent of in-band amplitude (and phase) distortion. Distortion of the spectrum amplitude produces a general degradation in the error performance over and above the penalty incurred by the general attenuation of the signal. In wideband digital radios, frequency selective fading (as opposed to flat fading) has proved to be the dominant source of multipath outages. Fortunately, better performance can be achieved if the spectrum amplitude is equalized (adjusted to a uniform level).

Since atmospheric induced multipath varies with time, removal of multipath induced amplitude distortion requires adaptive equalization. A common approach to adaptive amplitude equalization merely samples the energy at selected frequencies in the received signal spectrum. A compensating filter shape is then inserted into the signal path to adjust all energy samples to a common level. This basic technique is used in digital radios manufactured by Bell Northern [36] and Rockwell International [41]. Reference [41] reports a 7-dB improvement in effective fade margin can be achieved with an adaptive equalizer.

Because channel distortions produce intersymbol interference in the received baseband signal, time domain equalization with adaptive transversal filters or decision feedback equalizers is also used. Reference [42] contrasts the relative merits of both approaches and provides an extensive list of references. The use of high-density modulation formats such as 256-QAM places very stringent requirements on the adaptive equalizers so these radios commonly use both types of equalizers [38].

Frequency selective fading, which corrupts only a portion of a wide bandwidth channel, is particularly difficult to equalize because it changes

rapidly and requires complicated equalizer structures (in the frequency domain or the time domain). The problem can be greatly simplified if, instead of using a single carrier, multiple carriers within the allotted bandwidth are used to create multiple subchannels. With this approach, fading within any subchannel is relatively flat and easy to equalize. The disadvantage of this approach is the cost of replicated circuitry for each subchannel. As reported in [31], however, atmospheric induced outages are greatly reduced.

6.4.7 Route Design

The layout of a point-to-point microwave relay system involves numerous considerations of the local terrain, prevailing atmospheric conditions, radio frequency interference with other systems, and interference from one hop to another within a single system.

Path Length
The foremost consideration in setting up a single hop of a microwave system is that line-of-sight transmission is required. Using antenna heights of about 60 m, the curvature of the earth limits the transmission distance to approximately 50 km. Longer line-of-sight distances are possible if taller antenna towers are used. However, practical considerations of mechanical stability limit the height of a tower, particularly since longer distances imply larger and more directive antennas.

When antennas are rigidly mounted on the top of a building or the side of a mountain, mechanical stability may not be a problem but the stability of the path itself may become a limiting factor. Under some atmospheric conditions, radio waves can be refracted to the point that a narrow beam completely misses the receiving antenna. Bending of the propagation path can also cause a fixed obstacle to intermittently obstruct transmission, implying that clearance between the normal path and nearby physical obstructions is required. Thus there is always a practical limit on how narrow the beam can be.

Local terrain also influences the prevalence of multipath fading. Nearby bodies of water contribute significantly to multipath conditions during late evening or early morning hours when there is little wind. Direct transmission over water is usually very difficult because of reflections off the water surface. Transmission over water often requires higher antennas and some means of blocking direct reflections.

Cochannel Interference
Since there are a limited number of channels available for point-to-point microwave systems, the same channels must be used over and over again. Reuse of microwave frequency bands is enhanced by the directivity of the antennas and the general need for line-of-sight reception. In many metropolitan areas, however, there is so much traffic converging into one particular area it is impossible to completely isolate two systems using the same channel. This type

of interference is referred to as cochannel interference.

Cochannel interference results from converging routes or from overreach of one hop into another hop of the same system reusing a channel. Sometimes reflections or atmospheric refractions can contribute to overreach, even when direct line of sight is not present. Overreach is usually controlled by zigzagging the hops so that a beam from one transmitter misses all subsequent receive antennas in a route.

Cochannel interference of analog radios into digital radios is often the performance-limiting factor of digital radios being installed in AT&T junction stations where many routes converge [38]. To circumvent this problem, AT&T developed a FM carrier interference canceler (FMCIC) [43] that samples the FM signal at the colocated FM receiver and adaptively subtracts it from the received digital signal.

REFERENCES

1 Federal Communications Commission, *Report and Order*, Docket No. 19311, FCC 74-985, released September 27, 1974.

2 "Special Issue: AR6A Radio System," *Bell System Technical Journal*, December 1983.

3 W. A. Conner, R. J. Girvin, and W. R. McClellan, "Frequency Control of the MSR-6 Single Sideband System," *IEEE International Conference on Communications*, 1983, pp E2.5.1–5.

4 Y. Yoshida, Y. Kitahara, and S. Yokoyama, "6G-90 Mbps Digital Radio System with 16-QAM Modulation," *IEEE International Conference on Communications*, 1980, pp 52.4.1–52.4.5.

5 S. Pasupathy, "Minimum Shift Keying: A Spectrally Efficient Modulation," *IEEE Communications Magazine*, July 1979, pp 14–22.

6 C.-E. Sundberg, "Continuous Phase Modulation," *IEEE Communications Magazine*, April 1986, pp 25–38.

7 Y. Morihiro, S. Nakajima, and N. Furuya, "A 100 Mbit/s Prototype MSK Modem for Satellite Communications," *IEEE Transactions on Communications*, October 1979, pp 1512–1518.

8 R. E. Ziemer and W. H. Tranter, *Principles of Communications*, Houghton Mifflin Company, Boston, 1976.

9 F. M. Gardner, "Rapid Synchronization: Carrier and Clock Recovery for High Speed Digital Communication," *Microwave Systems News*, Vol. 6, No. 1, 1976, p 57.

10 C. R. Hogge, Jr. "Carrier and Clock Recovery for 8-PSK Synchronous Demodulation," *IEEE Transactions on Communications*, May 1978, pp 528–533.

11 P. R. Hartmann and J. A. Crosset, "A 90 MBS Digital Transmission System at 11 GHz using 8 PSK Modulation," *IEEE International Conference on Communications*, 1976, pp 18-8–18-13.

12 I. Horikawa, T. Murase, and Y. Saito, "Design and Performances of 200 Mbit/s 16 QAM Digital Radio System," *IEEE Transactions on Computers*, December 1979, pp 1953–1958.

13 J. A. Crosset and P. R. Hartmann, "64-QAM Digital Radio Transmission System Integration and Performance," *IEEE International Communications Conference,* 1984, pp 636–641.

14 W. A. Conner, "Direct RF Modulation 256QAM Microwave System," *IEEE Globecom Proceedings,* 1988, pp 52.7.1–6.

15 H. Matsue, T. Shirato, and K. Watanabe, "256 QAM 400 Mb/s Microwave Radio System with DSP Fading Countermeasures," *IEEE International Conference on Communications,* June 1988, pp 41.5.1–6.

16 K. Pahlavan and J. L. Holsinger, "A Model for the Effects of PCM Compandors on the Performance of High Speed Modems," *IEEE Globecom Proceedings,* 1985, pp 28.8.1–5.

17 T. Ryu, J. Uchibori, and Y. Yoshida, "A Stepped Square 256 QAM for Digital Radio System," *IEEE International Conference on Communications,* 1986, pp 1477–1481.

18 M. Tahara, T. Deguchi, S. Mizoguchi, and Y. Yoshida, "6 GHz 140 Mbps Digital Radio System with 256-SSQAM Modulation," *IEEE Globecom Proceedings,* 1987, pp 38.3.1–6.

19 S. A. Rhodes, "Performance of Offset-QPSK Communications With Partially Coherent Detection," *National Telecommunications Conference Record,* November 1973, pp 32A-1–32A-6.

20 M. K. Simon and J. G. Smith, "Offset Quadrature Communications with Decision-Feedback Carrier Synchronization," *IEEE Transactions on Communications,* October 1974, pp 1576–1584.

21 J. W. Bayless, R. D. Pedersen, and J. C. Bellamy, "High Density Digital Data Transmission," *National Telecommunications Conference Record,* 1976, pp 51.3-1–51.3-6.

22 S. A. Gronemeyer and A. L. McBride, "MSK and Offset QPSK Modulation," *IEEE Transactions on Communications,* August 1976, pp 809–820.

23 C. W. Anderson and S. G. Barber, "Modulation Considerations for a 91 Mbit/s Digital Radio," *IEEE Transactions on Communications,* May 1978, pp 523–528.

24 D. P. Taylor and M. Shafi, "Fade Margin and Outage Computation of 49-QPSR Radio Employing Decision Feedback Equalization," *IEEE International Conference on Communications,* 1983, pp 1453–1458.

25 A. J. Viterbi, "Error Bounds for Convolutional Codes and an Asymptotically Optimum Decoding Algorithm," *IEEE Transactions on Information Theory,* April 1967, pp 260–269.

26 G. Ungerboeck, "Trellis-Coded Modulation with Redundant Signal Sets Part I: Introduction," *IEEE Communications Magazine,* February 1987, pp 5–11.

27 G. Ungerboeck, "Trellis-Coded Modulation with Redundant Signal Sets Part II: State of the Art," *IEEE Communications Magazine,* February 1987, pp 12–21.

28 R. J. F. Fang, "A Coded 8-PSK System for 140-Mbit/s Information Rate Transmission over 80-MHz Nonlinear Transponders," *Proceedings 7th International Conference on Digital Satellite Communications,* 1986, pp 12–16.

29 T. Kamitake, K. Uehara, M. Abe, and S. Kawamura, "Trellis Coding 14.4 kb/s Data Modem Implemented with a Single-Chip High-Speed Digital Signal Processor," *IEEE Globecom Proceedings,* 1987, pp 12.8.1–6.

30 Ramin Sadr and J. K. Omura, "Generalized Minimum Shift-Keying Modulation Techniques," *IEEE Transactions of Communications,* Jan 1988, pp 32–40.

31 Y. Nakamura, H. Ohtsuka, S. Aikawa, and H. Takanashi, "Advanced Techniques for Super Multi-Carrier Digital Microwave Radio with Trellis-Coded 256 QAM Modulation," *IEEE Globecom Proceedings*, 1989, pp 11.3.1–6.

32 E. D. Sunde, "Ideal Binary Pulse Transmission by AM and FM," *Bell System Technical Journal*, November 1959, pp 1357–1426.

33 W. R. Bennet and J. R. Davey, *Data Transmission*, McGraw-Hill, New York, 1965.

34 R. W. Lucky, J. Salz, and E. J. Weldon, Jr., *Principles of Data Communications*, McGraw-Hill, New York, 1968.

35 Federal Communications Commission, Memorandum Opinion and Order modifying FCC *Report and Order* of Docket 19311, released January 29, 1975.

36 G. H. M. de Witte, "DRS-8: System Design of a Long Haul 91 Mb/s Digital Radio," *IEEE National Telecommunications Conference*, 1978, pp 38.1.1–38.1.6.

37 H. Kostal, D. R. Jeske, and V. K. Prabu, "Advanced Engineering Methods for Digital Radio Route Design," *IEEE International Communications Conference*, 1987, pp 19B.6.1–6.

38 C. P. Bates, P. L. Penney, and K. L. Seastrand, "Digital Radio Technology in the AT&T Network," *IEEE International Communications Conference*, 1987, pp 19B.5.1–7.

39 E. A. Sweedyk and P. Balaban, "A Multipath Fading Model for Terrestrial Microwave Radio," *IEEE Globecom Proceedings*, 1988, pp 52.6.1–7.

40 S. H. Lin, "Opening New Vistas for Microwave Radio," *Bellcore Exchange*, July/August 1988, pp 9–12.

41 P. R. Hartmann and E. W. Allen, "An Adaptive Equalizer for Correction of Multipath Distortion in a 90 MB/S 8-PSK System," *IEEE International Conference on Communications*, 1979, pp 5.6.1–5.6.4.

42 J. K. Chamberlain, F. M. Clayton, H. Sari, and P. Vandamme, "Receiver Techniques for Microwave Digital Radio," *IEEE Communications Magazine*, November 1986, pp 43–54.

43 D. J. Ramirez, M. V. Thompson, and D. J. Wisler, "Modifying FM Analog Radio to Carry Broadband Digital Signals," *IEEE Globecom Proceedings*, 1986, pp 15.5.1–8.

PROBLEMS

6.1 To prevent the transmission of line spectra, digital radio terminals use data scramblers to randomize the data patterns. Furthermore, differential encoding is normally required for proper data detection. Both functions introduce error multiplication. If the combined effect of these operations causes an average of five decoded errors for every channel error, what is the effective penalty in transmit power for a 4-PSK system at a $BER = 10^{-6}$? At a $BER = 10^{-3}$?

6.2 If a digital radio receiver is experiencing thermal noise induced errors at a rate of 1 per 10^6 bits, what is the new error rate if the path length is decreased from 30 to 25 miles?

6.3 Derive Equation 6.18.

6.4 What is the minimum theoretical bandwidth of an 8-PSK signal carrying 4800 bps?

6.5 A 32-QAM signal set is implemented by eliminating the four corner points of a 36-QAM signal. What is the minimum error distance in terms of the peak signal power? How does this answer compare to the error distance for antipodal signaling? (Express answers in decibels.)

6.6 What is the peak-to-average ratio of a 32-QAM signal (Problem 6.5) assuming transmission of random data?

6.7 How does 32-QAM (Problem 6.5) error performance compare to 32-PSK error performance?
(a) In terms of average signal powers?
(b) In terms of peak signal powers?

6.8 A carrier transmission system using 4-PSK modulation provides an error rate of 10^{-6}. If the modulation is changed to 16-PSK to narrow the transmission bandwidth (data rate unchanged), how much must the transmit power be increased to maintain the same error rate?

6.9 Repeat Problem 6.8 assuming that the bandwidth is unchanged and the data rate is doubled.

6.10 What is the error rate of an ideal 16-QAM signal with a signal to Gaussian noise ratio (SNR) of 18 dB?

6.11 What is the error rate of Problem 6.10 if interference is also present at a level that is 21 dB below the received signal? (Assume the effect of the interference is identical to Gaussian noise at the same rms power.)

6.12 How much must the transmit power in Problem 6.11 be increased to offset the effect of the interference? (Assume the interference level is fixed.)

6.13 Repeat Problem 6.12, but assume that the interference increases in proportion to the increase in transmit power. (Signal powers in the adjacent channels are increased along with the desired signal power.)

7

NETWORK SYNCHRONIZATION, CONTROL, AND MANAGEMENT

In Chapters 4 and 6 some synchronization requirements of transmission systems are discussed. These requirements involve carrier recovery for coherent detection of modulated signals, clock recovery for sampling incoming data, and framing procedures for identifying individual channels in a TDM signal format. All of these considerations are inherent in digital transmission systems and, for the most part, operate independently of other equipment in a network. An instance of one system's dependency on another was noted for Tl lines. Tl source data must include minimum densities of l's to maintain timing on the original tranmission link. In contrast, other line codes are described that maintain clock synchronization independently of the source data.

This chapter discusses more general synchronization considerations for interconnecting various digital transmission and switching equipment. Foremost among these considerations is coordinating the operations of the synchronous transmission and switching equipment. When individual synchronous equipments are interconnected to form a network, certain procedures need to be established that either synchronize the clocks to each other or provide for their interoperability when each subsystem uses independent clocks. Following the discussion of network clock synchronization, the concept of synchronization is extended to higher-level aspects of network control. These considerations include connection control, traffic routing, and flow control.

7.1 TIMING

All digital systems inherently require a frequency source or "clock" as a means of timing internal and external operations. Operations timed from a single frequency source do not require particularly stable sources since all commonly clocked elements experience timing variations in common. A different situation occurs when transfers are made from one synchronous equipment to another (as from a transmitter to a receiver). Even if the clock of the receiving terminal is

"synchronized" to the transmitting terminal on a long-term or average basis, short-term variations in either clock may jeopardize the integrity of the data transfer. Thus it is generally necessary to use frequency sources (oscillators) in both the transmitter and the receiver that are as stable as is economically feasible.

7.1.1 Timing Recovery: The Phase-Locked Loop

A common means of synchronizing a receiver clock to a transmitter clock uses a phase-locked loop (PLL) as shown in Figure 7.1. A phase detector continuously measures the phase difference between the incoming clock and a locally generated clock. The phase detector in Figure 7.1 merely measures the difference in the zero crossings between the two signals. When the zero crossing of the line clock precedes the zero crossing of the local clock a positive voltage is generated—otherwise, a negative voltage is produced. The output of the phase detector is filtered to eliminate as much receive noise as possible, and then the phase measurement adjusts the frequency of the voltage-controlled oscillator (VCO) to reduce the phase difference. Some amount of noise or interference inevitably passes through the phase detector and the filter causing erroneous adjustments in the VCO frequency. As time passes, however, a frequency offset produces ever-increasing phase shifts. When the phase difference builds up, it is easier to detect, and the appropriate changes in the VCO occur. Hence the local clock maintains the desired average frequency but inherently produces short-term variations of phase and frequency as it "hunts" the underlying frequency of the line clock.

The line clock in Figure 7.1 is shown to have a transition in every clock interval—a situation that does occur with Manchester or diphase-type line codes. With some line codes (bipolar or AMI in particular) there are no clock transitions during intervals of binary zeros. In these cases the clock extraction

Figure 7.1. Phase-locked loop clock recovery circuit.

circuitry either inserts artificial transitions extrapolated from previous intervals or the phase detector is disabled during intervals when no pulse is detected.

7.1.2 Clock Instability

The variation in the output frequency of the VCO described above is an example of clock instability. All clocks have a certain amount of instability—even free running oscillators. An important aspect of clock instability is its frequency: the rate at which the clock frequency changes from being too high to being too low. The frequency of the instability can be directly observed as the frequency spectrum of the VCO control voltage in a PLL clock recovery circuit. When the VCO control voltage varies slowly the variations are referred to as clock wander. When the variations are more rapid the instability of the clock is referred to as jitter. The most generally accepted dividing point between wander and jitter is 10 Hz. Pure wander on a T1 line therefore produces positive phase errors for more than 77,200 bit intervals (0.05 sec) followed by more than 77,200 bit intervals of only negative phase errors.* The main sources of clock instability (both wander and jitter) in a network are

1. Noise and interference
2. Changes in the length of transmission media
3. Changes in velocity of propagation
4. Doppler shifts from mobile terminals
5. Irregular timing information.

Noise and Interference
If the low-pass filter of a PLL is designed with a very low cutoff frequency it can filter out almost all of the noise and interference on a transmission link that would otherwise corrupt the timing recovery. There are three main reasons why arbitrarily low low-pass filters cannot be used. First, the ability of the PLL to acquire synchronization (e.g., lock onto to the underlying clock) is inversely related to the PLL bandwidth. If the VCO begins oscillating at the wrong frequency and the bandwidth is too narrow the PLL may never pull the oscillator to the frequency of the line clock. Sometimes, this problem is ameliorated by using two bandwidths: a wide one to acquire synchronization and a narrow one that is selected after lock is achieved.

A second consideration that generally precludes the use of very narrow filters, even after acquisition, is that the source may vary in frequency (i.e., have jitter) that cannot be tracked by a slowly responding PLL in a corresponding receiver.

*As discussed later, observing wander in this way requires extremely stable clocks as a reference. The normal PLL clock recovery circuitry cannot be used to observe wander because the VCO tracks the relatively long-term phase offsets. Thus the clock recovery circuit does not filter out the wander but passes it on.

In this case, the recovered clock does not track the ideal sample times leading to high error rates or, worse yet, the PLL loses synchronization altogether and has to reacquire lock.

The third limiting factor for low bandwidth PLLs is the instability of the VCO itself. If the VCO begins to drift in frequency very low-bandwidth filters preclude adjusting the VCO input voltage soon enough to prevent bit errors or possible loss of synchronization.

Because operational considerations dictate certain minimum PLL bandwidths, noise and interference on the transmission link always cause the recovered clock to be more impaired than the source clock. The PLL does, however, eliminate that portion of a disturbance with frequency content above the bandwidth of the PLL. Thus disturbances with low-frequency content are the most difficult to deal with. Systematic jitter, as produced by particular patterns of intersymbol interference, can have an arbitrarily low-frequency content and is discussed more fully in a later section.

An important consideration in the design of a digital transmission link is the build up of jitter at tandem clock recovery circuits. If a recovered clock is used to time the transmission of outgoing data, as in a regenerative repeater, some amount of incoming jitter is imbedded into the outgoing clock. The clock recovery circuit in the next receiver tracks its incoming clock but introduces even more jitter due to noise and interference on the second section. Thus jitter accumulates at every regenerative repeater using its received line clock as its transmit clock. If there is a large number of regenerative repeaters, the jitter can accumulate to a point where subsequent clock recovery circuits have difficulty tracking the receive clock, produce sampling errors, and possibly lose lock.

Changes in Length of Transmission Media

Path length changes occur as a result of thermal expansion or contraction of guided transmission media, or as a result of atmospheric bending of a radio path. While a path is increasing in length, the effective bit rate at the receiver is reduced because more and more bits are being "stored" in the medium. Similarly, as the path shortens, the bit rate at the receiver increases because the number of bits stored in the transmission link is decreasing. After the path has stabilized, the receive signal returns to the nominal data rate. The most significant changes in path length occur with communications satellites. Current geostationary satellites produce path length variations of approximately 200 miles, which cause propagation time variations of approximately 1 msec [1].

Example 7.1. Determine the change in path length, the change in the number of bits in the path, and the relative change in received data rate of a 500-mile-long T2 transmission link using 22-gauge copper wires. Assume the temperature changes by 20°C over a period of 1 hr.

Solution. The coefficient of thermal expansion for copper and the velocity of propagation of 22-gauge wire can be obtained from reference [2] as

$16.5(10)^{-6}/°C$ and 29,400 miles per second, respectively.* Thus the change in path length is determined as

$$\Delta d = (500)(16.5)(10^{-6})(20) = 0.165 \text{ miles}$$

The change in the number of bits in the path is determined as

$$\Delta B = \frac{(6.312)(10)^6(0.165)}{29,400} = 35.4 \text{ bits}$$

Assuming that the temperature changes at a constant rate, we determine the change in the data rate as

$$\Delta R = \frac{35.4}{3600} = 9.8(10)^{-3} \text{ bps}$$

Hence the relative change in the received data rate (instability) is

$$\frac{\Delta R}{R} = \frac{9.8(10)^{-3}}{6.312(10)^6} = 1.56(10)^{-9}$$

Although the value of received clock instability obtained in Example 7.1 is a very small number, it is actually greater than the accuracy requirements of 1 part in 10^{11} for primary references in the North American network [3]. Thus distribution of a primary reference requires careful attenuation to the nature of the transmission medium. It is important to note, however, that instability (wander) due to path length changes is a transient phenomenon, while accuracy objectives represent maximum offsets in steady-state clock frequencies.

Changes in the Velocity of Propagation

Temperature changes not only cause expansion and contraction of wireline transmission media, they can also change those propagation constants of the media that determine the velocity of propagation. The resulting change in received clocked stability, however, is much less than that produced by the change in path length [4].

The propagation velocity of radio waves in the atmosphere also changes with temperature and humidity. Although these velocity changes are more significant than those occurring in wire lines, they are still smaller than the path-length-induced variations derived in Example 7.1. Notice that a change in propagation velocity is effectively equivalent to a change in path length since the number of bits "stored" in the transmission path is changed.

*These parameter values are obtained for conventional 22-gauge cable pairs and do not necessarily represent exact values for low-capacitance cable first used in T2 systems.

Doppler Shifts

The most significant source of potential timing instability in a received clock occurs as a result of Doppler shifts from airplanes or satellites. For example, a Doppler shift induced by a 350 mph airplane amounts to an equivalent clock instability of $5(10)^{-7}$. Again, Doppler shifts occur, in essence, as a result of path changes.

Irregular Timing Information

As discussed in Chapter 4, a fundamental requirement of a digital line code is that it provide sufficient timing information to establish and maintain a receiver line clock. If the timing information is data dependent, jitter in the recovered clock increases during periods of relatively low-density timing marks. The magnitude of the jitter is not dependent solely on the density of timing marks, but also on the timing (data) patterns. In an ideal repeater, only the density would matter. In practice, however, various imperfections lead to pattern dependent jitter [5].
insert overhead bits into a composite data stream for various purposes. When the higher-rate data stream is demultiplexed, the arrival rate of data within individual channels is irregular. This irregularity produces timing jitter when generating new line clocks for the lower-rate signals. This source of jitter (waiting time jitter) is often the most troublesome and is discussed in more detail later.

7.1.3 Elastic Stores

The timing instabilities described in the preceding paragraphs essentially represent changes in the number of bits stored in a transmission link. In the case of noise and interference induced jitter, the change in "bits stored" occurs because data are sampled a little earlier or a little later than nominal. Since the outgoing data of a regenerative repeater are transmitted according to the recovered clock, a phase offset in the clock means the delay through the repeater is different from when there is no misalignment in timing.

If phase offsets in successive regenerative repeaters coincide, a net change of several bits of storage in a long repeatered transmission link occurs. Since these extra bits enter or leave the transmission link over relatively short periods of time, the accumulated jitter may represent a relatively large, but short lived, instability in the receive clock.

Because regenerative repeaters use incoming sample clocks as output clocks, sustained timing differences between inputs and outputs do not exist. The endpoints of a transmission link, however, may interface to a local clock. In this case a difference between a received and a relatively fixed local clock must be reconciled with an elastic store. An elastic store is a data buffer that is written into by one clock and read from by another. If short-term instabilities exist in either clock, the elastic store absorbs the differences in the amount of data transmitted and the amount of data received. An elastic store can compensate

only for short-term instabilities that produce a limited difference in the amounts of data transmitted and received. If sustained clock offsets exist, as with highly accurate but unsynchronized clocks, an elastic store will eventually underflow or overflow.

TDM–Switch Interface
A typical need for an elastic store occurs when a digital transmission link is interfaced to a digital time division switch. As shown in Figure 7.2, the elastic store is placed between the incoming digital transmission link and the inlet side of the switch. In most instances the digital switch provides timing for all outgoing TDM links so that no timing discrepancies exist between these links and the switch. For the time being, assume that the far end of the digital link derives its clock from the receive signal and uses that clock to time digital transmissions returning to the switch. This is the situation that arises when a remote channel bank is connected to a digital switch through T1 lines and is commonly referred to as channel bank "loop timing." When loop timing is used, the line clock on the incoming link of the switch is nominally synchronized to the switch clock. However, for reasons discussed previously, a certain amount of instability in the incoming clock necessarily exists. The elastic store absorbs these instabilities so that purely synchronized data are available for the switch.

In essence, the loop formed by the transmission links and the elastic store maintains a constant and integral number of clock intervals between the inlet and outlet of the switch. Thus, from a timing point of view, the inlets and outlets operate as though directly connected to each other using a common source of timing.

Removal of Accumulated Jitter
Another application for an elastic store is shown in Figure 7.3 where it is used to remove transmission-induced timing jitter in a regenerative repeater. Normally,

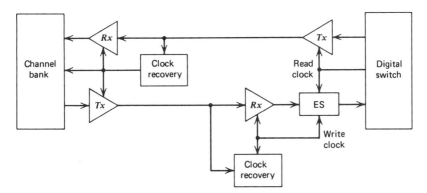

Figure 7.2. Interface between TDM transmission link and a digital switch using an elastic store.

Figure 7.3. Jitter-removing regenerative repeater.

a regenerative repeater establishes the transmit timing directly from the locally derived sample clock. In Figure 7.3, however, the transmit timing is defined by a separate local clock. The elastic store absorbs the short-term instabilities in the receive clock, but the long-term frequency of the transmit clock is controlled by maintaining a certain "average level of storage" in the elastic store. Thus the transmit clock is synchronized to the line clock on a long-term basis, but not on a short-term basis. If the elastic store is large enough to accommodate all transient variations in the data rate, high-frequency instability of the input clock is removed.

All regenerative repeaters, regardless of the mechanism used to recover timing, derive their output clocks by averaging the incoming timing information over a period of time. Tuned circuits average the incoming clock for relatively few signal intervals, phase-locked loops for many intervals. In all cases a certain amount of storage or delay is implied. An elastic store is merely a mechanism to increase the available delay so that output timing adjustments can be made more gradually. Notice that an elastic store always inserts significant amounts of artificial delay into the data path. To remove arbitrarily low-frequency instabilities (wander) arbitrarily large elastic stores (and arbitrarily stable VCOs) are required. Thus a jitter-removing elastic store should be inserted into a chain of repeaters only when jitter accumulation threatens the ability of a regular regenerative repeater to maintain synchronization.

Jitter is not just a phenomenon of a digital transmission link but also occurs in digital storage systems. For example, jitter removal like that shown in Figure 7.3 is used in laser disc players to eliminate blurring of the visual images [6].

Elastic Store Implementations

The required size of an elastic store varies from a few bits to several hundred bits for high-speed, long-distance communications links. Figure 7.4 shows one means of implementing a small elastic store utilizing a series-to-parallel converter, a register, and a parallel-to-series converter. As indicated, incoming data are transferred into the register as soon as each word is accumulated in the series-to-parallel converter. Some time later, data in the register are transferred to the output parallel-to-series converter as a complete word is shifted out. Notice that transfers to the parallel-to-series converter are independent of the

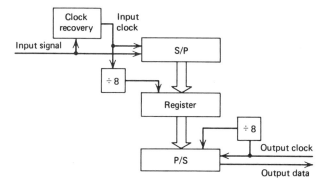

Figure 7.4. Basic implementation of an elastic store.

incoming clock. As long as output transfers occur between input transfers no data are lost, and short-term jitter is absorbed by varying delays through the elastic store.

Normally, some control circuitry (not shown) is needed to initialize the elastic store so that the first transfer into the register occurs midway between output transfers. This process means some incoming data are initially discarded by the series-to-parallel register until the desired transfer time occurs.

The relative times of the parallel transfers into and out of the holding register provide a direct indication of the relative phase of the input and output clocks. Thus the parallel transfer clocks contain the information needed to generate VCO control voltages if the elastic store is being used to remove accumulated transmission jitter.

The basic structure shown in Figure 7.4 can be extended to implement larger elastic stores as shown in Figure 7.5. The only change involves the substitution

Figure 7.5. FIFO implementation of an elastic store.

of a first-in, first-out (FIFO) buffer for the holding register of Figure 7.4. This data buffer is designed specifically to allow input transfers under the control of one clock while outputs are controlled by a different clock. Normally, the FIFO is initialized by inhibiting output transfers until the buffer is half full. In fact, some commercially available FIFOs have an output signal specifically indicating when the buffer is half full or greater.

7.1.4 Jitter Measurements

A simple circuit for measuring timing jitter is shown in Figure 7.6. As indicated, it is nothing more than a phase-locked loop (PLL) with the output of the phase comparator providing the measurement of the timing jitter. Normally, the bandwidth of the low-pass filter (LPF) is very small so the VCO ignores short-term jitter in the timing signal. If there is no jitter at all, the output of the phase comparator is constant and no signal is passed by the high-pass selection filter (HPF).

Very low-frequency jitter cannot be measured by the circuit in Figure 7.6 because the VCO tracks slowly changing phase shifts. Low-frequency jitter may be of no concern, however, because it can be tracked by a PLL. Higher-frequency jitter, on the other hand, is more apt to cause sampling errors or a loss of lock in the clock recovery of a repeater. Thus the spectral content of the jitter as well as its magnitude is of interest. Besides not being able to measure low-frequency jitter the circuit of Figure 7.6 cannot operate if it cannot lock onto the fundamental clock signal. Section 7.4.7 describes another way to measure jitter, and other timing impairments, within a network.

Phase jitter is commonly specified by communications theorists as a power measurement in units of radians squared or cycles squared. (One cycle $= 2\pi$ radians.) As indicated in Figure 7.7 phase jitter power is then a measure of the variance of the number of clock cycles or unit intervals (UIs) stored in the transmission link. In a physical sense jitter "power" has little meaning because it represents timing variations—not power. Some physical justification for expressing jitter as a power can be obtained by observing that the rms power σ_v^2

Figure 7.6. Circuit for measuring timing jitter.

Figure 7.7. Phase jitter modeled as the variance in the number of symbols "stored" in the transmission link.

of the phase detector output signal is proportional to the rms phase jitter σ_ϕ^2:

$$\sigma_v^2 = K_d^2 \sigma_\phi^2 \quad (V^2) \tag{7.1}$$

where K_d is the phase-detector gain factor in volts/radian.

Example 7.2. Given an rms phase jitter of 10.7 dB relative to one slot-squared, what is the standard deviation of the phase offset?

Solution. The variance of the signal phase is determined as $\sigma_\phi^2 = 10\exp(10.7/10) = 11.76$ slots-squared. Hence, the standard deviation is $(11.76)^{1/2} = 3.43$ symbol intervals. Since 68% of a normal probability distribution lies within one standard deviation, the phase of this signal is within ± 3.43 symbol intervals for 68% of the time. One percent of the time the signal phase will be outside 2.6 standard deviations or ± 8.9 symbol intervals.

If phase jitter arises as a result of additive Gaussian noise on a stable signal, the phase noise can be approximated as

$$\sigma_\phi^2 = \frac{\sigma_n^2}{2 \cdot P_s} \quad (rad^2) \tag{7.2}$$

where σ_n^2 is the additive noise power, and P_s is the signal power.

Equation 7.2 is the basic equation of phase jitter produced by additive noise on a continuous sinusoid [7]. When timing is extracted from a data signal, the timing information is usually not continuous. The distinction is not important because jitter produced by additive noise is normally insignificant compared to other sources [8]. For an analysis of phase jitter produced by regenerative repeaters operating on randomly occurring timing transitions see reference [9].

As an example of a particular instability specification, Table 7.1 lists the wander and jitter specifications of a DS1 digital carrier (T1) interface to the

TABLE 7.1 Maximum Instability of DS1 Customer Interface

Frequency Band	Maximum Peak-to-Peak Variation
Less than 10 Hz	28 UIs in 24 hr
Less than 10 Hz	5 UIs in 15 min
10 Hz–40 kHz	5 UIs
8 kHz–40 kHz	0.1 UIs

public network [10]. Notice that the higher frequency jitter spec allows deviations in pulse centers of only 5%—which effectively specifies the accuracy of the data sample clock. Larger variations are allowed for lower-frequency instabilities because the clock recovery circuits can track the changes to maintain a good sample clock.

7.1.5 Systematic Jitter

An original analysis of jitter in a chain of digital regenerators was reported by Byrne, Karafin, and Robinson in reference [5]. Figure 7.8 shows the basic model of their analysis. Each of the regenerative repeaters in a T-carrier line extracts timing from the received waveform and passes that timing on to the next regenerator as a transmit clock. Because of implementation imperfections (primarily intersymbol interference) in the timing recovery circuits, jitter produced by repeaters is dependent on the data patterns. One worst case pattern produces an extreme phase lag. Another pattern produces an extreme phase lead. When the data pattern shifts from one worst case to the other, a phase ramp occurs. Because every repeater has the same basic implementation, the jitter produced by individual repeaters tends to be coherent. The systematic

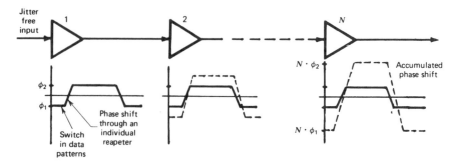

Figure 7.8. Model of systematic jitter in a string of regenerative repeaters; ϕ_2 = phase shift produced by worst case data pattern for phase lead; ϕ_1 = phase shift produced by worst case data for phase lag.

nature of this jitter makes it the most significant source of accumulated line clock jitter at the end of a chain of repeaters.

As indicated in Figure 7.8 the last repeater in the chain experiences a large phase ramp equal to the number of repeaters times the phase shift of each individual repeater. This phase ramp represents an abrupt change in the clock frequency that may cause bit errors or a complete loss of synchronization. Thus there is a limit to the number of repeaters that can be used without jitter removal. For more analyses of jitter accumulation including the combined effects of systematic jitter and random perturbations see references [11] through [13].

7.2 TIMING INACCURACIES

In the preceding section the nature of certain instabilities or transient variations in timing was discussed. Although these variations represent shifts in the frequency of a line clock, the shifts are only temporary and can be absorbed by elastic stores. In some instances digital communications equipment using autonomous frequency sources must be interconnected. When this happens, the clock rates of the two system are never exactly the same, no matter how much accuracy is designed into the frequency sources. An offset in the two clocks, no matter how small, cannot be reconciled by elastic stores alone.

In the preceding section, channel bank loop timing was mentioned as an example of how remote terminals are synchronized to a digital switch. When the remote terminal is another digital switch using its own frequency source as a reference, a different situation results. As shown in Figure 7.9, the outgoing clock for each direction of transmission is defined by the local switch clock. Thus the incoming clock at each switch interface not only contains transmission line induced jitter but also a small and unavoidable frequency offset.

7.2.1 Slips

As indicated in Figure 7.9, the interface of each incoming digital link necessarily contains an elastic store to remove transmission link timing jitter. The elastic store at the first digital switch is *written into* by the recovered line clock, but *read*

Figure 7.9. Communication between autonomously timed digital switches.

from at the local rate R_1. If the average rate of the recovered line clock R_2 is different from R_1, the elastic store will eventually underflow or overflow, depending on which rate is larger. When R_2 is greater than R_1, the elastic store at the first digital switch overflows causing a loss of data. If R_2 is less than R_1, the same elastic store underflows, causing extraneous data to be inserted into the bit stream entering the switch. Normally, the extraneous data are a repetition of data bits already transferred into the switch. Disruptions in the data stream caused by underflows or overflows of an elastic store are referred to as "slips."

Uncontrolled slips represent very significant impairments to a digital network because they generally cause a loss of frame synchronization. Therefore, slips are allowed to occur only in prescribed manners that do not upset framing. One general approach to controlling the slips is to ensure that they occur only in the form of a repetition or deletion of an entire frame. Thus the time slot counters and framing logic associated with the multiplex group remain synchronized. Controlled slips comprising entire frames can be assured by using elastic stores with at least one frame of storage. As a slip occurs, the storage level in the elastic store is effectively increased or decreased by a full frame. Rather than actually inserting or deleting a frame of information, the desired effect is achieved more easily by indexing address pointers in a random access memory. Such a system is shown in Figure 7.10.

The elastic store in Figure 7.10 operates by sequentially writing input information into memory addresses corresponding to individual TDM channels. Data for individual output channels are obtained by reading the same addresses in the same sequential manner. Ideally, if there is no offset between the clock rates, the read times of each channel occur midway between write times for corresponding channels. The elastic store then has the capability of absorbing transmission delay variations up to one-half of a frame time.

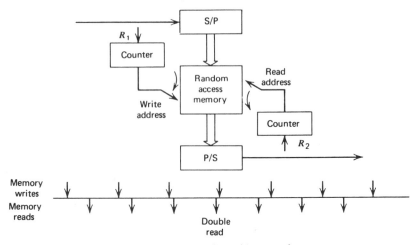

Figure 7.10. Elastic store operation with a one-frame memory.

The timing diagram in Figure 7.10 depicts an exaggerated timing offset in which the switch clock R_2 is greater than the incoming clock R_1. As indicated, the read times catch up gradually with the write times until a "double read" occurs. At that time the information retrieved for each channel is a repetition of the information retrieved for the previous outgoing frame. Although write and read times for only one channel are shown, the corresponding times for all other channels have the same relationship. Thus all channels slip together. Notice that if R_1 is greater than R_2 a slip occurs when a "double write" on all channels causes the information in the previous incoming frame to be overwritten.

The elastic store operation depicted in Figure 7.10 is very similar to the operation of a time slot interchange memory described in Chapter 5. This relationship is exploitable in a TST switch where the inlet memory can provide both the elastic store function and the time switching function [14]. When the two functions are combined, slips generally occur at different times for different channels. Nevertheless, individual channels maintain proper frame alignment since each channel is transferred through the inlet memory using dedicated memory addresses.

One attractive feature of using the inlet memory as an elastic store is that, when setting up a new connection, an internal switching time slot can be chosen so that the inlet memory read is halfway between inlet memory writes for the particular channel. Thus a slip in that connection will not occur for a long time—probably not until long after the connection is released. (With clock accuracies of only one part in 10^8, the time between slips in any one channel is 3.5 hr.)

One potential problem with the elastic store in Figure 7.10 occurs when write and read times nearly coincide. When both accesses to a single channel occur one after the other, transient timing instabilities can cause the two accesses to cross back and forth with respect to each other. Thus slips caused by double reads may follow slips caused by double writes and vice versa. To remedy this situation, some amount of hysteresis is needed in the counter adjustment process. The hysteresis, in turn, implies that additional storage is needed to defer the occurrence of one type of slip after a slip of the other type has recently occurred.

One means of implementing an elastic store with the desired hysteresis is to use two frames of storage as shown in Figure 7.11. For convenience the elastic store is divided into an A-frame memory and a B-frame memory. The counter logic again accesses the memories in sequential fashion except that frames are written alternately into the A and B memories. Under normal operation, the memories are accessed in the same way for output data. When a slip is imminent, however, control logic causes the output channel counter to be reset so that the A memory is read twice in a row. This situation is depicted in the timing diagram of Figure 7.11, which again assumes that R_2 is greater than R_1. The important point to be noticed in the timing diagram is that after the counter adjustment produces a double read of memory A, the write and read times of each individual memory are approximately one frame time apart. Thus another adjustment can

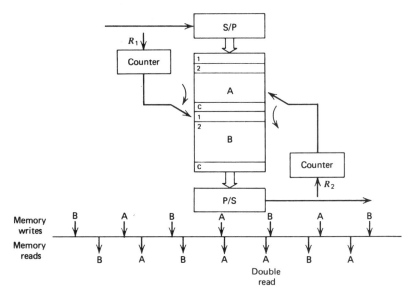

Figure 7.11. Elastic store with a two-frame memory.

be deferred until the write and read accesses again drift one full frame time with respect to each other. The structure and mode of operation shown in Figure 7.11 describe the elastic store used for DS1 signal interfaces to the Bell System No. 4 ESS [15].

Slip Rate Objectives

If the difference between an elastic store's input data rate and its output data rate is ΔR,* the time between slips is

$$\Delta T = \frac{N}{\Delta R} \qquad (7.3)$$

where N is the number of bits that get dropped or repeated whenever a slip occurs. Normally, a slip involves a full frame of data, in which case the time between slips is determined as

$$\Delta T = \frac{1}{\Delta F} \qquad (7.4)$$

where ΔF is the difference in frame rates.

*Typically, clock offsets are specified in relative terms (e.g., one part in 10^9). A clock that is accurate to P parts per million (ppm) has a maximum offset of $\Delta R = R \cdot P/10^6$.

As long as slips are controlled so they do not disrupt framing, their only effect is an infrequent repetition or deletion of the information within affected TDM channels. The audible effect of slips on a digitized voice signal is an occasional "click." Only one slip in 25 produces an audible click in PCM voice [1]. Voice signals can therefore tolerate several slips per minute [16].

Encrypted traffic (voice or data) is more susceptible to slips since the encryption/decryption process usually relies on bit synchronous scramblers and unscramblers. When the bit count is altered by insertion or deletion of bits in a time slot, counters in the source and destination become unsynchronized. At best, the decryption process causes every slip to be audible. At worst, unintelligible speech or data result until the unscrambler is resynchronized.

A more subtle, and troublesome, aspect of slips occurs when a digitized channel carries voiceband data. High-speed data modems for the analog telephone network use phase shift modulation with coherent detection in the receiver. Since these modems are particularly sensitive to phase shifts, they are particularly vulnerable to slips. An 8 bit slip in a digitized modem signal using a carrier of 1800 Hz causes an instantaneous phase shift of 81°. Obviously, a phase shift of this size causes a data error, but more importantly, it upsets the carrier recovery circuitry in the receiver and causes multiple errors. A single slip can upset the operation of some voiceband modems for several seconds [17, 18].

Characterizations of the effects of slips on Group 3 facsimile equipment [19] reveal that a single slip can cause the loss of four to eight scan lines without an error report. Sometimes the loss of the lines is not immediately evident to the reconstructed image. Diagonal lines, however, readily reveal missing vertical space.

When a digital transmission link is being used to transmit data directly, the effect of a slip may not be any more significant than a single channel error. Most data communications receiving equipment requests a complete retransmission of any block of data not satisfying certain redundancy checks. Thus one error is as bad as many errors or a complete loss of data. The effect of the slip, will be more significant, however, if the communications protocol (e.g., Digital Equipment Corporation's DDCMP) relies on byte count procedures to delimit message blocks. Insertion or deletion of data by the network causes the receive counter to become unsynchronized, and the normal exchange of information is disrupted until the loss of synchronization is recognized.

From the foregoing considerations for data transmission, the slip rate objective for the AT&T network in North America was set at one slip in 5 hr for an end-to-end connection [20]. Since slips can occur at multiple points within a network, the objective for slips at individual trunk and switching interfaces was set at one slip every 20 hr.

Example 7.3. Determine the relative accuracy requirements of two independent clocks to maintain a mutual slip rate objective of one slip in 20 hr. Assume a frame rate of 8 kHz as in PCM voice signals.

Solution. The slip rate objective implies that the frame rate produced by one

clock can be different than the frame rate produced by the other clock by no more than

$$\Delta F = \frac{1}{20(60)(60)} = 1.39(10)^{-5} \text{ slips per second}$$

Since there are 8000 frames per second, the relative accuracy is determined as

$$\frac{1.39(10)^{-5}}{8000} = 1.7(10)^{-9} \text{ slips/frame}$$

Hence the clocks must be accurate to 1.7 parts in 10^9.

Because Example 7.3 determines only a maximum relative inaccuracy, the absolute inaccuracy of each clock must be less than $(1.7)(10^{-9})/2$ or 0.85 parts in 10^9. As discussed later in this chapter, AT&T's revised approach to network synchronization [16] provides even better performance.

7.2.2 Asynchronous Multiplexing

In the preceding section certain aspects of network synchronization were discussed that implied the need for clock synchronization to prevent a loss of data by way of slips. In this section we discuss a procedure referred to as "pulse stuffing" that avoids both slips and clock synchronization. The term pulse stuffing can be somewhat misleading since it implies that pulses are inserted into the line code to make timing adjustments. Actually, pulse stuffing involves only the data stream and is independent of the line code or modulation system in use. Pulse stuffing is a term commonly used in North America while the same concept is referred to as "justification" in Europe.

The basic concept of pulse stuffing involves the use of an output channel whose rate is purposely higher than the input rate. Thus the output channel can carry all input data plus some variable number of "null" bits or "stuff bits." The null bits are not part of the incoming data. They are inserted, in a prescribed manner, to pad the input data stream to the higher output rate. Naturally, the extraneous null or "stuff" bits must be identifiable so that "destuffing" can recover the original data stream.

The practice of pulse stuffing arose when the initial digital TDM hierarchy (Table 1.10) was defined. At this time there were only isolated digital transmission links within the network that precluded synchronizing them to a common clock. When it came time to combine lower rate tributaries (e.g., DS1s) into higher level signals (e.g., DS2s) the multiplexing procedure necessarily had to accommodate tributaries operating at slightly different rates. The generic

term for combining these unsynchronized signals is asynchronous multiplexing. In this context, "asynchronous" refers to multiplexing of unsynchronized tributaries into a higher level signal (using pulse stuffing). It does not refer, in any way, to a means of transmission. The higher level signal is always carried on a synchronous transmission link.

As the amount of digital equipment in the network grew and more and more of it became interconnected both the means and the necessity for synchronization arose. Thus various forms of synchronous multiplexing have been proposed and developed. One of these techniques (SYNTRAN) is described in reference [21]. Another form of synchronous multiplexing (SONET) is described in conjunction with fiber systems in Chapter 8.

Pulse Stuffing Concepts

As a starting point for understanding the need for pulse stuffing consider the simple two-channel, bit-interleaved multiplexer in Figure 7.12. As indicated, within any string of even-numbered bits the number of bits carried in each subchannel is necessarily identical. Thus the rates of the subchannels are also identical. If the two input data streams are running at different rates the output can be synchronized to one of the channels but not both. Thus slips would necessarily occur in at least one of the tributaries.

As a simplified example of pulse stuffing Figure 7.13 shows a two-channel, bit-interleaved format as before but the additional detail needed to allow adjustments of the information flow within each subchannel. As indicated the multiplexed output is formatted into 10-bit master frames with 5 bits assigned to each subchannel. The first three bits in each subchannel of each masterframe always carry data from the respective tributaries. The fourth bit in each subchannel (C_1 and C_2) specify whether the last bits (S_1 and S_2) carry data or are stuff bits. When C_1 is a "1" S_1 is stuffed, otherwise S_1 carries tributary data. Hence each master frame can carry three or four bits from each tributary. If, on average, each tributary sends 3.5 bits during a masterframe variations of $\pm 14\%$ in the tributary clock rates can be accommodated.

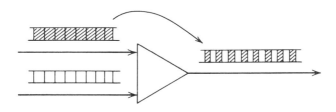

Figure 7.12. Two-channel multiplexer showing equal output data rates for each input.

Figure 7.13. Simplified pulse stuffing example.

An important point to notice about an asynchronous multiplexer is that the output frame structure is unrelated to the frame structure of the lower-level inputs. As far as the higher-level multiplexer is concerned, each input signal is merely a serial bit stream with no particular structure assumed. Framing bits in the lower-level multiplex signals are transmitted right along with the information bits. After the higher-level signal is demultiplexed (and destuffed), framing of the lower-level signals must be established for further demultiplexing.

Although pulse stuffing can be implemented with a variety of higher-level framing formats, the generally most desirable features of a pulse stuffing format can be identified as follows:

1. The use of fixed length master frames with each channel allowed to stuff or not to stuff a single bit in the master frame.
2. Redundant stuffing specifications.
3. Noninformation bits distributed across a master frame.

Timing offsets are generally quite small, so only small adjustments from a single occasional stuff bit are required. Thus large numbers of tributary bits can be combined into a master frame with one specific bit position identified as the S bit. Nominally, approximately one-half of the master frames contain the maximum number of information bits N_M, and the other half contain $N_M - 1$ information bits.

The purpose of a pulse stuffing operation is to prevent a loss of data when two interconnected digital transmission links are unsynchronized with respect to each other. If single bit errors can cause a stuff bit to be interpreted as information (or vice versa), the basic objective is lost. Furthermore, notice that an erroneous interpretation of a stuff code causes lower-level multiplex signals to lose framing. For these reasons, the interpretation of C bits must be encoded redundantly. If we assume that channel errors are random, the probability of misinterpreting a stuff code is

$$P_F = \sum \binom{2n + 1}{n + 1} p^{n+1}(1 - p)^n \tag{7.5}$$

where p is the probability of a channel error, and n is the number of correctable stuff code errors ($2n + 1$ bits in a stuff code).

Noninformation bits should be distributed across a master frame for several reasons. First, by separating these bits as much as possible, errors in redundant stuffing specification bits (C bits) are more likely to be independent. If the specification bits are too close together and burst errors are prevalent, the redundant encoding is of little use. Second, by distributing noninformation bits, the irregularity of information flow is minimized. When higher-level multiplex signals are demultiplexed, a clock for each individual lower-level signal needs to be derived from the irregular information rate in each channel. Generation of a suitably stable clock synchronized to the information rate is simplified if information bursts or gaps are minimized. Furthermore, elastic stores needed to smooth out the information rates are smaller when the tendency toward bursts is minimized.

M12 Multiplexer

An example of a higher level multiplexing format is provided in Figure 7.14. This is the format used for 6.312 Mbps DS2 signals in the North American digital hierarchy. A DS2 signal is derived by bit interleaving four DS1 signals and adding the appropriate overhead bits.

A DS2 masterframe is 1176 bits long. Of these there are 1148 information bits (287 per channel), 12 framing bits (M_0, M_1, F_0, F_1), 12 stuffing control bits (C_1, C_2, C_3), and 4S bits (S_1, S_2, S_3, S_4). Since an S bit can be a null bit or an information bit, each channel can send 287 or 288 bits in a master frame. An S bit is designated as an information bit if all three of the corresponding C bits are 1. The S bit is a null (stuff) bit if all three corresponding C bits are 0. Obviously, this encoding procedure allows for single error correction in the stuffing control bits.

The first level of framing is established by the alternating F_0, F_1, F_0... pattern. Notice that exactly 146 bits separate the F_0 and F_1 bits. Another level of framing for identifying the C and S bits is established by the M_0 and M_1 bits. A fourth M bit (M_x) is not used for framing and therefore can be used as an alarm service digit. Similar frame structures exist for other higher level digital signals. Figures 7.15 through 7.18 show the structures for DS3, DS4, DS1C, and the CEPT2 second level digital signal of CCITT, respectively.

Figure 7.14. Frame format of DS2 digital signal. Stuffing occurs in channel *i* when the previous C_i bits = 111. X is an alarm bit that equals 1 for no alarm condition. Framing is established by the $F_0 F_1 F_0$... sequence with 146 intervening bits.

Figure 7.15. Frame format of DS3 digital signal. P is even parity over all message bits in the previous master frame. Stuffing occurs in channel i when the previous C_i bits $= 111$. The X (alarm) bits and the P bits must be 11 or 00 so the $M_0 M_1 M_0$ sequence can identify the end of the master frame.

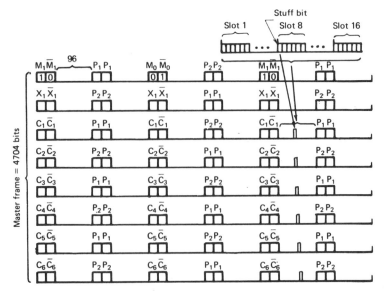

Figure 7.16. Frame format of DS4 digital signal. $C_i = 111$ implies stuff the eight message bit position for channel i following the last C_i. P_1 is even parity over the 192 previous odd-numbered message bits. P_0 is even parity over the 192 previous even-numbered message bits.

Figure 7.17. Frame format of DS1C digital signal. Stuffing occurs in channel *i* when the previous C_i bits = 111. X is an alarm bit that equals 1 for no alarm condition. Framing is established by the $F_0F_1F_0$... sequence with 158 intervening bits.

Figure 7.18. Frame format of second level digital signal of CCITT. Stuffing occurs in channel *i* when the previous C_i bits are 111.

Example 7.4. Determine the minimum and maximum input channel rates accommodated by an M12 multiplexer. Also determine the rate of DS1 misframes caused by an erroneous interpretation of a stuffed bit. Assume the bit error rate is 10^{-6}.

Solution. The maximum information rate per channel is determined as

$$\frac{6.312(288)}{1176} = 1.5458 \text{ Mbps}$$

The minimum information rate per channel is determined as

$$\frac{6.312(287)}{1176} = 1.5404 \text{ Mbps}$$

Since there are three possible combinations of two errors in the C bits, the probability of misinterpreting an S bit is closely approximated by

$$3(10^{-6})^2 = 3 \cdot 10^{-12}$$

The duration of each master frame is $1176/6.312 = 186 \, \mu\text{sec}$. Thus the rate of misframes per DS1 signal is

$$\frac{(3)10^{-12}}{(186)(10)^{-6}} = 0.016(10)^{-6} \text{ misframes per second}$$

which is equivalent to one misframe every 2 years.

Example 7.4 demonstrates that the tolerance of a 1.544-MHz DS1 clock is -3.572 to $+1.796$ kHz. Thus the relative accuracy between the DS1 and DS2 clocks must be $1.796/1544$ or only 1 part in 860. This relatively large timing tolerance is greater than what is normally required for reasonable clock and line instabilities. The timing adjustment capabilities and unsymmetric tolerance range were chosen out of a desire to (1) minimize DS2 reframe times, (2) provide a line clock that is a multiple of 8 kHz, and (3) minimize waiting time jitter [22].

Elastic Store Size Requirements
A functional model of an M12 multiplexer is shown in Figure 7.19. Associated with each lower level (DS1) input is an elastic store to hold incoming data until it is transferred to the higher level (DS2) output. The elastic stores serve two

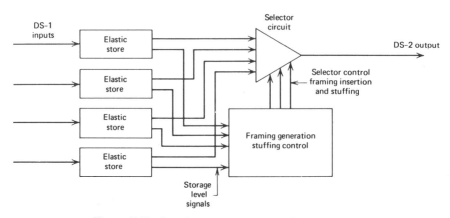

Figure 7.19. Functional diagram of a M12 multiplexer.

purposes: to remove the arrival jitter of the incoming data and hold data for the proper time slots. In addition to generating framing, the control logic of the multiplexer monitors the storage level (which serves as a phase comparator) of each elastic store and initiates a stuffing operation whenever the elastic store is less than half full. Conversely, no stuffing occurs when the elastic store is more than half full.

Since stuffing can occur at only certain times and only at a certain maximum rate, the elastic store must be at least as large as the peak jitter (peak phase offset) of the incoming signal. As discussed at the beginning of this chapter, jitter accumulates along the entire length of a repeatered transmission link. Thus longer links require larger elastic stores if slips are to be prevented.

As an example of the relationship between line length (number of repeaters) and number of bits of elastic storage needed by an M12 multiplexer refer to Figure 7.20. This figure was submitted by AT&T to the CCITT special study group on jitter [23]. The analysis is an extension of the systematic jitter analysis of reference [5].

Figure 7.20. Maximum number of regenerative repeaters as a function of elastic store size and jitter slope. A = number of cells in elastic store.

The abscissa of Figure 7.20 is the maximum phase slope produced by a clock recovery circuit in the presence of a worst case shift in data patterns (worst case systematic jitter). Since phase slope is nothing more than frequency offset, the required elastic store size can be determined as the maximum phase slope times its maximum duration. Since the total phase slope is proportional to the number of repeaters, Figure 7.20 displays the maximum number of repeaters per storage cell in an elastic store versus the jitter slope of an individual repeater when making a worst case timing transition.

As an example, first generation T_1 repeaters produce a worst case slope of 2.4 kHz. M12 multiplexers allocate 5 bits of storage to input phase jitter (3 more bits are included for implementation ease and waiting time jitter). From Figure 7.20 it can be seen that the ratio of N_{max} to A is 56, which implies that $N_{max} = (56)(5) = 280$ repeaters.

7.2.3 Waiting Time Jitter

When demultiplexing a higher-level TDM data stream, it is necessary to generate a clock for each derived subchannel. Because the subchannels are transferred (or transmitted) as synchronous data, the derived clock must be continuous. Derivation of synchronous subchannel clocks is complicated by the insertion into TDM data streams of overhead bits that create gaps in the bit arrival times. Irregularity in the data arrival rate caused by these gaps is referred to as "waiting time jitter."

Most of the overhead bits (e.g., framing bits, parity bits, stuffing control C bits) occur on a regular and predictable basis. The waiting time jitter caused by these gaps can be eliminated easily with an elastic store and a output clock derived from the incoming line clock. For example, a single PCM channel clock at 64 kbps can be derived from a 1.544 Mbps T1 line clock by multiplying by 8 and dividing by 193.

Figure 7.21. Functional diagram of M12 demultiplexer.

In contrast, waiting time jitter produced by pulse stuffing is significantly more difficult to deal with. The difficulty arises because waiting times produced by stuffed pulses are irregular and unpredictable. For this reason the subchannel output clocks derived from a pulse stuffed TDM line must be derived independently and only from the average arrival rate of each channel's data— not from the higher-level TDM rate!

Output clocks from M12 demultiplexers are generated using jitter removing elastic stores as shown in Figure 7.21. If large elastic stores and very slowly adjusted output clocks are used, most of the jitter can be removed. Unfortunately, waiting time jitter has frequency components down to zero frequency so the jitter can never be eliminated entirely. However, the jitter can be confined to as low a band of frequencies as desired by using a large enough elastic store.

Waiting time jitter is basically a function of how often pulses are stuffed, but it is also dependent on the ratio of actual stuffs to stuffing opportunities. If the input clock is jitter-free the output jitter peaks when one-half of the opportunities are used. From the point of view of maximum tolerance for clock offsets, a stuffing ratio of one-half is ideal. To reduce the waiting time jitter, however, stuffing ratios of approximately one-third are often used. For a thorough analysis of waiting time jitter see reference [24]. As an example of waiting time jitter dependence on stuffing (justification) ratios, see Figure 7.22 obtained from reference [23]. The abscissa of Figure 7.22 represents the ratio of stuffs to opportunities while the ordinate is jitter power produced by a single pulse stuffing process. The jitter power is expressed in decibels relative to one time slot squared (0 dB implies an rms jitter value of 1 bit-time squared, i.e., 2π radians squared). Curve A shows the output jitter produced when the input jitter is -20 dB (0.1 slot rms). Curve B shows the output jitter when the input jitter is -12 dB (0.25 slot rms). Figure 7.22 shows jitter produced by a single mult-

A: Tributary jitter = 0.10 slot r.m.s.
B: Tributary jitter = 0.25 slot r.m.s.

Figure 7.22. Waiting time jitter dependence on justification ratio.

iplexer. From measured data [23], a good order-of-magnitude estimate of the waiting time jitter accumulated by N_m tandem M12 stuffing/destuffing operations is

$$\sigma_w^2 = \frac{N_m}{100} \text{ slots}^2 \tag{7.6}$$

7.3 NETWORK SYNCHRONIZATION

As discussed in the preceding section, whenever a digital transmission link is connected to a digital switch it is desirable to synchronize the two systems by having the transmission link obtain its timing from the switch. An obvious exception to this mode of operation occurs when a digital transmission link is connected to a digital switch on both ends. Generally, a transmission link in an all-digital network derives its timing from just one of the switches to which it is connected. If the other switch is not synchronized to the first in some manner, an unsynchronized interface necessarily results. This section is concerned with network synchronization as a whole, not simply the synchronization of a single interface. Basically, network synchronization involves synchronizing the switches of the network. The transmission links can then be synchronized automatically by deriving timing directly from a switching node.

There are two basic reasons for paying assiduous attention to the timing requirements of a digital network. First, the network must prevent uncontrolled slips that could produce misframes, inadvertent disconnects, and crossconnects. It is generally very difficult, or very expensive, to prevent slips altogether. Thus a second aspect of a network timing plan requires establishing a maximum rate of controlled slips as part of the end-to-end circuit quality objectives.

Synchronizing private networks is sometimes difficult because the network topologies are not designed with network synchronization in mind and the switching equipment (PBXs) are not designed to provide synchronization to other nodes. Furthermore, the private networks often interface to multiple carriers in multiple locations. Determining which signals to synchronize to, particularly on a dynamic basis when a reference signal becomes unavailable, is exceptionally difficult.

There are six basic approaches used, or considered for use, in synchronizing a digital network:

1. Plesiochronous
2. Networkwide pulse stuffing
3. Mutual synchronization
4. Network master
5. Master–slave clocking
6. Packetization.

7.3.1 Plesiochronous

A plesiochronous network does not synchronize the network, but merely uses highly accurate clocks at all switching nodes so the slip rate between the nodes is acceptably low. This mode of operation is probably the simplest to implement since it avoids distributing timing throughout the network. A plesiochronous network, however, implies that the smaller switching nodes carry the cost burden of highly accurate and redundant timing sources. As a compromise, large networks can be divided into subnetworks for timing purposes and use plesiochronous operations for intersubnetwork synchronization and some other, more cost-effective, means of providing intrasubnetwork synchronization. As described in Section 7.5 the public telephone network in the United States uses plesiochronous synchronization at the upper levels.

Plesiochronous timing is also used to synchronize international digital network interconnections. In recommendation G.811 [25], the CCITT has established the stability objectives for clocks of all international gateway digital switches. The stability objective of one part in 10^{11} implies that slips between international gateway switches will occur at a rate of one per 70 days. (This assumes one clock is positive one part in 10^{11} and another clock is negative one part in 10^{11}.)

7.3.2 Networkwide Pulse Stuffing

If all internal links and switches of a network were designed to run at nominal rates slightly higher than the nominal rates of the voice digitization processes, all voice signals could propagate through the network without slips by stuffing the information rate up to the local channel rate. None of the clocks would have to be synchronized to each other, and relatively coarse clock accuracies could be tolerated. At every interface between systems running under different clocks, however, the individual channels would have to be unstuffed from the incoming rate and stuffed up to the local or outgoing rate. In essence, the TDM links of the network would provide TDM channels through which user data flows at lower and variable rates—the differences being absorbed by internal pulse stuffing.

In contrast to pulse stuffing operations of higher-level multiplexers where all channels in a lower-level digital signal are stuffed as a group, switching operations imply that each channel must be stuffed independently. The need for separate pulse stuffing operations is illustrated in Figure 7.23, which depicts two voice signals being switched into a common TDM outlet link. Obviously, the bit rate R_3 of both output channels is identical. If the two channels originate in portions of the network running under different clocks R_1 and R_2, the pulse stuffing adjustments must be made separately for each channel.

The need to independently stuff each channel at every switching interface precludes pulse stuffing as an economically feasible solution to networkwide timing problems. Pulse stuffing is used for higher-level multiplexers primarily because other synchronization procedures were not yet established. As the

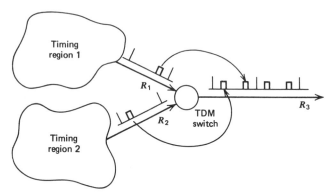

Figure 7.23. Switching two channels with different rates onto a common TDM output.

digital network evolves and the ability to synchronize the clocks of all lower-level digital signals arises, pulse stuffing may become unnecessary. One instance in which pulse stuffing might always be desirable is over digital satellite links that are timed by free running clocks in the satellite [26].

7.3.3 Mutual Synchronization

The two preceding sections discuss modes of operation for the network that do not involve synchronization of individual clocks. This section, and the next two, describe network timing plans that synchronize each individual clock to a common frequency. The first method, mutual synchronization, establishes a common network clock frequency by having all nodes in the network exchange frequency references as shown in Figure 7.24. Each node averages the incoming references and uses this for its local and transmitted clock. After an initialization period, the network clock normally converges to a single stable frequency. Under certain conditions, however, the averaging process can become unstable [27].

The main attractiveness of a mutually synchronized network is its ability to remain operational in spite of a clock failure in any node. The main disadvantages are the uncertainties of the exact average frequency and unknown transient behavior. For these reasons, mutual synchronization is not being considered for the North American telephone network. In Great Britain, however, a hierarchical timing structure has been investigated that utilizes mutual synchronization within some of the levels [28].

7.3.4 Network Master

Another method of synchronizing the network is shown in Figure 7.25. With this method a single master clock is transmitted to all nodes enabling them to lock onto a common frequency. As indicated, all network nodes are directly connected to the network master implying the need for a separate transmission

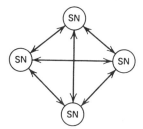

Figure 7.24. Mutual synchronization. SN = switching node.

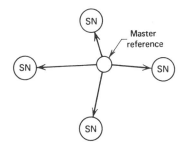

Figure 7.25. Network master synchronization. SN = switching node.

network dedicated to the distribution of the reference. Reliability considerations also imply that alternate paths be provided to each node. Because of cost considerations for the separate timing network and reliability problems with reference distribution a network master with direct transmission to each node is undesirable.

7.3.5 Master–Slave Synchronization

The main drawback to network master synchronization as described in the preceding section is its need for separate and reliable transmission facilities to every node. Figure 7.26 shows a network configuration that disseminates a master reference by way of the message links themselves. A network reference frequency is transmitted to a few selected higher-level switching nodes. After these nodes synchronize their clocks to the reference, and remove transmission link-induced timing jitter, the reference is passed on to lower-level switches by way of existing digital links. The next lower-level switches, in turn, synchronize to an incoming link from the higher level and pass timing on to another level of switches by way of their outgoing digital links. The process of passing the reference downward from one level to the next is referred to as "master–slave synchronization."

Since all switching nodes in the network are synchronized either directly or indirectly to the same reference, they all run at the same nominal clock rate. Thus slips should not occur normally. However, because of the different paths through which timing is disseminated, short-term frequency differences can

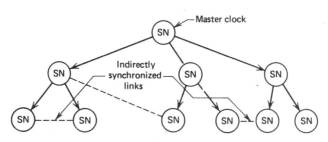

Figure 7.26. Master–slave synchronization.

occur between some nodes. If these nodes are synchronized indirectly, as shown in Figure 7.26, infrequent slips might occur. Furthermore, reliability considerations imply that backup clocks must be provided in all switches should the clock distribution system fail. When this happens, slips become more likely, but only after relatively stable backup clocks have had enough time to drift from the common reference frequency.

AT&T and the United States Independent Telephone Association (USITA) originally selected master–slave synchronization for the switched digital network in the United States [29]. The reference frequency is located in Hillsboro, Missouri from which selected No. 4 ESS switching centers receive their timing by way of dedicated transmission facilities. Synchronization of all other switches occurs by way of existing digital transmission links. As discussed in Section 7.3.7 the original plan has been modified to use plesiochronous synchronization at the highest level.

7.3.6 Packetization

The synchronization discussions of the five preceding sections have assumed implicitly that a synchronous, circuit-switched network was being considered, since prevailing digital voice networks operate in that manner. For completeness, however, another form of network must be mentioned—a packet-switched network.

As discussed in Chapter 9, packet-switched networks break up messages into identifiable blocks (packets) of data. In between the blocks, the transmission links carry either idle codes or control messages. If all messages (control and data) are separated by a nominal interval of idle transmission, elastic stores can be reset in preparation for the next block. As long as each block is limited in length, the elastic stores can absorb clock differences and avoid losses of data. (In essence, slips occur in the idle codes only.)

Packet-switched networks have been developed primarily for data communications applications although packet-switched voice networks have also been proposed. A significant proposal in these regards is the broadband ISDN (BISDN) network recommendation that specifies asynchronous transfer mode (ATM) for the network architecture for voice and data.

7.3.7 Network Timing Performance Measurements

After choosing a synchronization architecture for a network it is necessary to be able to measure the quality of the timing signals within the network to qualify equipment, determine performance margins, isolate faulty equipment or transmission links, and possibly evaluate design alternatives. ANSI and CCITT standards use the concept of time interval error (TIE) and maximum time interval error (MTIE) for these purposes. A TIE is the difference in delay between a selected timing signal and an ideal timing signal at the end of a particular time period (i.e., the error at the end of a time interval). An MTIE is the maximum error that occurs at any time during a particular interval. These concepts are illustrated in Figure 7.27.

Figure 7.27a displays the TIE that occurs when measuring a perfectly stable but inaccurate timing signal. Because the frequency of the signal under test differs from the reference frequency by a constant value (Δf) the TIE is directly proportional to the measurement interval S. The error is determined by counting clock cycles in both signals and expressing the difference as the time required by the test signal to catch up (or fall back) to the current reference count. Thus the TIE produced by a constant frequency offset is

$$\text{TIE} = \Delta n/f$$
$$= [S(f + \Delta f) - S(f)]/f$$
$$= S(\Delta f/f) \tag{7.7}$$

where Δn is the accumulated difference in clock cycles, Δf is the frequency offset, and S is the measurement interval.

Because the TIE in Figure 7.27a increases monotonically with S the MTIE and the TIE are identical. In contrast, Figure 7.27b depicts the TIE produced by a timing signal with no long-term frequency offset but some short-term instabilities (jitter and wander) as would occur if the reference was relayed through a network. As indicated, the TIE varies as a function of S but is bounded. The MTIE is the largest TIE in a measurement interval.

Figure 7.27c displays the more general case where the signal under test contains both instability and a frequency offset. In this case the choice of the value for S is critical. If S is too small jitter will mask the presence of the offset. If S is too large the frequency offset dominates the TIE and MTIE measurements. Thus periodic measurements need to be recorded for a time history to fully characterize the timing imperfections.

The first three examples in Figure 7.27 assume the use of a "perfect" reference with which to make the measurements. As long as the timing imperfections are significantly greater than the corresponding imperfections of an "imperfect" reference meaningful absolute measures are possible. Sometimes the relative performance between two timing signals is of interest. For example, a PBX with T1 connections to two different places (or carriers) is more dependent on the

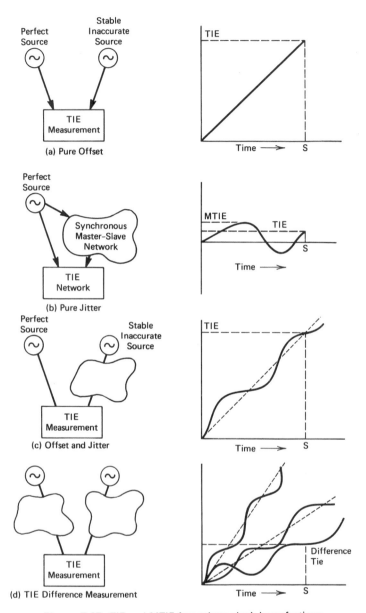

Figure 7.27. TIE and MTIE for various clock imperfections.

timing differences between the two transmission links than on the absolute performances. As long as the PBX can synchronize to both of the links slips occur only as a result of differences between the two signals. Figure 7.27d depicts measuring the relative performance of two such links. In this case, all measurements are meaningful—even arbitrarily low frequency wander which

can cause slips but are difficult or impossible to measure in an absolute sense. Of course, if relative measurements indicate timing problems exist absolute measurements are needed to isolate the problem source.

7.4 U.S. NETWORK SYNCHRONIZATION

The original plan for synchronization of the U.S. network formulated by the Bell System and the U.S. Independent Telephone Association was to use master-slave synchronization with a single master clock [29]. Due to the breakup of the network into multiple independent companies and to difficulties in reliably distributing a highly accurate master, the synchronization architecture for the U.S. was changed to a plesiochronous/hierarchical design [3, 16, 30].

7.4.1 Synchronization Regions

As shown in Figure 7.28 the public network is now partitioned into synchronization regions that are internally synchronized with a master–slave timing hierarchy that establishes different levels of timing quality: stratum 1 to stratum 4. Stratum 1 clocks have the highest quality while stratum 4 clocks have the lowest. Timing for each region is established by a primary reference source (PRS) at stratum 1. Stratum 1 clocks are free running clocks with inaccuracies no greater than one part in 10^{11}. Some regions may have their own PRS while others may use a synchronization signal from another party (e.g., an interexchange carrier such as AT&T).

For the most part, the synchronization regions correspond to LATAs. Every region must have at least one stratum 2 clock, which is typically associated with an access tandem switch. Toll offices within the LATAs may also have stratum 2 clocks. All toll switches within the AT&T network contain stratum 2 clocks [16]. Except when they use a common PRS, the synchronization regions are independently synchronized. Thus connections between the regions (using interexchange carriers like AT&T) are typically plesiochronously timed. However, all nodes within a region are synchronized in a master–slave hierarchy as indicated.

The accuracy requirements of the four levels of stratum clocks is provided in Table 7.2. These accuracies pertain only to situations in which the nodes are operating in a free-running mode. Normally the nodes are synchronized to higher-level clocks so the accuracy is traceable back to the respective PRS. In addition to listing free-running accuracies Table 7.2 lists accuracies that must be met during holdover states. The holdover mode for stratum 2 and 3 clocks requires these nodes to gradually transition to the free-running state when they lose their reference signals. A stratum 2 clock, for example, loses the previously established frequency at the rate of 1×10^{-10} per day—implying it would take 160 days to reach the worst case free-running accuracy. Thus free running accuracies at the stratum 2 and 3 levels of the hierarchy are hopefully relatively

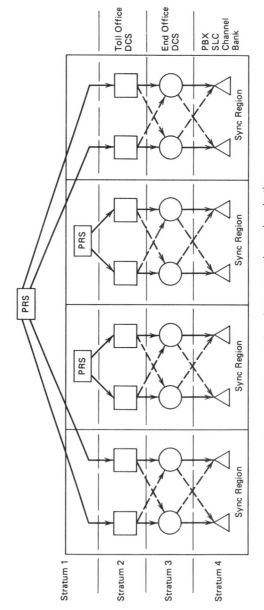

Figure 7.28. North American network synchronization.

TABLE 7.2 Stratum Clock Accuracy Requirements

Level	Minimum Free-Run Accuracy	24-hr Holdover Accuracy
Stratum 1	1×10^{-11}	Not applicable
Stratum 2	1.6×10^{-8}	1×10^{-10}
Stratum 3	4.6×10^{-6}	255 slips/24 hr
Stratum 4	32×10^{-6}	Not required

unimportant. Stratum 4 clocks, however, have no holdover requirements so they enter the free-running state immediately upon losing their reference(s).

Network synchronization requirements also specify how the various nodes respond to degradation or complete failure of their references. Stratum 2 and 3 clocks must bridge short interruptions in the reference with minimum specified time keeping errors [3]. Stratum 3 clocks, which typically have a primary reference and a secondary reference, must perform a very gradual switchover to the secondary reference when the primary fails. Abrupt switchovers, as often occur in stratum 4 nodes (PBXs), can cause phase transients in the output clock, which can in turn disrupt synchronization in all down stream devices (e.g., within a private network).

7.4.2 Primary Reference Sources

The ANSI MTIE specification for the accuracy of a primary reference source is shown in Figure 7.29. Notice that jitter (instability that can be observed in less than 0.05 sec) is not specified. Wander, as measured over 500-sec time intervals, is limited to 3000 nsec. The asymptote for long term timing errors corresponds to inaccuracies of 1×10^{-11}.

In addition to being designed with highly accurate (cesium beam or rubidium) clocks all PRSs must be continuously verified with universal coordinated time (UTC). Such verification can be achieved by using a UTC-based navigation system such as Loran-C or Global Positioning System (GPS). Loran-C is used by MCI to directly synchronize its PRS [3]. AT&T uses GPS to monitor (verify) the long-term accuracy of each of 14 PRS nodes established in North America. The typical accuracy of these nodes is much better than the ANSI or CCITT requirement [16].

7.5 NETWORK CONTROL

In essence, the synchronization procedures described in the preceding section represent methods for controlling the timing between transmission and switching systems. In this section we discuss synchronization in a more general sense. Instead of just time or frequency control, the synchronization concept is

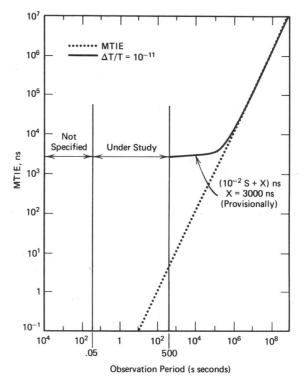

Figure 7.29. Permissible MTIE vs. observation period at the output of a primary reference source.

extended to higher-level functions of connection control and network control as a whole. Fundamental to the control concept is the interaction between two processes (e.g., the exchange of information from one switching machine to another to set up or monitor a connection).

A particularly useful means of defining the interaction of two processes is a state transition diagram. The main purpose of the state transition diagram is to abstract the operational states of a process and indicate what events cause transitions from one state to another. When these events are messages (signaling tones) from another process, the state transition diagram effectively defines how two communicating processes interact.

7.5.1 Hierarchical Synchronization Processes

As an example of a more generalized concept of synchronization, Figure 7.30 has been included to demonstrate three distinct levels of control for a conventional telephone connection using digital transmission and multiplexing. The lowest-level process shown in Figure 7.30a depicts nothing more than the clock synchronization process required to transmit and receive digital in-

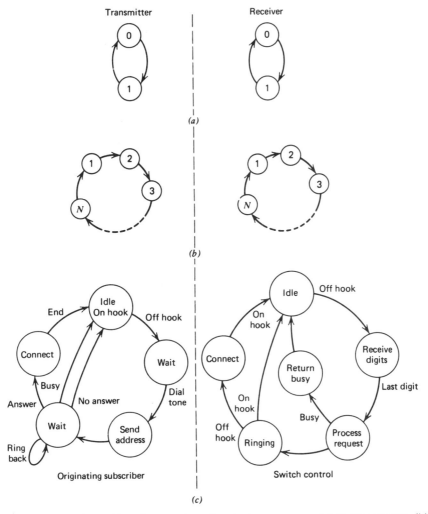

Figure 7.30. State transition diagram of synchronization processes. (a) Timing process. (b) Framing process. (c) Telephone connection process.

formation. There are only two states to the process in both the transmitter and the receiver. The purpose of clock synchronization is to cause transitions between the two states in the receiver to coincide with transitions in the clocking process of the transmitter. To accomplish this a certain amount of transmission capacity is required in the form of line code transitions.

Figure 7.30b depicts a higher-level synchronization process involving the framing of a time division multiplexer. Both processes represent a modulo-N counter where N is the number of channels in the TDM frame. The two processes are synchronized (framed) by utilizing some of the transmission

capacity to send framing patterns. Once the receiver acquires framing, the counter in the receiver counts in synchronism with the counter in the transmitter so that individual TDM channels are properly identified.

Figure 7.30c provides state transition diagrams of a somewhat more complicated, but easily understood process. The figure depicts the connection control of a conventional telephone call. The state transition diagram of the first process represents a subscriber placing a call (going off-hook). The second state transition diagram represents the sequence of states the control element in the local switch goes through to set up the connection.

As indicated, the process begins by the originating subscriber going off-hook and waiting for the dial tone. When the switch recognizes the off-hook signal (current flow in the line), it connects the subscriber line to a digit receiver that returns dial tone. The subscriber then dials the address of the desired telephone and enters another wait state. Upon receiving the last digit of the address, the switch control processes the request. If a local connection is desired, a busy tone or a ringback tone is returned to the originating subscriber. A busy tone prompts the subscriber to hang up (go on-hook) while a ringback signal causes the subscriber to stay in the wait state until the called party answers, or until the caller "times out" and abandons the call.

When the called party answers, both processes enter the connected state and communication between the end users begins. The end users then get involved in yet another level of "synchronization." Voice telephone users begin by exchanging greetings and identities to establish a "connection" between their thought processes to communicate on a mutually understood subject. The message exchange process also requires synchronization so that only one person talks at a time. Hence various forms of control signals are needed to "turn the line around." Although being somewhat subtle in nature, these control signals represent transmission overhead in the same sense as control signals within the network. A talker may indicate his end-of-transmission by asking a question, by inflections in the voice, by the message itself, or more commonly by a pause.

Data communications equipment goes through the same basic procedures in order to establish connections and exchange information. In this case, the procedures are defined more formally and, consequently, are more restrictive. The formal rules of communication between data communications equipment are usually refered to as a "protocol." Data communications protocols typically include a definition of certain control codes, code interpretations, message framing, turn-around procedures for half-duplex lines, error control, message sequencing, fault control, and recovery.

Automated fault control and recovery procedures for communications networks can become quite involved and difficult to implement reliably. When individual voice circuits malfunction (e.g., become noisy or disconnected), the recovery procedures are left to the users. They merely redial the connection and take up where they left off. However, large trunk groups or switching systems must be designed for higher levels of dependability and maintainability. The dependability criterion ensures that failures or malfunctions rarely occur, or that

they are circumvented automatically by protection switching. High levels of maintainability ensure that failures are repaired quickly when they occur. Within switching systems, most of the instructions and memory words of the processor are typically dedicated to hardware and software performance monitoring, recovery procedures, and maintenance diagnostics.

7.6 NETWORK MANAGEMENT

In addition to controlling individual connections and equipment, a communications network must also manage its facilities on more macroscopic levels. The basic goal of network management is to maintain efficient operations during equipment failures and traffic overloads. The main considerations are the following:

1. Routing control
2. Flow control.

7.6.1 Routing Control

Routing control refers to procedures that determine which paths in a network are assigned to particular connections. If possible, connections should use the most direct routes at the lowest levels of the network. The direct routes are obviously desirable because they use fewer network facilities and generally provide better transmission quality. However, economic considerations often limit the capacities of the direct routes so that alternate routes are needed to maintain suitably low-blocking probabilities between one end office and another.

If a trunk group between two switching offices contains enough circuits to provide an acceptably low-blocking probability, a significant number of the circuits in the group are idle during average traffic loads. A more economical design allocates a limited number of heavily utilized trunks in the direct route and provides alternate routes for overflow (alternately routed) traffic. In this manner the users are able to share larger portions of the network. Chapter 10 presents basic examples of how a network can be engineered to minimize the transmission facilities while providing a given grade of service (blocking probability). As discussed in Chapter 1 the use of centralized control for the network, with common channel signaling, provides significant efficiencies of operation in congested networks.

7.6.2 Flow Control

In the preceding section, alternate routing is discussed as one aspect of managing traffic in a communications network. Routing algorithms are concerned only with the utilization of paths or directions of travel within a network.

Another requirement of network management is to control the amount of traffic in a network. Managing the rate at which traffic enters a network is referred to as flow control. A network without effective flow control procedures becomes very inefficient or ceases to function entirely when presented with excessively heavy traffic conditions.

The generalized performance of a large, uncontrolled network is shown in Figure 7.31 as a function of the offered traffic. As indicated, when light traffic conditions exist, the network carries all traffic requests presented. As the load increases, however, some of the offered traffic is rejected because no appropriate circuits are available for particular connections, that is, blocking exists. As the input load increases even further, a network with no flow control eventually begins to carry less traffic than it does when lighter loads are presented. If the offered load increases even more, the network may even cease to carry any traffic at all.

The reason that the volume of carried traffic decreases when the offered traffic exceeds some critical value is that partially completed requests tie up network resources while trying to acquire other resources tied up by other partially completed requests. Thus a form of deadlock occurs. As a specific example, consider the public telephone network and the occurrence of extremely heavy traffic between two distant cities. Since each request requires an end-to-end connection, a sequence of intermediate transmission and switching facilities is needed to set up each circuit. In a network with distributed control both sources of traffic are serviced by successively seizing trunks to intermediate switching nodes until the destination is reached. If heavy traffic exists, requests emanating from both sides of the network encounter congestion somewhere near the middle. At that time, the partially completed connections are tying up facilities needed by other requests. In the limit, when extremely heavy traffic exists, all network resources are held by partially completed requests and no complete connections can be established!

Another example of the need for flow control, is automobile traffic at a metropolitan intersection. Have you ever encountered an intersection in which

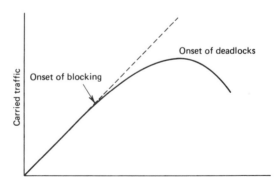

Figure 7.31. Traffic carried vs. traffic offered for a network with no flow control.

your lane of traffic was blocked by cross traffic backed up into the intersection? In essence, the driver blocking your direction of travel seized a common resource (the middle of the intersection) without being able to obtain the next resource required (the other side of the intersection). Bumper-to-bumper traffic in one direction can significantly degrade the throughput in other directions. With heavy traffic in all directions, total throughput can grind to a halt until the congestion is relieved from the periphery inward.

The fundamental principle demonstrated by these examples is that efficient use of the common resources of a heavily loaded network requires some form of flow control. In the automobile example, smooth operation of an intersection depends on each individual driver looking ahead and not entering an intersection unless he can get all the way across. A telecommunications network must use the same basic principle (it is hoped with more discipline). The control elements at the periphery of the network must be aware of the internal status of the network and control the flow of traffic from its sources.

More than one level of flow control is sometimes implemented within a network. In a data communications link, some form of flow control is usually required to keep a source terminal from overloading the terminal at the other end of the link. The receiving terminal uses a reverse channel (sometimes with a lower bit rate) to inform the source when to cease or begin transmissions. This level of flow control in a circuit-switched network involves the terminals themselves and is of no concern to the network since the traffic flows within an established connection. Of more concern to a circuit-switched network is how to control the flow of connection requests into the interior of the network. In setting up a long-distance connection the first few circuits required should not be seized unless there is a reasonable chance that all of the circuits necessary to complete the connection can be obtained. Partially completed circuits only degrade the network capacity by increasing congestion without satisfying a service request. Network flow control is greatly simplified with common channel signaling support for centralized network control. The following paragraphs describe basic, uncentralized, flow control techniques and how centralized control simplifies their implementation.

Dial Tone Delay

A natural consequence of a busy network at the local level is that common equipment such as digit receivers is much in demand. The effect of the heavy use is to produce queues for service and concomitant delays in returned dial tones. Although dial tone delays are not intended to do so, they provide an effective means of slowing the flow of requests into the interior of the network. Since flow control for local connections is normally unnecessary, dial tone delay is undesirable as an intentional means of delaying service requests at the periphery of the network. A preferred means of flow control is first to receive the digits of a service request, and then to reject the request, if necessary, by returning a reorder tone (busy tone) or recorded message to the originating subscriber. The subscriber can effectively place himself in a queue for service, if he continually

repeats the request. This approach has the one drawback that subscribers with serviceable requests experience increased dial tone and processing delays because of the load produced by unserviceable retries.

Trunk Directionalization

The operation of trunk circuits can be classified according to two different ways of controlling seizures for particular calls. Two-way trunks can be seized at either end. One-way trunks, on the other hand, can be seized only at one end. (Notice that this has nothing to do with the direction of message transfer on established connections, which is always in both directions.) When one-way trunking is used, the trunk group is usually partitioned into one group that can be seized at one end and one group that can be seized at the other end. Two-way trunk groups are obviously more flexible in servicing fluctuating traffic patterns, but they are more difficult to control since the possibility of simultaneous seizures (called glare) at both ends must be resolved.

A useful feature to incorporate into two-way trunks is the ability to directionalize them by marking them busy at one end and effectively creating one-way trunks. With this mechanism, a distant, overloaded switching node can be relieved of additional incoming traffic while providing it sole access to the trunk group for outgoing traffic. Thus the overloaded node relieves its congestion while inhibiting new arrivals.

When the network as a whole experiences heavy traffic loads, trunk directionalization can be used to reduce the flow of connect requests into the interior of the network while establishing one-way trunks from the interior to the periphery. Thus connect requests that manage to get to the interior have a much better chance of obtaining all facilities required for the complete connection.

Cancellation of Alternate Routing

Alternate routing of traffic is primarily useful for accommodating localized overloads by transferring traffic to underutilized routes. During networkwide overloads, however, alternate routing is undesirable for two reasons. First, alternate routes imply that a greater number of transmission and switching facilities are needed for a connection. If these same facilities could be assigned to two or more direct connections, the total number of links per call could be reduced and the network could carry more traffic.

Second, the probability that an alternately routed call can acquire all the necessary resources is relatively low. Trying to set up a connection with a large number of facilities is undesirable if the probability of getting all of the facilities is low (particularly so, if some facilities are fruitlessly tied up while less demanding requests are pending).

Code Blocking

Code blocking refers to artificially blocking calls intended for specific destination codes. If the calls are blocked at originating end offices before they

acquire internal network facilities, the destinations are relieved of incoming traffic without tying up facilities that may be needed for outgoing requests from the specified areas.

The method of flow control is particularly useful in times of natural disasters, which typically stimulate large numbers of calls both into and out of the area of the disaster. In these events a network control center can initiate code blocking for all, or a large percentage, of the calls into the area. The principle of giving preference to the outgoing calls serves two purposes. First, no network facilities are seized unless there is a reasonable chance of obtaining all facilities necessary. It is the trunks into or out of the disaster area that are the focal point of network congestion. Once one of these trunks is seized the rest of the connection can probably be established. Second, code blocking is useful because outgoing calls are probably more important than incoming calls.

Centralized Connection Control

All of the flow control procedures described previously are designed to eliminate seizures of common resources if the desired connection has a low completion probability. Because of the distributed nature of network control implied by these operations, these control procedures are necessarily probabilistic. In order to maintain a certain amount of network efficiency, the network is purposely operated at less than maximum capacity.

A more desirable mode of operation, from a throughput point of view, is to allocate network facilities from a single centralized control node. Since this central node has access to the status of all network resources, no facilities are assigned to a particular request unless all facilities needed for the complete connection are available. In this manner network transmission links are assigned in an analogous fashion as internal links of common control switches.

Complete centralized control of a network as large as the public telephone network is obviously infeasible from the point of view of maintaining status of all interstage links within end-office switches, and from the point of view of survivability of the network when the control node fails. However, many aspects of centralized control have been implemented in North America and around the world with common channel signaling (CCS). For example, INWATS call requests are routed to a central node that determines if the destination is busy or not. If the destination is busy, the originating end office is instructed to return the busy tone without any of the internal transmission links ever being seized [31]. This mode of operation is particularly useful for 800 numbers(INWATS) that occasionally experience very heavy traffic flow because of national television announcements.

Without CCS, the previous mode of operation was to return a busy tone all the way through the network from the place at which the busy circuit or subscriber is located. Thus the path through the network was tied up during the time the busy tone was being returned. CCS allows the originating office to return the busy tone so internal network facilities can be released and reassigned immediately upon detecting the busy condition.

REFERENCES

1 M. Decina and U. deJulio, "International Activities on Network Synchronization for Digital Communication," *IEEE International Communications Conference*, 1979.

2 *Reference Data for Radio Engineers*, Howard Sams Inc., Indianapolis, Indiana, 1972.

3 *Synchronization Interface Standards for Digital Networks: ANSI T1.101-1987*, American National Standards Institute, New York, 1987.

4 J. R. Pierce, "Synchronizing Digital Networks," *Bell System Technical Journal*, March 1969, pp. 615–636.

5 C. J. Byrne, B. J. Karafin, and D. B. Robinson, "Systematic Jitter in a Chain of Digital Repeaters," *Bell System Technical Journal*, November 1963, pp 2679–2714.

6 Digitally Dejittering Laser Disc Players," *IEEE Spectrum*, February 1990, p 14.

7 F. M. Gardner, *Phaselock Techniques*, 2nd ed., Wiley, New York, 1979.

8 E. D. Sunde, "Self-Timing Regenerative Repeaters," *Bell System Technical Journal*, July 1957, pp 891–938.

9 D. L. Duttweiler, "The Jitter Performance of Phase-Locked Loops Extracting Timing from Baseband Data Waveforms," *Bell System Technical Journal*, January 1976, pp 37–58.

10 *Carrier-to-Customer Installation–DSI Metallic Interface: ANSI T1.403-1989*, American National Standards Institute, New York, 1989.

11 H. Meyer, L. Popken, and H. R. Mueller, "Synchronization Failures in a Chain of PLL Synchronizers," *IEEE Transactions on Communications*, May 1986, pp 436–445.

12 M. Amemiya, M. Aiki, and T. Ito, "Jitter Accumulation for Periodic Pattern Signals," *IEEE Transactions on Communications*, May 1986, pp 475–480.

13 P. R. Trischitta and P. Sannuti, "The Accumulation of Pattern-Dependent Jitter for a Chain of Fiber Optic Regenerators," *IEEE Transactions on Communications*, June 1988, pp 761–765.

14 A. A. Collins and R. D. Pedersen, *Telecommunications Time for Innovation*, Merle Collins Foundation, Dallas, 1973.

15 J. F. Boyle, J. R. Colton, C. L. Dammann, B. J. Karafin, and H. Mann, "No. 4ESS: Transmission/Switching Interfaces and Toll Terminal Equipment," *Bell System Technical Journal*, September 1977, pp 1057–1097.

16 J. E. Abate, E. W. Butterline, R. A. Carley, P. Greendyk, A. M. Montenegro, C. D. Near, S. H. Richman, and G. P. Zampetti, "AT&T's New Approach to the Synchronization of Telecommunication Networks," *IEEE Communications Magazine*, April 1989, pp 35–45.

17 H. Drucker and A. C. Morton, "The Effects of Slips on Data Modems," *IEEE International Conference on Communications*, 1987, pp 12.4.1–3.

18 J. F. Ingle, "Identification of Digital Impairments in a Voiceband Channel," *IEEE International Conference on Communications*, 1989, pp 12.3.1–5.

19. J. E. Abate and H. Drucker, "The Effect of Slips on Facsimile Transmission," *IEEE International Conference on Communications*, 1988, pp 32.3.1–4.

20 J. E. Abate, L. H. Brandenburg, J. C. Lawson, and W. L. Ross, "The Switched Digital Network Plan," *Bell System Technical Journal*, September 1977, pp 1297–1320.

21 G. R. Ritchie, "SYNTRAN—A new Direction for Digital Transmission Terminals," *IEEE Communications Magazine*, November 1985, pp 20–25.

22 Members of Technical Staff, Bell Telephone Laboratories, *Transmission Systems for Communications*, Bell Telephone Laboratories, 1971.

23 Impact of Jitter on the Second Order Digital Multiplex at 6312 kbit/s," AT&T Submittal to CCITT study group on jitter, *Green Book*, Vol. 3, pp 861–869.

24 D. L. Duttweiler, "Waiting Time Jitter," *Bell System Technical Journal*, January 1972, pp 165–207.

25 "Plesiochronous Operation of International Digital Links," CCITT Recommendation G. 811, *Orange Book*, Geneva, Switzerland 1976.

26 T. Muratani and H. Saitoh, "Synchronization in TDMA Satellite Communications," *IEEE International Conference on Communications*, 1979, pp 11.4.1–11.4.6.

27 J. P. Moreland, "Performance of a System of Mutually Synchronized Clocks," *Bell System Technical Journal*, September 1971, pp 2449–2464.

28 P. A. Mitchell and R. A. Boutler "Synchronization of the Digital Network in the United Kindom," *IEEE International Conferences on Communications*, 1979, pp. 11.2.1–11.2.4.

29. C. A. Cooper, "Synchronization for Telecommunications in a Switched Digital Network," *IEEE Transactions on Communications*, July 1979, pp 1028–1033.

30 *Digital Synchronization Network Plan: Bellcore TA-NPL-000436*, Issue 1, Bell Communications Research, November 1986.

31 G. Ebner and L. Tomko, "CCIS: A Signaling System for the Stored Program Controlled Network," *Bell Laboratories Record*, February 1979, pp 53–59.

PROBLEMS

7.1 Determine the size of an elastic store needed to accommodate a velocity shift of ± 1000 km/hr that lasts for 10 sec if the data rate is 10 Mbps. [The speed of light is $3(10)^8$ m/sec.]

7.2 How many bits are needed in an elastic store designed to interface the CEPT1 digital signal of CCITT to a digital switch? What is the maximum slip rate if the line clock and switch clock differ by $+50$ parts per million to -50 parts per million (the maximum recommended offsets)?

7.3 What is the maximum (99% probability) phase offset (in signal intervals) produced by a jitter power of $+10$ dB relative to 1 radian squared?

7.4 Determine the rate at which DS1 signals in a DS2 multiplex lose framing because stuff codes of a DS2 signal are incorrectly interpreted. Assume the channel BER is 10^{-3}. Assume the BER is 10^{-6}.

7.5 What would the rate of incorrect DS2 stuff code occurrences be if a 5 bit stuff code were used instead of the 3 bit stuff code? Assume the channel BER is 10^{-3}

7.6 A digital transmission link is to be used to transmit blocks (packets) of data without slips. If the transmission link is autonomously timed with respect to the receiving terminal, what is the maximum allowable block length if the clocks vary by ± 50 parts per million each and an elastic store of 16 bits exists in the receiving terminal? Assume the elastic store is initialized to half full between each block.

7.7 Assume that worst case systematic jitter from a single repeater produces a phase slope of 300 radians per second for 1 msec. What is the peak-to-peak jitter in decibels relative to a signaling interval at the end of a line with 200 such repeaters?

7.8 Repeat Example 7.4 for DS2 signals in a DS3.

7.9 A jitter power of 100 dB relative to 1 rad^2 is observed at the receiving end of a digital microwave link. What is the probability that the phase offset will exceed ± 3.99 symbol intervals?

7.10 Determine the amount of phase shift injected into a 2400 Hz carrier signal by a slip of one PCM sample.

7.11 Determine the TIE and the MTIE at the end of a ten second interval produced by a DS3 signal that has a constant offset of one part in 10^6.

7.12 Repeat Problem 7.11 but assume a 1 MHz jitter component is added that has a peak-to-peak amplitude of 8 UIs. Assume the starting phase of the jitter component is $0°$.

8

FIBER OPTIC TRANSMISSION SYSTEMS

The evolution of the common equipment portions of the public telephone network from analog technology to almost completely digital technology began and ended with digital transmission systems: T1 systems for short-haul trunks and fiber optic systems for long-haul trunks. If fiber optic technology had not emerged, extensive digitization of the long-distance network would not have occurred as rapidly. High bandwidth coaxial cable transmission systems are too expensive to universally replace radio systems and analog radios are more efficient than digital radios in terms of voice circuits per preciously available bandwidth. Even though high-density digital modulation techniques and sophisticated voice compression algorithms can make up for the bandwidth inefficiencies of digitized voice, any ultimate advantage of digital radios over analog radios is not nearly great enough to warrant wholesale replacement of the analog radio backbone. Fiber transmission systems, on the other hand, provide such dramatic savings in equipment and operational costs that wholesale deployment of high-density routes with fiber systems was economically justified. Because many of the major routes in the AT&T network are now served by fiber, the analog radio systems along these routes have been converted to digital radio systems for spare capacity and protection for the digital fiber systems.

As a note of historical interest Figure 8.1 shows a diagram of a "photophone" patented by Alexander Graham Bell in 1880. Bell developed several techniques for modulating a light source (the sun) so as to directly vary the resistance of a selenium cell detector and thereby create an analog electrical current for the speaker. The modulation mechanism shown in Figure 8.1 is a mirror that is vibrated by acoustic energy to deflect more or less light to the receiver. It was almost exactly 100 years later that a suitable transmission medium and suitable sources made optical transmission practical.

The particular characteristics of optical fibers that make them so useful for transmission systems are low loss, high bandwidth, small physical cross section, EMI immunity, and security.

A. G. BELL.
Apparatus for Signaling and Communicating, called
Photophone.
No. 235,199.　　　　　　　　Patented Dec. 7, 1880.

Figure 8.1. Alexander Graham Bell's photophone.

Fiber Attenuation

As a result of the invention of the laser in 1960, materials scientists began searching for optical transmission media that could be usefully applied as a communication system. The first instance of a practical optical fiber was announced 10 years later in 1970 [1]. This announcement described a silica-based fiber with "only" 20 dB/km attenuation. In just a little less than 10 more years, commercially viable optical fibers with 0.2 dB/km attenuation had been developed [2]. Such remarkably low attenuation attracted immediate attention because it meant intercity transmission links could be traversed with few repeaters—implying a dramatic savings in equipment and maintenance.

As shown in Figure 8.2, the attenuation of an optical fiber is very dependent on the wavelength of the light signal in use. Two wavelengths that experience particularly low attenuation in contemporary fibers are 1300 and 1550 nm. Representative attenuations at these wavelengths are 0.35 and 0.2 dB/km, respectively. With an easily achievable net loss of 20 dB the latter attenuation value allows 100 km between repeaters. Because 0.2 dB/km is close to the

Figure 8.2. Fiber attenuation vs. wavelength.

minimum theoretical attenuation at 1550 nm, additional reductions in attenuation cannot be realized without going to higher wavelengths. If fibers and associated electronics can be developed to operate at higher wavelengths, much lower levels of attenuation may be achieved in the future. The main application of such systems would be in submarine cables where the avoidance of repeaters is most desirable.

Bandwidth

When compared to electromagnetic transmission media, the bandwidth of an optical fiber is mind boggling: a single optical fiber operating at 1300 or 1550 nm of wavelength has a potential bandwidth of 20 THz (20×10^{12} Hz), which is enough for 312 million 64-kps channels. Bandwidth limits of fiber optic transmission systems are mostly determined by the electrooptic drivers and receivers or the electronic interfaces to these devices. As described in Section 8.1.1, multimode fibers have an inherent distance-dependent bandwidth limitation but this particular limitation is avoided in single-mode fibers. Single-mode fiber systems also have a distance-dependent bandwidth limitation, but this limitation is as much a limitation of the optical sources as it is a limitation of the fiber.

Small Physical Cross Section

One of the most beneficial features of T1 systems is their ability to relieve overcrowded cable ducts with a 12-to-1 savings of copper pairs compared to two-wire analog transmission. Fiber systems provide the same advantage with respect to T1 systems—only to a greater degree. A 25-pair copper cable has approximately the same cross section as a fiber cable with 24 fibers and some copper pairs for line-powered repeaters. The latter can easily carry one hundred thousand voice circuits. Furthermore, fiber systems can often be upgraded for more capacity by merely installing higher speed electronics. Thus, after copper

cable is replaced by a fiber cable, duct congestion is unlikely to reappear unless extensive use of broadband services greatly expands the need for fiber facilities.

Electromagnetic Interference Immunity

Because glass has a very high resistance to electricity, the fiber itself can not pick up interfering signals or noise, or propagate harmful or damaging transients to personnel and equipment at the endpoints. The immunity to interference also means crosstalk is not a problem with multiple-fiber cables. In systems utilizing line-powered repeaters, however, some amount of copper is included in the cable to carry power—implying that the immunity, particularly with respect to hazardous voltages, is compromised.

The immunity of fibers to external interferences such as noise and crosstalk implies there is no crosstalk induced limit to the benefit of developing higher powered transmitters or more sensitive receivers. Receiver sensitivity is ultimately limited by internal noise in the photo detectors and transmit power has technology limits due to spectrum spreading in the sources and nonlinearities in the fiber. Until these limits are reached, however, the absence of crosstalk allows the capacity of a fiber system to be increased by merely upgrading the electronics—not the fiber.

Security

Because optical fibers radiate no energy, noninvasive ease dropping of the signal is impossible. Furthermore, invasive taps are more difficult to implement than are wireline taps, which merely require bridging across the conductors with a sufficiently high impedance to remove a usable but unnoticeable amount of signal energy. A similar process is possible with optical fibers, but it requires bending the fiber a very precise amount to allow a small amount of energy to escape and be amplified by a tap. Not only does this process allow passively tapping a fiber but signals can also be injected into the fiber through such a bend. This very technique has been used as a means of locally testing the effectiveness of a fiber splice and is being considered as a means of implementing distributed, passive taps for fiber distribution to the home [3].

8.1 FIBER OPTIC TRANSMISSION SYSTEM ELEMENTS

As shown in Figure 8.3, the basic elements of a fiber optic transmission system are the electrical-to-optical transducer in the transmitting terminal, the optical fiber itself, the optical-to-electrical transducer in the receiver, and signal processing circuitry for amplification, clock recovery, and data detection. If repeaters are necessary, currently available technology requires optical-to-electrical conversion for the signal processing functions and conversion back to optical for transmission. One area of intense research interest involves the development of direct optical amplification for repeaters and other applications.

The system depicted in Figure 8.3 is not necessarily a digital one. Although all

Figure 8.3. Fiber optic transmission system elements.

of the fiber applications for telephony have utilized digital signals, analog transmission is possible. Analog FM modulation of optical signals has been successfully applied to feeder applications of CATV systems [4]. Furthermore, the ability to carry 80 channels of AM video on fiber has been demonstrated—a capability that allows existing television sets to receive fiber-based cable TV without a digital video decoder.

8.1.1 Optical Fiber Fundamentals

The basic construction of an optical fiber is shown in Figure 8.4. Both the core and the cladding are transparent to the desired light signal but the cladding is designed with a lower index of refraction, which causes most light waves in the core to be reflected back into the core. As shown in Figure 8.5, whether or not a ray is reflected back into the core is dependent on the angle at which it strikes the core-cladding boundary. If the angle is too sharp, the ray is not reflected but passes through the cladding and is absorbed by the opaque, protective jacket. The sharp angles can occur at two places: (1) near the source where all of the source's output is not focused into the center of the fiber and (2) at bends, splices, or other imperfections in the fiber.

Figure 8.4. Optical fiber construction.

Figure 8.5. Fiber with multimode propagation.

Multimode Fibers

Although it is usually desirable to deliver as much signal energy as possible to the receiver, waves that reflect back and forth before reaching the receiver may be undesirable if they experience too much delay with respect to the primary ray traveling down the center of the fiber. An optical fiber that allows rays to arrive at the receiver via multiple paths is referred to as a multimode fiber. Multimode fibers have core diameters that are large compared to the wavelength of the signal. A typical multimode fiber will have a 50-μm core diameter and a 125-μm cladding diameter (such a fiber is designated as a 50/125 fiber). The delayed rays cause pulse spreading referred to as multimode dispersion. The significance of the spreading depends on the width of the pulse or, conversely, the data rate being used. Multimode dispersion therefore creates an inherent operational limit defined as a bandwidth-distance product (BDP). The bandwidth-distance product of a typical step-index multimode fiber is 13 Mbps-km [5].

The previously described multimode fiber is referred to as a step-index fiber because the index of refraction in the core is constant with a step change in the index occurring at the core-cladding boundary. Multimode dispersion can be significantly reduced by varying the index of refraction within the core so that a high value occurs in the center and a low value occurs at the edge. Because the speed of propagation of light is higher in lower indices of refraction, rays that reflect back and forth within the core travel at an average speed that is greater than a primary ray that remains entirely within the center. Thus, if the index of refraction is carefully graded within the core, all rays can be made to arrive at the receiver with the same amount of delay. Such a fiber is referred to as a graded-index fiber. A representative bandwidth-distance product of a graded-index fiber is 2 Gbps-km [6]. Notice that this is an improvement of more than two orders of magnitude over typical step-index multimode fibers.

Example 8.1. Determine the loss limit and the multimode dispersion limit of a graded-index multimode fiber system operating at 820 nm and providing a bandwidth of 90 Mbps (enough to carry two DS3 signals). Assume that the difference between the available output power from the source and the input power required by the receiver for an acceptable maximum error rate is 42 dB.

Solution. From Figure 8.2, the attenuation of a multimode fiber operating at 820 nm is approximately 3 dB/km. Thus, the loss limit can be determined as

$$\text{Loss limit} = 42/3.0 = 14\,\text{km}$$

Using 2 Gbps-km as a typical bandwidth-distance product of a graded-index multimode fiber, the multimode dispersion distance limit is determined as

$$\text{Dispersion limit} = 2000/90 = 22.2\,\text{km}$$

The results of Example 8.1 are representative capabilities of first generation fiber systems such as the FT3C system of AT&T [7]. In actual practice, repeater

spacings would be less than the 14 km loss limit to allow margin for component tolerances, splicing, aging, etc. The first FT3C system, which began service between New York and Washington in February 1983, had repeater spacings of 7 km—the location of repeaters in a coaxial cable system replaced by the fiber system.

Single-Mode Fibers

As the first generation fiber systems were being brought into service in the long-distance network, optical technology had already advanced to the point that deployment of second generation systems was underway. One key technology of the second generation was the use of single-mode fibers (SMFs) that provide two distinct advantages. First, SMFs have smaller diameter cores (8 μm typically) that restrict propagation to a fundamental mode and therefore eliminate multimode dispersion. Second, SMFs have less internal (Rayleigh) scattering, which reduces the attenuation by about 50% [6, 8]. For example, the attenuation of an SMF operating at 820 nm is about 2 dB/km as opposed to 3 dB/km for a multimode fiber.

Another key development of second generation systems involved the use of longer wavelengths (1300 nm), which, as shown in Figure 8.2 experience significantly less attenuation than the 800- to 900-nm wavelengths used in first generation systems. Use of these wavelengths required new technology for sources and receivers as described in Sections 8.1.2 and 8.1.3.

Chromatic Dispersion

The combination of eliminating multimode dispersion and using a wavelength with much lower signal attenuation revealed another transmission limitation referred to as chromatic dispersion. Chromatic dispersion arises when a photonic signal contains more than one wavelength and the individual wavelengths propagate at different speeds. Thus, chromatic dispersion is the photonic equivalent of phase distortion (also called delay distortion) in electromagnetic (wireline) propagation. The effect of chromatic dispersion is minimized by using one or both of two basic techniques. First, operation can be centered about a wavelength that exhibits a small amount of chromatic dispersion. Typical silica fibers, for example, produce approximately 15 times less chromatic dispersion at 1300 nm than at 1550 nm. Second, an optical source should be chosen that is as spectrally pure as possible (e.g., has a narrow spread of wavelengths). When both of these techniques are used bandwidth-distance products (BDPs) of 250 GHz-km are possible [6].

Example 8.2. Determine the loss limit and the chromatic dispersion limit of a high-performance SMF optical transmission system operating at 1300 nm and providing a bandwidth of 417 Mbps (enough to carry nine DS3 signals). Assume a narrowband source is used with an output power that is 42 dB greater than the minimum acceptable receive power (i.e., assume the system gain to be 42 dB)

Solution. As indicated in Figure 8.2, the attenuation of a single-mode fiber operating at 1300 nm is approximately 0.35 dB/km. Thus, the

$$\text{Loss limit} = 42/0.35 = 120\,\text{km}$$

Using 250 Gbps-km as the BDP of a silica, single-mode fiber, the chromatic dispersion distance limit is determined as

$$\text{Dispersion limit} = 250{,}000/417 = 600\,\text{km}$$

When the dispersion limit is much higher than the loss limit as is the case in Example 8.2, it indicates that the optical sources are providing a narrower spectrum than is necessary. Thus, the system design could be changed to use less expensive sources or the power output of the source could be increased to get a longer loss limit. (Increasing the power output of the source usually widens its spectrum and thereby reduces the bandwidth-distance product.)

The system parameters of Example 8.2 are representative of the second generation FT series G systems deployed by AT&T [9]. The FT series G systems have maximum repeater spacings of 48 km (29 miles)—indicating that the initial systems could be installed with significant margin for reliability and future upgrades. Many of the routes were, in fact, upgraded to 1.7 Gbps without changing the fiber [10, 11]. Even at the higher data rates, the FT series G routes have significant loss margin for splicing and the insertion of wavelength division multiplexers as another means of upgrading the system. Wavelength division multiplexing is described Section 8.3.

Example 8.2 shows that attenuation is the limiting distance factor in 1300-nm systems operating at low and intermediate data rates. To operate with even longer distances, Figure 8.2 indicates that 1550 nm should be used. With typical silica fibers, the limiting factor at 1550 nm would be chromatic dispersion. To overcome this limitation two other types of fiber have been developed: a dispersion shifted single-mode fiber (DS-SMF) that has minimal dispersion at 1550 nm and a dispersion flattened single-mode fiber (DF-SMF) that has less dispersion across a range of wavelengths. Figure 8.6 contrasts the dispersion of both of these types of fibers with a conventional SMF. Even though Figure 8.6 indicates zero dispersion exists at 1300 nm on the SMF and at 1550 nm on the DS-SMF, some amount of dispersion has to be assumed due to fiber manufacturing variations and uncertainty of the actual center wavelength produced by a source. Depending on the accuracy of the source, values of 1 to 3.5 psec/km-nm are typically used as the dispersion coefficient at the "zero" dispersion points.

8.1.2 Electrical-to-Optical Transducers

Two basic types of semiconductor devices convert electrical signals into optical signals and can have the output coupled into an optical fiber: laser diodes (LDs)

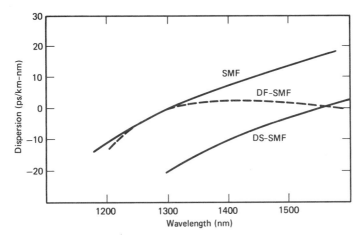

Figure 8.6. Dispersion of SMF and DS-SMF fibers.

and light-emitting diodes (LEDs). LDs generally provide better performance in terms of higher output power, greater bandwidth, and narrower signal spectrum. LEDs, on the other hand, are less expensive, require only simple interface circuitry, are more tolerant of environmental conditions, and are generally more reliable. Thus LDs are used for long transmission distances and LEDs are used when ruggedness and interface costs are more important than performance.

To achieve maximum distances between repeaters it is necessary to minimize chromatic dispersion by using narrowband sources. Various forms of laser diodes have been developed that come close to achieving a spectrally pure output. Figure 8.7 shows the typical output spectrum of one of the most popular types of "single frequency" devices: a Fabry–Perot laser diode. Another type of diode with a very narrow spectrum is a distributed feedback ID (DFB-ID). As indicated, the dominant mode of a Fabry–Perot diode is typically 10 dB stronger than the adjacent modes. To maintain the spectrum indicated it is

Figure 8.7. Representative laser output spectrum.

TABLE 8.1 Representative Parameters of Various Optical Sources

Device Type	Wavelength (nm)	Launched Output Power (dBm)	FWHM Spectrum Width (nm)
Si LED	850	−16	50
Ge LED	1300	−19	70
InGaAsP LED	1300	−10	120
DFB LD	1300	−5	1.0
DFB LD	1550	−5	0.4
IL/DFB LD	1550	+2	0.8

generally necessary to tightly control the operational environment (bias points, modulation level, and temperature). Otherwise, modulation-dependent effects broaden the spectrum or even cause very short duration modes shifts that produce a dominant mode at one of the adjacent wavelengths. This latter phenomenon is referred to as mode partition noise [12], which can cause a halving of the bandwidth-distance product of a system [6]. The spectral width of an optical source is commonly specified as the full-width half maximum (FWHM) value, which represents the spread between the wavelengths at which the spectrum is at half the magnitude of the peak.

To achieve greater distances in loss-limited systems, either a more sensitive receiver or a higher powered source is necessary.* Laser diodes provide the highest available powers as indicated in Table 8.1, which lists representative parameters of various LD and LED sources. Systems that are dispersion limited cannot be improved by higher output powers because the intersymbol interference dominates the imperfections in the received signal.

The effective transmit power depends on the coupling efficiency between the source and the fiber. The coupling efficiency may be as low as 1% in the case of an unfocused LED connected to a single mode fiber. High efficiency couplings of 60% are possible with focused sources. The launched power values provided in Table 8.1 include the coupling efficiencies.

Bandwidth-Distance Products
The bandwidth-distance factor of a fiber system resulting from chromatic dispersion is determined from the fiber dispersion coefficient and the spectral width of the source as follows. The maximum acceptable amount of pulse spreading with an NRZ line code is typically specified to be one-fourth of a bit interval T. Thus,

$$(D)(L)(\Delta\lambda) \leqslant 0.25(T) \tag{8.1}$$

*Transmit power cannot be increased arbitrarily without encountering nonlinear phenomenon in the fiber itself. The optical power threshold where nonlinear effects begin to occur may be as low as 10 mW [6].

where D = the dispersion coefficient of the fiber in psec/km-nm
L= the distance in km
$\Delta\lambda$ = the spectrum width of the source in nm
T= the duration of a bit interval in psec

Using the relationship that the bandwidth (B) is the reciprocal of the bit interval, the bandwidth-distance product (BDP) is determined as

$$\text{NRZ-BDP} = (B)(L)$$

$$\leqslant \frac{250}{(D)(\Delta\lambda)} \text{Gbps-km} \qquad (8.2)$$

An RZ line code allows twice as much pulse spreading because energy in one interval is half a bit period from energy in the next interval. Thus the RZ bandwidth-distance product is determined as

$$\text{RZ-BDP} < \frac{500}{(D)(\Delta\lambda)} \text{Gbps-km} \qquad (8.3)$$

RZ line codes are obviously preferred in dispersion-limited systems but are inferior to NRZ line codes in loss-limited systems because the receiver must detect pulses with half as much energy (assuming the peak transmit power is held constant).

Example 8.3. Determine the bandwidth-distance product of a 1550-nm SMF system and a 1550-nm DS-SMF system using the 1550-nm DFB LD source identified in Table 8.1. Assume an NRZ line code and the "zero" dispersion value is 3.5 psec/km-nm to allow relaxed tolerances for the fiber and the source operating wavelength.

Solution. From Figure 8.6, the dispersion coefficient of an SMF fiber at 1550 nm is 17 psec/km-nm. From Table 8.1, the spectral width of the source is 0.4 nm. Thus, the

$$\text{NRZ-BDP} = 250/(17)(0.4)$$

$$= 36.8 \text{ Gbps-km}$$

Using the given "zero" dispersion value, the BDP of the DS-SMF system is determined as

$$\text{NRZ-BDP} = 250/(3.5)(0.4)$$

$$= 179 \text{ Gbps-km}$$

8.1.3 Optical-to-Electrical Transducers

Two basic types of photodetectors are available as transducers to convert the optical energy at the receiver to electrical energy for amplification and other processing such as clock recovery and data detection. Silicon based p-i-n diodes were the first photodetectors to be used, operating in systems using wavelengths of 800–900 nm. These diodes are inexpensive, reliable, and provide good performance. Their major drawback is that they do not operate at higher wavelengths where fiber loss is minimized. Germanium devices function at the higher wavelengths and are more sensitive, but are more temperature dependent and less reliable.

The second basic type of photodetector is an Avalanche PhotoDiode (APD), which enhances the receiver sensitivity because it operates with internal gain. (A p-i-n diode does not have internal gain and therefore requires all amplification to be externally applied, which raises the noise level.) The amplification inherent in the APD's conversion from an optical signal to an electrical signal is useful because it means the APD can be 10–15 dB more sensitive in detecting low level signals at a given error rate. A major drawback of an APD is that it needs a high supply voltage to operate and is very sensitive to temperature. APDs tend to have somewhat lower gain-bandwidths than p-i-n diodes, which restricts their use in very high data rate systems. Table 8.2 lists various types of photodetectors and provides representative performance parameters for each.

Both types of receivers utilize direct detection, which means that they merely measure the presence or absence of optical energy to detect data. Because higher data rates imply less energy per bit, the sensitivity decreases with data rate. As a first approximation, a doubling of the data rate implies the receiver becomes 3 dB less sensitive. The general formula for determining the change of sensitivity as a function of a change in the data rate is

$$\text{Sensitivity } (R) = \text{Sensitivity } (R_0) + 10 \log_{10}(R/R_0) \text{ dB} \qquad (8.4)$$

where R = the new data rate
R_0 = the data rate at the given sensitivity

TABLE 8.2 Representatiave Parameters of Optical Detectors (BER = 10^{-11})[a]

Device Type	Wavelength (nm)	Receive Power (dBm)	Data Rate (Mbps)
Si p-i-n	850	−48.0	50
Si APD	850	−58.0	50
InGaAs p-i-n	1310	−35.0	420
InGaAs APD	1310	−43.0	420
InGaAs APD	1310	−26.0	8000
InGaAs APD	1550	−37.5	678
InGaAs p-i-n	1550	−37.0	1200
InGaAs APD	1550	−33.0	4000

[a]The detector sensitivities assume an NRZ line code. The sensitivities decrease by 3 dB for (50%) RZ codes.

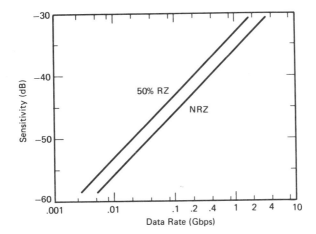

Figure 8.8. Receiver sensitivity vs. data rate.

Figure 8.8 plots the receiver sensitivity of representative p-i-n diode detectors and APD detectors as a function of the data rate for both NRZ and RZ line codes.

8.2 LINE CODES FOR FIBER OPTIC TRANSMISSION

Line codes for optical fiber transmission systems must satisfy the same basic requirements as line codes in electrical (wireline) transmission systems described in Chapter 4. Some fiber systems use line codes described previously. A few new line codes developed specifically for fiber applications are described here. Optical sources and detectors are primarily used in nonlinear modes of operation with significant gain and threshold variations, which implies they are best suited to operating in only two states: on or off. Hence, a two-level line code (on–off keying) is most natural. Considering the wide bandwidth available, multilevel line codes are usually unnecessary and extra timing transitions can be included in the line code without significant penalty. An exception to the availability of "free" bandwidth occurs in submarine systems where maximum repeater spacing is achieved by minimizing the bandwidth of the optical signal. These systems are typically dispersion limited because the need for ultrahigh reliability precludes the latest single-frequency lasers with unproven longevity.

A particularly useful property of direct detection optical receivers is the absence of any polarity ambiguity, which means the transmitter and receiver are inherently dc coupled so differential encoding is unnecessary. Even though the optical signal is inherently dc coupled, the dc balance of the line code may still be a consideration because the gain of some photodiodes (particularly APDs) is sometimes dependent on the dc level of the line code. Furthermore, it is usually

desirable to ac couple amplifier stages in the receivers [13], which means electrical pulse amplitudes at the detecting comparators are dependent on optical pulse densities. In all, these constraints are typically easier to deal with than their counterparts in wireline systems so line codes for fiber systems may tolerate greater amounts of dc variation. Notice that, in fiber systems, both a Manchester and a CMI line code have unvarying dc levels equal to one-half of the pulse amplitude. A number of other line codes used in fiber systems allow some variation in the average dc level. Because these variations are controlled, the line codes are sometimes referred to as "dc-constrained" line codes as opposed to being perfect or near-perfect "dc-balanced" line codes. The dc variations that do exist in these line codes are either too small to cause problems or are accommodated by providing some form of dc restoration [14].

FOT applications that do not require maximum repeater spacing or maximum data rates typically use Manchester (diphase) or CMI line codes because both of these codes provide strong timing content and a completely constrained dc level (50% on). An example of the use of CMI is a high-speed intraoffice link developed by NTT of Japan [15].

Manchester and CMI line codes experience the worst case of both the loss limit and the dispersion limit. The worst case loss limit arises because the receiver must detect optical pulses with a duration equal to half the bit interval (i.e., with a sensitivity of an RZ code). The dispersion limit of a Manchester or CMI line code is actually half of the NRZ dispersion limit because the symbol rate is essentially doubled. (Each bit interval is divided in half with the need to detect a pulse or no pulse in both halves.) Thus, the Manchester or CMI bandwidth-distance product can be determined as

$$\text{MC-BDP} = \frac{125}{(D)(\Delta\lambda)} \text{ Gbps} \qquad (8.5)$$

From the previous discussion it can be seen that when performance is more important than low cost, some other line code is needed. Three basic line codes are commonly used: scrambled NRZ, scrambled RZ, or some variation of a new code referred to as an mBnB line code. The choice of an NRZ or RZ line code is primarily based on whether the system is loss limited or dispersion limited. Loss-limited systems use NRZ line codes whereas dispersion-limited systems use RZ codes. A scrambler is typically used to enhance timing recovery and control dc wander in a statistical sense (but not in an absolute sense). Often, overhead in the data stream provides some assurance that the data stream cannot exactly match the scrambler sequence so a certain minimum number of transitions is assured. This is the case in SONET transmission systems described in Section 8.5.

8.2.1 mBnB Line Codes

As an alternative to relying on scramblers for ensuring timing transitions a new class of line codes has been developed that encodes *m* binary data bits into

blocks of n binary line bits (mBnB) where $m < n$. Because only 2^m data codes must be selected from 2^n codes in each block, there is coding flexibility for controlling timing and dc wander. For example, if $m = 4$ and $n = 5$ 16 data codes must be chosen from the set of 32 5-bit line codes. A specific choice of these codes is provided in Table 8.3, which depicts the line codes selected for the fiber distributed data interface (FDDI) standard [16].

The significant aspects of the coding assignments in Table 8.3 are as follows:

1. The code space is used for control as well as data.
2. When transmitting data, the minimum pulse density is 40% and there can never be more than three intervals without a pulse.

TABLE 8.3 FDDI 4B5B Line Codes

Decimal	Line Code Binary	Line Code Symbol	Name	Functional Assignment
0	00000	Q	QUIET	Line state symbol
1	00001	V	Violation	Disallowed
2	00010	V	Violation	Disallowed
3	00011	V	Violation	Disallowed
4	00100	H	HALT	Line state symbol
5	00101	L		Start delimiter
6	00110	V	Violation	Disallowed
7	00111	R	RESET	Control indicator
8	01000	V	Violation	Disallowed
9	01001	1		Data symbol 0001
10	01010	4		Data symbol 0100
11	01011	5		Data symbol 0101
12	01100	V	Violation	Disallowed
13	01101	T		End delimiter
14	01110	6		Data symbol 0110
15	01111	7		Data symbol 0111
16	10000	V	Violation	Disallowed
17	10001	K		Start delimiter
18	10010	8		Data symbol 1000
19	10011	9		Data symbol 1001
20	10100	2		Data symbol 0010
21	10101	3		Data symbol 0011
22	10110	A		Data symbol 1010
23	10111	B		Data symbol 1011
24	11000	J	SET	Start delimiter
25	11001	S	SET	Control indicator
26	11010	C		Data symbol 1100
27	11011	D		Data symbol 1101
28	11100	E		Data symbol 1110
29	11101	F		Data symbol 1111
30	11110	0		Data symbol 0000
31	11111	I	IDLE	Line state symbol

3. The dc component is constrained to fall between a minimum of 40% pulses and a maximum of 60% pulses—a range that is one-fifth of an unconstrained random data stream.

The features of the 4B5B line code described above are achieved at the expense of a 25% increase in the line data rate. By way of comparison, a digital biphase (Manchester) line code and the CMI line code described in Chapter 4 can both be represented as 1B2B line codes with the coding assignments provided in Tables 8.4 and 8.5, respectively. Notice that both of these codes come at the expense of a 100% increase in the line data rate.

Because the 4B5B line code defined for FDDI is intended for a specific application it contains codes for control as well as for data. The 5B6B codes given in Table 8.6 are intended for transmission only and therefore do not allocate code space for control. Notice that the 5B6B code assignments are made in such a way that the dc level is fixed at 50% pulses but the maximum run length of no pulses is six. Thus, tighter control of the dc level comes at the expense of increasing the worst case duration between pulses. 5B6B and 7B8B line codes have been used extensively in Europe. Examples of 7B8B usage are a 565-Mbps terrestrial system developed by British Telecom [17] and a 280-Mbps NL1 submarine system developed by STC of Great Britain [18].

The 4B5B and 5B6B examples use output blocks that are only 1 bit greater than the input blocks. Considerably more coding flexibility is achieved when the output block is more than 1 bit longer than the input block. As an example, a

TABLE 8.4 Digital Biphase (Manchester) 1B2B Line Codes

Line Code		
Decimal	Binary	Functional Assignment
0	00	Disallowed
1	01	Data symbol 0
2	10	Data symbol 1
3	11	Disallowed

TABLE 8.5 Coded Mark Inversion (CMI) 1B2B Line Codes

Line Code		
Decimal	Binary	Functional Assignment
0	00	Data symbol 1 (if 11 previously sent)
1	01	Data symbol 0
2	10	Disallowed
3	11	Data symbol 1 (if 00 previously sent)

TABLE 8.6 5B6B Line Code Assignments[a]

5B Input	6B Output	5B Input	6B Output
00000	011101/100010	10000	111010/000101
00001	101110/010001	10001	100011
00010	010111/101000	10010	100101
00011	000111	10011	100110
00100	101011/010100	10100	101001
00101	001011	10101	101010
00110	001101	10110	101100
00111	001110	10111	011011/100100
01000	110101/001010	11000	110001
01001	010011	11001	110010
01010	010101	11010	110100
01011	010110	11011	101101/010010
01100	011001	11100	111000
01101	011010	11101	110011/001100
01110	011100	11110	101101/010010
01111	110110/001001	11111	100111/011000

[a]A code 1/code 0 output must alternate between the code with four ones and the code with two ones to maintain dc balance.

6B8B code allows all 64 input codes to be encoded with an output code containing exactly four ones in every code. (The number of combinations of four ones in 8 bits is 70.) Thus dc wander is maintained without having to alternate between low-density and high-density codes.

8.2.2 Bit-Insertion Codes

The mBnB block codes described in the previous section have one significant disadvantage: they are difficult to implement on very high-speed data streams. (In lower speed applications, table lookup in a small read only memory is trivial.) Very high-speed links therefore use another type of code referred to as bit-insertion codes. These codes are actually special cases of mBnB codes with $n = m + 1$ and the codes selected so the coding/decoding logic is greatly simplified.

The first bit-insertion code to be considered is the mB1P code, which merely inserts an odd parity bit after every m bits of source data. Odd parity ensures that at least one "one" is included in the parity field of $m + 1$ bits. As a point of reference, notice that the digital biphase (Manchester) line code is a 1B1P line code. An example of the use of a 24B1P line code is the Trans Pacific Submarine Cable system (TPC-3) brought into service in December 1988 by AT&T and KDD of Japan [19].

An even simpler bit-insertion code described by engineers of NTT in Japan [20] is the mB1C code, which merely inserts a bit after every m source bits—the value of which is the complement of the immediately preceding bit. Thus, the

added bit always forces a data transition and is very easy to implement. An obvious disadvantage of the simplicity is the loss of performance monitorability over the first $m - 1$ bits of a block. Again notice that a diphase line code is a degenerate case of a mB1C line code with $m = 1$. An example of the use of a 10B1C line code is the F-1.6G System of NTT in Japan [21].

8.3 WAVELENGTH DIVISION MULTIPLEXING

Wavelength division multiplexing (WDM) is the basic technique used to establish multiple, independent optical channels on a single fiber. The concept of WDM is illustrated in Figure 8.9 showing the coupling (multiplexing) of separate sources into a fiber and the separation (demultiplexing) of the signals out of the fiber. Prismatic refraction is depicted as the basic coupling mechanism at both ends of the link. Refraction is usable only for channels with relatively large wavelength separations. Diffractive grating devices [22], on the other hand, can operate with channel spacing on the order of 1–2 nm [23]. All WDM devices share the common characteristics of being purely passive and being reversible so any particular device can perform either multiplexing or de-multiplexing functions. Other than some optical insertion loss and some crosstalk from imperfect separation, the multiplexing/demultiplexing operations are transparent to the individual channel signals. The insertion losses of the diffraction grating devices can vary from 1 to 7 dB depending on fiber size (multimode vs single-mode) and number of channels [23].

WDM is commonly used to upgrade an existing transmission system to expand its total capacity. For example, the original 90-Mbps, 820-nm FT3C system installed in 1983 by AT&T in the Northeast corridor was upgraded in 1984 by adding an additional 180 Mbps on a 1300-nm carrier [24]. In addition, references [9] and [10] describe how the FT Series G can have its capacity increased by multiplexing a 1550-nm system onto an existing 1300-nm system and/or adding higher speed electronics without taking the existing system out of service.

When a fiber route is first installed WDM is typically not used because a single high-speed transmitter and receiver is generally less expensive than separate pairs of transmitters and receivers and accompanying WDM devices. However, if the desired data rates exceed practical limits of proven technology it may be necessary to use WDM at the outset. For example, an experimental system described in [23] achieved a bandwidth-distance product equal to 2070 km-Gbps using 18 channels spaced 2 nm apart.

Figure 8.9. Wavelength division multiplexing.

The passive property of WDM devices is a dominant attraction for some applications such as fiber to the home. Allocating individual fibers from a central office to each home is expensive—implying the need for some form of fiber sharing. The outside plant environment (-40–$70°$C) and the need for remote power for active components imply that passive WDM devices are preferred. The use of WDM in this application is sometimes referred to as the passive photonic loop (PPL) [25].

WDM is fundamentally identical to FDM as used on electrical (copper) or electromagnetic (radio) transmission systems. In fact, using the relationship

$$v = f\lambda \tag{8.6}$$

where $v =$ the velocity of light $= 3(10^8)$ m/sec
$\quad f =$ the frequency in Hz
$\quad \lambda =$ the wavelength in meters

allows wavelengths to be related to frequency. Optical systems are traditionally defined in terms of wavelength as opposed to frequency because optical sources are traditionally specified in terms of the wavelengths of their outputs. Nevertheless, systems with extremely close WDM channel spacings (on the order of 0.04 nm) are sometimes referred to as FDM systems [26].

Systems that use electronic multiplexing to produce electrical FDM signals for modulation of a single optical carrier are referred to as subcarrier multiplexing systems. Because the individual channels of these systems are typically close together in frequency, and hence in wavelength, passive separation is usually infeasible.

Example 8.4. Determine the difference in wavelength of two optical signals separated by 2 GHz and centered at 1500 nm.

Solution: The frequency (in free space) of a 1500 nm signal is

$$f = 3(10^8)/1500(10^{-9})$$

$$= 200,000 \text{ GHz}$$

Thus the upper and lower frequencies are determined as 200,001 and 199,999 GHz, respectively. The corresponding wavelengths are

$$= 3(10^8)/199,999(10^9)$$

$$= 1500.007 \text{ nm}$$

$$= 3(10^8)/200,001(10^9)$$

$$= 1499.992 \text{ nm}$$

and the difference in wavelengths is 0.015 nm.

Figure 8.10. WDM with power splitting/filtering.

Example 8.4 shows that even when subcarriers are separated by 2 GHz the resulting wavelength difference is small—too small for passive demultiplexing.

Wavelength demultiplexing with passive diffraction does not inherently cause a loss of signal power in the individual channels. An alternative approach to demultiplexing is shown in Figure 8.10, which involves power splitting of the received signal followed by wavelength filtering to extract individual channels. This approach is passive and functionally identical to diffractive separation but is wasteful of optical power. Its main advantage is that power splitting can be implemented as passive taps distributed along the fiber route. Thus, it is a useful technique in applications (like local area networks) where distances are less important than flexibility in network topology.

8.4 FIBER SYSTEM DESIGN

The examples in the previous sections highlight only the most basic aspects of fiber optic technology. This section provides system level considerations for the design of practical FOT systems. As is the case with any transmission technology, the system design is strongly dependent on system availability requirements and cost of maintenance. At the low end of these requirements are easily accessible fiber connections to individual workstations within a single building. At the other end of the spectrum of reliability requirements are submarine cables that are obviously very expensive to repair. Repairs are minimized by using only ultrareliable components. Even though submarine cables are immune to backhoes* and gophers, they must contend with oceanic counterparts: fishing trawlers and sharks. AT&T has developed a "seaplow" for burying submarine cables in areas of large-scale fishing activity [27]. About the only advantage that submarine applications have in their favor is the constant, relatively low-temperature environment for repeaters.

8.4.1 Fiber Connectors and Splices

Three basic techniques of joining fibers are connectors, mechanical splices, and fusion splices [28]. Connectors are used for terminal connections, patch panels,

*Some outside plant engineers of the telephone companies would probably not be too surprised if a backhoe could somehow get to a submarine cable.

or other instances where loss (0.4–1 dB) is less important than ease of maintenance and reconfiguration.

Mechanical splices involve mechanical alignment and clamping of two adjoining fibers. There are many different types of mechanical splicing equipment that provide various levels of loss depending on the fiber size (multimode or single-mode) and installation time. Splices can be made on individual fibers or in bulk on fiber bundles or ribbons. Two basic splicing procedures are possible: passive or active. Passive splices are faster because the craftsperson merely places the fibers in the alignment sleeves and clamps the fibers in place. Active alignment involves passing a signal through the splice and making fine mechanical adjustments to minimize the loss before the final clamp is applied. A single fiber splice described in reference [29] provides an average of 0.2 dB loss when passively installed and 0.05 dB loss when actively installed. The installation times are 5 and 7 min, respectively.

Fusion splices provide the lowest losses because they essentially form one continuous fiber that is almost as good as a single drawn fiber. In fact, submarine fibers are pretested at the factory for tensile strength and fused at the weak points where breaks occur [27]. This process ensures that the fibers can withstand the tensile stress encountered when laying and recovering the fiber cable but introduces slightly higher levels of average attenuation in the fiber. For example, the 104-km repeaterless system installed between Taiwan and the island of Peng Hu has 0.24 dB/km at 1550 nm including all splices [27]. Fusion splicing equipment is available that is completely automated with high-resolution TV used to align the fibers and estimate the splice loss by measuring the misalignment after fusion [28]. Fusion losses can vary between 0.01 and 0.1 dB. For undersea applications, only values close to 0.01 dB are accepted.

8.4.2 Protection Switching

Protection switching within an FOT system is basically the same as in other high-capacity transmission systems wherein one spare line, including transmitters, repeaters, and receivers, can be switched into service if one of N main lines fails (1:N protection). There are, however, a few unique considerations for protecting fiber systems—arising primarily because of the high capacity of the systems. First, a system can be installed with lower cost (lower performance) optoelectronic devices with an upgrade for higher performance anticipated at a later date. In this case, the protection switching procedures and packaging concepts should allow one-at-a-time upgrading of electronics without disrupting service [11].

A more significant and difficult aspect of protection switching arises because fiber routes tend to carry large traffic volumes representing aggregations of traffic that would otherwise be carried by separate facilities on separate routes. One-for-N protection of most fiber systems is of no use when a backhoe cuts an entire cable. Protection from the loss of a complete facility requires network

level routing diversity—a feature of a network that tends to disappear when using fiber and large switching systems to their maximum economic advantage: large hubs fed by large trunk groups. A network architecture that provides 1:1 route diversity and no hub dependence is the self-healing ring (SHR) [30, 31], the basic concepts of which are described in Chapter 4. Network level protection switching (for hubs or rings) may use add-drop multiplexers, cross-connect systems, or photonic switches that protect routes but not equipment [32].

8.4.3 System Gain

The first step in determining the repeater spacing of an FOT section is to determine the difference between the launched output power of a transmitter and the receive power required for a designated minimum BER. In analagous fashion to the same term introduced for radio systems this parameter is referred to as the system gain. Notice that the system gain includes coupling losses at the optical source and optical detector. The system gain must be greater than the sum of all the losses in the path. Sources of loss include inherent attenuation of the fiber, splicing/connector losses, bending losses, WDM device losses, and photonic switch losses if any. Splicing losses that occur at regular intervals during the installation of the fiber are often added into the fiber attenuation so there is only one distance-dependent parameter involved in determining repeater spacing. The difference between the system gain and the sum of all the nominal losses is the loss margin. In any particular installation, the loss margin allows for equipment manufacturing tolerances, power penalties due to dispersion, unanticipated splices, component aging, and possible WDM upgrades.

Example 8.5. Determine the system gain, the bandwidth-distance product, the dispersion-limited repeater spacing, and the loss margin for a FOT system with the following parameters: data rate = 565 Mbps, line code = 5B6B RZ, wavelength = 1550 nm, source = −5 dBm DFB-LD with 0.4 nm FWHM, fiber = SMF, detector = InGaAs APD, repeater spacing = 65 km, and 0.2 dB/km splicing losses.

Solution. The use of the 5B6B line code implies the line data rate = 565(6/5) = 678 Mbps. The receiver sensitivity for 678 Mbps is determined from Figure 8.8 or Table 8.2 as −34.5 dBm. Thus,

$$\text{System gain} = -5 - (-34.5) = 29.5\,\text{dB}$$

Because an RZ line code is in use, the

$$\text{BDP} = 500/(17)(0.4) = 73.6\,\text{Gbps-km}$$

$$\text{BDP spacing} = 73.6/0.678 = 109\,\text{km}$$

$$\text{Path losses} = (0.2 + 0.2)(65) = 26\,\text{dB}$$

$$\text{Loss margin} = 29.5 - 26 = 3.5\,\text{dB}$$

Example 8.5 is representative of the 565-Mbps system developed by Fujitsu of Japan for carrying a CEPT5 digital signal (565.148 Mbps) [33]. The fact that the repeater spacing is 60% of the BDP limit implies that dispersion penalties are small (estimated to be 1.2 dB).

8.5 SONET

The first generations of fiber optic systems in the public telephone network used proprietary architectures, equipment, line codes, multiplexing formats, and maintenance procedures. Some commonality with other systems in the network came from suppliers who also supplied digital radio systems. In these cases, the multiplexing formats and maintenance protocols emulated counterparts in the radio systems—which also had proprietary architectures. The only thing in common with all of the radio and fiber systems from all of the suppliers was that the interface to the network was some number of DSX3 cross-connect signals. Proprietary multiplexing formats for multiple DS3 signals evolved because there was no higher level standard compatible with the applications. A DS4 signal, which is composed of six DS3 signals, requires too much bandwidth for radio systems and carries a larger cross section of channels (4032) than needed in many applications.

The Regional Bell Operating Companies and interexchange carriers (IXCs), the users of the equipment, naturally wanted standards so they could mix and match equipment from different suppliers. This became particularly important as a result of competition among the interexchange carriers who desired fiber interfaces to the local exchange carriers (LECs), but did not want to necessarily buy from the same suppliers as the LECs. (It might be necessary for an IXC to interface with a different supplier at each LEC.) To solve these problems, and others, Bellcore initiated an effort that was later taken up by the T1X1 committee of the Exchange Carriers Standards Association (ECSA) to establish a standard for connecting one fiber system to another at the optical level (i.e., "in the glass"). This standard is referred to as the Synchronous Optical NETwork: SONET [34, 35]. In the late stages of the development of this standard CCITT became involved so that a single international standard exists for fiber interconnect between telephone networks of different countries. Internationally, the standard is known as the Synchronous Digital Hierarchy [36]. The SONET standard addresses the following specific issues:

1. Establishes a standard multiplexing format using any number of 51.84-Mbps signals as building blocks. Because each building block can carry a DS3 signal a standard rate is defined for any high-bandwidth transmission system that might be developed.
2. Establishes an optical signal standard for interconnecting equipment from different suppliers.
3. Establishes extensive operations, administration, and maintenance (OAM) capabilities as part of the standard.

4. Defines a synchronous multiplexing format for carrying lower level digital signals (DS1, DS1C, DS2, CEPT1). The synchronous structure greatly simplifies the interface to digital switches, digital cross-connect switches, and add-drop multiplexers.
5. Supports CCITT (CEPT) digital signal hierarchy.
6. Establishes a flexible architecture capable of accommodating future applications such as broadband ISDN with a variety of transmission rates.

The data rates and signal designations of the SONET hierarchy are provided in Table 8.7. At the lowest level is the basic SONET signal referred to as the synchronous transport signal level-1 (STS-1). Higher level signals are referred to as STS-N signals. An STS-N signal is composed of N byte-interleaved STS-1 signals. The optical counterpart of each STS-N signal is an optical carrier level-N signal (OC-N). Table 8.7 also includes CCITT nomenclature for the SONET hierarchy, which refers to signals as synchronous transport modules-N (STM-N). Because common usage of the CCITT signal hierarchy cannot effectively use a 51.84-Mbps signal, the lowest level STM signal is a 155.52-Mbps (STS-3c) signal.

Although the SONET specification is primarily concerned with OC-N interconnect standards, electrical signals within the SONET hierarchy are useful within a switching office for interconnecting network elements (e.g., multiplexers, switching machines, and cross-connect systems) [37].

TABLE 8.7 SONET Signal Hierarchy

North American Designation			
Electrical Signal	Optical Signal	Data Rate (Mbps)	CCITT Designation
STS-1	OC-1	51.84	
STS-3	OC-3	155.52	STM-1
STS-9[a]	OC-9	466.56	STM-3
STS-12[a]	OC-12	622.08	STM-4
STS-18[a]	OC-18	933.12	STM-6
STS-24[a]	OC-24	1244.16	STM-8
STS-36[a]	OC-36	1866.24	STM-12
STS-48[a]	OC-48	2488.32	STM-16

[a]Not defined as electrical standards.

8.5.1 SONET Multiplexing Overview

The first step in the SONET multiplexing process (shown in Figure 8.11) involves generation of a 51.840-Mbps STS-1 signal for each tributary. The STS-1 signal contains the tributary (payload) traffic plus transport overhead. As indicated in the figure, a variety of tributary types are accommodated:

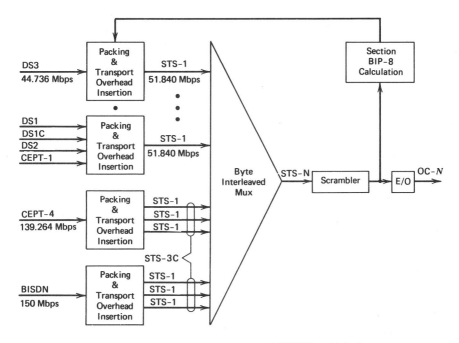

Figure 8.11. Functional block diagram of SONET multiplexing.

1. A single DS3 per STS-1 that can be a standard asynchronous DS3 signal generated by a M13 or M23 multiplexer, a synchronous formatted (SYNTRAN) DS3, or a DS3 payload equivalent composed of 28 synchronous DS1s. Asynchronous DS3 inputs are passed transparently through the system to a DS3 output. Because this transparent option exists, any 44.736-Mbps signal can be carried within the payload envelope.

2. A group of lower rate tributaries such as DS1, DS1C, DS2, or CEPT1 signals can be packed into the STS-1 payload. These signals can also be carried as asynchronous or synchronous signals. In the latter case, direct access to DS0 channels are possible.

3. A higher rate (wideband) signal can be packed into a multiple number of concatenated STS-1 signals. Prevalent examples of higher rate signals are 139.264-Mbps fourth-level multiplexes of CCITT or a broadband ISDN signal at 150 Mbps. Each of these applications require three STS-1 signals concatenated together to form an STS-3c signal. Higher levels of concatenation (to form STS-Nc signals) are possible for higher rate tributaries. Concatenated STS-1 signals contain internal control bytes that identify the signal as a component of a higher speed channel so the integrity of the concatenated data can be maintained as it passes through a network.

An STS-N signal is created by interleaving bytes from N STS-1 signals that are mutually synchronized. Thus there is no timing adjustment required when

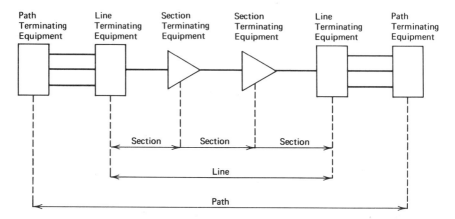

Figure 8.12. SONET system hierarchy.

going to higher level multiplex formats. All timing (frequency) adjustment is done in generating each of the individual STS-1 signals. No matter what the nature of the tributary traffic is, the STS-1 signal is formatted in the same way and has a common synchronized rate of 51.840 MHz.

Optical carrier level-N signals are generated by first scrambling the STS-N signal (except for framing bytes and STS-ID bytes) and then converting the electrical signal to an optical signal. Other than scrambling, the OC-N signal is derived by bit-for-bit conversion to an optical signal. Thus, the data rates, formats, and framing of the STS-N and OC-N signals are identical.

A SONET system is defined as a hierarchy of three levels: sections, lines, and paths as indicated in Figure 8.12. Each of these levels has overhead bandwidth dedicated to administering and maintaining the respective level. As indicated in Figure 8.11, one of the overhead functions provided within an STS-N signal involves calculation and transmission of a parity byte for the entire STS-N signal. Parity is also defined for the other levels of the architecture as described in the following section.

8.5.2 SONET Frame Formats

The frame format of an STS-1 signal is shown in Figure 8.13. As indicated, each frame consists of 9 rows of 90 bytes each. The first three bytes of each row are allocated to transport overhead with the balance available for path overhead and payload mapping. The transport overhead is itself composed of section overhead and line overhead. Path overhead is contained within the information payload as indicated.

The 9 rows of 87 bytes (783 bytes in all) in the information payload block is

Transport Overhead			STS-1 Information Payload		

	A1	A2	C1	J1		
Section	B1	E1	F1	B3		
Overhead	D1	D2	D3	C2		
	H1	H2	H3	G1	Path	9 Rows
	B2	K1	K2	F2	Overhead	
Line	D4	D5	D6	H4		
Overhead	D7	D8	D9	Z3		
	D10	D11	D12	Z4		
	Z1	Z2	E2	Z5		

3 Columns 87 Columns

Figure 8.13. STS-1 frame format.

referred to as the envelope capacity. Because the frame rate is 8 kHz, the composite data rate of each STS-1 signal can be represented as the sum of the transport overhead rate and the information envelope capacity:

$$STS\text{-}1 \text{ rate} = \text{Overhead rate} + \text{Information envelope rate}$$

$$= (9)(3)(8)(8000) + (9)(87)(8)(8000)$$

$$= 1.728(10^6) + 50.112(10^6)$$

$$= 51.840 \text{ Mbps} \tag{8.7}$$

The internal format of the envelope capacity is dependent on the type of tributary traffic being carried. One aspect of the envelope format that is common to all types of traffic is the nine bytes of path overhead indicated in Figure 8.13. The actual location and purpose of this overhead are described in the next two sections.

As a specific example of a higher level (STS-N) signal, Figure 8.14 depicts the details of an STS-3 signal which also represents the STM-1 signal format in CCITT terminology. Transmission of the bytes occurs row-by-row and left-to-right. Thus, the first three bytes of an STS-3 frame are the three framing bytes A1, A1, A1. Most of the Section and Line overhead functions within an STS-3 signal are carried in the STS-1 number 1 overhead. Thus many of the corresponding bytes of the other STS-1 signals are unused and are so designated with an asterisk "*". Notice, however, that path overhead is included in the information envelope for each of the STS-1 signals.

After a frame of an STS-N signal is scrambled, a parity byte (BIP-8) is generated which provides even parity over corresponding bits in all bytes of the STS-N frame. This parity byte is inserted into the section overhead of the first STS-1 signal of the next STS-N frame.

Transport Overhead									STS-3 Information Payload		
A1	A1	A1	A2	A2	A2	C1	C1	C1	J1	J1	J1
B1	*	*	E1	*	*	F1	*	*	B3	B3	B3
D1	*	*	D2	*	*	D3	*	*	C2	C2	C2
H1	H1	H1	H2	H2	H2	H3	H3	H3	G1	G1	G1
B2	B2	B2	K1	*	*	K2	*	*	F2	F2	F2
D4	*	*	D5	*	*	D6	*	*	H4	H4	H4
D7	*	*	D8	*	*	D9	*	*	Z3	Z3	Z3
D10	*	*	D11	*	*	D12	*	*	Z4	Z4	Z4
Z1	Z1	Z1	Z2	Z2	Z2	E2	*	*	Z5	Z5	Z5
(1)	(2)	(3)	(1)	(2)	(3)	(1)	(2)	(3)	(1) (2) (3) ⋯ (1) ⋯ (2)		
9 Columns									261 Columns		

Figure 8.14. STS-3 frame format.

8.5.3 SONET Operations, Administration, and Maintenance

The SONET standard places significant emphasis on the need for operations, administration, and maintenance (OAM) of an end-to-end system. As shown in Figure 8.15, the OAM architecture is based on the section, line, and path layers described previously. OAM standardization is a requirement for mixing equipment from multiple vendors and ease of management of all levels of a system (an individual repeater section or an end-to-end path).

Section Overhead
The functional allocation of the nine bytes of section overhead in each STS-1 frame shown in Figure 8.13 are as follows:

A1, A2: Framing bytes = F6, 28 hex (1111 0110, 0010 1000)
 C1: STS-1 ID identifies the STS-1 number (1 to N) for each STS-1 within an STS-N multiplex
 B1: Bit-interleaved parity byte providing even parity over previous STS-N frame after scrambling
 E1: Section level 64-kbps PCM orderwire (local orderwire)
 F1: 64-kbps channel set aside for user purposes
D1–D3: 192-kbps data communications channel for alarms, maintenance, control, and administration between sections

The fact that there is such a richness of maintenance support at the section level (from one repeater to another) is indicative of the recognized need for extensive OAM facilities and the availability of economical technology to provide it.

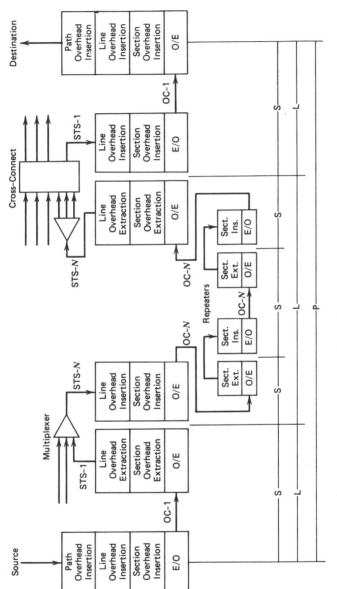

Figure 8.15. SONET overhead layers.

409

Line Overhead

The functional allocation of the 18 bytes of line overhead in each STS-1 frame shown in Figure 8.13 are as follows:

H1–H3: Pointer bytes used in frame alignment and frequency adjustment of payload data; the functions of these bytes are defined in detail in Section 8.5.4

B2: Bit-interleaved parity for line level error monitoring

K1, K2: Two bytes allocated for signaling between line level automatic protection switching equipment

D4–D12: 576-kbps data communications channel for alarms, maintenance, control, monitoring, and administration at the line level

Z1, Z2: Reserved for future use

E2: 64-kbps PCM voice channel for line level orderwire

Notice that the line level OAM facilities are similar to those available at the section level with the addition of the protection switching signaling channel and H1, H2, and H3 pointer bytes use for payload framing and frequency adjustment.

Path Overhead

As indicated in Figure 8.13 there are nine bytes of path overhead included in every block (9 × 87 bytes) of information payload. The important aspect of this overhead is that it is inserted when the tributary data are packed into the synchronous payload envelope (SPE) and not removed (processed) until the tributary data are unpacked. Thus, it provides end-to-end OAM support independent of the path through the synchronous network—which may involve numerous intermediate multiplexers, cross-connect switches, or add-drop multiplexers. The exact location of these nine bytes within the payload envelope is dependent on the pointer values defined in the next section. The functions of the path overhead bytes are as follows:

J1: 64-kbps channel used to repetitively send a 64-byte fixed-length string so a receiving terminal can continuously verify the integrity of a path; the contents of the message are user programmable

B3: Bit-interleaved parity at the path level

C2: STS path signal label to designate equipped versus unequipped STS signals and, for equipped signals, the specific STS payload mapping that might be needed in receiving terminals to interpret the payloads

G1: Status byte sent from path terminating equipment back to path originating equipment to convey status of terminating equipment and path error performance (received BIP error counts)

F2: 64-kbps channel for path user

H4: Multiframe indicator for payloads needing frames that are longer than a single STS frame; multiframe indicators are used when packing lower rate channels (virtual tributaries) into the SPE

Z3–Z5: Reserved for future use

8.5.4 Payload Framing and Frequency Justification

Payload Framing

The location of the nine bytes of path overhead in the STS-1 envelope is not defined in terms of the STS-1 transport framing. Instead, the path overhead is considered to be the first column of a frame of data referred to as the synchronous payload envelope (SPE), which can begin in any byte position within the STS-1 payload envelope (see Figure 8.16). The exact location of the beginning of the SPE (byte J1 of the path overhead) is specified by a pointer in bytes H1 and H2 of the STS line overhead. Notice that this means that an SPE typically overlaps two STS-1 frames.

The use of a pointer to define the location of the SPE frame location has two significant advantages. First, SPE frames do not have to be aligned with higher level multiplex frames. It may be that when first generated, an SPE is aligned with the line overhead at the originating node (i.e., the pointer value is 0). As the frame is carried through a network, however, it arrives at intermediate nodes (e.g., multiplexers or cross-connects) having an arbitrary phase with respect to the outgoing transport framing. If the SPE had to be frame aligned with the outgoing signal, a full SPE frame of storage and delay would be necessary. Thus, the avoidance of frame alignment allows SPEs on incoming links to be immediately relayed to outgoing links without artificial delay. The location of the SPE in the outgoing payload envelope is specified by setting the H1, H2 pointer to the proper value (0 to 782).

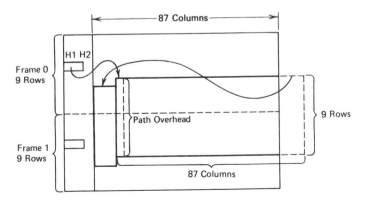

Figure 8.16. Representative location of SPE.

The second advantage of the pointer approach to framing SPE signals is realized when direct access to subchannels such as DS1s is desired. Because the pointer provides immediate access to the start of an SPE frame, any other position or time slot within the SPE is also immediately accessible. This capability should be compared to the procedures required to demultiplex an asynchronous DS3. In a DS3 signal there is no relationship between the higher level framing and the lower level DS2 and DS1 framing positions. In essence, two more frame recovery processes are needed to identify a DS0 time slot. The use of pointers* in the SONET architecture eliminates the need for more than one frame recovery process when accessing lower level signals.

Frequency Justification

Although it is generally intended that SONET equipment be synchronized to each other or to a common clock, allowances must be made for the interworking of SONET equipment that operates with slightly different clocks. Frequency offsets imply that an SPE may be generated with one clock rate but be carried by a SONET transport running at a different rate. The means of accommodating a frequency offset is to accept variable SPE frame rates using dynamic adjustments in the SPE pointers. Pointer adjustments allow SPE frames to float with respect to the transport overhead to maintain a nominal level of storage in interface elastic stores. Figure 8.17 shows the basic means of accommodating a slow SPE. If the elastic store begins to empty, positive byte stuffing can be accomplished by skipping one information time slot (the slot immediately following the H3 byte) and incrementing the pointer to delay the SPE frame by one byte.

Negative byte stuffing, to accommodate a fast SPE clock, requires sending an extra SPE byte whenever the elastic store begins to fill. As indicated in Figure 8.18, the H3 slot carries the extra byte of data, which requires the pointer to be simultaneously decremented—thereby advancing the SPE frame by one slot. To protect against errors in misinterpreting byte-stuffing operations, the need to increment or decrement a pointer is redundantly encoded and the new pointer value must be transmitted in the three frames following the frame in which the pointer adjustment occurs. This implies that a one-byte adjustment can be made once every four frames (once every 500 μsec). For an analysis of the effects of channel errors on SONET payload pointers see reference [38].

Example 8.6. Determine the range of SPE data rates that can be accommodated by the byte-stuffing operation described above.

Solution. Four SPE frames nominally contain $(4)(9)(87) = 3132$ bytes of data. Thus, the nominal SPE rate is $(8)(3132)(2000) = 50.112$ Mbps. When positive byte stuffing is used to accommodate a slow SPE rate, 3131 bytes of data are

*As discussed in Section 8.55, not all SONET multiplexing formats use pointers to identify subchannel framing.

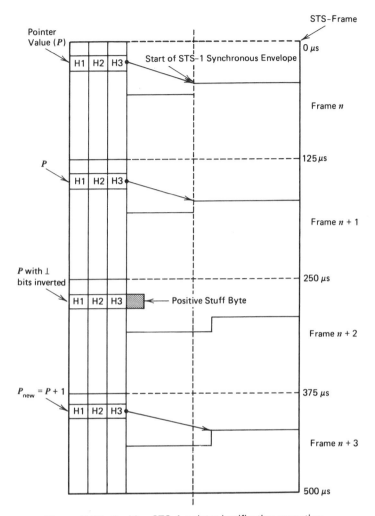

Figure 8.17. Positive STS-1 pointer justification operation.

transmitted in four frames. Thus, the lowest, slip-free rate is

$$\text{Minimum SPE rate} = (8)(3131)(2000)$$
$$= 50.096 \, \text{Mbps}$$

When negative byte stuffing is used to accommodate a fast SPE rate, 3133 bytes of data are transmitted in four frames. Thus, the highest, slip-free rate is

$$\text{Maximum SPE rate} = (8)(3133)(2000)$$
$$= 50.128 \, \text{Mbps}$$

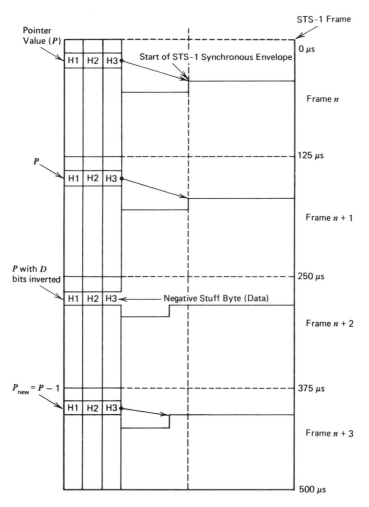

Figure 8.18. Negative STS-1 pointer justification operation.

Example 8.6 demonstrates that the SONET clock accuracy required for maintaining SPE data is $50.112 \pm .016$ Mbps—a very wide tolerance of ± 320 ppm. In comparison, a SONET node is specified to maintain a minimum timing accuracy of 20 ppm if it loses its reference. Thus, the frequency of timing adjustments was chosen more from a desire to simplify the process than from just assuring a margin for slip-free operations.

The use of byte stuffing to accommodate timing differences introduces waiting time jitter into SPE payloads—just as bit stuffing introduces waiting time jitter into DS1 signals being asynchronously multiplexed into DS2 or DS3 signals. If the SPE is carrying DS1 traffic, the effect of byte stuffing at the SPE rate is about one-fourth of a bit interval at the DS1 rate. (Because the SPE rate is

32 times the DS1 rate, the duration of an SPE byte is one-fourth the duration of a DS1 bit.) Notice that this amount of phase jitter is comparable to the amount introduced by bit stuffing a DS2 signal into a DS3 signal. Hence, if a SONET transmission link is used to replace a DS3 transmission link, the underlying traffic will experience no additional degradation when recovering the DS1 clock at the far end of a network.

8.5.5 Virtual Tributaries

To facilitate the transport of lower rate digital signals, the SONET standard uses sub-STS-1 payload mappings referred to as virtual tributary (VT) structures as shown in Figure 8.19. This mapping divides the SPE frame into seven equal sized subframes or VT blocks with 12 columns (108 bytes) in each. Thus, the subframes account for $(7)(12) = 84$ columns with the path overhead and two unused columns (reserved bytes R) accounting for the remainder of the 87 columns in an SPE. The rate of each VT structure is determined as $(108)(8)(8000) = 6.912$ Mbps.

The VT structures can be individually assigned to carry one of four types of signals. Depending on the data rate of a particular signal, more than one signal may be carried within a VT structure as a VT group. All signals within a VT group must be of the same type, but VT groups within a single SPE can be different types. The particular lower rate signals accommodated as virtual tributaries are listed in Table 8.8. The last column indicates how many of the lower rate signals are carried in a single SPE if all seven VT groups are the same type.

Two different modes of operation are defined for the VT structures: locked and floating. The locked byte-synchronous mode assigns every byte (DS0 channel) of a tributary to a specific byte position in the SPE. If all seven VTs are locked DS1s, all 672 DS0 channels are assigned to specific SPE time slots. In essence, this establishes a single-level digital signal with 672 channels. The locked mode of operation is the simplest interface to switching nodes but is inappropriate for interfacing to digital signals of the asynchronous hierarchy.

The floating mode of operation defines pointers to a VT-SPE payload in the same fashion as pointers to SPE payloads are defined at the STS-1 level. Thus, the floating VT-SPE mode allows for minimal framing delays at intermediate

TABLE 8.8 SONET VIRTUAL TRIBUTARIES

Tributary Type	VT Designation	No. Columns per VT	Number in VT Group	Maximum Number in SPE
DS1	VT1.5	3	4	28
CEPT1	VT2.0	4	3	21
DS1C	VT3.0	6	2	14
DS2	VT6.0	12	1	7

SPE Column Number

	1	2	3	4	5	6	7	8	9	...	30	31	32	...	59	60	61	...	84	85	86	87
J1																						
B3																						
C2											u				u							
G1											n				n							
F2											u				u							
H4											s				s							
Z3											e				e							
Z4											d				d							
Z5		1	1	1	1	1	1	1	2	...		5	5	...		9	9	...	12	12	12	12
		(1)	(2)	(3)	(4)	(5)	(6)	(7)	(1)	...		(1)	(2)	...		(1)	(2)	...	(4)	(5)	(6)	(7)

VT Column Number

VT Group Number

Figure 8.19. SPE mapping for virtual tributaries.

Figure 8.20. VT1.5 envelope and SPE structure.

Notes: V_1, V_2, V_3, V_4 is the VT transport overhead.
V_1, V_2 point to the path overhead V_5, the first byte of the VT SPE.
∗ is the positive stuff byte for floating VT-SPE.
I represents information bits in the asynchronous mapping.
R represents unused (fixed stuff) bits.
C_1, C_2 are stuff control bits for asynchronous mapping.
S_1, S_2 are stuff opportunities for asynchronous mapping.
O are reserved for future purposes.
P_1, P_0 identify the signaling bits in floating byte-synchronous mapping.
S_1, S_2, S_3, S_4 carry 24 signaling bits in byte-synchronous mapping.
F is a DS0 framing bit in byte-synchronous mapping.

417

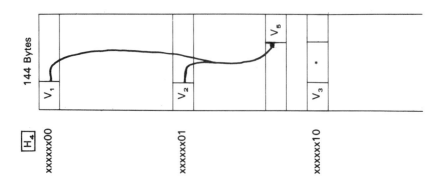

Figure 8.21 VT2 envelope and SPE structure.

Channels 16–30

VT2-SPE
Byte-Synchronous
Mapping (CAS)

Unused or Slot 0

Channels 1–15

Signaling

Channels 16–30

P_1 P_0

VT2-SPE
Asynchronous
Mapping

C_1 C_2

S_2

S_1

224 I Bits
(31×8)

xxxxxx11

V_4

VT2
Superframe
Structure

Notes:
V_1, V_2, V_3, V_4 is the VT transport overhead.
V_1, V_2 point to the path overhead V_5; the first byte of the VT-SPE.
* is the positive stuff byte for floating VT-SPE.
I represents information bits in the asynchronous mapping.
R represents unused (fixed stuff) bits.
C_1, C_2 are stuff control bits for asynchronous mapping.
S_1, S_2 are stuff opportunities for asynchronous mapping.
O are reserved for future purposes.
P_1, P_0 identify the signaling bits in floating byte-synchronous mapping.

419

nodes and for frequency justification of VT-SPEs undergoing transitions between timing boundaries. The floating VT-SPE structures for DS1 and CEPT1 signals are shown in Figures 8.20 and 8.21, respectively. High rate VT-SPEs are accommodated by inserting an information byte into V3 while slow rate VT-SPEs are accommodated by stuffing into the information byte immediately following V3 when necessary. Figures 8.20 and 8.21 both show two different VT-SPE payload structures: one for asynchronous multiplexing and one for byte-synchronous multiplexing.

Asynchronous Multiplexing

The asynchronous operation is identical in concept to the bit-stuffing operation described in Chapter 7. The DS1 or CEPT1 bit stream is inserted into the information bits (I) with no relationship to the VT-SPE frame or byte boundaries. As indicated, there are two stuffing opportunities (S1 and S2) available in every four-frame superframe. Thus, the VT1.5 superframe carries 771, 772, or 773 information bits depending on the value of the stuff control bits C1 and C2. Similarly, the VT2 superframe carries 1023, 1024, or 1025 information bits depending on the value of the C bits. The nominal number of information bits in each frame is $(193)(4) = 772$ for DS1s and $(256)(4) = 1024$ for CEPT1s. Nominal frame sizes carry information in S2 while stuffing in S1.

Because the asynchronous operation is compatible with the asynchronous network it is the format used in most SONET applications. The major advantage of the asynchronous mode of operation is that it provides for totally transparent transmission of the tributary signal in terms of information and in terms of information rate. The major disadvantage of the asynchronous mode is that 64-kbps DS0 channels and signaling bits are not readily extracted.

It is important to note that a floating mode of operation is necessary even with the asynchronous mapping because an individual VT-SPE generated with a clock rate at one SONET node may be cross-connected with other VT-SPEs generated with a different clock. In this case the VT-SPEs must be individually frequency justified to share a common rate, higher-level channel (the STS-1). The asynchronous multiplexing format provides frequency justification at the SONET network interface but the floating mode of operation is needed to accommodate possible timing perturbations within the SONET network itself.

Byte-Synchronous Multiplexing

In contrast to the asynchronous mode of operation, the byte-synchronous payload mappings shown in Figures 8.20 and 8.21 allocate specific bytes of the payload to specific bytes (channels) of the tributary signals. Hence, this mode of operation overcomes the main drawback of the asynchronous mode in that 64-kbps channels and signaling bits within the payload are easily identified. In fact, the signaling bits of a DS1 are moved from the lsb of every sixth frame of respective channels and placed in dedicated signaling bit positions within the VT-SPE. Thus byte-synchronous multiplexing offers an additional feature of converting from in-slot signaling to out-slot signaling for DS1 signals.

Figure 8.21 depicts a mapping for 30-channel CEPT1 wherein time slot 16 carries channel-associated signaling (CAS)—the form of out-slot signaling designed into CEPT1 signals at their inception. The same basic format supports common channel signaling, which is sometimes referred to as a 31-channel CEPT1 format. In this case channel 16 is the CCS channel and channels 1–15 and 17–31 are the bearer channels. Thus, the muliplex mapping is not changed—just the nomenclature of the channels and the SPE type designation in the VT path overhead byte V5.

The major disadvantages of the byte-synchronous mode of operation are that transparency and timing adjustment capabilities are compromised. Transparency is diminished because the tributary signals must be channelized and contain framing to identify the channels. In contrast, the tributaries in the asynchronous mapping are merely bit streams (at 1.544 or 2.048 Mbps) with no specific internal structure required.

Timing flexibility is severely compromised because either the tributary needs to be synchronized to the VT-SPE frame rate or the VT-SPE (at the originating SONET interface) must be allowed to float with pointer changes for frequency adjustments. The floating mode of operation eliminates slips at the originating interface but may produce timing problems at the far-end tributary interface as a result of waiting-time jitter. The jitter is particularly significant because it arises from stuffing full *bytes* of data at the tributary rate as compared to *bit* stuffing at the tributary rate in the asynchronous digital hierarchy. Normal clock recovery circuits (phase-locked loops) perform poorly when subjected to such large amounts of waiting-time jitter. Without special clock recovery circuitry to ameliorate the effects of byte stuffing, the floating mode of operation cannot be used when interfacing to an unsynchronized tributary. The alternative is to require the tributary to be synchronized, at least in a plesiochronous sense, to the SONET network. Because synchronous interfaces require slip buffers to prevent uncontrolled slips when the synchronization procedure fails, artificial delay of up to two frames of data is undesirably inserted into the signal path.

Bit-Synchronous Multiplexing

The SONET standard defines a third mapping format for DS1 and CEPT1 signals that are actually special cases of the respective asynchronous mappings shown in Figures 8.20 and 8.21. The special provision is that the tributary rates are fixed at the nominal rates of 772 bits/frame for DS1 and 1024 bits/frame for CEPT1. The nominal rates are assured by always stuffing in S1 and never stuffing in S2. This mapping is very simple to implement because it merely involves a bit-by-bit transfer from the tributary into the VT-SPE. Thus, the mapping is transparent but there is no relationship between bits or bytes in the tributary and bits or bytes in the VT-SPE (despite the physical appearance that there might be).

The absence of timing adjustments in the interface limits this mode of operation to synchronized tributaries. If the synchronization path fails for some reason, the resulting slips are not controlled slips—forcing reframes in down-

stream DS1 or CEPT1 receivers. The slips are uncontrolled even if slip buffers are used because framing bits are also contained in the slip buffer. In this case the slip buffer can only control how often a slip occurs—not preserve the position and value of the framing bits.

Again, slips can be avoided by using floating VT-SPEs at the originating node for frequency adjustments, but this introduces large amounts of waiting-time jitter, which may not be acceptable.

8.5.6 DS3 Payload Mapping

The previous section describes several alternatives for packing virtual tributaries into an STS-1 envelope. When all seven VTs in an envelope are VT1.5s, a total capacity of 28 DS1s is provided—the same as a DS3 signal. Thus one method of carrying a DS3 signal involves demultiplexing it into its constituent DS1 (or DS2 signals) and packing the constituents as virtual tributaries. This approach is attractive in that the virtual tributaries are individually accessible for cross-connect or add/drop multiplexer systems. If the application does not not need to access the individual tributaries it is simpler to pack the DS3 signal directly into an STS-1 as indicated in Figure 8.22. The payload mapping in Figure 8.22 treats the DS3 signal simply as a 44.736-Mbps data stream with no implied internal structure. Thus, this mapping is required for transparent transport of DS3-rate data streams.

Each row of a nine row SPE envelope contains 87(8) = 696 bits, which can carry 621 or 622 DS3 data bits depending on the value of the C bits. Notice that this format has five C bits, which allows for single and double bit error correction. The path overhead bytes (POH) carry the nine bytes of path

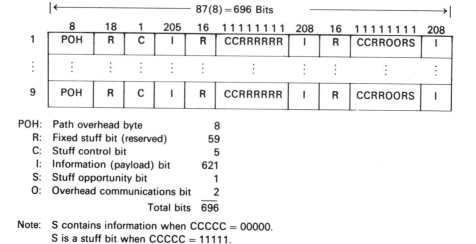

POH: Path overhead byte — 8
R: Fixed stuff bit (reserved) — 59
C: Stuff control bit — 5
I: Information (payload) bit — 621
S: Stuff opportunity bit — 1
O: Overhead communications bit — 2

Total bits 696

Note: S contains information when CCCCC = 00000.
S is a stuff bit when CCCCC = 11111.

Figure 8.22. Asynchronous 44.736-Mbps (DS3) payload mapping.

overhead as defined in Section 8.5.3. Determination of the minimum and maximum DS3 rates is left as an exercise at the end of the chapter.

8.5.7 CEPT4 Payload Mapping

One example of a SONET super-rate mapping is shown in Figure 8.23 for a 139.264-Mbps fourth level CCITT signal (CEPT4). This signal is packed into a 155.52-Mbps STS-3c (or STM-1) signal. Figure 8.23 shows only the synchronous payload envelope (SPE-3c)—not the 9 bytes of section and line overhead in each row. Notice that there is only one column of path overhead within the SPE-3c envelope. The path overhead (POH) bytes carry the 9 bytes of path overhead as defined in Section 8.5.3.

The payload mapping in Figure 8.23 treats the 139.264-Mbps signal as a transparent data stream with no implied internal structure. Each row of a nine row SPE-3c envelope contains 87(3) = 261 bytes, which can carry 1934 or 1935 data bits depending on the value of the C bits. Notice that this format also has five C bits, which allows for single and double bit error correction.

POH: Path overhead byte — 1
I: Information (payload) byte — 241
X: CRRRRROO — 5
Y: RRRRRRRR — 13
Z: iiiiiiSR — 1
Total bytes: 261

C: Stuff control bit
R: Fixed stuff bit
O: Overhead communications bit
S: Stuff opportunity bit
i: Information bit

Note: S contains information when CCCCC = 00000.
S is a stuff bit when CCCCC = 11111.

Figure 8.23. Asynchronous 139.264-Mbps (CEPT-4) payload mapping.

8.5.8 SONET Optical Standards

The optical interface standard [35] defined for "mid-span-meet" of SONET equipment allows for either NRZ or RZ line codes on single-mode fibers. Generation of the OC-N signal from the STS-N signal requires a scrambler as shown in Figure 8.24. The scrambler is synchronized to each STS-N frame by presetting the shift register to all ones immediately after transmitting the last C1 byte of the STS-N section overhead. Thus, the frame codes (A1, A2) and STS-1 ID (C1) code are not scrambled. Furthermore, the static overhead bits of the STS-N frame that are anticoincident with the scrambler sequence provide sufficient timing content for SONET receivers. (Because the scrambler is preset at the same point of every frame, every bit position in successive frames experience the same scrambler value.)

The BER objective is $1 \times (10^{-10})$ for optical sections of 40 km or less. Equipment from separate manufacturers can be freely interchanged for applications with distances up to 25 km. Longer distances may require joint engineering. (Transmitters and receivers from separate suppliers must be jointly specified to support longer distances.)

SONET systems are specified to operate with central wavelengths at 1310 nm

Figure 8.24. SONET scrambler.

TABLE 8.9 Representative Maximum Spectral Widths of SONET Sources[a,b]

		<25 km	<40 km
OC-1	51.84 Mbps	30 nm	25.0 nm
OC-3	155.52 Mbps	15 nm	10.0 nm
OC-9	466.56 Mbps	10 nm	7.5 nm
OC-12	622.08	8 nm	6.0 nm

[a]The spectral width of a source is determined as the wavelength difference between the peak mode and the farthest mode that is 10 dB below the peak.
[b]These specifications are for lasers operating within 10 nm of the central wavelengths (1310 and 1550 nm). Lasers with greater deviation from the central wavelength are allowed but have narrower spectral width specifications to compensate for greater fiber dispersion.

with SMF fibers or at 1550 nm with DS-SMF fibers. Operation at 1310 nm with DS-SMF fibers or at 1550 nm with SMF fibers is not disallowed but must be jointly engineered. A range of laser wavelength tolerances and maximum allowable spectral widths is specified for both 1310 or 1550 nm. Table 8.9 provides representative values of the specifications.

8.5.9 SONET Applications

The evolution of the digital telephone network necessarily occurred with analog equipment being replaced one at a time by compatible digital counterparts. Because much of the digital equipment initially existed as isolated islands, the ultimate advantages of an all digital network could not be realized. One aspect of the evolution was the deployment of asynchronous higher-level multiplexers (e.g., M23, M13, M34, and various proprietary digital radios and fiber systems). The deployment of SONET equipment represents a second evolution of the telephone network: from an asynchronous digital hierarchy to a synchronous digital hierarchy. Again, the introduction of new equipment requires compatibility with the existing equipment. Thus a key feature of the SONET standard is support of asynchronous interfaces to existing digital signals so the new equipment can transparently replace existing higher-level multiplexers and transmission systems. A smooth migration to a synchronous network is possible because a single SONET link can simultaneously transport asynchronous and synchronous signals as indicated in Figure 8.25.

Because synchronous multiplexing advantages of SONET equipment are initially unattainable, other benefits are needed to justify its initial deployment. Beyond the main purpose of interconnecting fiber transmission equipment from separate suppliers there are the following ancillary benefits of SONET deployment:

1. Centralized, remote, and integrated OAM interface for the multiplexing and transmission equipment
2. Rapid fault isolation
3. End-to-end performance monitoring
4. Support of new high-rate transmission services.

Figure 8.25. SONET transmission link supporting asynchronous and synchronous interfaces.

When extensive amounts of SONET equipments are interconnected, there are the following additional advantages of a synchronous network:

1. Higher-level multiplex signals can be integrated into switching equipment
2. Less expensive add/drop multiplexers allow more distributed network structures.
3. SONET-based digital cross-connect systems support network-level protection, rapid and remote provisioning, virtual private networks, and higher-rate switched services such as SMDS or BISDN
4. Fewer pieces of equipment are required because back-to-back multiplexing and manual cross-connects are eliminated.
5. Centralized and network-wide OAM operations are possible.

REFERENCES

1 F. P. Kapron, D. B. Keck, and R. D. Maurer, "Radiation Losses in Glass Optical Waveguides," *Applied Physics Letters*, November 1970, pp 423–425.
2 T. Miya, Y. Terunuma, T. Hosaka, and T. Miyashita, "Ultimate Low-Loss Single-Mode Fibre at 1.55 μm," *Electronic Letters*, February 1979, pp 106–108.
3 D. Large, "The Star-Bus Network: Fiber Optics to the Home," *CED Magazine*, January 1989. Raynet Corporation Reprint.
4 R. Olshansky, V. A. Lanzisera, and P. M. Hill, "Subcarrier Multiplexed Lightwave Systems for Broadband Distribution," *Journal of Lightwave Technology*, September 1989, pp 1329–1342.
5 T. Li, "Advances in Optical Fiber Communications: An Historical Perpective," *IEEE Journal on Selected Areas in Communications*, April 1983, pp 356–372.
6 P. S. Henry, "Lightwave Primer," *IEEE Journal of Quantum Electronics*, December 1985, pp 1862–1877.
7 J. R. Stauffer, "FT3C-A Lightwave System for Metropolitan and Intercity Applications," *IEEE Journal on Selected Areas in Communications*, April 1983, pp 413–419.
8 P. S. Henry, R. A. Linke, and A. H. Gnauck, "Introduction to Lightwave Systems," in *Optical Fiber Transmission II*, S. Miller and I. P. Kaminow, Eds., Academic Press, San Diego, 1988, pp 781–832.
9 D. C. Gloge and I. Jacobs, "Terrestial Intercity Transmission Systems," in *Optical Fiber Transmission II*, S. Miller and I. P. Kaminow, Eds., Academic Press, San Diego, 1988.
10 D. L. Howells, "High Capacity Lightwave Technology Comes of Age," *AT&T Technology*, Vol 3, No 4, 1988, pp 10–15.
11 W. C. Marra, J. S. Linnell, N. M. Denkin, and K. Ogawa, "FT Series G Lightwave Digital Transmission System Architecture and Upgrade Capability," *IEEE International Conference on Communications*, 1987, pp 19B.3.1–19B.3.5.
12 K. Ogawa, "Analysis of Mode Partition Noise in Laser Transmission Systems," *IEEE Journal of Quantum Electronics*, May 1982, pp 849–855.

13　J. E. Goell, "An Optical Repeater with High-Impedance Input Amplifier," *Bell System Technical Journal*, April 1974, pp 629–643.

14　N. Yoshikai, K. I. Katagiri, and T. Ito, "mB1C Code and Its Performance in an Optical Communications System," *IEEE Transactions on Communications*, February 1984, pp 163–168.

15　K. Hagishima, Y. Kobayashi, and K. Aida, "High-Speed CMI Optical Intraoffice Transmission System: Design and Performance," *IEEE Journal on Selected Areas in Communications*, December 1986, pp 1451–1457.

16　F. Ross, "Overview of FDDI: The Fiber Distributed Data Interface," *IEEE Journal on Selected Areas in Communications*, September 1989, pp 1043–1051.

17　P. Cochrane, R. Brooks, and R. Dawes, "A High-Reliability 565 Mbit/s Trunk Transmission System," *IEEE Journal on Selected Areas in Communications*, December 1986, pp 1396–1403.

18.　R. L. Williamson and M. Chown, "NL1 Submarine System," *IEEE Journal on Selected Areas in Communications*, April 1983, pp 454–458.

19　Y. Niro, Y. Ejiri, and H. Yamamoto, "The First Transpacific Optical Fiber Submarine Cable System," *IEEE International Conference on Communications*, June 1989 pp 50.1.1–50.1.5.

20　N. Yoshikai, S. Nishi, and J. I. Yamada, "Line Code and Terminal Configuration for Very Large-Capacity Optical Transmission System," *IEEE Journal on Selected Areas in Communications*, December 1986, pp 1432–1437.

21　H. Kimura, and K. Nakagawa, "F-1.6G System Overview," *Review of the Electrical Communications Laboratories NTT Japan*, Vol 35, No 3, 1987, pp 219–225.

22　G. Winzer, "Wavelength Multiplexing Components—A Review of Single Mode Devices and Their Applications," *Journal of Lightwave Technology*, August, 1984, pp 369–378.

23　S. S. Wagner and H. Kobrinski, "WDM Applications in Broadband Telecommunications Networks," *IEEE Communications Magazine*, March 1989, pp 22–30.

24　R. J. Sanferrare, "Terrestrial Lightwave Systems," *AT&T Technical Journal*, January/February, 1987.

25　S. S. Wagner, H. L. Lemberg, H. Kobrinski, L. S. Smoot, and T. J. Robe, "A Passive Photonic Loop Architecture Employing Wavelength-Division Multiplexing," *IEEE Globecom Proceedings*, 1988, pp 48.1.1–48.1.5.

26　N. K. Dutta, "III-V Device Technologies for Lightwave Applications," *AT&T Technical Journal*, January/February 1989, pp 5–18.

27　P. R. Trischitta and D. T. S. Chen, "Repeaterless Undersea Lightwave Systems," *IEEE Communications Magazine*, March 1989, pp 16–21.

28　S. C. Mettler and C. M. Miller, "Optical Fiber Splicing," in *Optical Fiber Transmission II*, S. Miller and I. P. Kaminow, Eds., Academic Press, San Diego, 1988, pp 263–300.

29　J. B. Haber and J. W. Rogers, "Fiber in the Loop," *AT&T Technology*, Vol 3, No 4, 1988, pp 2–9.

30　T. H. Wu, D. J. Koolar, and R. H. Cardwell, "High-Speed Self-Healing Ring Architectures for Future Interoffice Networks," *IEEE Globecom Proceedings*, November 1989, pp 23.1.1–23.1.7.

31 T. Flanagan, S. Oxner, and D. Elkaim, "Principles and Technologies for Planning Survivability—A Metropolitan Case Study," *IEEE Globecom Proceedings,* November 1989, pp 23.3.1–23.3.8.

32 T.-H. Wu, R. H. Cardwell, and W. E. Woodall, "Decreasing Survivable Fiber Network Cost Using Optical Switches," *IEEE Globecom Conference,* 1988, pp 3.6.1–3.6.5.

33 N. Yamaguchi, K. Yamane, and T. Kihara, "565-Mbit/s Optical Fiber Transmission System," *Fujitsu Science and Technology Journal,* March 1989, pp 60–63.

34 ANSI T1.105–1988 American National Standard for Telecommunications: *Digital Hierarchy–Optical Interface Rates and Formats Specifications.*

35 ANSI T1.106–1988 American National Standard for Telecommunications: *Digital Hierarchy–Optical Interface Specifications (Single Mode).*

36 "Global Deployment of SDH-Compliant Networks," *IEEE Communications Magazine,* 1990, August.

37 N. B. Sandesara, T. H. Jones, and A. G. Edwards, " SONET Intra-Office Interconnect Signal," *IEEE Globecom Proceedings,* 1988 pp 30.5.1–30.5.7.

38 N. Zhang, K.-H. Liu, and E. C. Posner, "Reliable Payload Pointer Protocol for Synchronous Optical Network," *IEEE International Communications Conference Proceedings,* 1989, pp 14.2.1–14.2.5.

PROBLEMS

8.1 Determine the attenuation α in dB/km such that the loss limit of a 500 Mbps fiber system is exactly equal to the dispersion limit. Assume the transmitter output is 30 dB above the receiver threshold for the desired error rate and that the system has BDP of 80 Gbps-km.

8.2 An 850 nm, graded-index, multimode, 50-Mbps fiber system with 2 dB/km loss in the fiber is to be used for a token passing bus local area network. Assume the system uses the first entries in Tables 8.1 and 8.2 as a source/detector pair. If the BDP of the fiber is 500 Mbps-km how many passive taps with 0.4 dB of loss can be inserted per kilometer without affecting the distance between transmitters and receivers.

8.3 What is the distance limit (without taps) of the system in Problem 8.2 if the data rate is reduced to 10 Mbps?

8.4 A graded, multimode fiber exhibits 100 ps/km-nm of chromatic dispersion at 850 nm. Determine the BDP of a NRZ system using an LED having 40 nm of spectral width.

8.5 A WDM SMF system using 1300 and 1550 nm is dispersion limited at both wavelengths. How must the spectral bandwidths of the two sources be related? (Assume D = 3.5 PS/km-nm at 1300 nm.)

8.6 What is the minimum and maximum pulse density of mB1P and mB1C line codes?

8.7 What is the longest string of no pulses that can occur in mB1P and mB1C line codes?

8.8 A 4B6B line code can encode all 4 bits of source data with 6-bit line codes having exactly three pulses in every code. How many dc-balanced codes are left over for link control?

8.9 Determine the minimum and maximum data rate of a DS1 signal that can be carried within a floating mode VT1.5 envelope.

8.10 Determine the minimum and maximum data rates of CEPT1 signals that can be carried within floating mode VT2 envelopes.

8.11 Determine the minimum and maximum data rates of an asynchronous DS3 payload that can be carried within a STS-1 envelope.

8.12 Determine the minimum and maximum CEPT4 data rates that can be accommodated within a STS-3c envelope.

8.13 Determine the average frame acquisition time of an STS-1 signal assuming all non-framing bits have equally likely random values of "0" or "1".

8.14 Repeat Problem 8.13 for an STS-3c signal.

9

DATA AND INTEGRATED SERVICES DIGITAL NETWORKS

Although this book emphasizes the use of modern digital electronics as a means of supplanting traditional analog implementations of voice communications networks, it should be remembered that communications by digital techniques is actually an ancient practice. In early times communication over distances greater than the range of the human voice was provided by sight or sound of discrete signals (e.g., heliographs, smoke signals, flags, drum beats, and bugles). Furthermore, the first practical communication system to use electricity, the telegraph, is inherently digital. As telegraphy evolved from the original manually based systems to fully automated systems, they developed into what are commonly referred to as message switching networks. Modern message switching is discussed in the first section of this chapter.

More recently, digital communications networks have evolved to satisfy data communications needs. Because of its widespread availability it is only natural that the public telephone network is being used to provide most data transmission services. The availability overshadows numerous technical shortcomings of a network designed primarily for voice communications services. The main deficiencies of the telephone network for data transmission are

1. Need for signal transducers (modems)
2. Limited data rates
3. High-error rates
4. Inefficient circuit utilizations.

As data communications requirements increased, so did the justification for more cost-effective data communications networks. One solution to reducing data transmission costs is to improve circuit utilizations through the use of packet switching networks. The technology of packet switching was pioneered by the Advanced Research Projects Agency of the U.S. government. This agency developed a network referred to as the ARPANET [1]. In addition to ARPANET, which is used only by government educational, and industrial

research institutions, a number of public packet switching networks have been developed throughout the world.

Packet switching networks are an outgrowth of the need to utilize the capacity of a data communications circuit more efficiently. Circuit inefficiencies arise because the statistics of most data communications traffic vary significantly from voice traffic statistics. A subscriber voice terminal is typically involved in one or two calls per busiest hour of the day. The average telephone call lasts for 3 to 4 min. Thus, a typical voice terminal is in use only for 10% of the busiest hour of the day. In contrast, data communications terminals generally produce more frequent and shorter messages. In common applications a data terminal uses a voice-grade circuit for extended periods of time, even though actual use of the transmission capacity is very low. Data communications usage patterns represent abnormal loads to the telephone network whether infrequent and long holding times are used or whether frequent, short holding times occur: Short calls overload the call processing elements of a switch, and a few calls with extended holding times overload interoffice trunks and paths through switching matrices. Because of increasing amounts of "abnormal" traffic, and because of computational capabilities of stored program control switches, telephone operating companies are beginning to employ usage sensitive accounting for local calls in lieu of traditional flat rate service.

From a customer's point of view, the most inefficient aspect of data communications occurs on long-distance connections (or measured local service) that incur charges based on total holding times and not on actual utilization of channel capacity. In a typical interactive computer session, the transmission capacity is less than 1% utilized.

A second approach to improving data communications involves developing separate networks specifically designed for digital transmission (no analog circuits with modems). The first major enterprise of this type in the United States was a nationwide digital microwave network developed by Digital Transmission Corporation (DATRAN). Unfortunately for data communications users, DATRAN ran into financial difficulty and declared bankruptcy in 1977.

With the exception of the dataphone digital service (DDS) offering of AT&T, other attempts to develop digital networks for data communications were unsuccessful—partly because of insufficient demand and partly because the capabilities and service offerings of the telephone common carriers improved. The advent of digital access to the digital facilities of the telephone network provides the ultimate benefit for data communications users: the integrated services digital network (ISDN) described later in this chapter.

9.1 MESSAGE SWITCHING

As one telegraph system after another was installed in the countries around the world, nationwide communications networks evolved. A message could be sent from one point to another even if the two points were not serviced by a common

telegraph line. In this case, telegraph operators at intermediate points would receive a message on one line and retransmit it on another. When a telegraph office had several lines emanating from it, the process of transferring a message from one line to another was, in essence, a switching function.

The process of relaying, or switching, a message from one telegraph line to another became semiautomated when teletypes with paper tape punches and readers were developed. An incoming message could be punched automatically onto a paper tape by one teletype and subsequently read by another teletype for transmission on the appropriate outgoing line. The process of transferring a message from one line to another in this manner led to these systems being referred to as torn-tape message switches.

One of the world's largest message switches was completely automated in 1963 when Collins Radio Company of Cedar Rapids, Iowa installed a computer-based message switch for the airline companies of North America. This system, and its more recent successors, eliminates paper tape transfers by storing incoming messages directly into a computer memory (disk file) and forwarding them automatically to the appropriate output line when available. Hence this mode of operation is often referred to as "store-and-forward" message switching.

Included with each message is a header containing an address and possibly routing information so the message processor at each node can determine which output line to switch the message to. As indicated in Figure 9.1, the processor in each node maintains message queues for each outgoing link. These queues are normally serviced on the first-come first-served basis. However, priority in-

Figure 9.1. Message switching network.

formation can sometimes be included in each header to establish different classes, or grades of service—thereby allowing time-critical messages to be placed at the head of a queue.

A message-switching network is fundamentally different from a circuit-switching network in that the source and destination do not interact in real time. In fact, most message-switching networks can deliver a message on a delayed bases if a destination node is busy or otherwise unable to accept traffic. There is no need to determine the status of the destination node before sending a message, as there is in circuit switching.

Message-switching networks are also fundamentally different from circuit-switching networks in their response to traffic overloads. A circuit-switching network blocks or rejects excess traffic, while a message-switching network normally accepts all traffic but provides longer delivery times as a result of greater average queue lengths.

Another important distinction of a message-switching network is that the transmission links are never idle while traffic is waiting to use them. In a circuit-switching network, a circuit may be assigned to a particular connection but not actually carrying traffic. Thus some of the transmission capacity may be idle while some users are denied service. In contrast, utilization of the transmission links of a message-switching network is directly related to the actual flow of information. Arbitrarily high-utilization efficiencies are possible if increased store-and-forward queueing delays are acceptable. Chapter 10 provides basic results of queueing theory that relate utilization efficiency to queueing delay.

9.2 PACKET SWITCHING

The circuit-switched telephone network is ill-suited to interactive data traffic because it is fundamentally designed for less frequent service requests with comparatively long holding times (3 to 4 min on average). Neither the control elements in the switches nor the capacity of the signaling channels are capable of accommodating frequent requests for very short messages. The result is that connection setup time is many times greater than the holding time of a data message. Obviously, more efficient utilization of the network requires greater control channel bandwidth and increased call processing capacities in the switches. Beyond this, however, interactive data traffic with low-activity factors requires a network operation that is fundamentally different from a conventional circuit-switched network. The most appropriate mode of operation for traffic that comes in bursts is more closely related to a message-switched network than to a circuit-switched network.

Figure 9.2 depicts both the conceptual structure and the conceptual operation of a packet-switched network. A single message at the source is broken up into "packets" for transmission through the network. Included in each packet is a header containing address and other control information. Each packet is relayed through the network in a store-and-forward fashion similar to a message

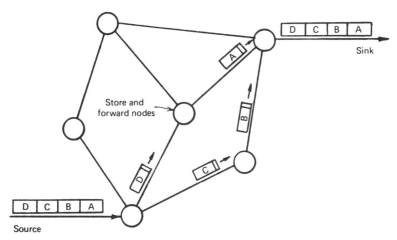

Figure 9.2. Packet switching network.

switching network. At the destination node, the packets are reassembled into the original contiguous message and delivered to the sink.

The main feature of a packet switching operation is the manner in which the transmission links are shared on an as-needed basis. Each packet is transmitted as soon as the appropriate link is available, but no transmission facilities are held by a source when it has nothing to send. In this manner, a large number of relatively inactive sources can share the transmission links. In essence, link utilization is improved at the expense of storage and control complexity in the nodes. A circuit-switched network has control overhead associated with connection setup, but very little control thereafter. In contrast, packet-switching nodes must process the header information in each packet as it arrives. Thus a long message in a packet-switched network requires more control overhead than if it were serviced in circuit-switched network. Considering the declining cost of digital memory and processing, the increased control complexity becomes less and less significant as time goes on.

As the traffic load in a packet-switched network increases, the average transmission delay increases correspondingly. In contrast, a circuit-switched network either grants service or rejects it. There is no graceful degradation in service. Conversely, when only a few circuits are in use in a circuit-switched network, much network transmission capacity is idle. When there is a light load on a packet-switched network, the active users benefit by shorter than usual delay times. Hence from a user's grade-of-service point of view, the two network types are fundamentally different.

Using automatic repeat request (ARQ) error control, packet-switching networks normally provide essentially error-free transmission for each node-to-node transfer. This process requires the receiving nodes to monitor redundant

check bits appended to each packet to determine if the packet was received correctly. When errors are detected, a retransmission is requested. Hence transmitting nodes must hold all transmitted packets in memory until a positive response (ACK) is returned by the receiving terminal. Furthermore, an entire packet is usually received and checked for errors before forwarding it to another node.

Customers access packet networks by way of local leased lines or dial-up connections. Dial-up connections are used by infrequent users, while leased lines are preferred by heavy users to achieve constant availability, higher data rates, and often lower error rates.

If flat rate accounting is in effect for local dial-up access, packet switching customers have long-distance data communications capabilities at standard monthly rates from the local operating company. A packet switching company typically charges only for the amount of data actually transmitted. Thus customers realize considerable savings if they transmit and receive data using low-activity factors.

Despite the similarity to a message-switching operation, a packet-switching network is different in two important respects:

1. The store-and-forward delay through a packet-switched network is relatively short. Thus interactive communications can occur in much the same manner as if a dedicated, end-to-end circuit is established
2. A packet-switched network does not provide storage of messages—except in an incidental manner while relaying packets from one node to another. The network is designed to provide switched communication between two nodes, both of which are actively involved in the communications process. A basic packet-switching network does not normally store a message for later delivery to an inactive or busy terminal.

One reason for breaking messages into packets is to allow transmission of the first segment of a long message while other segments are in transit. If the entire message had to be received at each node before forwarding it to the next node, the delays through the nodes might be too large. Another reason for breaking the messages into packets arises from operational simplifications derived from storing, processing, and transmitting fixed-length blocks of data. In addition, if long messages are transmitted intact, short messages experience excessive delays when queued behind long messages. Packetization allows short messages to get through a transmission link without waiting behind long messages. This same principle occurs in multiprogrammed computers, which use time slicing to allow short jobs the opportunity of finishing before previously started long jobs. One more motivation for packetization is that when a transmission block is too long, it is unlikely that the entire message will be received correctly. Packetization provides a means of retransmitting only those portions of a message that need to be retransmitted.

9.2.1 Packet Formats

The format of a packet in a packet-switching network can vary significantly from one network to another. Some formats include numerous fields for control information while other systems rely more heavily on special control packets to transmit control information. Generally speaking, the control information associated with a particular message or link is included in the header of a message packet. Less frequent, networkwide control information is communicated through special control packets.

As indicated in Figure 9.3, a packet contains three major fields: the header, the message, and the redundancy check bits. Some packets may not contain a message field if they are being used strictly for control purposes. Although a variety of techniques for generating redundancy checks are possible, the most popular technique uses cyclic redundancy codes (CRCs). Basically, a CRC is nothing more than a set of parity bits that cover overlapping fields of message bits. The fields overlap in such a way that small numbers of errors are always detected and large numbers of errors are detected with a probability of 1 in 2^M where M is the number of bits in the check code. Standard CRCs can now be generated and checked using cyclic shift registers implemented in a single intergrated circuit.

A header typically contains numerous subfields in addition to the necessary address field. Additional fields sometimes included in a header are the following:

1. An operation code to designate whether the packet is a message (text) packet or a control packet. In a sense this field is a part of the destination address, with the address specifying the control element of a switching node.
2. A source address for recovery purposes or identification of packets at a destination node that is capable of simultaneously accepting more than one message.
3. Sequence number to reassemble messages at the destination node, detect faults, and facilitate recovery procedures.
4. Length code to indicate the length of a packet when less than a standard size packet is transmitted. Some protocols insert special delimiters (flags) at the end of a packet and therefore do not use a length count.

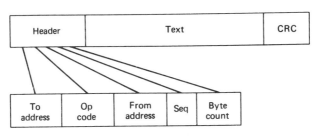

Figure 9.3. Typical packet format.

9.2.2 Statistical Multiplexing

The digital time division multiplexing techniques described in the previous chapters provide multiple channels by periodically assigning a time slot to each channel. The time slots are assigned whether or not the respective sources have anything to send. Channel assignments may be altered on a connection-by-connection basis, but for the duration of a "call" a particular time slot is dedicated to a respective connection. Because of the continuous manner in which time slots occur for each channel, this form of multiplexing is sometimes referred to as synchronous time division multiplexing (STDM).* In this section we describe another form of multiplexing, variously referred to as asynchronous time division multiplexing (ATDM) or statistical time division multiplexing (STATDM or STAT-MUX). This form of multiplexing is mentioned here because of its close relationship to packet-switching techniques.

Asynchronous time division multiplexers operate with framing formats that are basically identical to STDM framing formats. The major difference is that an ATDM system periodically redefines the length of its frames to change the number of time slots and, hence, the number of channels. Whereas an STDM system permanently assigns a time slot to each of its sources, an ATDM system assigns a time slot only when a source becomes active. A time slot is eliminated (the frame shortened) when the respective source becomes inactive. Notice that the terms "synchronous" and "asynchronous" in this context have nothing to do with the mode of operation of the transmission link. The transmission links of either an STDM or an ATDM system normally operate with synchronous transmission and detection but can conceivably operate with asynchronous (start/stop) transmission.

ATDM or STAT-MUX systems are typically used to provide line sharing for a multiple number of interactive terminals communicating with a host computer. When only a few sources (terminals) are active, each source has a relatively high data rate channel available to it. As the number of active sources increases, the frame length increases so individual channel rates decrease. Some systems limit the number of active sources to ensure certain minimum data rates.

The purpose and performance of STAT-MUX systems are very similar to the purpose and performance of a packet-switching link. The main difference is that a packet-switching link transmits larger blocks of data with a header included in each block. Each time slot of a STAT-MUX system is shorter and contains only source data. Figure 9.4 contrasts the basic operation of message switching,

*Here is yet another conflicting use of the term "synchronous." In the context of digital telephony, "synchronous multiplexing" refers to combining tributary signals that have been synchronized to each other as in the SONET multiplexing described in the previous chapter. Correspondingly, "asynchronous multiplexing" refers to the use of pulse stuffing to accommodate unsynchronized tributaries. As discussed at the end of this chapter the CCITT standards committee has complicated matters further by adopting the term "asynchronous transfer mode" (ATM) to refer to high-speed (broadband ISDN) packet switching.

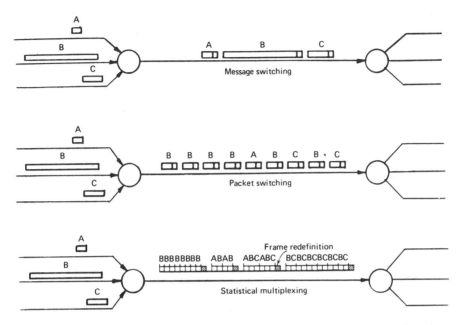

Figure 9.4. Comparison of message switching, packet switching, and statistical multiplexing.

packet switching, and statistical time division multiplexing. The message switch transmits each message in its entirety in a first-come first-served manner. Packet switching breaks messages up to allow interleaving of packets from other sources. Thus short messages never get queued behind long messages such as file transfers. The STAT-MUX system breaks the messages up into even finer blocks (words) of data and adds periodic frame definition messages so that receiving terminals can properly identify the individual time slots and switch the incoming data accordingly.

As indicated in Figure 9.4, a packet-switching operation becomes very similar to a STAT-MUX operation if the size of the packets is small. In fact, commercially available statistical time division multiplexers can be used to build up a packet-switching network. In addition to the multiplexers, however, a packet-switching node needs sufficient storage to accumulate an entire packet, check it for errors, process it, and retransmit it. Strictly speaking, a statistical multiplexer does not accumulate an entire message or packet. It switches the incoming data as they are received. Error control functions (redundancy checking and requests for retransmission) are implemented between endpoints of a "connection" instead of between nodes. In summary, a basic ATDM or STAT-MUX system is strictly a multiplexer/demultiplexer. A packet-switching node provides multiplexing-like functions, but also provides message-level and network-level control functions as well.

9.2.3 Routing Control

Much discussion and experimentation have been undertaken regarding various procedures for routing packets from sources to sinks through a packet-switching network [2]. All routing techniques allow for a certain amount of adaption or alternate routing to circumvent line or node failures. The various techniques differ, however, in how fast they respond to failures and whether or not they circumvent network congestion as well as equipment failures. The basic techniques are the following:

1. Dynamic routing
2. Virtual circuit routing
3. Fixed path routing.

Each of these techniques can be implemented in a variety of ways and can assume some of the characteristics of the other routing control procedures.

Dynamic Routing

Dynamic routing is implemented on a distributed basis with network nodes examining the destination address of each received packet to determine the appropriate outgoing link. The outgoing link is selected by processing locally stored information to determine which path provides minimum delay to the destination. The routing criteria are routinely updated to include both the operational status (health) and the size of the queues in the neighboring nodes. The routing decisions are adjusted rapidly enough that individual packets of a single message may follow *different* paths through the network. Dynamic routing, with its ability to respond quickly to changes in network topology or traffic conditions, is one of the original features espoused for packet switching. In fact, dynamic routing is sometimes considered to be inherent in the definition of a packet-switching network.

In spite of the obvious attraction of being able to adjust to rapid fluctuations in traffic patterns, dynamic routing has a number of significant drawbacks. One implication of allowing successive packets in a message to follow different routes through the network is that packets may arrive at a destination out of sequence. Although sequence numbers are used to rearrange the packets properly, the reassembly process is complicated—particularly since the destination does not know if a missing packet is merely delayed, or lost entirely.

Another drawback to dynamic routing is the possibility of oscillation occurring in the routing decisions. If the bandwidth assigned to updating the routing control algorithms is too small, a lightly loaded node will attract more traffic than it can handle before neighboring nodes are informed of the change in the traffic flow. In this instance, a packet might even wind up at a node from which it has previously been sent.

Purely distributed control, in general, and dynamic routing, in particular, also present difficulties with respect to flow control in a packet network. As

mentioned in Chapter 7, alternate routing in the switched telephone network is sometimes suspended when the network as a whole becomes overly congested (alternate routes require more resources). Obviously, the same principle applies to a packet-switching network. Flow control in packet networks is discussed in later sections.

Dynamic routing is most appropriate for small networks or in a military environment where survivability of the network in the presence of multiple-node failures is a requirement. A military network is typically more concerned with reliable and timely completion of a few important messages than with achieving the highest possible throughput from a given amount of resources.

Virtual Circuits

A virtual circuit network embodies some of the basic aspects of both circuit switching and packet switching. The transmission capacity is dynamically assigned on an "as needed" basis, but all packets of a multipacket message follow the *same* route through the network. Before interactive communication begins, a route is established through the network, and all participating nodes are informed of the "connection" and how to route the individual packets that follow. From then on, all packets flowing between the two end points follow the same route through the network. In essence, a virtual circuit is a logical concept involving addresses and pointers in the nodes of the network, but no dedicated transmission facilities. At the end of a connection (or "session" in data communications terminology), a virtual circuit is released by a "disconnect" message propagating through the network.

Except in the case of permanent virtual circuits separate "connections" or sessions involving the same two end points do not necessarily use identical paths through the network. Each virtual circuit is established during the call establishment phase depending on the traffic patterns at the time. Thus a virtual circuit network can respond to network failures or changing traffic patterns, but in a longer time frame than a dynamically routed network. When virtual circuits are changed from one "connection" to the next, the mode of operation is sometimes referred to as a "switched virtual circuit network" by direct analogy to conventional circuit switching.

Virtual circuits can be established using either distributed or centralized control. When distributed control is used, the call establishment message propagates through the network with each node making a local decision as to which outgoing link should be selected.

As discussed in Chapter 7 concerning conventional circuit switched networks, centralized call establishment has the basic advantage of being able to set up circuits with a networkwide view of network status and traffic patterns. The TYMNET network of the United States [3] and the DATAPAC network of Canada [4] are examples of virtual circuit, packet-switched networks with centralized control.

Since the call establishment phase of a virtual circuit represents excessive overhead for single-packet messages, a virtual circuit mode of operation is

obviously most useful when a network services a relatively large number of multiple-packet messages or sessions. Thus a dual mode network is suggested requiring virtual circuits for longer messages, and immediate transmission with dynamic routing for single-packet messages. In this instance, the single-packet messages are usually referred to as "datagrams."

One of the main advantages of a virtual circuit operation is its ability to provide more orderly control of packet delivery. If a node in a virtual circuit never forwards a packet pertaining to a particular "connection" until the previous packet has been acknowledged, packets cannot arrive out of order. A second advantage of a virtual circuit is the reduced addressing requirements of individual packets. Once a virtual circuit has been established, complete destination addresses are no longer needed in the packets. In place of destination addresses, virtual circuit identifiers can be used that are local to each link. In essence, virtual circuit identifiers are pointers to memory addresses in the call processors of the packet-switching nodes. The designated memory addresses contain the pertinent information concerning the actual source, destination, and routing of the packets. Another important feature of a virtual circuit operation is its inherent ability to aid in flow control, as discussed in the next section.

The main disadvantage of a virtual circuit operation is the possibility of greater transmission delays. When a path for a virtual circuit is established, it is chosen to minimize the delay through the network under the traffic patterns at that time. If the traffic patterns change, packets pertaining to a particular virtual circuit may experience long queueing delays on some links while alternate links are more lightly loaded. Yum and Schwartz [5] report that analyses of routing techniques for several small network configurations indicate a packet delay improvement of 10 to 27% is possible when adaptive routing is used in lieu of a fixed routing rule.

When first considered, call establishment might seem to be a disadvantage of a virtual circuit network. Actually, however, flow control considerations require some type of query/response message to determine the status of the destination nodes before a source is allowed to begin sending a packet stream. Thus the control overhead and the delay associated with setting up a "connection" is usually a fundamental requirement, even in a dynamically routed network.

Fixed Path Routing

Fixed path routing embodies the same basic concepts of a virtual circuit network except successive "connections" or sessions between any two end points always use the same path. In essence a fixed path network is one that assigns permanent virtual circuits to each pair of end points.

One attractive feature of a fixed path network is the absence of the call establishment phase of a virtual circuit network. However, unless the necessary resources are permanently allocated, a "clear-to-transmit" message is needed before sending packets. Inactive virtual circuits do not tie up network resources in the same manner as conventional circuits, but each virtual circuit does require

transmission capacity and store-and-forward buffers in a probabilistic sense. If minimum grades of service (delay times) are to be guaranteed, a network must limit the number of virtual circuits existing at any particular time. Hence clear-to-transmit signals are needed before a fixed path circuit becomes active. Of course, a network can provide two modes of operation: permanently active (hot line) virtual circuits and virtual circuits activated and deactivated as needed.

An obvious disadvantage of a purely fixed path network, as described so far, is its vulnerability to node or link failures. To circumvent this problem, a network control center usually assigns semipermanent paths through the network that are changed only when necessary for failure survivability or maintenance. Some message-switching networks are implemented with fixed path routing in the strict sense. This mode of operation is more appropriate to message-switching networks since message delivery is less time critical and can be deferred while repairs are undertaken for inoperative equipment.

9.2.4 Flow Control

As discussed in Chapter 7 concerning conventional circuit-switched networks, routing and flow control are two closely related operational requirements of any communications network. The same basic principle for controlling the flow in circuit-switched networks also applies to packet-switched networks. Namely, whenever the network is experiencing excessive stress, either from a loss of capacity due to failures or from an abnormally heavy demand for services, new service requests must be blocked at the periphery of the network before they tie up common resources and compound the congestion. In a packet-switched network, the common resources are store-and-forward buffers and transmission links.

Flow control in a packet network is primarily concerned with buffer management. For example, if all store-and-forward buffers in adjacent nodes become filled with packets destined to each other, the nodes are unable to receive additional packets, and a deadlock exists. Recall that a packet-switching node does not release a buffer as soon as it transmits a packet. The buffer is released when an acknowledgment is returned from the destination node. If a receiving node has no available buffers, it cannot accept a new packet and therefore cannot acknowledge it.

Flow control requirements imply that all interface nodes in a packet-switched network are aware of overload conditions and refuse new requests for service until the congestion is relieved. A particularly attractive feature of a virtual circuit network is that the call establishment phase provides an automatic means of determining whether or not a particular request should be serviced. If no path through the network can be established because the interior nodes are too congested, the request is rejected. On the other hand, if a virual circuit is established, there is reasonable expectation that the entire request will be serviced in a timely manner.

Unless network nodes are very conservative in accepting requests for new

virtual circuits, the ability to set up a circuit does not guarantee the avoidance of excessive congestion or deadlocks. A node accepts a new virtual circuit based on an expected load and its capacity to service that load. If the traffic volume and patterns happen to exceed the expected load, excessive delay or congestion is possible.

Virtual circuits are an effective means of controlling the flow of multiple-packet messages, but they represent too much overhead for single-packet or datagram flow control. If a virtual circuit network must support a significant number of single-packet messages, it can allow immediate transmission of these messages and forego the call establishment phase. In terms of transmission overhead, a datagram is not much greater than a call-establishment packet. Thus, from this point of view, the packet might as well be sent immediately and be considered its own circuit-setup message. In terms of store-and-forward buffers, however, a single-packet message is much different from a call-establishment packet. A call-establishment packet requires a certain amount of storage and processing by the call processor of each network node it reaches, but it does not compete for store-and-forward buffers as does a message packet. If necessary, a call-establishment message can be ignored by an overloaded node, and no ill effects result. The originating node merely times-out waiting for the network response and reissues the request. The time out should be long enough that the network has had a chance to relax.

In contrast to call-establishment packets, if a message packet is ignored by an overloaded node, congestion migrates to the node that last transmitted the packet, since this node is holding a copy of the rejected packet in a buffer. The buffer can not be released until an acknowledgment is received. Hence datagrams cannot be allowed to enter the network unless a reasonable chance for complete passage exists. When the network is heavily congested, attempts to set up virtual circuits might also be suspended.

A conventional circuit-switched network is unconcerned with flow control between the endpoints of a connection since, once the circuit is established, the activity or inactivity of the endpoints has no effect on other connections or on the network as a whole. End users necessarily administer flow control between themselves so the source does not ove. un the receive buffers of a sink. These procedures concern only the endpoints.

In contrast, the very nature of a packet-switching network implies direct involvement with endpoint activity. If a source pumps excessive traffic into a network, other users experience degraded performance. Hence interfaces to a packet-switching network necessarily include flow control for respective sources. Source flow control establishes a maximum data rate for a network. If a sink accepts data at a lower rate for a sustained period of time, this fact must be communicated to the network node serving the source in order to slow the source down.

Store-and-forward buffers of a packet-switching network are needed for communications purposes and are not used as a storage medium for messages. If

a packet-switching network also provides message-switching services, message storage functions should be implemented separately from the communications buffers. Then, the packet-switching network is used to transfer data to the message storage facility.

Another implication of maximizing store-and-forward buffer utilization is the need to wait until a sink is ready to accept data before a source begins sending. If a sink is not ready, packets get stuck in store-and-forward buffers at the far end of a network and cause congestion. Thus some form of request-to-transmit/clear-to-transmit sequence is needed before message transmission begins. This requirement is independent of the routing algorithms employed in the network. Hence in actual practice the setting up of a virtual circuit does not represent a time penalty.

Single-packet messages can be an exception. If they are transmitted without a clear-to-transmit signal, they may be discarded at the destination node if the sink is inactive or has no receive buffers available. The destination node then needs to return a rejection message to the source indicating the status of the sink. In this manner, the message itself is a request-to-transmit signal. The ARPANET uses a flow control strategy wherein single-packet messages serve as their own requests for buffer storage, but multiple-packet messages require preallocated buffers in groups of eight at the destination [6, 7].

Flow control in TYMENET is implemented in a different manner because of its exclusive use of virtual circuits for all messages. Before any node in the network can send a packet to a neighboring node, it must receive a clear-to-transmit signal from the neighboring node for the particular virtual circuit. The clear-to-transmit signal is an indication that a specified number of store-and-forward buffers are being held in reserve for that particular virtual circuit. After a node sends the specified number of packets, it sends no more until another clear-to-transmit signal is received (indicating the previous packets have been forwarded or more buffers have been allocated to the virtual circuit). By using the same node-to-node method of flow control at the interface between the network and the users, a very solid end-to-end flow control strategy is established.

The TYMENET flow control strategy is somewhat conservative in that it may allocate store-and-forward buffers to one virtual circuit while another virtual circuit has more use for them (possibly causing a decrease in line utilization). This conservatism, however, provides a number of useful features:

1. Networkwide flow control is established automatically by the flow control within each virtual circuit.
2. Under light traffic conditions, packet flow within each virtual circuit adjusts automatically to the maximum rate that the source and sink can support. If only a few virtual circuits exist, a relatively large number of buffers can be assigned to each circuit—allowing the return of more frequent clear-to-transmit signals.

3. If a sink stops accepting packets for some reason, this fact propagates back to the source by way of cessation of clear-to-transmit signals. Thus the source stops sending when all allocated buffers are full.

4. As long as a node never overcommits its buffers, store-and-forward lockups cannot occur. If several sinks stop accepting data, the store-and-forward buffers assigned to the particular virtual circuits get filled and become unavailable. Other virtual circuits, however, can maintain transmission through their own assigned buffers.

5. The mechanism is fail-safe in the sense that positive indications (clear-to-transmit signals) are needed before packets are forwarded to neighboring nodes. When network links are overloaded, flow control bandwidth requirements are minimal. If flow control signals stop altogether, packet transmission stops.

9.3 INTEGRATED SERVICES DIGITAL NETWORK

As described in the preceding chapters, the 1970s and 1980s produced two dramatic changes in the public telephone networks in the United States and around the world. The first of these was the much heralded use of digital technology for transmission and switching facilities that evolved into integrated digital networks (IDNs). The second, lesser known but also significant change involves the development of common channel signaling networks for network control. In both cases, these technologies started at internal portions of the network and migrated toward the periphery. Except for some special data service offerings and a few network-based features derived from the signaling network, these facilities provided no direct benefit to the end users. As shown in Figure 9.5 the integrated services digital network (ISDN) is a service offering that extends access to both of these facilities to the end user. Access to the digital transport facilities occurs on 64-kbps bearer (B) channels while access to the signaling network occurs on 16-kbps or 64-kbps signaling (D) channels. Major features or benefits made available by these channels are listed in Tables 9.1 and 9.2, respectively.

TABLE 9.1 Features of ISDN B Channels

1. End-to-end 4-wire digital circuits: no loss or echoes for voice circuits using digital instruments
2. Shared network access for voice, data, and leased lines
3. Relatively high bandwidth data channels (64 kbps)
4. Lower error rates than typical voiceband modems
5. In-service performance monitoring
6. Possible expansion of speech bandwidth because elimination of tandem encodings allows greater one-time quantization errors

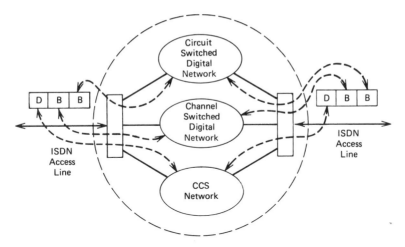

Figure 9.5. Integrated services digital network access to circuits, channels (leased lines), and common channel signaling.

TABLE 9.2 Features of ISDN D Channels

1. Signaling simultaneous with active connections
2. Calling number identification
3. Far-end supervision
4. User-to-user message transfer
5. Telemetry for fire alarms, security, meter reading, etc.
6. Access to packet switching network
7. Support for network services such as multiple directory numbers sharing one or more B channels, trunk group blocking statistics, and identification of calling number for abandoned or blocked calls

Two levels of digital access to the ISDN network have been standardized: basic rate access and primary rate access. As shown in Figure 9.6 the (worldwide) basic rate interface (BRI) standard is also referred to as a 2B + D interface. In North America, the primary rate interface (PRI) standard is sometimes referred to as 23B + D while the CEPT counterpart is 30B + D. The North American PRI is fundamentally a 1.544-Mbps DS1 signal with the D channel replacing one of the 24 message channels (usually the last one). To achieve a 64-kbps clear channel capability a B8ZS line code is used to eliminate ones density requirements and common channel signaling frees up the signaling bits so the full 64-kbps bandwidth is available for user data. The CEPT PRI is a 2.048-Mbps CEPT1 digital signal with the D channel occupying the signaling channel (time slot 16). Because a single D channel can support more than one PRI, 24B and 31B interfaces are allowed for all but one member of a group of PRIs.

Primary Rate Access

23B + D

ISDN
Network

2B + D
Basic Rate Access

Figure 9.6. Basic rate and primary rate access to ISDN.

9.3.1 ISDN Basic Rate Access Architecture

Transmission technology required for basic rate access is generally referred to as the digital subscriber loop (DSL), the main requirement of which is to use a single pair of wires for both directions of transmission. When displacing existing analog loops it is furthermore necessary that the DSL operate with bridged taps and wire gauge changes (both of which can cause reflections.) To allow flexibility in the selection and deployment of the DSL, the CCITT basic rate specification [8] does not define a two-wire transmission standard. Instead, it establishes an interface standard that assumes the presence of a network termination module that converts any chosen transmission system to the standard interface. In the interest of supporting deregulated customer premises equipment, the Exchange Carriers Standard Association in the United States established a basic rate transmission standard [9] so CPE equipment could connect directly to the transmission link or select network termination modules from alternate vendors. Figure 9.7 depicts the architecture and associated terminology of a basic rate interface (BRI).

Module Definitions

NT1: A network termination module for layer one functions that provides physical and electrical termination of the transmission link only. In essence, the NT1 isolates the user from the transmission technology but does not demultiplex or process D channel messages.

NT2: A second level of network termination that implements functions associated with layers two and three of the OSI protocol stack. Thus, NT2 equipment extracts and processes D channel messages. Representative NT2 equipment includes PBXs, multiplexers, or LAN gateways.

TE1: Type 1 terminal equipment such as a digital telephone that complies with the ISDN S interface recommendation.

Figure 9.7. Basic rate interface architecture.

TA: Terminal adaptor used to convert from an arbitrary (R) interface to the ISDN S interface.

TE2: A non-ISDN terminal that requires a terminal adaptor to interface to the ISDN S interface . Prevalent examples of a TE2 equipment are analog telephones or asynchronous (RS-232) data terminals.

Reference Points

U: Interface to the two-wire transmission line.
T: CCITT ISDN interface defined in recommendation I.430.
S: Interface to NT2 equipment identical to a T interface.
R: A non-ISDN interface such as an analog tip and ring.

9.3.2 S/T Interface

The S/T interface is defined in CCITT recommendation I.430 to be supplied by network termination equipment (NT2/NT1). It is intended for customer premises installations only. (No overvoltage protection is prescribed.) The most significant aspects of the S/T interface are

1. Four-wire facility (one pair for each direction)
2. One kilometer maximum required distance
3. Alternate space inversion line code (which is the inverse of an AMI line code: see Figure 9.8)
4. Point-to-point or point-to-multipoint configurations
5. Data rate is 192 kbps with 48 kbps of framing, control, and synchronization.

The frame structure at reference points S and T is shown in Figure 9.9. As indicated, each 250-µsec frame contains 48 bits. Thirty-eight of these bits are

Figure 9.8. Alternate space inversion line code.

Figure 9.9. S/T frame structure.

common to both directions of transmission and defined as follows:

 16 bits in first B channel (B1)
 16 bits in second B channel (B2)
 4 bits in the D channel
 1 bit in the framing channel F
 1 bit in the auxiliary framing channel F_A.

The remaining 10 bits are assigned different functions depending on the direction of transmission. From the TE to the NT all remaining 10 bits are defined as L bits, which are used to maintain dc balance. The definition of the 10 bits from the NT to the TE are

 2 L bits for maintaining dc balance
 4 E bits that echo D bits received from the TE(s)
 1 A bit for activation
 1 N bit, which is the complement of the F_A bit
 1 M bit for multiframe identification
 1 S bit for S-channel.

Figure 9.9 indicates that the framing bit F is always a binary ZERO. Even though a positive voltage level is indicated, either a positive or negative voltage

is allowed so the receivers are not sensitive to wiring polarity. A transmitter always produces the same level, however, so the receiver always receives the same polarity in every framing bit. As an aid in rapid acquisition of the framing pattern, the framing bit always represents a line code violation (it is the same polarity as the previous ZERO). To maintain dc balance, an L bit with the opposite polarity of the F bit always follows the F bit. The first ZERO in a data block following the L bit is encoded with the same polarity as the L bit, which implies another line code violation. Dc balancing of this violation is the purpose of the L bit at the end of each data block, which also assures that the next (fixed polarity) F bit produces a line code violation.

The reason for the additional L bits in the frame from the TE to the NT arises because more than one TE can be connected to the S interface as a passive bus (Figure 9.10). Because the TEs transmit independently of each other, each individual transmission (D channel bits and B channel bytes) is individually dc balanced.

Passive bus operations are also the reason for the existence of the NT-to-TE E bits. Multiple station access to the D channel is controlled by having a terminal wait for an idle code on the NT-to-TE D channel before transmitting on the TE-to-NT D channel. When a terminal begins D channel transmission it monitors the incoming E bits. If an incoming E bit does not match the previously transmitted D bit that terminal stops transmitting and waits to seize the channel at a later time. Two levels of priority are defined for accessing the D channel. Signaling information is the highest priority while user packet messages are the lower priority. All terminals on a passive bus have equal priorities within each level.

Example 9.1. Determine the distance limit imposed by the need to echo E bits in a BRI S/T interface. Assume the speed of transmission of a signal on pair of

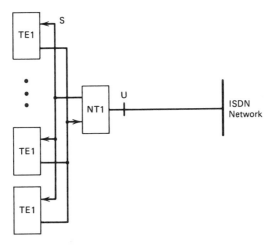

Figure 9.10. S-bus connections.

wires is one-third the speed of light in a vacuum. Ignore transmitter and receiver filter delays and assume no delays in the NT logic circuitry.

Solution. From Figure 9.9 it can be seen that the minimum delay between a terminal transmitting a D bit and receiving it back in the following E bit is seven bit times (this is the D bit following the first B2 byte). At 192 kbps data rate the duration of a bit is 5.2 μsec.

Thus, the total roundtrip propagation time is $(7)(5.2) = 36.4\,\mu sec$. Assuming there are no logic circuitry delays in the NT, the maximum length of wire is determined as

$$\text{Maximum wire length} = 36.4(10^{-6}) \times [3(10^8)]/3$$
$$= 3.64\,\text{km}$$

Because roundtrip propagation involves both directions of transmission the maximum distance is

$$\text{Maximum distance} = 3.64/2$$
$$= 1.82\,\text{km}$$

Example 9.1 shows that the BRI standard has a fundamental distance limitation that is not far above the minimum transmission distance specification of 1 km.

9.3.3 ISDN U Interface

Prior to the establishment of a standard U interface by the Exchange Carriers Standards Association in North America a number of basic rate transmission systems were developed by telephone equipment suppliers around the world and put into service. One of these involves the use of time compression multiplex (TCM). TCM, developed by NEC in Japan [10], provides full duplex transmission on a single pair of wires by alternately transmitting bursts of data in each direction. For this reason it is sometimes referred to as "ping-pong" transmission. One big advantage of TCM transmission is that near-end crosstalk is avoided because a station is never receiving while transmitting. The big disadvantage is that the burst data rate must be more than twice the desired data throughput.

AT&T in the United States also developed a BRI transmission system for the No. 5 ESS end office switching system available with generic releases 5E4 and 5E5 (11). This system has a 160-kbps data rate utilizing a 50% duty cycle AMI line code. Full duplex transmission is achieved by simultaneous transmission in both directions using hybrids and echo cancellors (ECs) to separate the two signals as indicated in Figure 9.11. Beginning with generic 5E6 the No. 5 ESS supports both the AT&T U interface (referred to as a 5E4/5E5 U interface) and the ANSI U interface.

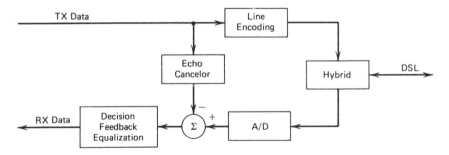

Figure 9.11. ISDN DSL TX/RX block diagram.

Like the 5E4/5E5 U interface, the ANSI U interface uses simultaneous transmission in both directions with echo cancelors and a data rate of 160 kbps. The major difference is the use of a four-level line code referred to as 2B1Q (two binary digits in one quaternary digit). Thus, the symbol (baud) rate on the line is 80,000 symbols/sec. Because the line code itself does not prevent dc wander, dc restoration is necessary. Reference [12] describes an implementation with adaptive quantized feedback as being the best approach. A 2B1Q line code was chosen primarily because the lower symbol rate minimizes the two dominant transmission limitations in this application: intersymbol interference and near-end crosstalk [13].

The frame format and superframe structure of the ANSI U interface are shown in Figure 9.12. Each frame consists of 240 bits containing 18 framing bits, 216 payload bits (12 fields of 18 2B + D data bits), and 6 overhead bits. Because the frame rate is 667 frames/sec the data rate is 160 kbps. The 6 overhead bits are organized as a block of 48 bits in a 8-frame superframe. Functions included in

	18	18	18	...	18	6
1	ISW	2B + D	2B + D		2B + D	$M_1 - M_6$
2	SW	2B + D	2B + D		2B + D	$M_1 - M_6$
3	SW	2B + D	2B + D		2B + D	$M_1 - M_6$
4	SW	2B + D	2B + D	...	2B + D	$M_1 - M_6$
5	SW	2B + D	2B + D		2B + D	$M_1 - M_6$
6	SW	2B + D	2B + D		2B + D	$M_1 - M_6$
7	SW	2B + D	2B + D		2B + D	$M_1 - M_6$
8	SW	2B + D	2B + D		2B + D	$M_1 - M_6$

$\overset{\longleftarrow\quad\quad\quad 240 \quad\quad\quad\longrightarrow}{}$

SW = Sync Word = $+3+3-3-3-3+3+3-3+3+3$
ISW = Inverted SW = $-3-3+3+3+3-3+3-3-3-3$
2B + D = | B_1 | B_2 | D | (|8|8|2)
$M_1 - M_6$ = Overhead bits
Data are encoded as 00 = -3, 01 = -1, 11 =1, 11 = $+1$, 10 = $+3$

Figure 9.12 ANSI U Interface Frame and Superframe Structure.

the overhead bits are 24 bits of embedded operations channel, 1 activation bit, 1 deactivation bit, 1 far-end block error bit, 12 CRC bits, and 9 fixed ONE bits. All bits except the framing bits are scrambled for transmission.

9.3.4 ISDN D Channel Protocol

The D channel protocol is defined in two separate series of CCITT recommendations: the I series and the Q series. The data link layer (LAPD) is defined in I.441 or Q.921. This protocol is similar to LAPB of the X.25 standard except Q.921 allows more than one logical link. (Therefore separate "connections" can exist for signaling, packet network, or far-end terminals.) The main functions of the data link layer are message sequencing, error checking and retransmission, and data layer link recovery. The network layer of the D channel protocol is defined in I.451 or Q.931. This layer provides connection set up, alerting, routing, and release of ISDN calls. When a B channel accesses a packet network the X.25 protocol is used while in the connected state.

9.3.5 ISDN Applications

Of the two main aspects of ISDN, switched digital circuits and connectivity to the signaling network, the latter is the most significant in terms of introducing new services. Digital connectivity at lower rates is readily available with voiceband modems or at rates comparable to 64 kbps through special access circuits to adjunct public networks. In fact, because of the large installed base of voiceband hardware and software many initial uses of ISDN B channels involve only lower rate interfaces. Rate adaptation of common voiceband synchronous signals into 64-kbps B channels is defined in CCITT recommendation I.463.

In contrast to the B channels, the D channels present opportunities for entirely new features and services. The most significant of the D channel features is delivery of calling number identification (CNI). Although initial offerings of CNI generated controversy with consumer rights groups and the Public Utility Commissions (PUCs) regarding rights to privacy, CNI service offerings are generally approved—sometimes with restrictions. The following identify some of the most obvious applications of CNI:

Security for dial-up computer access or for telephone order taking.

Immediate, automated database retrieval for service bureaus, answering services, brokerage firms, telemarketing organizations, etc.

PBX or ACD routing based on CNI as opposed to called number. In an ACD application, for example, a call can be automatically routed to the same attendant that handled a previous call from the same number.

Automatic callback to calling parties that abandon a call after encountering a busy or unanswering station.

Support for centrex-like operations in multisite private networks.

Other features provided by the D channel but not derived from CNI are as follows:

Multiline telephones with single button feature activation similar to proprietary electronic telephones commonly available with PBXs.

Traveling class marks to request special services or maintain compatibility between terminals (e.g., identify voice or data calls, encrypted traffic, or special encodings for speakerphones or teleconferencing).

Call-by-call service selection for access to circuit switched services, packet switched services, or private line services.

In all, access to the signaling network tends to blur the distinction between private and public networks because the end users have the flexibility to dynamically reconfigure virtual private networks within the public network and have them interoperate with private facilities without a loss of features.

9.4 BROADBAND ISDN

Due to the emergence of high bandwidth fiber optic transmission facilities, digital connectivity at rates well above the 1.544- or 2.048-Mbps primary rates are economically viable for end-user applications. High-speed interconnect using public network facilities is referred to as broadband ISDN (BISDN). CCITT recommendations call for BISDN to be offered at speeds up to 155 Mbps (OC-3) or possibly 620 Mbps (OC-12). The main applications for BISDN are LAN interconnection for private networks, LAN services via public networks referred to as metropolitan area networks (MANs) or wide area networks (WANs), and video services for residential markets. In addition, the advantages of maintaining a single transmission system suggest that plain old telephone service (POTS), BRI, and PRI (narrowband) ISDN should also be carried on the BISDN facilities.

Because many of the potential applications for BISDN are unpredictable, a flexible transport/multiplexing architecture is needed that supports a variety of channel rates as listed in Table 9.3. Notice that the channel rates defined for ISDN and BISDN are the payload rates available to the end user before framing or other overhead is added.

A fundamental goal of the BISDN architecture is to be an all-purpose digital network. As such, it must support a wide variety of applications with a wide range of data rates. Some applications such as voice have balanced bandwidth requirements in both directions but others such as interactive graphics or remote database queries require high speed, bursty transmission in one direction and very low data rates in the other. Video distribution applications require yet another type of service: continuous (stream) transmission at relatively high rates in only one direction. To complicate matters even more the various applications

TABLE 9.3 ISDN and BISDN Channel Rates

Designation	Data Rate
B	64 kbps
H0	384 kbps
H11	1.536 Mbps
H12	1.920 Mbps
H21	32.768 Mbps
H22	44.160 Mbps
H4	135 Mbps

have different grade of service requirements in terms of connection setup times, blocking probabilities, error rate tolerances, absolute propagation delays, delay variations, signaling services, etc.

Conventional, hierarchical, and channelized multiplexing and switching architectures designed primarily for voice services do not support bursty, high bandwidth applications well. A major difficulty with hierarchical TDM structures is achieving dynamic bandwidth allocation and switching of bursty, high-speed services. Because of this, CCITT recommendations (I.121) suggest the use of a network architecture based on asynchronous transfer mode (ATM) multiplexing and switching [14]. ATM utilizes labeled multiplexing and switching as opposed to positional multiplexing and switching in synchronous TDM systems. Thus bandwidth allocation automatically adapts according to the size of the data block (cell) associated with each label.

Because ATM is basically a fast packet switching network it meets the needs of most high-speed data communications applications quite well. However, the introduction of ATM into the present network is complicated by the artificial delays inherent in packetized voice and the effects of this delay on echoes and singing. In an all-digital network, with end-to-end four-wire circuits, absolute delay is not a problem (within reason) but delay variability is. Furthermore, voice channels can tolerate relatively high bit error rates but not lost or discarded packets caused by corrupted headers (labels) or congested nodes. The exact opposite is true of most data applications—a lost packet is of little more consequence than an errored packet. In contrast to ATM, synchronous TDM virtually assures that no information is lost because TDM (positional) multiplexing is much more robust than ATM (labeled) multiplexing. Reference [15] describes some basic considerations for packetizing various types of digitized voice signals. Whether ATM, or any "pure" network architecture, can become a universal services network is a matter of definition. Any particular layer of an architecture can be made to be application independent, but then layers above and below that need to provide specialization to support the diverse requirements.

REFERENCES

1 R. E. Kahn, "Resource-Sharing Computer Communications Networks," *Proceedings of IEEE*, November 1972, pp 1397–1407.

2 H. Rudin and H. Müller, "More on Routing and Flow Control," *National Telecommunications Conference*, 1979, pp 34.5.1–34.5.9.

3 J. Rinde, "Routing and Control in a Centrally Directed Network," *National Computer Conference*, 1977, pp 603–608.

4 S. C. K. Young and C. I. McGibbon, "The Control System of the Datapac Network," *International Conference on Communications*, 1976, pp 137–141.

5 T. S. Yum and M. Schwartz, "Comparison of Adaptive Routing Algorithms for Computer Communications Networks," *National Telecommunications Conference*, 1978, pp 4.1.1–4.1.5.

6 F. E. Heart, R. E. Kahn, S. M. Ornstein, W. R. Crowther, and D. C. Walden, "The Interface Message Processor for the ARPA Computer Network," *Spring Joint Computer Conference*, 1970, pp 551–556.

7 L. Kleinrock, "Principles and Lessons in Packet Communications," *Proceedings of IEEE*, November 1978, pp 1320–1329.

8 *Recommendation I.430 ISDN User-Network Interface: Layer 1 Recommendations*, Fascicle III.8, Blue Book, 1988, pp 171–240.

9 American National Standards for Telecommunications, "Integrated Services Digital Network—Basic Access Interface for Use on Metallic Loops for Application on the Network Side of the NT-Layer 1 Specification," *ANSI T1.601-1988*.

10 R. Komiya, K. Yoshida, and N. Tamaki, "The Loop Coverage Comparison Between TCM and Echo Cancellor under Various Noise Considerations," *IEEE Transactions on Communications*, November 1986, pp 1058–1067.

11 *5ESS^R Switch ISDN Basic Rate Interface Specification, 5E5 Generic Program Issue 1.00*, AT&T, December 1987, pp II-B-1/15.

12 S. Sugimoto, K. Hayashi, and F. Mano, "Design of 2B1Q Transceiver for ISDN Subscriber Loops," *IEEE International Conference on Communication*, June 1989, pp 8.1.1–8.1.5.

13 J. W. Lechleider, "Line Codes for Digital Subscriber Lines," *IEEE Communications Magazine*, September 1989, pp 25–32.

14 S. E. Minzer, "Broadband ISDN and Asynchronous Transfer Mode (ATM)," *IEEE Communications Magazine*, September 1989, pp 17–24.

15 N. Kitawaki, H. Nagabuchi, M. Taka, and K. Takahashi, "Speech Coding Technology for ATM Networks," *IEEE Communications Magazine*, January 1990, pp 21–27.

10

TRAFFIC ANALYSIS

Except for station sets and their associated loops, a telephone network is composed of a variety of common equipment such as digit receivers, call processors, interstage switching links, and interoffice trunks. The amount of common equipment designed into a network is determined under an assumption that not all users of the network need service at one time. The exact amount of common equipment required is unpredictable because of the random nature of the service requests. Networks conceivably could be designed with enough common equipment to instantly service all requests except for occurrences of very rare or unanticipated peaks. However, this solution is uneconomical because much of the common equipment is unused during normal network loads. The basic goal of traffic analysis is to provide a method for determining the cost-effectiveness of various sizes and configurations of networks.

Traffic in a communications network refers to the aggregate of all user requests being serviced by the network. As far as the network is concerned, the service requests arrive randomly and usually require unpredictable service times. The first step of traffic analysis is the characterization of traffic arrivals and service times in a probabilistic framework. Then the effectiveness of a network can be evaluated in terms of how much traffic it carries under normal or average loads and how often the traffic volume exceeds the capacity of the network.

The techniques of traffic analysis can be divided into two general categories: loss systems and delay systems. The appropriate analysis category for a particular system depends on the system's treatment of overload traffic. In a loss system overload traffic is rejected without being serviced. In a delay system overload traffic is held in a queue until the facilities become available to service it. Conventional circuit switching operates as a loss system since excess traffic is blocked and not serviced without a retry on the part of the user. In some instances "lost" calls actually represent a loss of revenue to the carriers by virtue of their not being completed.

Store-and-forward message or packet switching obviously possesses the basic characteristics of a delay system. Sometimes, however, a packet-switching

operation can also contain certain aspects of a loss system. Limited queue sizes and virtual circuits both imply loss operations during traffic overloads. Circuit-switching networks also incorporate certain operations of a delay nature in addition to the loss operation of the circuits themselves. For example, access to a digit receiver, operator, or a call processor is normally controlled by a queuing process.

The basic measure of performance for a loss system is the probability of rejection (blocking probability). A delay system, on the other hand, is measured in terms of service delays. Sometimes the average delay is desired, while at other times the probability of the delay exceeding some specified value is of more interest.

Some of the analyses presented in this chapter are similar to those presented in Chapter 5 for the blocking probabilities of a switch. Chapter 5 is concerned mostly with matching loss: the probability of not being able to set up a connection through a switch under normal or average traffic volumes. This chapter, however, is mostly concerned with the probability that the number of active sources exceeds some specified value. Typically, the specified value is the number of trunk circuits in a route.

10.1 TRAFFIC CHARACTERIZATION

Because of the random nature of network traffic, the following analyses involve certain fundamentals of probability theory and stochastic processes. In this treatment only the most basic assumptions and results of traffic analysis are presented. The intent is to provide an indication of how to apply results of traffic analysis—not to delve deeply into analytical formulations. However, a few basic derivations are presented, to acquaint the user with assumptions in the models so they can be appropriately applied.

In the realm of applied mathematics, where these subjects are treated more formally, blocking probability analyses are referred to as congestion theory and delay analyses are refered to as queuing theory. These topics are also commonly referred to as traffic flow analysis. In a circuit-switched network, the "flow" of messages is not so much of a concern as are the holding times of common equipment. A circuit-switched network establishes an end-to-end circuit involving various network facilities (transmission links and switching stages) that are held for the duration of a call. From a network point of view, it is the holding of these resources that is important, not the flow of information within individual circuits.

On the other hand, message-switching and packet-switching networks are directly concerned with the actual flow of information, since in these systems, traffic on the transmission links is directly related to the activity of the sources.

As mentioned in Chapter 7, circuit switching does involve certain aspects of traffic flow in the process of setting up a connection. Connect requests flow from the sources to the destinations acquiring, holding, and releasing certain

resources in the process. As was discussed, controlling the flow of connect-requests during network overloads is a vital function of network management.

The unpredictable nature of communications traffic arises as a result of two underlying random processes: call arrivals and holding times. An arrival from any particular user is generally assumed to occur purely by chance and be totally independent of arrivals from other users. Thus the number of arrivals during any particular time interval is indeterminant. In most cases holding times are also distributed randomly. In some applications this element of randomness can be removed by assuming constant holding times (e.g., fixed length packets). In either case the traffic load presented to a network is fundamentally dependent on both the frequency of arrivals and the average holding time for each arrival. Figure 10.1 depicts a representative situation in which both the arrivals and the holding times of 20 different sources are unpredictable. The bottom of the figure depicts activity of each individual source while the top displays the instantaneous total of all activity. If we assume that the 20 sources are to be connected to a trunk group, the activity curve displays the number of circuits in use at any particular time. Notice that the maximum number of circuits in use at any one time is 16 and that the average utilization is a little under 11 circuits. In general terms, the trunks are referred to as servers, and a trunk group is a server group.

Traffic Measurements
One measure of network capacity is the volume of traffic carried over a period of time. Traffic volume is essentially the sum of all holding times carried during the interval. The traffic volume represented in Figure 10.1 is the area under the activity curve (approximately 84 call minutes).

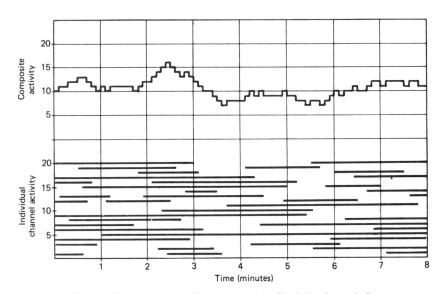

Figure 10.1. Activity profile of network traffic (all calls carried).

A more useful measure of traffic is the traffic intensity (also called traffic flow). Traffic intensity is obtained by dividing the traffic volume by the length of time during which it is measured. Thus traffic intensity represents the average activity during a period of time (10.5 in Figure 10.1). Although traffic intensity is fundamentally dimensionless (time divided by time), it is usually expressed in units of erlangs, in honor of the Danish pioneer traffic theorist A. K. Erlang, or in terms of hundred (century) call seconds per hour (CCS). The relationship between erlangs and CCS units can be derived by observing that there are 3600 sec in an hour:

$$1 \text{ erlang} = 36 \text{ CCS}$$

The maximum capacity of a single server (channel) is 1 erlang, which is to say that the server is always busy. Thus the maximum capacity in erlangs of a group of servers is merely equal to the number of servers. Because traffic in a loss system experiences infinite blocking probabilities when the traffic intensity is equal to the number of servers, the average activity is necessarily less than the number of servers. Similarly, delay systems operate at less than full capacity, on average, because infinite delays occur when the average load approaches the number of servers.

Two important parameters used to characterize traffic are the average arrival rate λ and the average holding time t_m. If the traffic intensity A is expressed in erlangs, then

$$A = \lambda t_m \tag{10.1}$$

where λ and t_m are expressed in like units of time (e.g., calls per second and seconds per call, respectively).

Notice that traffic intensity is only a measure of average utilization during a time period and does not reflect the relationship between arrivals and holding times. That is, many short calls can produce the same traffic intensity as a few long ones. In many of the analyses that follow the results are dependent only on the traffic intensity. In some cases, however, the results are also dependent on the individual arrival patterns and holding time distributions.

Public telephone networks are typically analyzed in terms of the average activity during the busiest hour of a day. The use of busy hour traffic measurements to design and analyze telephone networks represents a compromise between designing for the overall average utilization (which includes virtually unused nighttime hours) and designing for short duration peaks that may occur by chance or as a result of TV commercial breaks, radio call-in contests, and so on.

Busy hour traffic measurements indicate that an individual residential telephone is typically in use between 5 and 10% of the busy hour. Thus each telephone represents a traffic load of between 0.05 and 0.10 erlangs. The average holding time is between 3 and 4 min, indicating that a typical telephone is involved in one or two phone calls during the busy hour.

Business telephones usually produce loading patterns different from residential phones. In the first place, a business phone is generally utilized more heavily. Second, the busy hour of business traffic is often different from the busy hour of residential traffic. Figure 10.2 shows a typical hourly variation for both sources of traffic. The trunks of a telephone network are sometimes designed to take advantage of variations in calling patterns from different offices. Toll connecting trunks from residential areas are often busiest during evening hours, and trunks from business areas are obviously busiest during midmorning or midafternoon. Traffic engineering depends not only on overall traffic volume, but also on time–volume traffic patterns within the network.

A certain amount of care must be exercised when determining the total traffic load of a system from the loadings of individual lines or trunks. For example, since two telephones are involved in each connection, the total load on a switching system is exactly one-half the total of all traffic on the lines connected to the switch. In addition, it may be important to include certain setup and release times into the average holding times of some common equipment. A 10-sec setup time is not particularly significant for a 4-min voice call, but can actually dominate holding time of equipment used for short data messages. Common equipment setup times also become more significant in the presence of

Figure 10.2. Traffic volume dependence on time of day.

voice traffic overloads. A greater percentage of the overall load is represented by call attempts since they increase at a faster rate than completions.

An important distinction to be made when discussing traffic in a communications network is the difference between the offered traffic and the carried traffic. The offered traffic is the total traffic that would be carried by a network capable of servicing all requests as they arise. Since economics generally precludes designing a network to immediately carry the maximum offered traffic, a small percentage of offered traffic typically experiences network blocking or delay. When the blocked calls are rejected by the network, the mode of operation is referred to as blocked calls cleared or lost calls cleared. In essence, blocked calls are assumed to disappear and never return. This assumption is most appropriate for trunk groups with alternate routes. In this case a blocked call is normally serviced by another trunk group and does not, in fact, return.

The carried traffic of a loss system is always less than the offered traffic. A delay system, on the other hand, does not reject blocked calls but holds them until the necessary facilities are available. With the assumption that the long-term average of offered traffic is less than the capacity of the network, a delay system carries all offered traffic. If the number of requests that can be waiting for service is limited, however, a delay system also takes on properties of a loss system. For example, if the queue for holding blocked arrivals is finite, requests arriving when the queue is full are cleared.

10.1.1 Arrival Distributions

The most fundamental assumption of classical traffic analysis is that call arrivals are independent. That is, an arrival from one source is unrelated to an arrival from any other source. Even though this assumption may be invalid in some instances, it has general usefulness for most applications. In those cases where call arrivals tend to be correlated, useful results can still be obtained by modifying a random arrival analysis. In this manner the random arrival assumption provides a mathematical formulation that can be adjusted to produce approximate solutions to problems that are otherwise mathematically intractable.

Negative Exponential Interarrival Times
Designate the average call arrival rate from a large group of independent sources (subscriber lines) as λ. Use the following assumptions:

1. Only one arrival can occur in any sufficiently small interval.
2. The probability of an arrival in any sufficiently small interval is directly proportional to the length of the interval. (The probability of an arrival is $\lambda \Delta t$ where Δt is the interval length.)
3. The probability of an arrival in any particular interval is independent of what has occurred in other intervals.

It is straightforward [1] to show that the probability distribution of interarrival times is

$$P_0(\lambda t) = e^{-\lambda t} \tag{10.2}$$

Equation 10.2 defines the probability that no arrivals occur in a randomly selected interval t. This is identical to the probability that t sec elapse from one arrival to the next.

Example 10.1. Assuming each of 10,000 subscriber lines originate one call per hour, how often do two calls arrive with less than 0.01 sec between them?

Solution. The average arrival rate

$$\lambda = (10,000)\left(\frac{1}{3600}\right)$$

$$= 2.78 \text{ arrivals per second}$$

From Equation 10.2, the probability of no arrival in a 0.01-sec interval is

$$P_0(0.0278) = e^{-0.0278}$$

$$= 0.973$$

Thus 2.7% of the arrivals occur within 0.01 sec of the previous arrival. Since the arrival rate is 2.78 arrivals per second, the rate of occurrence of interarrival times less than 0.01 sec is

$$(2.78)(0.027) = 0.075 \text{ times a second}$$

The first two assumptions made in deriving the negative exponential arrival distribution can be intuitively justified for most applications. The third assumption, however, implies certain aspects of the sources that cannot always be supported. First, certain events, such as television commercial breaks, might stimulate the sources to place their calls at nearly the same time. In this case the negative exponential distribution may still hold, but for a much higher calling rate during the commercial.

A more subtle implication of the independent arrival assumption involves the number of sources, and not just their calling patterns. When the probability of an arrival in any small time interval is independent of other arrivals, it implies that the number of sources available to generate requests is constant. If a number of arrivals occur immediately before any subinterval in question, some of the sources become busy and cannot generate requests. The effect of busy sources is to reduce the average arrival rate. Thus the interarrival times are always somewhat larger than what Equation 10.2 predicts them to be. The only

time the arrival rate is truly independent of source activity is when an infinite number of sources exist.

If the number of sources is large and their average activity is relatively low, busy sources do not appreciably reduce the arrival rate. For example, consider an end office that services 10,000 subscribers with 0.1 erlangs of activity each. Normally, there are 1000 active links and 9000 subscribers available to generate new arrivals. If the number of active subscribers increases by an unlikely 50% to 1500 active lines, the number of idle subscribers reduces to 8500, a change of only 5.6%. Thus the arrival rate is relatively constant over a wide range of source activity. Whenever the arrival rate is fairly constant for the entire range of normal source activity, an infinite source assumption is justified.

Actually, some effects of finite sources have already been discussed in Chapter 5 when analyzing blocking probabilities of a switch. It is pointed out that Lee graph analyses overestimate the blocking probability because, if some number of interstage links in a group are known to be busy, the remaining links in the group are less likely to be busy. A Jacobaeus analysis produces a more rigorous and accurate solution to the blocking probability—particularly when space expansion is used. Accurate analyses of interarrival times for finite sources are also possible. These are included in the blocking analyses to follow.

Poisson Arrival Distribution

Equation 10.2 merely provides a means of determining the distribution of interarrival times. It does not, by itself, provide the generally more desirable information of how many arrivals can be expected to occur in some arbitrary time interval. Using the same assumptions presented, however, the probability of j arrivals in an interval t can be determined [1] as

$$P_j(\lambda t) = \frac{(\lambda t)^j}{j!} e^{-\lambda t} \qquad (10.3)$$

Equation 10.3 is the well-known Poisson probability law. Notice that when $j = 0$, the probability of a no arrivals in an interval t is $P_0(t)$, as obtained in Equation 10.2.

Again, Equation 10.3 assumes arrivals are independent and occur at a given average rate λ, irrespective of the number of arrivals occurring just prior to an interval in question. Thus the Poisson probability distribution should only be used for arrivals from a large number of independent sources.

Equation 10.3 defines the probability of experiencing exactly j arrivals in t sec. Usually there is more interest in determining the probability of j or more arrivals in t sec:

$$P_{\geq j}(\lambda t) = \sum_{i=j}^{\infty} P_i(\lambda t)$$

$$= 1 - \sum_{i=0}^{j-1} P_i(\lambda t) \qquad (10.4)$$

$$= 1 - P_{<j}(\lambda t)$$

where $P_i(\lambda t)$ is defined in Equation 10.3.

Example 10.2. Given a message-switching node that normally experiences four arrivals per minute, what is the probability that eight or more arrivals occur in an arbitrarily chosen 30-sec interval?

Solution. The average number of arrivals in a 30-sec interval is

$$\lambda t = (4)\left(\frac{30}{60}\right) = 2$$

The probability of eight or more arrivals (when the average is 2) is

$$P_{\geq 8}(2) = \sum_{i=8}^{\infty} P_i(2)$$

$$= 1 - \sum_{i=0}^{7} P_i(2)$$

$$= 1 - e^{-2}\left(1 + \frac{2^1}{1!} + \frac{2^2}{2!} + \frac{2^3}{3!} + \cdots + \frac{2^7}{7!}\right)$$

$$= 0.0011.$$

Example 10.3. What is the probability that a 1000 bit data block experiences exactly four errors while being transmitted over a transmission link with a bit error rate (BER) $= 10^{-5}$?

Solution. Assuming independent errors (a questionable assumption on many transmission links), we can obtain the probability of exactly four errors directly from the Poisson distribution. The average number of errors (arrivals) $\lambda t = (10^3)10^{-5} = 0.01$. Thus

$$\text{prob}(4 \text{ errors}) = P_4(0.01)$$

$$= \frac{(0.01)^4}{4!} e^{-0.01}$$

$$= 4.125(10)^{-10}$$

Alternative solution can be obtained from the binomial probability law:

$$\text{Prob}(4\,\text{errors}) = \binom{1000}{4} p^4 (1-p)^{996} \qquad \text{where } p = 10^{-5}$$

$$= 4.101\,(10)^{-10}$$

As can be seen, the two solutions of Example 10.3 are nearly identical. The closeness of the two answers reflects the fact that the Poisson probability distribution is often derived as a limiting case of a binomial probability distribution. Because it is easier to calculate, a Poisson distribution is often used as an approximation to a binomial distribution.

10.1.2 Holding Time Distributions

The second factor of traffic intensity as specified in Equation 10.1 is the average holding time t_m. In some cases the average of the holding times is all that needs to be known about holding times to determine blocking probabilities in a loss system or delays in a delay system. In other cases it is necessary to know the probability distribution of the holding times to obtain the desired results. This section describes the two most commonly assumed holding time distributions: constant holding times and exponential holding times.

Constant Holding Times

Although constant holding times cannot be assumed for conventional voice conversations, it is a reasonable assumption for such activities as per-call call processing requirements, interoffice address signaling, operator assistance, and recorded message playback. Furthermore, constant holding times are obviously valid for transmission times in fixed length packet networks.

When constant holding time messages are in effect, it is straightforward to use Equation 10.3 to determine the probability distribution of active channels. Assume, for the time being, that all requests are serviced, then the probability of j channels being busy at any particular time is merely the probability that j arrivals occurred in the time interval of length t_m immediately preceding the instant in question. Since the average number of active circuits over all time is the traffic intensity $A = \lambda t_m$, the probability of j circuits being busy is dependent only on the traffic intensity:

$$P_j(\lambda t_m) = P_j(A)$$

$$= \frac{A^j}{j!} e^{-A} \tag{10.5}$$

where λ = the arrival rate
 t_m = the constant holding time
 A = the traffic intensity in erlangs

Exponential Holding Times

The most commonly assumed holding time distribution for conventional telephone conversations is the exponential holding time distribution:

$$P(>t) = e^{-t/t_m} \qquad (10.6)$$

where t_m is the average holding time. Equation 10.6 specifies the probability that a holding time exceeds the value t. This relationship can be derived from a few simple assumptions concerning the nature of the call termination process. Its basic justification, however, lies in the fact that observations of actual voice conversations exhibit a remarkably close correspondence to an exponential distribution.

The exponential distribution possesses the curious property that the probability of a termination is independent of how long a call has been in progress. That is, no matter how long a call has been in existence, the probability of it lasting another t sec is defined by Equation 10.6. In this sense exponential holding times represent the most random process possible. Not even knowledge of how long a call has been in progress provides any information as to when the call will terminate.

Combining a Poisson arrival process with an exponential holding time process to obtain the probability distribution of active circuits is more complicated than it was for constant holding times because calls can last indefinitely. The final result, however, proves to be dependent on only the average holding time. Thus Equation 10.5 is valid for exponential holding times as well as for constant holding times (or any holding time distribution). Equation 10.5 is therefore repeated for emphasis: The probability of j circuits being busy at any particular instant, assuming a Poisson arrival process and that all requests are serviced immediately is

$$P_j(A) = \frac{A^j}{j!} e^{-A} \qquad (10.7)$$

where A is the traffic intensity in erlangs. This result is true for any distribution of holding times.

Example 10.4. Assume that a trunk group has enough channels to immediately carry all of the traffic offered to it by a Poisson process with an arrival rate of one call per minute. Assume that the average holding time is 2 min. What percentage of the total traffic is carried by the first five circuits, and how much traffic is carried by all remaining circuits? (Assume that the traffic is always packed into the lowest-numbered circuits.)

Solution. The traffic intensity (offered load) of the system is $A = (1)(2) = 2$ erlangs. The traffic intensity carried by i active circuits is exactly i erlangs.

Hence the traffic carried by the first five circuits can be determined as follows:

$$A_5 = 1\,P_1(2) + 2P_2(2) + 3P_3(2) + 4P_4(2) + 5P_5(2)$$

$$= e^{-2}\left[2 + \frac{2(2)^2}{2!} + \frac{3(2)^3}{3!} + \frac{4(2)^4}{4!} + \frac{5(2)^5}{5!}\right]$$

$$= 1.89 \text{ erlangs}$$

All of the remaining circuits carry

$$2 - 1.89 = 0.11 \text{ erlangs}$$

The result of Example 10.4 demonstrates the principle of diminishing returns as the capacity of a system is increased to carry greater and greater percentages of the offered traffic. The first five circuits in Example 10.4 carry 94.5% of the traffic while all remaining circuits carry only 5.5% of the traffic. If there are 100 sources, 95 extra circuits are needed to carry the 5.5%.

10.2 LOSS SYSTEMS

Example 10.4 provides an indication of the blocking probabilities that arise when the number of servers (circuits) is less than the maximum possible traffic load (number of sources). The example demonstrates that 94.5% of the traffic is carried by only five circuits. The implication is that the blocking probability, if only five circuits are available to carry the traffic, is 5.5%. Actually, Example 10.4 is carefully worded to indicate that all of the offered traffic is carried but that only the traffic carried by the first five circuits is of interest. There is a subtle but important distinction between the probability that six or more circuits are busy (as can be obtained from Equation 10.7) and the blocking probability that arises when only five circuits exist.

The basic reason for the discrepancy is indicated in Figure 10.3, which depicts the same traffic pattern arising from 20 sources as is shown previously in Figure 10.1. Figure 10.3, however, assumes that only 13 circuits are available to carry the traffic. Thus the three arrivals at $t = 2.2$, 2.3, and 2.4 min are blocked and assumed to have left the system. The total amount of traffic volume lost is indicated by the shaded area, which is the difference between all traffic being serviced as it arrives and traffic being carried by a blocked calls cleared system with 13 circuits. The most important feature to notice in Figure 10.3 is that the call arriving at $t = 2.8$ is not blocked, even though the original profile indicates that it arrives when all 13 circuits are busy. The reason it is not blocked is that the previously blocked calls left the system and therefore reduced the congestion for subsequent arrivals. Hence the percentage of time that the original traffic profile is at or above 13 is not the same as the blocking probability when only 13 circuits are available.

Figure 10.3. Activity profile of blocked calls cleared (13 channels).

10.2.1 Lost Calls Cleared

The first person to account fully and accurately for the effect of cleared calls in the calculation of blocking probabilities was A. K. Erlang in 1917. In this section we discuss Erlang's most often used result: his formulation of the blocking probability for a lost calls cleared system with Poisson arrivals. Recall that the Poisson arrival assumption implies infinite sources. This result is variously referred to as Erlang's formula of the first kind $E_{1,N}(A)$, the Erlang-B formula, or Erlang's loss formula.

A fundamental aspect of Erlang's formulation, and a key contribution to modern stochastic process theory, is the concept of statistical equilibrium. Basically, statistical equilibrium implies that the probability of a system's being in a particular state (number of busy circuits in a trunk group) is independent of the time at which the system is examined. For a system to be in statistical equilibrium, a long time must pass (several average holding times) from when the system is in a known state until it is again examined. For example, when a trunk group first begins to accept traffic it has no busy circuits. For a short time thereafter, the system is most likely to have only a few busy circuits. As time passes, however, the system reaches equilibrium. At this point the most likely state of the system is to have $A = \lambda t_m$ busy circuits.

When in equilibrium, a system is as likely to have an arrival as it is to have a termination. If the number of active circuits happens to increase above the average A, departures become more likely than arrivals. Similarly, if the number of active circuits happens to drop below A, an arrival is more likely than a

departure. Thus if a system is perturbed by chance from its average state, it tends to return.

Although Erlang's elegant formulation is not particularly complicated, it is not presented here because we are mostly interested in application of the results. The interested reader is invited to see reference [2] or [3] for a derivation of the result:

$$B = E_{1,N}(A) = \frac{A^N}{N!\Sigma_{i=0}^{N}\left(\frac{A^i}{i!}\right)} \tag{10.8}$$

where N is the number of servers (channels) and $A = \lambda t_m$ is the offered traffic intensity in erlangs.

Equation 10.8 specifies the probability of blocking for a system with random arrivals from an infinite source and arbitrary holding time distributions. The blocking probability of Equation 10.8 is plotted in Figure 10.4 as a function of offered traffic intensity for various numbers of channels. An often more useful presentation of Erlang's results is provided in Figure 10.5, which presents the output channel utilization for various blocking probabilities and numbers of servers. The output utilization ρ represents the traffic carried by each circuit:

$$\rho = \frac{(1 - B)A}{N} \tag{10.9}$$

where A = the offered traffic
 B = the blocking probability
 $(1 - B)A$ = the carried traffic

Blocking probabilities are also provided in tabular form in Appendix D.

Example 10.5. A T1 line is to be used as a high-usage trunk group between two end offices. How much traffic can the trunk group carry if the blocking probability is to be 0.1? What is the offered traffic intensity? (The blocking probability between switching offices is normally designed to be less than 0.01. Because alternate routes are available, however, direct trunks can be blocked more than 10%.)

Solution. From Figure 10.5 it can be seen that the output circuit utilization for $B = 0.1$ and $N = 24$ is 0.8. Thus the carried traffic intensity is $(0.8)(24) = 19.2$ erlangs. Since the blocking probability is 0.1, the maximum level of offered traffic is

$$A = \frac{19.2}{1 - 0.1}$$

$$= 21.3 \text{ erlangs}$$

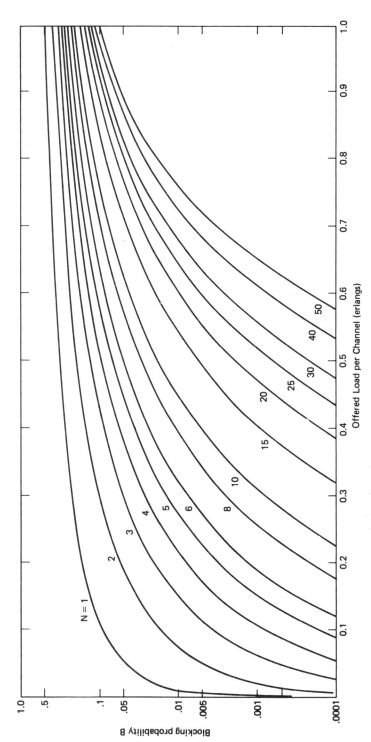

Figure 10.4. Blocking probability of lost calls cleared system.

Figure 10.5. Output channel utilization of lost calls cleared system.

Example 10.6. Four clusters of data terminals are to be connected to a computer by way of leased circuits as shown in Figure 10.6. In Figure 10.6a the traffic from the clusters uses separate groups of shared circuits. In Figure 10.6b the traffic from all clusters is concentrated onto one common group of circuits. Determine the total number of circuits required in both cases when the maximum desired blocking probability is 5%. Assume that 22 terminals are in each cluster and each terminal is active 10% of the time. (Use a blocked calls cleared analysis.)

Solution. The offered traffic from each cluster is 22(0.1) = 2.2 erlangs. Since the average number of active circuits is much smaller than the number of sources, an infinite source analysis can be used. Using Table D.1, the number of circuits required for $B = 5\%$ at a loading of 2.2 erlangs is 5. Thus the configuration of Figure 10.6a requires a total of 20 circuits.

The total offered traffic to the concentrator of the configuration of Figure 10.6b is 4(2.2) = 8.8 erlangs. From Table D.1, 13 circuits are required to support the given traffic load.

Example 10.6 demonstrates that consolidation of small traffic groups into one large traffic group can provide significant savings in total circuit requirements. Large groups are more efficient than multiple small groups because it is unlikely that the small groups will become overloaded at the same time (assuming independent arrivals). In effect, excess traffic in one group can use idle circuits in another group. Thus those circuits that are needed to accommodate traffic peaks, but are normally idle, are utilized more efficiently when the traffic is combined into one group. This feature is one of the motivations mentioned in Chapter 8 for integrating voice and data traffic into a common network. The total savings in transmission costs is most significant when the individual traffic intensities are low. Hence it is the peripheral area of a network that benefits the most by concentrating the traffic.

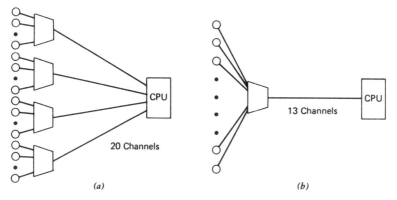

Figure 10.6. Data terminal network of Example 10.6. (a) Four separate groups. (b) All traffic concentrated into one group.

The greater circuit efficiency obtained by combining traffic into large groups is often referred to as the advantage of large group sizes. This efficiency of circuit utilization is the basic motivation for hierarchical switching structures. Instead of interconnecting a large number of nodes with rather small trunk groups between each pair, it is more economical to combine all traffic from individual nodes into one large trunk group and route the traffic through a tandem switching node. Figure 10.7 contrasts a mesh versus a star network with a centralized switching node at the center. Obviously, the cost of the tandem switch becomes justified when the savings in total circuit miles is large enough.

Example 10.7. What happens to the blocking probabilities in Figure 10.6a and b discussed in Example 10.6 when the traffic intensity increases by 50%?

Solution. If the traffic intensity of each group increases from 2.2 to 3.3 erlangs, the blocking probability of the configuration of Figure 10.6a increases from 5% to almost 14%.

In the configuration of Figure 10.6b a 50% increase in the traffic intensity causes a 400% increase in the blocking probability (from 5% to 20%).

Example 10.7 demonstrates some important considerations in network design. As indicated, blocking probabilities are very sensitive to increases in traffic intensities, particularly when the channels are heavily utilized. Thus because large trunk groups utilize their channels more efficiently they are more vulnerable to traffic increases than are a number of smaller groups designed to provide the same grade of service. Furthermore, failures of equal percentages of transmission capacity affect the performance of a large group more than the performance of several small groups. In both cases the vulnerability of the large groups arises because large groups operate with less spare capacity than do multiple small groups.

A second aspect of blocking analyses demonstrated in Example 10.7 is that the calculated results are highly dependent on the accuracy of the traffic intensities. Accurate values of traffic intensities are not always available.

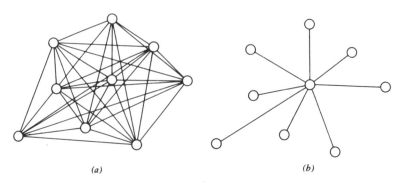

(a) *(b)*

Figure 10.7. Use of trandem switching to concentrate traffic. (a) Mesh. (b) Star.

Furthermore, even when accurate traffic measurements are obtainable, they do not provide an absolute indication of how much growth to expect. Thus only limited confidence can be attached to calculations of blocking probabilities in an absolute sense. The main value of these analyses is that they provide an objective means of comparing various network sizes and configurations. The most cost-effective design for a given grade of service is the one that should be chosen, even if the traffic statistics are hypothetical. If a network is liable to experience wildly varying traffic patterns or rapid growth, these factors must be considered when comparing design alternatives. A network with a somewhat larger initial cost may be more desirable if it can absorb or grow to accommodate unanticipated traffic volumes more easily.

10.2.2 Lost Calls Returning

In the lost calls cleared analyses just presented, it is assumed that unserviceable requests leave the system and never return. As mentioned, this assumption is most appropriate for trunk groups whose blocked requests overflow to another route and are usually serviced elsewhere. However, lost calls cleared analyses are also used in instances where blocked calls do not get serviced elsewhere. In many of these cases, blocked calls tend to return to the system in the form of retries. Some examples are subscriber concentrator systems, corporate tie lines and PBX trunks, calls to busy telephone numbers, and access to WATS lines (if DDD alternatives are not used). This section derives blocking probability relationships for lost calls cleared systems with random retries.

The following analysis involves three fundamental assumptions regarding the nature of the returning calls:

1. All blocked calls return to the system and eventually get serviced, even if multiple retries are required.
2. The elapsed times between call blocking occurrences and the generation of retries are random and statistically independent of each other. (This assumption allows the analysis to avoid complications arising when retries are correlated to each other and tend to cause recurring traffic peaks at a particular waiting time interval.)
3. The typical waiting time before retries occur is somewhat longer than the average holding time of a connection. This assumption essentially states that the system is allowed to reach statistical equilibrium before a retry occurs. Obviously, if retries occur too soon, they are very likely to encounter congestion since the system has not had a chance to "relax." In the limit, if all retries are immediate and continuous the network operation becomes similar to a delay system discussed in latter sections of this chapter. In this case, however, the system does not queue requests—the sources do so by continually "redialing."

When considered in their entirety, these assumptions characterize retries as being statistically indistinguishable from first-attempt traffic.* Hence blocked calls merely add to the first-attempt call arrival rate.

Consider a system with a first-attempt call arrival rate of λ. If a percentage B of the calls is blocked, B times λ retries will occur in the future. Of these retries, however, a percentage B will be blocked again. Continuing in this manner, the total arrival rate λ', after the system has reached statistical equilibrium, can be determined as the infinite series:

$$\lambda' = \lambda + B\lambda + B^2\lambda + B^3\lambda + \cdots$$

$$= \frac{\lambda}{(1 - B)} \tag{10.10}$$

where B is the blocking probability from a lost calls cleared analysis with traffic intensity $A' = \lambda't_m$.

Equation 10.10 relates the average arrival rate λ', including the retries, to the first-attempt arrival rate and the blocking probability in terms of λ'. Thus this relationship does not provide a direct means of determining λ' or B since each is expressed in terms of the other. However, the desired result can be obtained by iterating the lost calls cleared analysis of Equation 10.8. First, determine an estimate of B using λ and then calculate λ'. Next, use λ' to obtain a new value of B and an updated value of λ'. Continue in this manner until values of λ' and B are obtained.

Example 10.8. What is the blocking probability of a PBX to central office trunk group with 10 circuits and an offered traffic load of 7 erlangs? What is the blocking probability if the number of circuits is increased to 13? Assume random retries for all blocked calls.

Solution. It can be assumed that the 7 erlangs of traffic arise from a large number of PBX stations. Thus an infinite source analysis is justified. The blocking probability for $A = 7$ erlangs and $N = 10$ servers is about 8%. Thus the total offered load, including retries, is approximately 7.6 erlangs. With $N = 10$ and $A = 7.6$, the blocking probability is 10%. Two more iterations effectively produce convergence at $A = 8$ erlangs and $B = 12\%$. If the number of circuits in the trunk group is increased to 13, the blocking probability of a lost calls cleared system is 1.5%. Thus a first approximation to the returning traffic intensity is $7/0.985 = 7.1$ erlangs. Hence the blocking probability including all returning traffic increases only slightly above the 1.5%.

Example 10.8 demonstrates that the effect of returning traffic is insignificant when operating at low blocking probabilities. At high blocking probabilities,

*First-attempt traffic is also referred to as demand traffic: the service demands assuming all arrivals are serviced immediately. The offered traffic is the demand traffic plus the retries.

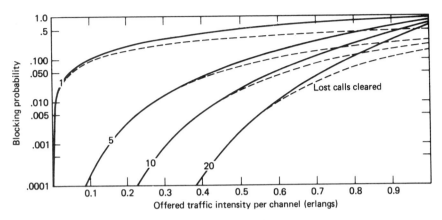

Figure 10.8. Blocking probability of lost calls returning.

however, it is necessary to incorporate the effects of the returning traffic into the analysis. This relationship between lost calls cleared and lost calls returning is shown in Figure 10.8.

When measurements are made to determine the blocking probability of an outgoing trunk group, the measurements cannot distinguish between first-attempt calls (demand traffic) and retries. Thus if a significant number of retries are contained in the measurements, this fact should be incorporated into an analysis of how many circuits must be added to reduce the blocking of an overloaded trunk group. The apparent offered load will decrease as the number of servers increases because the number of retries decreases. Thus fewer additional circuits are needed than if no retries are contained in the measurements.

10.2.3 Lost Calls Held

In a lost calls held system, blocked calls are held by the system and serviced when the necessary facilities become available. Lost calls held systems are distinctly different from the delay systems discussed later in one important respect: the total elapsed time of a call in the system, including waiting time and service time, is independent of the waiting time. In essence, each arrival requires service for a continuous period of time and terminates its request independently of its being serviced or not. Figure 10.9 demonstrates the basic operation of a lost calls held system. Notice that most blocked calls eventually get some service, but only for a portion of the time that the respective sources are busy.

Although a switched telephone network does not operate in a lost calls held manner, some systems do. Lost calls held systems generally arise in real time applications in which the sources are continuously in need of service, whether or not the facilities are available. When operating under conditions of heavy traffic,

Figure 10.9. Activity profile of lost calls held.

a lost calls held system typically provides service for only a portion of the time a particular source is active.

Even though conventional circuit switching does not operate according to the theoretical model of lost calls held, Bell System traffic engineers have used it to calculate blocking probabilities for trunk groups [4]. A lost calls held analysis always produces a larger value for blocking than does Erlang's loss formula. Thus the lost calls held analysis produces a conservative design that helps account for retries and day-to-day variations in the busy hour calling intensities. In contrast, CCITT recommendations [5] stipulate Erlang-B formulas should be used in determining blocking probabilities.

One example of a system that closely fits the lost calls held model is time assignment speech interpolation (TASI). A TASI system concentrates some number of voice sources onto a smaller number of transmission channels. A source receives service (is connected to a channel) only when it is active. If a source becomes active when all channels are busy, it is blocked and speech clipping occurs. Each speech segment starts and stops independently of whether it is serviced or not. In contrast to conventional TASI systems, a system developed by STC communications adds a small amount of delay to speech segments, when necessary, to minimize the clipping. In this case, a lost calls held analysis is not rigorously justified because the total time a speech segment is "in the system" increases as the delay for service increases. However, if the average delay is a small percentage of the holding time, or if the coding rate of delayed

speech is reduced to allow the transmission channel time to "catch up," a lost calls held analysis is still justified.

Lost calls held systems are easily analyzed to determine the probability of the total number of calls in the system at any one time. Since the duration of a source's activity is independent of whether it is being serviced, the number in the system at any time is identical to the number of active sources in a system capable of carrying all traffic as it arises. Thus the distribution of the number in the system is the Poisson distribution provided earlier in Equation 10.3. The probability that i sources requesting service are being blocked is simply the probability that $i + N$ sources are active when N is the number of servers. Recall that the Poisson distribution essentially determines the desired probability as the probability that $i + N$ arrivals occurred in the preceding t_m sec. The distribution is dependent only on the product of the average arrival rate λ and the average holding time t_m.

Example 10.9. What is the probability that a talkspurt experiences clipping in a TASI system with 10 sources and five channels? With 100 sources and 50 channels? Assume that the activity factor of each talker is 0.4. (Ignore finite sources.)

Solution. For the first case, the clipping probability (time congestion) can be determined as the probability that five or more sources are busy in a Poisson process with an average of $A = (0.4)(10) = 4$ busy servers. Using Equation 10.7.

$$\text{prob (clipping)} = \sum_{j=5}^{9} P_j(4)$$

$$= e^{-4}\left(\frac{4^5}{5!} + \frac{4^6}{6!} + \frac{4^7}{7!} + \frac{4^8}{8!} + \frac{4^9}{9!}\right)$$

$$= 0.36$$

With 100 sources, the average number of busy circuits $A = (0.4)(100) = 40$. A speech segment is clipped if 50 or more talkers are active at once. Thus the clipping probability can be determined as

$$\text{prob (clipping)} = \sum_{j=50}^{99} P_j(40)$$

$$= 0.04.$$

Example 10.9 demonstrates that TASI systems are much more effective for large group sizes than for small ones. The 36% clipping factor occurring with five channels produces unacceptable voice quality. On the other hand, the 4% clipping probability for 50 channels can be tolerated when the line costs are high enough.

In reality, the values for blocking probabilities obtained in Example 10.9 are higher than actual because an infinite source assumption was used. A more accurate solution is obtained in a later section using a finite source analysis.

10.2.4 Lost Calls Cleared—Finite Sources

As mentioned previously, a fundamental assumption in the derivation of the Poisson arrival distribution, and consequently Erlang's loss formula, is that call arrivals occur independently of the number of active callers. Obviously, this assumption can be justified only when the number of sources is much larger than the number of servers. This section presents some fundamental relationships for determining blocking probabilities of lost calls cleared systems when the number of sources is not much larger than the number of servers. The blocking probabilities in these cases are always less than those for infinite source systems since the arrival rate decreases as the number of busy sources increases.

When considering finite source systems, traffic theorists introduce another parameter of interest called time congestion. Time congestion is the percentage of time that all servers in a group are busy. It is identical to the probability that all servers are busy at randomly selected times. However, time congestion is not necessarily identical to blocking probability (which is sometimes referred to as call congestion). Time congestion merely specifies the probability that all servers are busy. Before blocking can occur, there must be an arrival.

In an infinite source system, time congestion and call congestion are identical because the percentage of arrivals encountering all servers busy is exactly equal to the time congestion. (The fact that all servers are busy has no bearing on whether or not an arrival occurs.) In a finite source system, however, the percentage of arrivals encountering congestion is smaller because fewer arrivals occur during periods when all servers are busy. Thus in a finite source system, call congestion (blocking probability) is always less than the time congestion. As an extreme example, consider equal numbers of sources and servers. The time congestion is the probability that all servers are busy. The blocking probability is obviously zero.

The same basic techniques introduced by Erlang when he determined the loss formula for infinite sources can be used to derive loss formulas for finite sources [3]. Using these techniques, we find the probability of n servers being busy in a system with M sources and N servers is

$$P_n = \frac{\binom{M}{n}(\lambda' t_m)^n}{\sum_{i=0}^{N} \binom{M}{i}(\lambda' t_m)^i} \tag{10.11}$$

where λ' is the calling rate per idle source and t_m is the average holding time. Equation 10.11 is known as the truncated Bernoullian distribution and also as the Engset distribution.

Setting $n = N$ in Equation 10.11 produces an expression for the time congestion:

$$P_N = \frac{\binom{M}{N}(\lambda' t_m)^N}{\sum_{i=0}^{N} \binom{M}{i}(\lambda' t_m)^i} \tag{10.12}$$

Using the fact that the arrival rate when N servers are busy is $(M-N)/M$ times the arrival rate when no servers are busy, we can determine the blocking probability for lost calls cleared with a finite source as follows:

$$B = \frac{\binom{M-1}{N}(\lambda' t_m)^N}{\sum_{i=0}^{N} \binom{M-1}{i}(\lambda' t_m)^i} \tag{10.13}$$

which is identical to P_N for $M-1$ sources.

Equations 10.11, 10.12, and 10.13 are easily evaluated in terms of the parameters λ' and t_m. However, λ' and t_m do not, by themselves, specify the average activity of a source. If the average activity of a source, assuming no traffic is cleared, is designated as ρ, the value of $\lambda' t_m$ can be determined as

$$\lambda' t_m = \frac{\rho}{1 - \rho(1 - B)} \tag{10.14}$$

where B is the blocking probability defined by Equation 10.13.

The difficulty with using the unblocked source activity factor ρ to character-ize a source's offered load is now apparent. The value of $\lambda' t_m$ depends on B, which, in turn, depends on $\lambda' t_m$. Thus some form of iteration is needed to determine B when the sources are characterized by ρ (an easily measured parameter) instead of λ'. If the total offered load is considered to be $M\rho$, the carried traffic is

$$A_{\text{carried}} = M\rho(1 - B) \tag{10.15}$$

A table of traffic capacities for finite sources is provided in Appendix D where the offered load $A = M\rho$ is listed for various combinations of M, N, and B. Some of the results are plotted in Figure 10.10 where they can be compared to blocking probabilities of infinite source systems. As expected, infinite source analyses (Erlang-B) are acceptable when the number of sources M is large.

Example 10.10. A group of callers generate requests at a rate of five calls per hour per telephone (including incoming and outgoing calls). Assuming the average holding time is 4 min, what is the average calling rate of each idle source? How many callers can be supported by a 12 channel concentrator/multiplexer if the maximum acceptable blocking probability is 1%?

Solution. Since each caller typically is active for 20 min of every hour, the calling rate for idle sources $\lambda' = 5/40 = 0.125$ calls per minute. The offered load for M sources, assuming all traffic is carried, is $0.33M$. Thus Table D.2 must be

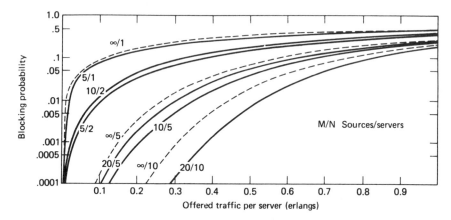

Figure 10.10. Blocking probability of lost calls cleared with finite sources.

searched to find the largest M such that $0.33M$ is less than or equal to the maximum offered load for $B = 1\%$ and $N = 12$. Using interpolation for $M = 21$ reveals that 12 servers can support 7.11 erlangs at $B = 1\%$. Since $21(0.33) = 6.93$ is the offered load, 21 sources is an acceptable solution. If 22 sources are used, the offered load of 7.26 erlangs is higher than the 7.04 erlangs obtainable from interpolation in Table D.2 as the maximum offered load for $B = 1\%$.

It is worthwhile comparing the result of Example 10.10 to a result obtained from an infinite source analysis (Erlang-B). For a blocking probability of 1%, Table D.1 reveals that the maximum offered load for 12 servers is 5.88 erlangs. Thus the maximum number of sources can be determined as $5.88/0.333 = 17.64$. Hence in this case an infinite source analysis produces a result that is conservative by 15%.

10.2.5 Lost Calls Held—Finite Sources

A lost calls held system with finite sources is analyzed in the same basic manner as a lost calls held systems with infinite sources. At all times the number of calls "in the system" is defined to be identical to the number of calls that would be serviced by a strictly nonblocking server group. Thus Equation 10.11 is used to determine the probability that exactly n calls are in the system:

$$P_n = \frac{\binom{M}{n}(\lambda' t_m)^n}{\sum_{i=0}^{M}\binom{M}{i}(\lambda' t_m)}$$

$$= \frac{\binom{M}{n}(\lambda' t_m)^n}{(1 + \lambda' t_m)^M}$$

(10.16)

Since no calls are cleared, the source activity (offered load per source) can be determined as

$$P = \frac{\lambda' t_m}{1 + \lambda' t_m.} \tag{10.17}$$

Combining Equations 10.16 and 10.17 produces a more useful expression for the probability that n calls are "in the system":

$$P_n = \binom{M}{n} \rho^n (1 - \rho)^{M-n} \tag{10.18}$$

If there are N servers, the time congestion is merely the probability that N or more servers are busy:

$$P_N = \sum_{n=N}^{M} P_n \tag{10.19}$$

The blocking probability, in a lost calls held sense, is the probability of an arrival encountering N or more calls in the system:

$$
\begin{aligned}
B_h &= \frac{\sum_{n=N}^{M} P_n \cdot \text{prob (arrival} \mid n \text{ sources are busy)}}{(\text{average arrival rate})} \\
&= \sum_{n=N}^{M-1} \binom{M-1}{n} \rho^n (1 - \rho)^{M-1-n}
\end{aligned}
\tag{10.20}
$$

where ρ = the offered load per source
 M = the number of sources
 N = the number of servers

Example 10.11. Determine the probability of clipping for the TASI systems described in Example 10.9. In this case, however, use a lost calls held analysis for finite sources.

Solution. In this example we are concerned only with the probability that a speech utterance is clipped for some period of time until a channel becomes available. Thus Equation 10.20 provides the desired answer using $\rho = 0.4$ for the offered load per source. In the first case, with 10 sources and five channels,

$$B_h = \sum_{n=5}^{9} \binom{9}{n} (0.4)^n (0.6)^{9-n}$$

$$= 0.27$$

In the second case with 100 sources and 50 servers,

$$B_h = \sum_{n=50}^{99} \binom{99}{n} (0.4)^n (0.6)^{99-n}$$

$$= 0.023$$

The results of Example 10.11 show again that a TASI system requires large group sizes to provide low clipping probabilities. When compared to the results of Example 10.9, these results also indicate that an infinite source analysis overestimates the clipping probability in both cases (0.36 versus 0.27, and 0.04 versus 0.023). Notice that the percentage error in the infinite source analysis is almost identical for the 10 source system and the 100 source system. Hence the validity of choosing an infinite source model is more dependent on the ratio of sources to servers than it is on the number of sources.

Example 10.11 merely determines the probability that a speech segment encounters congestion and is subsequently clipped. A complete analysis of a TASI system must consider the time duration of clips in addition to their frequency of occurrence. In essence, the desired information is represented by the amount of traffic volume in the clipped segments. Weinstein in reference [6] refers to the clipped segments as "fractional speech loss" or simply "cutout fraction." This is not the same as the lost traffic, since a conventional lost calls held analysis considers any arrival that encounters congestion as being completely "lost"—even if it eventually receives some service. The cutout fraction is determined as the ratio of untransmitted traffic intensity to offered traffic intensity:

$$B_{cf} = \frac{1}{A} \sum_{n=N+1}^{M} (n-N)P_n \tag{10.21}$$

where M = the number of sources
$A = M\rho$ is the offered load
N = the number of servers
P_n = the probability of n calls "in the system" (Equation 10.18)

Example 10.12. Determine the average duration of a clip in the two TASI systems of Example 10.11. Assume the average duration of a speech segment is 300 msec. (The average length of a speech segment is dependent on the volume threshold, which also influences the activity factor ρ.)

Solution. In the first case,

$$B_{cf} = \frac{1}{4} \sum_{n=6}^{10} (n-5) \binom{10}{n} (0.4)^n (0.6)^{10-n}$$

$$= 0.059$$

Thus, on average, 5.9% or 17.7 msec of every 300-msec speech segment is clipped. Since 27% of the segments experience clipping (Example 10.11), the average duration of a clip, for clipped segments, is $0.059/0.27 = 22\%$, or 66 msec—an obviously intolerable amount. In the second case for 100 sources and 50 channels,

$$B_{cf} = \frac{1}{40} \sum_{n=51}^{100} (n - 50) \binom{100}{n} (0.4)^n (0.6)^{100-n}$$

$$= 0.001$$

Figure 10.11. Clipping probability and clipping duration of TASI.

Thus in this case only 0.1% of all speech is clipped, which implies that when clipping occurs $(300)\,(0.001)/0.023 = 13\,\text{msec}$ of the segment is lost.

Example 10.12 shows that not only do large group sizes greatly reduce the clipping probability of TASI systems, but also they reduce the duration of the clips. The relationship of clipping probabilities and clipping durations (fractional speech loss) to group size and source activity is provided in Figure 10.11. As shown, the clipping probability is extremely sensitive to source activity (offered load). For a discussion of the effects of clipping on speech quality see references [7] and [8].

10.3 NETWORK BLOCKING PROBABILITIES

In the preceding sections basic techniques of congestion theory are presented to determine blocking probabilities of individual trunk groups. In this section techniques of calculating end-to-end blocking probabilities of a network with more than one route between endpoints is considered. In conjunction with calculating the end-to-end blocking probabilities, it is necessary to consider the interaction of traffic on various routes of a network. Foremost among these considerations is the effect of overflow traffic from one route onto another. The following sections discuss simplified analyses only. More sophisticated techniques for more complex networks can be obtained in references [9], [10], and [11].

10.3.1 End-To-End Blocking Probabilities

Generally, a connection through a large network involves a series of transmission links, each one of which is selected from a set of alternatives. Thus an end-to-end blocking probability analysis usually involves a composite of series and parallel probabilities. The simplest procedure is identical to the blocking probability (matching loss) analyses presented in Chapter 5 for switching networks. For example, Figure 10.12 depicts a representative set of alternative connections through a network and the resulting composite blocking probability.

The blocking probability equation in Figure 10.12 contains several simplifying assumptions. First, the blocking probability (matching loss) of the switches is not included. In a digital time division switch, matching loss can be low enough that it is easily eliminated from the analysis. In other switches, however, the matching loss may not be insignificant. When necessary, switch blocking is included in the analysis by considering it a source of blocking in series with the associated trunk groups.

When more than one route passes through the same switch, as in node C of Figure 10.12, proper treatment of correlation between matching losses is an

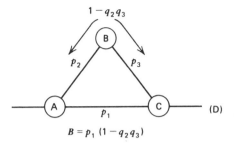

$$B = p_1 (1 - q_2 q_3)$$

Figure 10.12. Probability graph for end-to-end blocking analysis.

additional complication. A conservative approach considers the matching loss to be completely correlated. In this case the matching loss is in series with the common link. On the other hand, an optimistic analysis assumes that the matching losses are independent, which implies that they are in series with the individual links. Figure 10.13 depicts these two approaches for including the matching loss of switch C into the end-to-end blocking probability equation of Figure 10.12. In this case, the link from C to D is the common link.

A second simplifying assumption used in deriving the blocking probability equation in Figure 10.12 involves assuming independence for the blocking probabilities of the trunk groups. Thus the composite blocking of two parallel routes is merely the product of the respective probabilities (Equation 5.4). Similarly, independence implies that the blocking probability of two paths—in series—is 1 minus the product of the respective availabilities (Equation 5.5). In actual practice individual blocking probabilities are never completely independent. This is particularly true when a large amount of traffic on one route results as overflow from another route. Whenever the first route is busy, it is likely that more than the average amount of overflow is being diverted to the second route. Thus an alternate route is more likely to be busy when a primary route is busy.

In a large public network, trunks to tandem or toll switches normally carry traffic to many destinations. Thus no one direct route contributes an

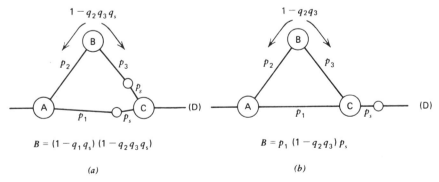

Figure 10.13. Incorporating switch matching loss into end-to-end blocking analysis. (a) Independent switch blocking. (b) Correlated switch blocking.

overwhelming amount of overflow traffic to a particular trunk group. In this case independent blocking probabilities on alternate routes are justified. In some instances of the public network, and often in private networks, overflow traffic from one route dominates the traffic on tandem routes. In these cases failure to account for the correlation in blocking probabilities can lead to overly optimistic results.

Example 10.13. Two trunk groups are to be used as direct routes between two switching offices. The first group has 12 channels and the second group has 6 channels. Assume 10.8 erlangs of traffic is offered to the 12-channel group, and overflows are offered to the 6-channel group when the first group is busy. What is the blocking probability of the first group, and how much traffic overflows to the second group? Using the overflow traffic volume as an offered load, determine the blocking probability of the second trunk group. What is the probability that both trunk groups are busy? Compare this answer to the blocking probability of one 18-channel trunk group.

Solution. Using a lost calls cleared analysis with an infinite source, we determine that blocking of the first group is 15% ($A = 10.8$, $N = 12$). Therefore the overflow traffic is $(10.8)(0.15) = 1.62$ erlangs. The blocking probability (assuming random arrivals?) of the second group is 0.5% ($A = 1.62$, $N = 6$). The probability that both trunk groups are busy simultaneously can be determined (assuming independence?) as

$$B = (0.15)(0.005)$$
$$= 0.00075$$

In contrast, the correct blocking probability of an 18-channel trunk group with an offered load of 10.8 erlangs is

$$B = 0.013$$

The question marks in the solution of Example 10.13 point to two sources of error in the determination of the first blocking probability value. The first error is the assumption of independence of blocking in the two trunk groups. The second error results from the use of an analysis predicated on purely random (Poisson) arrivals for overflow traffic into the second trunk group. Resolution of this error is discussed in the next section.

Separating the 18 channels of Example 10.13 into two groups is obviously an artifice. This example is useful in that it demonstrates an extreme case of correlation between blocking probabilities of two trunk groups. When correlation exists, the composite blocking probability of a direct route and an alternate route should be determined as follows:

$$B = (B_1)(B_2 \mid 1) \tag{10.22}$$

where B_1 = the blocking probability of group 1
 $B_2|1$ = the blocking probability of group 2 given that group 1 is busy

In the artificial case of dividing a trunk group into two subgroups, the conditional blocking probability can be determined as

$$B_2|1 = B(N|N_1)$$

 $= \text{prob} \ (N \text{ servers are busy when } N_1 \text{ are known to be busy})$

$$= \frac{P_N}{\sum_{n=N_1}^{N} P_n} \qquad (10.23)$$

$$= \frac{(A^N/N!)}{\sum_{n=N_1}^{N} (A^n/n!)}$$

where P_n is the probability that exactly n of N servers are busy (Equation 10.3).

Evaluating Equation 10.23 for $A = 10.8$, $N_1 = 12$, and $N = 18$ reveals that the appropriate conditional probability $B_2|1$ for Example 10.13 is 0.033. Thus the composite blocking probability per Equation 10.22 is $B = (0.15)(0.033) = 0.005$. The remaining inaccuracy (0.005 versus 0.013 actually) is due to nonrandom characteristics of overflow traffic.

Equation 10.23 is valid only for the contrived case of an alternate route carrying overflow traffic from only one primary route. It can be used, however, as a worst case solution to situations where overflow from one route tends to dominate the traffic on an alternate route.

The correlations between the blocking probabilities of individual routes arise because congestion on one route produces overflows that tend to cause congestion on other routes. External events stimulating networkwide overloads also cause the blocking probabilities to be correlated. Thus a third assumption in the end-to-end blocking probability equation of Figure 10.12 is that traffic throughout the network is independent. If fluctuations in the traffic volume on individual links tend to be correlated (presumably because of external events such as television commercials, etc.), significant degradation in overall performance results.

10.3.2 Overflow Traffic

The second source of error in Example 10.13 occurred because an Erlang-B analysis used the average volume of overflow traffic from the first group to determine the blocking probability of the second trunk group. An Erlang-B analysis assumes traffic arrivals are purely random, that is, they are modeled by a Poisson distribution. However, a Poisson arrival distribution is an erroneous assumption for the traffic offered to the second trunk group. Even though arrivals to the first group may be random, the overflow process tends to select groups of these arrivals and pass them on to the second trunk group. Thus

instead of being random the arrivals to the second group occur in bursts. This overflow effect is illustrated in Figure 10.14, which portrays a typical random arrival pattern to one trunk group and the overflow pattern to a second group. If a significant amount of the traffic flowing onto a trunk group results as overflow from other trunk groups, overly optimistic values of blocking probability arise when all of the traffic is assumed to be purely random.

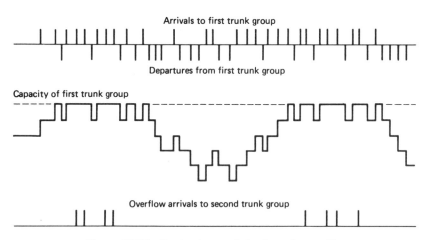

Figure 10.14. Bursty characteristic of overflow traffic.

The most common technique of dealing with overflow traffic is to relate the overflow traffic volume to an equivalent amount of random traffic in a blocking probability sense. For example, if the 1.62 erlangs of overflow traffic in Example 10.12 is equated to 2.04 erlangs of random traffic, a blocking probability of 1.3% is obtained for the second trunk group. (This is the correct probability of blocking for the second group since both groups are busy if and only if the second group is busy.)

This method of treating overflow traffic is referred to as the equivalent random theory [12]. Tables of traffic capacity are available [13] that incorporate the overflow effects directly into the maximum offered loads. The Neal–Wilkinson tables used by Bell System traffic engineers comprise one such set of tables. The Neal–Wilkinson tables, however, also incorporate the effects of day-to-day variations in the traffic load. (Forty erlangs on one day and 30 erlangs on another is not the same as 35 erlangs on both days.) These tables are also used for trunk groups that neither generate nor receive overflow traffic. The fact that cleared traffic does not get serviced by an alternate route implies that retries are likely. The effect of the retries, however, is effectively incorporated into the value of B by equivalent randomness.

10.4 DELAY SYSTEMS

The second category of teletraffic analysis concerns systems that delay non-serviceable requests until the necessary facilities become available. These systems are variously referred to as delay systems, waiting-call systems, and queuing systems. Call arrivals occurring when all servers are busy are placed in a queue and held until service commences. The queue might consist of storage facilities in a physical sense, such as blocks of memory in a message-switching node, or the queue might consist only of a list of sources waiting for service. In the latter case, storage of the messages is the responsibility of the sources themselves.

Using the more general term, queueing theory, we can apply the following analyses to a wide variety of applications outside of telecommunications. Some of the more common applications are data processing, supermarket check-out counters, aircraft landings, inventory control, and various forms of service bureaus. These and many other applications are considered in the relatively new field of operations research. The foundations of queuing theory, however, rest on fundamental techniques developed by early telecommunications traffic re-searchers. In fact, Erlang is credited with the first solution to the most basic type of delay system. Examples of delay system analysis applications in tele-communications are message switching, packet switching, statistical time division multiplexing, multipoint data communications, automatic call distri-bution, digit receiver access, signaling equipment usage, and call processing. Furthermore, many of the newer, stored-program control PBXs have features allowing queued access to corporate tie lines or WATS lines. Thus some systems formerly operating as loss systems now operate as delay systems.

In general, a delay operation allows for greater utilization of servers (transmission facilities) than does a loss system. Basically, the improved utilization is achieved because peaks in the arrival process are "smoothed" by the queue. Even though arrivals to the system are random, the servers see a somewhat regular arrival pattern. The effect of the queuing process on overload traffic is illustrated in Figure 10.15. This figure displays the same traffic patterns presented earlier in Figures 10.1, 10.3, and 10.9. In this case, however, overload traffic is delayed until call terminations produce available channels.

In most of the following analyses it is assumed that all traffic offered to the system eventually gets serviced. One implication of this assumption is that the offered traffic intensity A is less than the number of servers N. Even when A is less than N, there are two cases in which the carried traffic might be less than the offered traffic. First, some sources might tire of waiting in a long queue and abandon the request. Second, the capacity for storing requests may be finite. Hence requests may occasionally be rejected by the system.

A second assumption in the following analyses is that infinite sources exist. In a delay system, there may be a finite number of sources in a physical sense, but an infinite number of sources in an operational sense because each source may

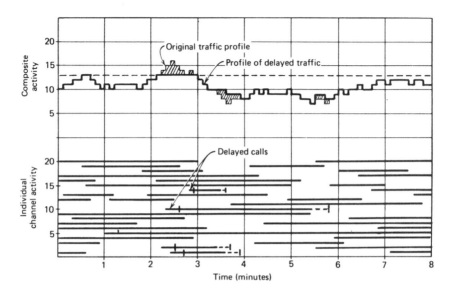

Figure 10.15. Activity profile of blocked calls delayed (13 servers).

have an arbitrary number of requests outstanding (e.g., a packet-switching node). There are instances in which a finite source analysis is necessary, but not in the applications considered here.

An additional implication of servicing all offered traffic arises when infinite sources exist. This implication is the need for infinite queuing capabilities. Even though the offered traffic intensity is less than the number of servers, no statistical limit exists on the number of arrivals occurring in a short period of time. Thus the queue of a purely lossless system must be arbitrarily long. In a practical sense, only finite queues can be realized, so either a statistical chance of blocking is always present, or all sources can be busy and not offer additional traffic.

When analyzing delay systems it is convenient to separate the total time that a request is in the system into the waiting time and the holding time. In delay systems analysis the holding time is more commonly referred to as the service time. In contrast to loss systems, delay system performance is generally dependent on the distribution of service times, and not just the mean value t_m. Two service time distributions are considered here: constant service times and exponential service times. Respectively, these distributions represent the most deterministic and the most random service times possible. Thus a system that operates with some other distribution of service times performs somewhere between the performance produced by these two distributions.

The basic purpose of the following analyses is to determine the probability distribution of waiting times. From the distribution, the average waiting time is easily determined. Sometimes only the average waiting time is of interest. More

generally, however, the probability that the waiting time exceeds some specified value is of interest. In either case, the waiting times are dependent on the following factors:

1. The intensity and the probabilistic nature of the offered traffic
2. The distribution of service times
3. The number of servers
4. The number of sources
5. The service discipline of the queue.

The service discipline of the queue can involve a number of factors. The first of these concerns the manner in which waiting calls are selected. Commonly, waiting calls are selected on a first come, first served (FCFS) basis, which is also referred to as first in, first out (FIFO) service. Sometimes, however, the server system itself does not maintain a queue, but merely polls its sources in a round robin fashion to determine which ones are waiting for service. Thus the queue may be serviced in sequential order of the waiting sources. In some applications waiting requests may even be selected at random. Furthermore, additional service variations arise if any of these schemes are augmented with a priority discipline that allows some calls to move ahead of others in the queue.

A second aspect of the service discipline that must be considered is the length of the queue. If the maximum queue size is smaller than the effective number of sources, blocking can occur in a lost calls sense. The result is that two characteristics of the grade of service must be considered: the delay probability and the blocking probability. A common example of a system with both delay and loss characteristics is an automatic call distributor with more access circuits than attendants (operators or reservationists). Normally, incoming calls are queued for service. Under heavy loads, however, blocking occurs before the ACD is even reached. Reference [14] contains an analysis of a delay system with finite queues and finite servers.

To simplify the characterization of particular systems, queuing theorists have adopted a concise notation for classifying various types of delay systems. This notation, which was introduced by D. G. Kendall, uses letter abbreviations to identify alternatives in each of the categories listed. Although the discussions in this book do not rely on this notation, it is introduced and used occasionally so the reader can relate the following discussions to classical queuing theory models. The interpretation of each letter is specified in Figure 10.16.

The specification format presented in Figure 10.16 actually represents an extension of the format commonly used by most queuing theorists. Thus this format is sometimes abbreviated by eliminating the last one or two entries. When these entries are eliminated, infinite case specifications are assumed. For example, a single-server system with random input and negative exponential service times is usually specified as M/M/1. Both the number of sources and the permissible queue length are assumed infinite.

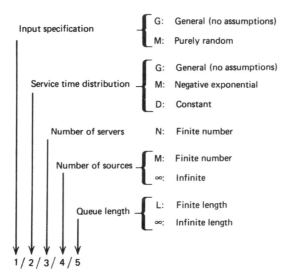

Figure 10.16. Queueing system notation.

10.4.1 Exponential Service Times

The simplest delay system to analyze is a system with random arrivals and negative exponential service times: M/M/N. Recall that a random arrival distribution is one with negative exponential interarrival times. Thus in the shorthand notation of queuing theorists, the letter M always refers to negative exponential distributions (an M is used because a purely random distribution is memoryless).

In the M/M/1 system, and all other systems considered here, it is assumed that calls are serviced in the order of their arrival. The following analyses also assume that the probability of an arrival is independent of the number of requests already in the queue (infinite sources). From these assumptions, the probability that a call experiences congestion and is therefore delayed was derived by Erlang:

$$\text{prob}(\text{delay}) = p(>0) = \frac{NB}{N - A(1 - B)} \qquad (10.24)$$

where N = the number of servers
A = the offered load in erlangs
B = the blocking probability for a lost calls cleared system (Equation 10.8)

The probability of delay $p(>0)$ is variously referred to as Erlang's second formula, $E_{2,N}(A)$ Erlang's delay formula; or the Erlang-C formula. For single-server systems $N = 1$, the probability of delay reduces to ρ, which is simply the

output utilization or traffic carried by the server. Thus the probability of delay for a single-server system is also equal to the offered load λt_m (assuming $\lambda t_m < 1$).

The distribution of waiting times for random arrivals, random service times, and a FIFO service discipline is

$$p(>t) = p(>0)e^{-(N-A)t/t_m} \qquad (10.25)$$

where $p(>0)$ = the probability of delay given in Equation 10.24,
t_m = the average service time of the negative exponential service time distribution

Equation 10.25 defines the probability that a call arriving at a randomly chosen instant is delayed for more than t/t_m service times. Figure 10.17 presents the

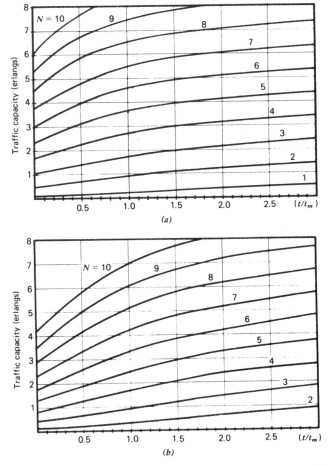

Figure 10.17. Traffic capacity of multiple-server delay systems with exponential serivce times. (a) Probability of exceeding t, $p(>t) = 10\%$. (b) Probability of exceeding t, $p(>t) = 1\%$.

relationship of Equation 10.25 by displaying the traffic capacities of various numbers of servers as a function of acceptable delay times. Given a delay time objective t/t_m, Figure 10.17a displays the maximum traffic intensity if the delay objective is to be exceeded for only 10% of the arrivals. Similarly, Figure 10.17b displays the maximum traffic intensity if the delay objective is exceeded for only 1% of the arrivals. Notice that at $p(>t) = 0.01$, the server systems do not approach their maximum capacity (number of servers) unless the acceptable delay is several times larger than t_m.

By integrating Equation 10.25 over all time, the average waiting time for all arrivals can be determined as

$$\bar{t} = \frac{p(>0)t_m}{N - A} \tag{10.26}$$

Notice that \bar{t} is the expected delay for all arrivals. The average delay of only those arrivals that get delayed is commonly denoted as

$$t_w = \frac{t_m}{N - A} \tag{10.27}$$

Example 10.14. A message-switching network is to be designed for 95% utilization of its transmission links. Assuming exponentially distributed message lengths and an arrival rate of 10 messages per minute, what is the average waiting time, and what is the probability that the waiting time exceeds 5 min?

Solution. Assume that the message-switching network uses a single channel between each pair of nodes. Thus there is a single server and a single queue for each transmission link. Since p is given to be 0.95 and $\lambda = 10$ arrivals per minute, the average service time can be determined as $t_m = 0.95/10 = 0.095$ min. The average waiting time (not including the service time) is easily determined as

$$\begin{aligned}
\bar{t} &= \frac{p(>0)t_m}{N - A} \\
&= \frac{(0.95)(0.095)}{1 - 0.95} \\
&= 1.805 \text{ min}
\end{aligned}$$

Using Equation 10.25, we can determine the probability of the waiting time exceeding 5 min as

$$p(>5) = (0.95)\,e^{-(1-0.95)5/0.095}$$

$$= 0.068$$

Thus 6.8% of the messages experience queuing delays of more than 5 min.

Example 10.15. Determine the number of digit receivers required to support 1000 telephones with an average calling rate of two calls per hour. Assume the dialing time is exponentially distributed with an average service time of 6 sec. The grade of service objective is to return dial tone within 1 sec of the off-hook signal for 99% of the call attempts. Compare the answer obtained from a delay system analysis to an answer obtained from a loss system analysis at $B = 1\%$. If the blocking probability is less than 1%, fewer than 1% of the calls are delayed.

Solution. The calling rate λ and the offered traffic intensity A are easily determined as 0.555 calls per second and 3.33 erlangs, respectively. Since the number of servers N cannot be solved for directly from the equations, Figure 10.17b is used to obtain a value of eight servers for $t/t_m = \frac{1}{6}$.

Examination of Table B.1 reveals that 99% of the call attempts can be serviced immediately if there are nine digit receivers. Thus in this case the ability to delay service provides a savings of only one server.

 Example 10.15 demonstrates that a blocking probability analysis produces approximately the same results as a delay system analysis when the maximum acceptable delay is a small percentage of the average service time. The two results are almost identical because, if a digit receiver is not immediately available, there is only a small probability that one will become available within a short time period. (With an average service time of 6 sec, the expected time for one of eight digit receivers to be released is $\frac{6}{8} = 0.75$ sec. Hence the delay operation in this case allows a savings of one digit receiver.)
 Because a digit receiver must be available within a relatively short time period after a request is generated, digit receiver group sizing is often determined strictly from a blocking probability analysis. The fact that digit receiver access is actually operated as a delay system implies the grade of service is always better than that calculated.

10.4.2 Constant Service Times

This section considers delay systems with random arrivals, constant service times, and a single server (M/D/1). Again, FIFO service disciplines and infinite sources are assumed. The case for multiple servers has been solved [3] but is too involved to include here. Graphs of multiple-server systems with constant service times are available in reference [15].
 The average waiting time for a single server with constant service times is determined as

$$\bar{t} = \frac{\rho t_m}{2(1 - \rho)} \tag{10.28}$$

where $\rho = A$ is the server utilization. Notice that Equation 10.28 produces an average waiting time that is exactly one-half of that for a single-server system

with exponential service times. Exponential service times cause greater average delays because there are two random processes involved in creating the delay. In both types of systems, delays occur when a large burst of arrivals exceeds the capacity of the servers. With exponential service times, however, long delays also arise because of excessive service times of just a few arrivals. (Recall that this aspect of conventional message switching systems is one of the motivations for breaking messages up into packets in a packet-switching network.)

If the activity profile of a constant service time system (M/D/1) is compared with the activity profile of an exponential service time system (M/M/1), the M/D/1 system is seen to be active for shorter and more frequent periods of time. That is, the M/M/1 system has a higher variance in the duration of its busy periods. The average activity of both systems is, of course, equal to the server utilization ρ. Hence the probability of delay for a single-server system with constant service times is identical to that for exponential service times: $p(>0) = \lambda t_m$.

The probability of congestion for larger N is relatively close to that for exponential service times. Thus Equation 10.25 can be used as a close approximation for $p(>0)$ for multiple-server systems with arbitrary service time distributions.

For single-server systems with constant holding times, the probability of delay greater than an arbitrary value t is

$$p(>t) = p[>(k + r)t_m]$$

$$= 1 - (1 - \rho) \sum_{i=0}^{k} \frac{\rho^i (i - t/t_m)^i e^{-\rho(i - t/t_m)}}{i!} \tag{10.29}$$

where k = the largest integral quotient of t/t_m
 r = the remainder of t/t_m
 $\rho = \lambda t_m$ is the server utilization

Comparisons of the waiting time distributions for single-server systems with exponential and constant service times are shown in Figure 10.18. For each pair of curves, the upper one is for exponential service times and the lower one is for constant service times. Since all other service time distributions produce delay probabilities between these extremes, Figure 10.18 provides a direct indication of the range of possible delays.

Example 10.16. A packet-switching node operates with fixed length packets of 300 bits on 9600 bps lines. If the link utilization is to be 90%, what is the average delay through a node? What percentage of packets encounter more than 0.35 sec of delay? What is the average delay if the offered load increases by 10%?

Solution. Message lengths of 300 bits and a data rate of 9600 bps imply that the fixed length service time is $300/9600 = 0.031$ sec. From Equation 10.28, the

Figure 10.18. Delay probabilities of single-server systems (exponential and constant service times).

average waiting time is

$$t = \frac{(0.9)(0.031)}{2(1 - 0.9)} = 0.140 \, \text{sec}$$

The total average delay through the node, excluding processing, is obtained by adding the average waiting time to the service time:

$$\text{average delay} = 0.140 + 0.031 = 0.171 \, \text{sec}$$

Since the service time is 0.031 sec, 0.35 sec of delay occur when the waiting time is $0.35 - 0.031 = 0.319$. This corresponds to $0.319/0.031 = 10$ service times. From Figure 10.18, the probability of delay for $t/t_m = 10$ is approximately 0.12. Thus 12% of the packets experience delays of greater than 0.35 sec. An increase of 10% in the traffic intensity implies that the new offered load is 0.99 erlangs. From Equation 10.30, the average waiting time becomes

$$t = \frac{(0.9)(0.031)}{2(1 - 0.99)} = 1.40 \text{ sec}$$

Thus when the offered load increases by only 10%, the average delay through the node increases more than eightfold to 1.43 sec.

Example 10.16 demonstrates the same characteristic for heavily utilized delay systems that was demonstrated for loss systems: the performance is very sensitive to increases in traffic intensity. Thus, as discussed in Chapter 7, flow control is a critical aspect of a packet-switching operation—particularly when there are real time delivery objectives.

10.4.3 Finite Queues

All of the delay system analyses presented so far have assumed that an arbitrarily large number of delayed requests could be placed in a queue. In many applications this assumption is invalid. Examples of systems that sometimes have significantly limited queue sizes are store-and-forward switching nodes, automatic call distributors, and various types of computer input/output devices. These systems treat arrivals in three different ways, depending on the number "in the system" at the time of an arrival:

1. Immediate service if one or more of N servers is idle.
2. Delayed service if all servers are busy and less than L requests are waiting.
3. Blocked or no service if the queue of length L is full.

In finite queue systems the arrivals getting blocked are those that would otherwise experience long delays in a pure delay system. Thus an indication of the blocking probability of a combined delay and loss system can be determined from the probability that arrivals in pure delay systems experience delays in excess of some specified value. However, there are two basic inaccuracies in such an analysis. First, the effect of blocked or lost calls cleared is to reduce congestion for a period of time and thereby to reduce the delay probabilities for subsequent arrivals. Second, delay times do not necessarily indicate how many calls are "in the system." Normally, queue lengths and blocking probabilities are determined in terms of the number of waiting requests, not the amount of work or total service time represented by the requests. With constant service times, there is no ambiguity between the size of a queue and its implied delay. With

exponential service times, however, a given size can represent a wide range of delay times.

A message-switching node is an example of a system in which the queue length is most appropriately determined by implied service time and not by the number of pending requests. That is, the maximum queue length may be determined by the amount of store-and-forward memory available for variable length messages, and not by some fixed number of messages.

Whenever a message-switching or packet-switching node rejects an arriving message, the source of the message must become aware of the rejection in some manner. First, the receiving node can remain passive and ignore the arrival. The source then determines that the transmission was unsuccessful by timing-out for an acknowledgment. This approach is attractive because time-outs are generally needed anyway, and the overloaded node or network does not need to respond actively to the arrival. The main drawback to the passive approach is that the source node does not learn why the transmission was unsuccessful and may continue to transmit undeliverable messages—thereby continuing to congest the network.

Obviously, a second technique is to respond to all blocked requests with a short control message that indicates why and where the rejection occurred and what the appropriate response should be.

For a system with random input, exponential service times, N servers, an infinite source, and a maximum queue length of $L(M/M/N/\infty/L)$, the probability of j calls in the system is

$$P_j(A) = P_0(A)\frac{A^j}{j!} \qquad (0 \leqslant j \leqslant N)$$

$$= \frac{P_0(A)A^j}{N!N^{j-N}} \qquad (N \leqslant j \leqslant N + L) \tag{10.30}$$

$$P_0(A) = \left\{\sum_{j=0}^{N+L}\left(\frac{A^j}{j!}\right)\right\}^{-1}$$

where $A = \lambda t_m$ = offered load in erlangs
N = the number of servers
L = maximum number in the queue

The time congestion, or probability of delay, can be determined from Equation 10.30 as

$$p(>0) = \sum_{j=N}^{N+L} P_j(A) \tag{10.31}$$

$$= P_N(A)\frac{1 - \rho^{L+1}}{1 - \rho}$$

where $\rho = A/N$ is the offered load per server.

The loss, or blocking probability, is determined as

$$B = P_{N+L}(A) = \frac{P_0(A)A^{N+L}}{N!N^L} \qquad (10.32)$$

It is worth noting that if there is no queue $(L = 0)$, these equations reduce to those of the Erlang loss equation (10.8). If L is infinite, Equation 10.31 reduces to Erlang's delay formula, Equation 10.24. Thus these equations represent a general formulation that produces the pure loss and pure delay formulas as special cases.

The waiting time distribution [3] is

$$p(>t) = P_N(A) \sum_{j=0}^{L-1} \frac{p^j}{j!} \int_{Nt/t_m}^{\infty} x^j e^{-x} dx \qquad (10.33)$$

from which the average delay can be determined as

$$\bar{t} = \frac{[p(>0) - L P_{N+L}(A)]t_m}{N - A} \qquad (10.34)$$

Again, Equation 10.34 is identical to Equation 10.26 for an infinite queue $(L = \infty)$.

The blocking probability of a single-server system $(N = 1)$ is plotted in Figure 10.19. When using Figure 10.19, keep in mind that the blocking probability (Equation 10.32) is determined by the number of waiting calls and not by the associated service time. Furthermore, since the curves of Figure 10.19 are based on exponential service times, they overestimate the blocking probabilities of constant holding time systems (e.g., fixed length packet networks). However, if fixed length packets arise primarily from longer, exponentially distributed messages, the arrivals are no longer independent, and the use of Figure 10.19 (or Equation 10.32) as a conservative analysis is more appropriate.

10.4.4 Tandem Queues

All of the equations provided in previous sections for delay system analysis have dealt with the performance of a single queue. In many applications a service request undergoes several stages of processing—each one of which involves queuing. Thus it is often desirable to analyze the performance of a system with a number of queues in series. Figure 10.20 depicts a series of queues that receive, as inputs, locally generated requests and outputs from other queues. Two principal examples of applications with tandem queues are data processing systems and store-and-forward switching networks.

Researchers in queuing theory have not been generally successful in deriving formulas for the performance of tandem queues. Often, simulation is used to

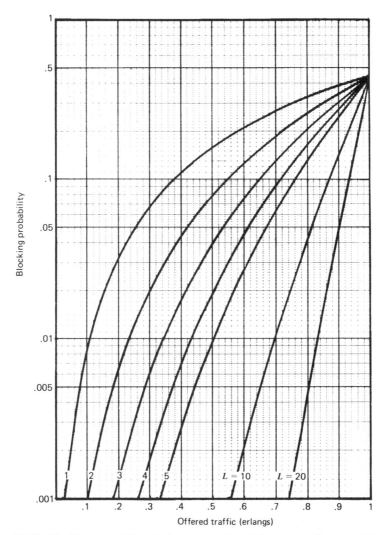

Figure 10.19. Blocking probability of single server loss/delay system (exponential service times).

analyze a complex arrangement of interdependent queues arising in systems like store-and-forward networks. Simulation has the advantage that special aspects of a network's operation—like routing and flow control—can be included in the simulation model. The main disadvantages of simulation are expense and, often, less visibility into the dependence of system performance on various design parameters.

One tandem queuing problem that has been solved [16] is for random inputs and random (negative exponential) holding times for all queues. The solution of

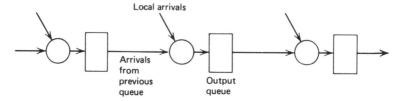

Figure 10.20. Tandem queues.

this system is based on the following theorem:

In a delay system with purely random arrivals and negative exponential holding times, the instants at which calls terminate is also a negative exponential distribution.

The significance of this theorem is that outputs from an M/M/R system have statistical properties that are identical to its inputs. Thus a queuing process in one stage does not affect the arrival process in a subsequent stage, and all queues can be analyzed independently. Specifically, if a delay system with N servers has exponentially distributed interarrival times with average $1/\lambda$, and if the average service time is t_m, calls leave each of the servers according to exponentially distributed intercompletion times with average $1/\lambda N$.

Although independent analysis of tandem queues can be rigorously justified only for purely random arrivals and service times, independence is often assumed in other cases. Before using such assumptions, however, the systems in question should be examined closely to determine if the state of one queue can influence the operation of another queue in the system.

Consider the connection setup process of a distributed control circuit-switched network. This process is basically a tandem queuing process because connect requests propagate from one switching node to another with each node processing a request before forwarding it to the next node. Depending on the manner in which signaling information is transmitted between nodes, congestion in one node can influence the operation of another node.

Prior to CCS switching offices commonly exchanged information by way of senders using in-channel multifrequency or dial-pulse signaling. The senders are used on a switched basis so they can be shared across a number of circuits. When the stored program control in one office wants to transfer a connect request to another office, a sender is assigned to the appropriate output trunk. However, before the information can be transferred, the destination office must assign a sender (also used for receiving) to the corresponding input trunk. In the meantime the sender in the first office is allocated and unavailable to service other requests. Hence if congestion exists in the second office, the first office is affected. In essence, the holding time of the sender is as much, or more, a function of the response time at the destination node as a function of the

underlying message length. One advantage of common channel interoffice signaling is that signaling information goes directly from one call processor to another and does not involve selection of common equipment like senders. (It is also an advantage to the traffic analyst since the independence assumption is easier to justify.)

Another example of interacting queues arises in packet-switching networks that do not preallocate storage space for all packets. Since the amount of available storage space is finite, the queue is finite and blocking can occur. When blocking does occur, a blocked packet returns (in a functional sense) to the node that sent it. Thus congestion in a destination node prevents the transmitting node from releasing internal storage space assigned to the packet and causes the packet to reenter the same or a different output queue for retransmission. Notice that in this case queue interaction arises because the maximum queue length of the destination node is finite instead of infinite, as assumed in the model for independent queues.

REFERENCES

1 W. Feller, *An Introduction to Probability Theory and Its Applications*, Wiley, New York, 1968.
2 A. A. Collins and R. D. Pederson, *Telecommunications a Time for Innovation*, Merle Collins Foundation, Dallas, Texas, 1973.
3 R. Syski, *Introduction to Congestion Theory in Telephone Systems*, Oliver and Boyd, London, 1965.
4 Members of Technical Staff, Bell Telephone Laboratories, *Engineering and Operations in the Bell System*, Western Electric, 1977.
5 "Determination of the Number of Circuits in Automatic and Semi-Automatic Operation," CCITT Recommendation E. 520, *Orange Book*, Vol. II.2, 1976, p 211.
6 C. J. Weinstein, "Fractional Speech Loss and Talker Activity Model for Packet-Switched Speech," *IEEE Transactions on Communications Technology*, August 1978, pp 1253–1256.
7 H. Miedema and M. Schachtman, "TASI Quality-Effect of Speech Detection and Interpolation," *Bell System Technical Journal*, July 1962, pp 1455–1473.
8 G. Szarvas and H. Suyderhoud, "Voice-Activated-Switch Performance Criteria," *Comsat Technical Review*, Spring 1980, pp 151–177.
9 S. S. Katz, "Improved Network Administration Process Utilizing End-to-End Service Considerations," *International Teletraffic Conference*, 1979.
10 R. Dayem, "Alternate Routing in High Blocking Communications Networks," *National Telecommunications Conference*, 1979, pp 28.4.1–28.4.6.
11 P. R. Boorstyn and H. Frank, "Large-Scale Network Topological Optimization," *IEEE Transactions on Communications*, January 1977, pp 29–47.
12 R. I. Wilkinson, "Theories for Toll Traffic Engineering in U.S.A.," *Bell System Technical Journal*, March 1956.

13 "Calculation of the Number of Circuits in a Group Carrying Overflow Traffic," CCITT Recommendation E.521, *Orange Book*, Vol. 2, No. 2, p 218.

14 J. A. Morrison, "Analysis of Some Overflow Problems with Queueing," *Bell System Technical Journal*, October 1980, pp 1427–1462.

15 *Telephone Traffic Theory, Tables, and Charts*, Siemens Aktiengesellschaft, Munich, 1970.

16 L. Kleinrock, *Queueing Systems Volume 1: Theory*, Wiley, New York, 1975.

PROBLEMS

10.1 A central office to PBX trunk group contains four circuits. If the average call duration is 3 min and the busy hour offered traffic intensity is 2 erlangs, determine each of the following:

(a) The busy hour calling rate.

(b) The probability that two arrivals occur less than 1 sec apart.

(c) The blocking probability assuming a lost calls cleared operation.

(d) The amount of lost traffic.

(e) The proportion of time the fourth circuit is in use (assuming fixed order selection).

10.2 A T1 line is used to carry traffic from a remote concentrator to a central office. How many 10 CCS subscribers can the concentrator system support at 0.5% blocking. Compare an infinite source analysis to a finite source analysis.

10.3 Two switching offices experience 20 erlangs of average busy hour traffic load between them. Assume a single T1 line provides 24 direct trunks between the offices. How much busy hour traffic overflows to a tandem switch?

10.4 A PBX with 200 stations has five trunks to the public network. What is the blocking probability if each station is involved in three external calls per 8-hr working day with an average duration of 2 min per call? Assume blocked calls return with random retries. What is the offered load? What is the demand traffic?

10.5 How many dial-up I/O ports are needed for a computer center to support 40 users with a blocking probability limit of 5%? Assume each user averages four calls per day with an average session duration of 30 min. If three users remain connected all day, what is the grade of service for the remaining 37 users?

10.6 A 24 channel trunk group is divided into two groups of 12 one-way trunks in each direction. (A one-way trunk is one that can only be seized at one end.) How many erlangs of traffic can this system support at 0.5%

blocking? How many erlangs can be supported if all 24 trunks are two-way trunks? (That is, every trunk can be seized at either end.)

10.7 The following 10 A.M. to 11 A.M. busy hour statistics have been observed on a 32-channel interoffice trunk group. What is the overall blocking probability? What is the blocking probability for the same busy hour, if day-to-day fluctuations are averaged together?

Monday	Tuesday	Wednesday	Thursday	Friday
20E	19E	22E	19E	30E

10.8 Traffic measurements on a PBX to central office trunk group indicate that during the busiest hour of the day the trunks are 80% utilized. If there are eight trunks in the group, what is the blocking probability, assuming blocked calls do not return? How many trunks must be added to achieve a maximum blocking probability of 5%?

10.9 Repeat Problem 10.8 assuming blocked calls return with random retries.

10.10 A small community with 400 subscribers is to be serviced with a community dial office switch. Assume that the average subscriber originates 0.1 erlangs of traffic. Also assume that 20% of the originations are local (intracommunity) calls and that 80% are transit calls to the serving central office. How many erlangs of traffic are offered to the community dial office to central office trunk group? How many trunks are needed for 0.5% blocking of the transit traffic?

10.11 For the community of Problem 10.10 determine the number of concentrator channels required if local calls are not switched locally but are merely concentrated into pair-gain systems and switched at the central office.

10.12 Repeat Problems 10.10 and 10.11 if 80% of the originations are intracommunity calls and 20% are transit calls.

10.13 A group of eight remote farm houses are serviced by four lines. If each of the eight families utilizes their telephones for 10% of the busy hour, compare the blocking probabilities of the following configurations:

(a) Four party lines with two stations per line.

(b) An 8 to 4 concentration system.

10.14 A PBX provides queuing and automatic call back for access to outgoing WATS lines. If there are 20 requests per hour for the WATS lines, and if the average call is 3 min in length, how many WATS lines are needed to provide delays of less than 1 hr for 90% of the requests?

10.15 A call processor has 50% of its time available for servicing requests. If each request requires 50 msec of processing time, what arrival rate can be supported if only 1% of the service requests are delayed by more than 1 sec? Assume that processor time is sliced into 500-msec time slots. (That is, 500 msec are allocated to call processing and then 500 msec to overhead functions etc.)

10.16 A group of 100 sources offers messages with exponentially distributed lengths to a 1200 bps line. The average message length is 200 bits, including overhead, and each source generates one message every 20 sec. Access to the line is controlled by message-switching concentration with an infinite queue. Determine the following:

(a) The probability of entering the queue.

(b) The average queueing delay for all arrivals.

(c) The probability of being in the queue for more than 1 sec.

(d) The utilization of the transmission link.

10.17 An airline company uses an automatic call distributor to service reservations and ticket purchases. Assume that the processing time of each inquiry is randomly distributed with a 40 sec average. Also assume that if customers are put on hold for more than 2 min, they hang up and call another airline. If each of 200 inquiries per hour produces $30 worth of sales, on average, what is the optimum number of reservationists? Assume each reservationist costs the company $20/hr (including overhead).

10.18 A radio station talk show solicits the listening public for comments on the ineptness of government (I assume this will be a topical subject for the life of this book). Assume that each caller talks for a random length of time with an average duration of 1 min. (Either the show is unpopular or the public has given up on the government.) How many incoming lines must the radio station have to keep the idle time below 5%?

APPENDIX A
DERIVATION OF EQUATIONS

A.1 QUANTIZING NOISE POWER: EQUATION 3.2

The probability density function of a noise sample η is assumed to be uniform:

$$p(\eta) = \frac{1}{q} \qquad \frac{-q}{2} \leqslant \eta \leqslant \frac{q}{2}$$

$$= 0 \qquad \text{otherwise}$$

The average or expected value of noise power is determined as

$$\text{quantization noise power} = \int_{-q/2}^{q/2} \left(\frac{1}{q}\right) \eta^2 \, dn$$

$$= \frac{q^2}{12}$$

A.2 NRZ LINE CODE: EQUATION 4.1

$$f(t) = 1 \qquad |t| \leqslant \frac{T}{2}$$

$$= 0 \qquad \text{otherwise}$$

$$F(j\omega) = \int_{-\infty}^{\infty} f(t) e^{-j\omega t} \, dt$$

$$= \int_{-T/2}^{T/2} e^{-j\omega t} \, dt$$

$$= \left(\frac{1}{j\omega}\right)(e^{j\omega T/2} - e^{-j\omega T/2})$$

$$= (T)\frac{\sin(\omega T/2)}{(\omega T/2)}$$

Note. $F(j\omega)$ is the spectrum of a single pulse. $|F(j\omega)|^2 (1/T)$ is the power spectral density of a random pulse train assuming positive and negative pulses are equally likely and occur independently.

A.3 DIGITAL BIPHASE: FIGURE 4.13

$$f(t) = 1 \qquad -\tfrac{1}{2}T \leqslant t < 0$$
$$= -1 \qquad 0 \leqslant t < \tfrac{1}{2}T$$
$$= 0 \qquad \text{otherwise}$$

$$F(j\omega) = \int_{-(1/2)T}^{0} e^{-j\omega t}\, dt - \int_{0}^{(1/2)T} e^{-j\omega t}\, dt$$

$$= \left(\frac{-1}{j\omega}\right)\{1 - e^{j(1/2)\omega T} - e^{-j(1/2)\omega T} + 1\}$$

$$= \left(\frac{j4}{\omega}\right)\sin^2\left(\frac{\omega T}{4}\right)$$

A.4 FRAME ACQUISITION TIME OF SINGLE-BIT FRAME CODE: EQUATION 4-10

Framing is established by successively examining one bit position after another until a sufficiently long framing pattern is detected. In this derivation it is assumed that the framing pattern alternates 1's and 0's. Furthermore, it is assumed that when beginning to test a particular bit position for framing, the value of the first appearance is saved and compared to the second appearance. Thus the minimum time to reject an invalid framing position is one frame time. If we denote by p the probability of a 1 and by $q = 1 - p$ the probability that a 0 is received first, the average number of frames required to receive a mismatch is

$A_0 =$ (1) (probability of mismatch at end of first frame)

$\qquad +$ (2) (probability of mismatch at end of second frame)

$\qquad +$ (3) (probability of mismatch at end of third frame) \cdots

$= (1)q + (2)(1-q)p + (3)(1-q)(1-p)q$

$\qquad + (4)(1-q)^2(1-p)p + \cdots$

$= (1)q + (2)p^2 + (3)pq^2 + (4)p^3q + (5)p^2q^3 + \cdots$

$= (q + 2p^2 + pq^2)[1 + (2)pq + (3)p^2q^2 + (4)p^3q^3 + \cdots]$

$= [1 - p + 2p^2 + p(1-p)^2](1 + pq + p^2q^2 + p^3q^3 + \cdots)^2$

$= \dfrac{(1 + p^3)}{(1 - pq)^2}$

$= \dfrac{(1 + p)}{(1 - pq)}$

Similarly, if a 1 is received first, the expected number of frames before receiving a mismatch is

$$A_1 = \frac{(1 + q)}{(1 - pq)}$$

The overall average number of frames required to detect a mismatch is

$$A = qA_0 + pA_1$$
$$= \frac{(1 + 2pq)}{(1 - pq)}$$

If we assume a random starting point in a frame with N bits, the average number of bits that must be tested before the true framing bit is encountered is

$$\text{frame time} = (\tfrac{1}{2}N)(A \cdot N) + \tfrac{1}{2}N$$
$$= \tfrac{1}{2}N(A \cdot N + 1) \text{ bit times}$$

If 1's and 0's are equally likely ($p = q = \tfrac{1}{2}$), $A = 2$ so frame time = $N^2 + \tfrac{1}{2}N$ bit times (Equation 4.10).

A.5 FRAME ACQUISITION TIME OF SINGLE-BIT FRAMING CODE USING A PARALLEL SEARCH: EQUATION 4.11

This framing algorithm assumes that all bit positions in a frame are simultaneously scanned for the framing pattern. If we assume that an alternating-bit frame code is used and that 1's and 0's in the information bits are equally likely, the probability that a particular information bit does not produce a framing violation in n frames is

$$p_n = (\tfrac{1}{2})^n$$

The probability that a framing violation has been received in n or less frames is $1 - p_n$. The probability that all $N - 1$ information bits in a frame produce a framing violation in n or less frames is

$$P_n = [1 - (\tfrac{1}{2})^n]^{N-1} \qquad (4.11)$$

A.6 FRAME ACQUISITION TIME OF MULTIBIT FRAME CODE: EQUATION 4.13

N = length of frame including framing code

L = length of framing code

$p = (\frac{1}{2})^L$ = probability of matching frame code with random data

The expected number of frames examined before a particular frame position mismatches the frame code follows:

$$A = (0)(1 - p) + (1)p(1 - p) + (2)p^2(1 - p) + \cdots$$
$$= (1 - p)p(1 + 2p + 3p^2 + 4p^3 + \cdots)$$
$$= (1 - p)p(1 + p + p^2 + p^3 + \cdots)^2$$
$$= (1 - p)p\left(\frac{1}{1 - p}\right)^2$$
$$= \frac{p}{1 - p}$$

The average number of bits that pass before the frame position is detected (assuming a random starting point and the test field is moved one bit position when a mismatch occurs) is

$$T = \left(\frac{N}{2}\right)(A)(N) + \frac{1}{2}N \qquad (4.13)$$
$$= \frac{N^2(1/2)^{L+1}}{1 - (1/2)^L} + \frac{1}{2}N$$

Note. Equation 4.13 with $L = 1$ is not identical to Equation 4.10 because Equation 4.13 assumes a fixed frame code while Equation 4.10 assumes an alternating code.

A.7 PATHFINDING TIME: EQUATION 5.11

Assume that all paths through a switch are independently busy with probability p. Let the probability that a path is not busy be denoted by $q = 1 - p$. The probability p_i that exactly i paths are tested before an idle one is found is the probability that the first $i - 1$ are busy and the ith is not:

$$p_i = p^{(i-1)}q$$

The expected number of paths tested before an idle path is found is

$$N_p = (1)q + (2)pq + (3)p^2q + \cdots + (k)p^{k-1}q + (k)p^k$$

where the last term represents the expectation that all possible paths k are unavailable. A closed form for A is determined as

$$N_p = (1 - p)(1 + 2p + 3p^2 + \cdots + kp^{k-1} + kp^k)$$

$$= 1 + p + p^2 + p^3 + \cdots + p^{k-1}$$

$$= \frac{1}{1 - p} - p^k \left(\frac{1}{1 - p} \right) \qquad (5.11)$$

$$= \frac{1 - p^k}{1 - p}$$

A.8 4 ESS SPACE STAGE BLOCKING PROBABILITY 5.21

The most straightforward way to calculate the blocking probability of the four-stage space structure shown in the probability graph of Figure A.1 is to list all elementary, mutually exclusive probability terms, determine their probabilities of occurrence, and generate the sum of those that represent blocking. Because there are eight links, which are assumed to be independently busy or idle, there are $2^8 = 256$ elementary terms. Rather than laboriously list them all the analysis can be greatly simplified, with a risk of miscounting blocking terms, by grouping the terms according to the number of busy links. The following table lists the groups and the corresponding numbers of terms that block and do not block.

Number of Busy Links i	Combinations $\binom{8}{i}$	Number That Block	Number that Do Not Block
0	1	0	1
1	8	0	8
2	28	2	26
3	56	16	40
4	70	50	20
5	56	52	4
6	28	28	0
7	8	8	0
8	1	1	0
	256	157	99

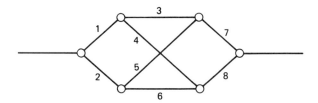

Figure A.1. Probability graph of No. 4 ESS four-stage space switch.

Entries in the last two columns are determined by analyzing the topology of the network. For example, when two links are busy only two of the 28 combinations produce blocking (1 and 2 and 7 and 8). When three links are busy 16 of the combinations produce blocking. To determine the remaining entries in column three it is easier to determine the number of combinations that do not block and subtract from the total. For example, when only three links are idle there are 4 of 56 combinations that do not block. Using the entries in column three the blocking probability is determined as

$$B = 2p^2q^6 + 16p^3q^5 + 50p^4q^4 + 52p^5q^3 + 28p^6q^2 + 8p^7q + p^8$$

where p is the probability that a link is busy and $q = 1 - p$ is the probability that it is idle. All of the interstage links are equally loaded because there is no concentration or expansion in the stages.

APPENDIX B
ENCODING/DECODING ALGORITHMS FOR SEGMENTED PCM

B.1 EIGHT BIT μ255 CODE

The encoded representations of μ255 PCM code words use a sign-magnitude format wherein one bit identifies the sample polarity and the remaining bits specify the magnitude of the sample. The seven magnitude bits are conveniently partitioned into a 3-bit segment identifier S and a 4-bit quantizing step identifier Q. Thus the basic structure of an 8-bit μ255 PCM code word is shown in Figure B.1.

In the following descriptions of encoding and decoding algorithms, it is assumed, for convenience in using integer representations, that analog input signals are scaled to a maximum amplitude of 8159. Furthermore, all amplitudes and segment identifiers are assumed to be encoded using conventional binary representations. The actual encoders used in T1 transmission systems, however, complement the code words to increase the density of 1's in a transmitted bit stream.

Figure B.1. Eight bit μ225 PCM code format.

B.1.1 Algorithm 1: Direct Encoding (Table B.1)

Polarity bit $P = 0$ for positive sample values

$\qquad\quad\; = 1$ for negative sample values

TABLE B.1 Piecewise Linear Approximation to μ255 Companding[a]

	Segment Code S								Quantization Code Q	
	000	001	010	011	100	101	110	111		
Quantization	0	31	95	223	479	991	2015	4063	0000	0
endpoints	1	35	103	239	511	1055	2143	4319	0001	1
	3	39	111	255	543	1119	2271	4575	0010	2
	5	43	119	271	575	1183	2399	4831	0011	3
	7	47	127	287	607	1247	2527	5087	0100	4
	9	51	135	303	639	1311	2655	5343	0101	5
	11	55	143	319	671	1375	2783	5599	0110	6
	13	59	151	335	703	1439	2911	5855	0111	7
	15	63	159	351	735	1503	3039	6111	1000	8
	17	67	167	367	767	1567	3167	6367	1001	9
	19	71	175	383	799	1631	3295	6623	1010	10
	21	75	183	399	831	1695	3423	6879	1011	11
	23	79	191	415	863	1759	3551	7135	1100	12
	25	83	199	431	895	1823	3679	7391	1101	13
	27	87	207	447	927	1887	3807	7647	1110	14
	29	91	215	463	959	1951	3935	7903	1111	15
	31	95	223	479	991	2015	4063	8159		

[a] Sample values are referenced to a full-scale value of 8159. Negative samples are encoded in sign-magnitude format with a polarity bit of 1. In actual transmission the codes are inverted to increase the density of 1's when low signal amplitudes are encoded. Analog output samples are decoded as the center of the encoded quantization interval. Quantization error is the difference between the reconstructed output value and the original input sample value.

Given a sample value with a magnitude x, the first step in the magnitude encoding process is to determine the segment identifier S. The major segments are identified by the segment endpoints: 31, 95, 223, 479, 991, 2015, 4063, and 8159. Thus S can be determined as the smallest endpoint that is greater than the sample value x. S is equal to the smallest a such that

$$x < 64 \cdot 2^a - 33 \qquad a = 0, 1, \ldots, 7$$

After the major segment containing the sample value has been determined, the particular quantization interval within the major segment must be identified. As a first step, a residue r is determined as the difference between the input amplitude and the lower endpoint of the segment:

$$r = x \qquad\qquad\qquad S = 0$$
$$r = x - (32 \cdot 2^S - 33) \qquad S = 1, 2, \ldots, 7$$

The value of Q can now be determined as the quantization interval containing the residue r. Q is equal to the smallest b such that

$$r < 2b + 1 \qquad\qquad S = 0$$
$$r < (2^{S+1})(b + 1) \qquad S = 1, 2, \ldots, 7$$

where $b = 0, 1, \ldots, 15$. Notice that this process identifies quantization intervals in segment $S = 0$ as having upper endpoints at $1, 3, 5, \ldots, 31$ while the other segments have quantization endpoints that are multiples of 4, 8, 16, 32, 64, 128, 256 for $S = 1, 2, 3, 4, 5, 6, 7$, respectively.

After S and Q have been determined, they are binary encoded into 3 and 4 bits, respectively. The concatenation of S and Q produces a 7-bit word that can be conveniently represented as an integer between 0 and 127. In essence, this integer identifies one of the 128 quantization intervals of a compressed signal amplitude.

The decoding process involves assigning the designated polarity to an analog output sample at the midpoint of the nth quantization interval $n = 0, 1, \ldots, 127$. Using the values of S and Q directly, we can determine a discrete output sample value y_n as

$$y_n = (2Q + 33)(2^S) - 33$$

where n is the integer obtained by concatenating the binary representations of S and Q.

Example

An input sample of $+242$ produces the following code word:

$$0, 3, 1 = \boxed{0} \; \boxed{011} \; \boxed{0001}$$

The decoder output becomes

$$y_{49} = (2 \cdot 1 + 33)(2^3) - 33$$
$$= 247$$

which is the midpoint of the forty-ninth quantization interval from 239 to 255.

B.1.2 Algorithm 2: Linear Code Conversion

The fundamental reason for using a μ-law companding characteristic with μ255 is the ease with which the segmented approximation can be digitally converted to and from a uniform code. This section describes the basic algorithms that implement the conversions. The first algorithm provides the means of implementing a μ255 PCM encoder using a 13-bit uniform encoder followed by digital logic to provide the compression function. The second algorithm indicates how to implement the decoder function of first expanding a compressed code into a 13-bit linear code to be used in generating the output samples.

Just as in algorithm 1, the polarity bit P is determined as

$$P = 0 \qquad \text{for positive sample values}$$
$$P = 1 \qquad \text{for negative sample values}$$

The simplicity of converting from a linear code to a compressed code is most evident if the linear code is biased by adding the value 33 to the magnitude of all samples. Notice that this bias shifts the encoding range from (0 to 8159) to (33 to 8192). The addition process can be performed directly on the analog samples before encoding or with digital logic after encoding. In either case, the general form of all biased linear code patterns and the corresponding compressed codes are shown in the following table.

μ255 Encoding Table

Biased Linear Input Code													Compressed Code						
0	0	0	0	0	0	0	1	w	x	y	z	a	0	0	0	w	x	y	z
0	0	0	0	0	0	1	w	x	y	z	a	b	0	0	1	w	x	y	z
0	0	0	0	0	1	w	x	y	z	a	b	c	0	1	0	w	x	y	z
0	0	0	0	1	w	x	y	z	a	b	c	d	0	1	1	w	x	y	z
0	0	0	1	w	x	y	z	a	b	c	d	e	1	0	0	w	x	y	z
0	0	1	w	x	y	z	a	b	c	d	e	f	1	0	1	w	x	y	z
0	1	w	x	y	z	a	b	c	d	e	f	g	1	1	0	w	x	y	z
1	w	x	y	z	a	b	c	d	e	f	g	h	1	1	1	w	x	y	z

From the foregoing table it can be seen that all biased linear codes have a leading 1 that indicates the value of the segment number S. Specifically, the value of S is equal to 7 minus the number of leading 0's before the 1. The value of Q is directly available as the four bits (w, x, y, z) immediately following the leading 1. All trailing bits (a through h) are merely ignored when generating a compressed code.

In reverse fashion the following table indicates how to generate a biased linear code from a compressed code. An unbiased output can be obtained by subtracting 33 from the biased code.

μ255 Decoding Table

Compressed Code							Biased Linear Output Code												
0	0	0	w	x	y	z	0	0	0	0	0	0	0	1	w	x	y	z	1
0	0	1	w	x	y	z	0	0	0	0	0	0	1	w	x	y	z	1	0
0	1	0	w	x	y	z	0	0	0	0	0	1	w	x	y	z	1	0	0
0	1	1	w	x	y	z	0	0	0	0	1	w	x	y	z	1	0	0	0
1	0	0	w	x	y	z	0	0	0	1	w	x	y	z	1	0	0	0	0
1	0	1	w	x	y	z	0	0	1	w	x	y	z	1	0	0	0	0	0
1	1	0	w	x	y	z	0	1	w	x	y	z	1	0	0	0	0	0	0
1	1	1	w	x	y	z	1	w	x	y	z	1	0	0	0	0	0	0	0

Both of these tables indicate that 13 bits of the linear code are used to represent the magnitude of the signal. In Chapter 3 it is mentioned that a μ255

PCM coder has an amplitude range equivalent to 12 bits. The discrepancy occurs because the first quantization interval has length one while all others in the first segment are of length two. Thus the extra bit is needed only to specify the first quantization interval. Notice further that the least significant bit in the tables carries no information but is included only to facilitate the integer relationships. In particular, the least significant bit in the encoding table is completely ignored when determining a compressed code (assuming that the bias is added to the analog sample). Furthermore, the least significant bit in the output codes is completely specified by the segment number S. It is a 0 for segment zero and a 1 for all other segments.

Example
An input code word of $+242$ is biased to produce a value of 275. The binary representation of 275 is

$$0\ 0\ 0\ 0\ 1\ 0\ 0\ 0\ 1\ 0\ 0\ 1\ 1 \quad \text{(biased linear code)}$$

Hence from the encoding table $S = 011$ and wxyz = 0001, and the compressed code is

| 0 | 011 | 0001 | (compressed code)

Using the decoding table, this compressed code produces the following biased linear output code:

$$0\ 0\ 0\ 0\ 1\ 0\ 0\ 0\ 1\ 1\ 0\ 0\ 0 \quad \text{(biased linear output)}$$

The decimal representation of the foregoing code is 280, which corresponds to an unbiased output equal to $+247$.

B.2 EIGHT BIT A-LAW CODE

The following companding algorithms for segmented A-law codes use the same basic procedures as those presented for the $\mu255$ codes. One difference that does occur, however, involves the elimination of a bias in the linear code for conversion to and from a compressed code. Another difference occurs in the use of the integer 4096 as the maximum amplitude of a sample in an A-law representation. If desired, the scale factors for the two systems can be brought into close agreement by doubling the A-law scale to 8192.

As mentioned in Chapter 3, the segmented A-law code is usually referred to as a 13-segment code owing to the existence of seven positive and seven negative segments with the two segments near the origin being colinear. In the following descriptions, however, the first segment for each polarity is divided into two parts to produce eight positive and eight negative segments. This point of view

TABLE B.2 Segmented A-Law Encoding Table

				Segment Code					Quantization Code	
	000	001	010	011	100	101	110	111		
Quantization	0	32	64	128	256	512	1024	2048	0000	0
endpoints	2	34	68	136	272	544	1088	2176	0001	1
	4	36	72	144	288	576	1152	2304	0010	2
	6	38	76	152	304	608	1216	2432	0011	3
	8	40	80	160	320	640	1280	2560	0100	4
	10	42	84	168	336	672	1344	2688	0101	5
	12	44	88	176	352	704	1408	2816	0110	6
	14	46	92	184	368	736	1472	2944	0111	7
	16	48	96	192	384	768	1536	3072	1000	8
	18	50	100	200	400	800	1600	3200	1001	9
	20	52	104	208	416	832	1664	3328	1010	10
	22	54	108	216	432	864	1728	3456	1011	11
	24	56	112	224	448	896	1792	3584	1100	12
	26	58	116	232	464	928	1856	3712	1101	13
	28	60	120	240	480	960	1920	3840	1110	14
	30	62	124	248	496	992	1984	3968	1111	15
	32	64	128	256	512	1024	2048	4096		

Note: Every other bit is inverted for transmission.

permits a code format that is identical to the $\mu255$ code format. Thus, a compressed code word consists of a sign bit P followed by 3 bits of a segment identifier S and 4 bits of quantizer level Q.

B.2.1 Algorithm 1: Direct Encoding

The segment endpoints of an A-law code are 32, 64, 128, 256, 512, 1024, 2048, and 4096. Thus for a sample with magnitude x the major segment identifier S can be determined as the smallest a such that

$$x < 32 \cdot 2^a \qquad a = 0, 1, \ldots, 7$$

After S has been determined, the residue r can be obtained as

$$r = x \qquad\qquad\quad S = 0$$
$$= x - 16 \cdot 2^S \qquad S = 1, 2, \ldots, 7$$

The value of Q can then be determined as the smallest b such that

$$r < 2(b + 1) \qquad S = 0$$
$$r < 2^S(b + 1) \qquad S = 1, 2, \ldots, 7$$

Just as in the case for $\mu255$ coding, an A-law magnitude can be represented as

an integer $n = 0, 1, \ldots, 127$ derived from the concatenation of 3 S-bits and 4 Q-bits. An output magnitude y_n can then be expressed as

$$
\begin{aligned}
y_n &= 2Q + 1 & S &= 0 \\
&= 2^S(Q + 16\tfrac{1}{2}) & S &= 1, 2, \ldots, 7
\end{aligned}
$$

Example

An input sample of $+121$ produces the following code word:

$$
\boxed{0 \;\mid\; 010 \;\mid\; 1110} = +46 \text{ in decimal}
$$

The decoder output becomes

$$
\begin{aligned}
y_{46} &= 2^S(14 + 16\tfrac{1}{2}) \\
&= 122
\end{aligned}
$$

which is the midpoint of the quantization interval from 120 to 124.

B.2.2 Algorithm 2: Linear Code Conversion

The following tables indicate how to convert a 12-bit linear code directly into a compressed A-law code. The algorithm is basically the same as for the $\mu 255$ conversion except that biasing the linear code is unnecessary and a first segment code does not have a leading 1. Thus the segment number S can be determined as 7 minus the number of leading zeros as before. The Q field data are obtained as the four bits (wxyz) immediately following the leading 1, except when $S = 0$, in which case the Q field is contained in the 4 bits following the 7 leading 0's.

A-Law Encoding Table

Linear Code																Compressed Code						
0	0	0	0	0	0	0	w	x	y	z	a					0	0	0	w	x	y	z
0	0	0	0	0	0	1	w	x	y	z	a					0	0	1	w	x	y	z
0	0	0	0	0	1	w	x	y	z	a	b					0	1	0	w	x	y	z
0	0	0	0	1	w	x	y	z	a	b	c					0	1	1	w	x	y	z
0	0	0	1	w	x	y	z	a	b	c	d					1	0	0	w	x	y	z
0	0	1	w	x	y	z	a	b	c	d	e					1	0	1	w	x	y	z
0	1	w	x	y	z	a	b	c	d	e	f					1	1	0	w	x	y	z
1	w	x	y	z	a	b	c	d	e	f	g					1	1	1	w	x	y	z

The following table provides the means of generating a linear code word directly from a compressed code word. The output value corresponds to the middle of the quantization interval designated by S and Q.

A-Law Decoding Table

Compressed Code							Linear Output Code											
0	0	0	w	x	y	z	0	0	0	0	0	0	0	w	x	y	z	1
0	0	1	w	x	y	z	0	0	0	0	0	0	1	w	x	y	z	1
0	1	0	w	x	y	z	0	0	0	0	0	1	w	x	y	z	1	0
0	1	1	w	x	y	z	0	0	0	0	1	w	x	y	z	1	0	0
1	0	0	w	x	y	z	0	0	0	1	w	x	y	z	1	0	0	0
1	0	1	w	x	y	z	0	0	1	w	x	y	z	1	0	0	0	0
1	1	0	w	x	y	z	0	1	w	x	y	z	1	0	0	0	0	0
1	1	1	w	x	y	z	1	w	x	y	z	1	0	0	0	0	0	0

Each of these tables relates 12 bits of magnitude in a linear code to a compressed code with 7 bits of magnitude. Notice, however, that the least significant bit of the encoder is always ignored. Thus the encoder needs only 11 bits of resolution if all of its outputs are immediately compressed. If any signal processing (such as adding two signals together) is to take place before compression, however, the extra bit is useful in reducing the composite quantization error.

Example
The previously used sample value 121 is represented in binary form as 000001111001. From the encoding table $S = 010$ and $Q = 1110$. Thus the compressed code word is

$$\boxed{0 \mid 010 \mid 1110}$$

Using the decoding table, the linear output can be determined as 000001111010, which is 122 decimal.

APPENDIX C
ANALYTIC FUNDAMENTALS OF DIGITAL TRANSMISSION

C.1 PULSE SPECTRA

This section presents the frequency spectra of common pulse waveforms used for digital transmission. These are square pulses as generated at a source. Since the spectra of square pulses have infinite frequency content, the spectra presented here do not correspond to pulse responses at the output of a channel where the pulse shapes have been altered by band limiting filters. In the next section channel output pulse responses are described. Then the necessary combinations of input pulse shapes and filter designs to produce particular output pulses are considered.

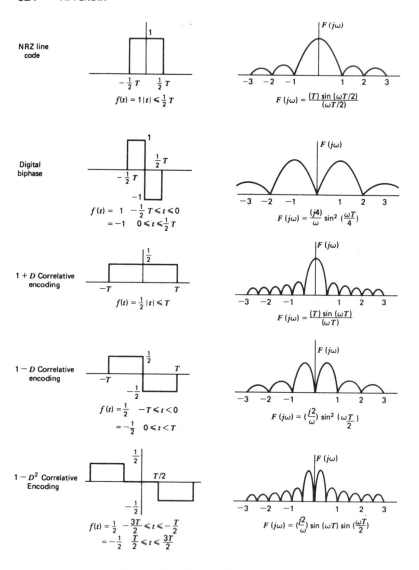

Figure C.1. Spectra of common pulse shapes.

The various pulse shapes and corresponding frequency spectra are presented in Figure C.1. In deriving the spectra, the following conditions and assumptions were made:

1. All pulses have equal energy.
2. All systems signal at rate $1/T$.
3. The waveforms shown are used to encode a 1.
4. Opposite polarities are used for a 0.
5. It is equally likely for 1's and 0's to occur, and occur at random (complete independence).

C.2 CHANNEL OUTPUT PULSE RESPONSES

Although a digital transmission system can be designed to produce a variety of output pulse responses, the most common is defined as

$$y_{rc}(t) = \frac{\sin(\pi t/T)}{(\pi t/T)} \frac{\cos(\alpha \pi t/T)}{1 - (2\alpha t/T)^2} \tag{C.1}$$

where $1/T$ is the signaling rate and α is an excess bandwidth factor between 0 and 1. Equation C.1 represents the response of a raised-cosine channel—so called because the frequency spectrum corresponding to $y_{rc}(t)$ in C.1 is

$$\begin{aligned} Y_{rc}(f) &= 1 & |f| &\leqslant \frac{(1-\alpha)}{2T} \\[2mm] &= \frac{1}{2}\left\{1 + \cos\left[\frac{\pi|f|T}{\alpha} - \frac{\pi(1-\alpha)}{2\alpha}\right]\right\} & \frac{(1-\alpha)}{2T} &\leqslant |f| \leqslant \frac{(1+\alpha)}{2T} \\[2mm] &= \cos^2\left[\frac{\pi|f|T}{2\alpha} - \frac{\pi(1-\alpha)}{4\alpha}\right] \\[2mm] &= 0 & \text{otherwise} \end{aligned} \tag{C.2}$$

The origin of the apellation "raised cosine" is apparent in the third line of Equation C.2.

The parameter α in Equations C.1 and C.2 is referred to as an excess bandwidth parameter. If $\alpha = 0$, the spectrum defined in Equation C.2 is exactly equal to the theoretical minimum bandwidth $1/2T$ for signaling rate $1/T$. As α increases from 0 to 1, the excess spectrum width increases to 100%. Raised-cosine channel spectra are illustrated in Figure C.2 for several values of α.

Practical systems are typically designed for excess bandwidths of 30% or more for several reasons. First, "brick wall" filters needed to produce the infinite attenuation slopes implied by $\alpha = 0$ are physically unrealizable. Second, as shown in Figure C.3, the time domain pulse response for small values of α exhibits large amounts of ringing. Slight errors in the sample times cause significant degradations in performance due to intersymbol interference. Third, a slight deviation in the signaling rate from the design rate also produces significant intersymbol interference.

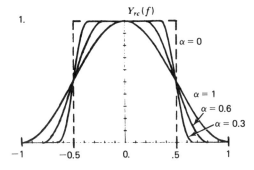

Figure C.2. Raised-cosine spectrums for various values of α.

Figure C.3. Raised-cosine pulse responses for various values of α.

It must be emphasized that Equation C.2 defines the desired spectrum at the output of the channel (the input to the decision circuit). Thus the desired response results from a combination of the input pulse spectrum and the channel filter responses. Often, the input pulse spectrum arises from a square pulse of duration T:

$$x_s(t) = 1 \qquad |t| \leqslant \frac{T}{2}$$

$$= 0 \qquad \text{otherwise}$$

(C.3)

The frequency spectrum corresponding to $x_s(t)$ in Equation C.3 is the "$\sin(x)/x$" spectrum also referred to as a "sinc" function:

$$X_s(f) = \frac{(T)\sin(\pi f T)}{(\pi f T)}$$

$$= T \operatorname{sinc}(\pi f T)$$

(C.4)

When the channel input spectrum is as defined in Equation C.4, the filter function of a channel to produce a raised-cosine output is determined as

$$H(f) = \frac{Y_{rc}(f)}{X_s(f)}$$

(C.5)

The channel filter functions implied in Equation C.5 are shown in Figure C.4 for the same values of α shown in Figure C.2. Figure C.2 displays channel output spectra while Figure C.4 displays corresponding frequency domain transfer functions for the channel.

As mentioned in Chapter 3, the design of a smoothing filter in a digital voice decoder sometimes requires a modification like that defined in Equation C.5. In the case mentioned, the ideal "flat" filter for reconstructing speech from narrow impulselike samples should be modified by dividing the flat response by the spectrum of the finite width samples (Equation C.4). When the samples are

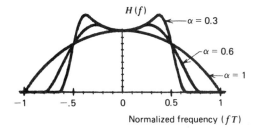

Figure C.4. Channel filter responses needed to produce raised-cosine outputs when excited by $\sin(\omega T/2)/(\omega T/2)$ pulses.

narrower than the signaling interval, the $\sin(x)/x$ response is essentially flat across the bandwidth of the filter. However, when the samples are made to last for the entire duration of the sample interval, the filter response should be "peaked" to compensate for $\sin(x)/x$ roll off.

As discussed in Chapter 6, a channel filter function is usually partitioned between the transmitter and the receiver. The so-called "optimum" filter partitioning occurs when the receive filter response is the square root of the desired output response $Y_{rc}(f)$, and the transmit filter response is whatever is necessary to transform the channel input spectrum into the complex conjugate of the receive filter response. (The output spectrum of the transmitter is also the square root of the desired channel response.) When the channel input is a pulse of duration T, the "optimum" filter functions are defined as

$$H_{RX}(f) = \{Y_{rc}(f)\}^{1/2} \tag{C.6}$$

$$H_{TX}(f) = \frac{\{Y_{rc}(f)\}^{1/2}}{X_s(f)} \tag{C.7}$$

Notice that the transmit spectrum resulting from $H_{TX}(f)$ in Equation C.7 is equal to the square root of the desired output response—no matter what the input pulse shape is. Hence, when optimum partitioning is used, the transmit spectrum and the receiver design are independent of the channel excitation.

The transmit and receive filter functions for square wave excitations (Equa-

Figure C.5. "Optimum" transmit and receive filter functions for raised-cosine response with $\alpha = 0.3$ and $\sin(\omega T/2)/(\omega T/2)$ excitation.

tions C.6 and C.7) are shown in Figure C.5.* Notice that the transmit filter function has peaking at frequencies other than direct current. The implication for midband attenuation is one of the reasons why "optimum" partitioning may not be optimum in a system with source power limitations.

C.2.1 Optimum Filtering for Minimum Shift Keying

As discussed in Chapter 6, minimum shift keyed (MSK) modulation can be represented as quadrature channel modulation with baseband excitation defined as cosine pulse shapes:

$$x_c(t) = \cos\left(\frac{\pi t}{T}\right) \quad |t| \leq \frac{T}{2}$$
$$= 0 \quad \text{otherwise}$$

(C.8)

The transform of $x_c(t)$ is

$$X_c(f) = \left(\frac{\pi}{2T}\right)\frac{\cos \pi f T}{1 - (2fT)^2}$$

(C.9)

When a raised-cosine output response and optimum filter partitioning is desired, the MSK filter functions are

$$H_{RX}(f) = \{Y_{rc}(f)\}^{1/2}$$

(C.10)

$$H'_{TX}(f) = \frac{\{Y_{rc}(f)\}^{1/2}}{X_c(f)}$$

(C.11)

The transmit spectrum and the receiver of an optimally partitioned MSK system are identical to counterparts in an optimally partitioned, offset-keyed 4-PSK system!

C.2.2 Partial Response Systems

As another digital transmission system design example, consider a $1 + D$ partial response system. The desired time response of the channel is defined in Equation C.12 and the associated frequency spectrum in Equation C.13.

$$y_c(t) = \frac{4 \cos(\pi t/T)}{\pi[1 - (2t/T)^2]}$$

(C.12)

*This discussion assumes baseband transmission. The principles are easily extended to carrier-based systems by translating the filter functions to the carrier frequency.

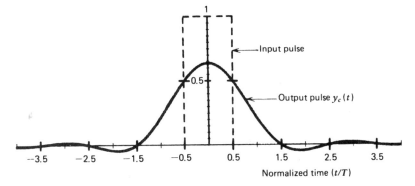

Figure C.6. Output pulse of $1 + D$ partial response channel.

$$Y_c(f) = \cos \pi f T \qquad |f| \leqslant \frac{1}{2T}$$

$$= 0 \qquad\qquad \text{otherwise}$$

(C.13)

The pulse response of a $1 + D$ partial response channel is shown in Figure C.6. Notice that a single pulse contributes equally to the response at two successive sample times but crosses zero at all other sample times.

Optimum filter partitioning is again achieved with a receive filter having an amplitude response equal to the square root of the desired output response $Y_c(f)$. If the channel is excited by square pulses of duration T, as defined in Equation C.3, the optimum filter functions are

$$H_{RX}(f) = \{Y_c(f)\}^{1/2} \qquad |f| \leqslant \frac{1}{2T}$$

(C.14)

$$H_{TX}(f) = \frac{\{Y_c(f)\}^{1/2}}{X_s(f)} \qquad |f| \leqslant \frac{1}{2T}$$

(C.15)

where $X_s(f)$ is defined in Equation C.4.

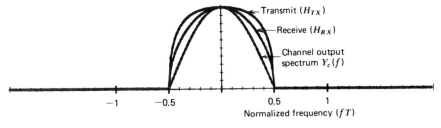

Figure C.7. Spectra of $1 + D$ PRS channel and "optimum" filter responses for $\sin(\omega T/2)/(\omega T/2)$ excitation.

The optimum filter functions defined in Equations C.14 and C.15 are shown in Figure C.7 along with the desired output response of the channel. Notice that, unlike the full response (raised-cosine) systems, optimum partitioning of a partial response system does not require peaking of the transmit filter.

C.3 ERROR RATE ANALYSES—BASEBAND SYSTEMS

C.3.1 Binary Transmission

Consider the receiver model of a digital transmission system shown in Figure C.8. The receiver consists of two parts: signal processing circuitry and a data detector (decision circuit). For the time being, assume that the output of the signal processing circuit produces a pulse of amplitude $+V$ when a 1 is transmitted and a pulse of amplitude $-V$ when a 0 is transmitted. Obviously, the detector merely examines the polarity of its input at the sample times defined by the sample clock. A decision error occurs if noise at the sample times has an amplitude greater than V and a polarity opposite to the transmitted pulse.

The most commonly analyzed type of noise is assumed to have a Gaussian or normal probability distribution. Thus the probability of error P_E can be determined as

$$P_E = \frac{1}{\sqrt{2\pi}\,\sigma} \int_V^\infty e^{-t^2/2\sigma^2}\, dt \qquad (C.16)$$

where σ^2 is the rms noise power at the detector. Using the error function:

$$\text{erf } z = \frac{2}{\sqrt{\pi}} \int_0^z e^{-t^2}\, dt \qquad (C.17)$$

Equation C.16 is sometimes rewritten as

$$P_E = \tfrac{1}{2}[1 - \text{erf } z] \qquad (C.18)$$

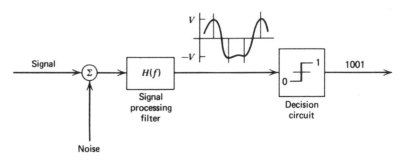

Figure C.8. Digital receiver model.

where $z = V/\sqrt{2}\sigma$. The error probability P_E can also be expressed in terms of the complimentary error function:

$$P_E = \tfrac{1}{2} \text{ erfc } z \qquad (C.19)$$

where erfc $z = 1 - \text{erf } z$

$$z = V/\sqrt{2}\sigma$$

In lieu of evaluating the integral in Equation C.17 or C.19 (which has no closed form solution), the error functions can be approximated as

$$\text{erfc } z \doteq \frac{e^{-z^2}}{(z\sqrt{\pi})} \qquad (z \gg 1) \qquad (C.20)$$

C.3.2 Multilevel Transmission

The error rate of a multilevel baseband system is easily determined by an appropriate reduction in the error distance. If the maximum amplitude is V, the error distance d between equally spaced levels at the detector is

$$d = \frac{V}{L-1} \qquad (C.21)$$

where L is the number of levels. Adjusting the error distance V of a binary system to that defined in C.21 provides the error rate of a multilevel system as

$$P_E = \left(\frac{1}{\log_2 L}\right)\left(\frac{L-1}{L}\right)\text{erfc}\left\{\frac{V}{(L-1)\sqrt{2}\sigma}\right\} \qquad (C.22)$$

where the factor $(L-1)/L$ reflects the fact that interior signal levels are vulnerable to both positive and negative noise, and the factor $1/\log_2 L$ arises because the multilevel system is assumed to be coded so symbol errors produce single bit errors ($\log_2 L$ is the number of bits per symbol).

Equation C.22 reveals that, with respect to peak signal-to-noise ratios at the detector, L-level transmission incurs a penalty of $20 \log_{10}(L-1)$ dB. If V is increased by a factor $(L-1)$, the error rate of the L-level system is identical to the error rate of the binary system. [Except for the factors $1/\log_2 L$ and $(L-1)/L$, which typically represent only a few tenths of a decibel.]

Equation C.22 relates error rate to the peak signal power V^2. To determine the error rate with respect to average power, the average power of an L-level system is determined by averaging the power associated with the various pulse amplitude levels:

$$(V^2)_{avg} = \frac{2}{L}\left\{\left(\frac{V}{L-1}\right)^2 + \left(\frac{3V}{L-1}\right)^2 + \cdots + V^2\right\} \tag{C.23}$$

$$= \frac{2V^2}{L(L-1)^2}\sum_{i=1}^{L/2}(2i-1)^2$$

where the levels

$$\frac{V}{L-1}\{\pm 1, \pm 3, \cdots \pm (L-1)\}$$

are assumed to be equally likely.

C.3.3 Energy-per-Bit to Noise Density Ratios

The foregoing error rate equations relate P_E to the signal energy at the sample times and the rms noise power at the detector. When comparing various digital modulation formats, multilevel systems in particular, it is more relevant to relate error performance to signal power and noise power at the input to the receiver (in front of the signal processing circuitry). As a first step in developing an error rate equation based on signal-to-noise ratios at the receiver input, the noise power at the detector is determined.

The variance of the noise σ^2 used in the previous equations is exactly equal to the rms power that would be measured at the detector in the absence of a signal. This noise power can be determined analytically as

$$\sigma^2 = \int_{-\infty}^{\infty}|H(f)(\tfrac{1}{2}N_0)|^2\,df \tag{C.24}$$

$$= N_0\int_0^{\infty}|H(f)|^2\,df$$

$$= N_0\cdot B_N \tag{C.25}$$

where N_0 is the one-sided noise power spectral density in watts per hertz and $B_N = \int_0^{\infty}|H(f)|^2\,df$ is the noise-equivalent bandwidth or simply the noise bandwidth of the receiver filter function $H(f)$.

In Equation C.25, the noise source is assumed to be white. That is, a uniform spectral density exists across the entire band of interest. This noise may exist in the transmission medium itself, or it may occur in the "front end" amplifier of the receiver. If the rms noise power passing through a square (ideal) filter of bandwidth B is measured, a reading of $(N_0)(B)$ W would be obtained. Thus B_N represents the bandwidth of a perfectly square filter that passes the same amount of noise as the receiver filter $H(f)$. [$H(f)$ may be decidedly nonsquare in its amplitude response.]

The receiver function $H(f)$ necessarily provides a compromise between two conflicting objectives. First, it must minimize the amount of noise passed to the detector (i.e., minimize B_N). Second, the difference between sample values ($+V$ and $-V$) must be maximized. Obviously, the signal-to-noise ratio at the detector (V^2/σ^2) should be maximized to minimize the error probability. A classical result of digital communication theory states that V/σ is maximized when $H(f)$ is "matched" to the received signal.

When viewed in the time domain, a "matched filter" is implemented by correlating (multiplying) the received signal with each of the receivable (noise-free) pulse shapes. The outputs of the correlators are integrated across a signal interval to determine the overall average correlation during the interval. The integrator with the maximum output indicates the most likely symbol to have been transmitted.

In most systems all signals or pulses have the same shape but differ only in amplitude and polarity. Thus a single matched filter can be used. Detection merely involves comparing the matched filter output to appropriate decision levels. The output of a single correlator $h(t)$ and its integrator is expressed as

$$V = \int_0^T s(t)h(t)\, dt$$

$$= \int_0^T |s(t)|^2\, dt \tag{C.26}$$

where $s(t)$ is the signal or pulse shape being measured. Notice that V is, in essence, a measure of the energy in the signal over the signaling interval T.

When analyzed in the frequency domain, a matched filter response $H(f)$ is the complex conjugate of the channel pulse spectrum $S(f)$. Thus the matched filter output $Y(f)$ can be expressed in the frequency domain as

$$Y(f) = H(f) \cdot S(f)$$

$$= S^*(f) \cdot S(f) \tag{C.27}$$

Frequency domain representations are most convenient when the transmit signal is strictly bandlimited—implying that the duration of the pulse response is theoretically unlimited. In this case the energy in a pulse (e.g., raised cosine) is directly proportional to the detector voltage at the optimum sample time. Hence optimum detection is achieved by merely sampling the output of the receive filter at the proper time.

Using the relationship of Equation C.25 and the parameter E_s to represent symbol energy, we express the binary error rate Equation C.19 as

$$P_E = \tfrac{1}{2}\operatorname{erfc}(z) \tag{C.28}$$

where $z^2 = E_s/(N_0 B_N)$.

Notice that for a given system (fixed B_N), the error rate is dependent only on the ratio of the symbol energy E_s and the noise density N_0. This ratio is commonly referred to as a signal-to-noise ratio although it is not a signal-power to noise-power measurement. Equation C.28 is the preferred form of the error rate equation for comparing different modulation schemes. In a binary scheme, the symbol energy E_s is equal to the energy per bit E_b.

As an example of a specific system, consider a baseband raised-cosine channel with optimum partitioning. The output spectrum of the transmitter is the square root of the raised cosine spectrum $Y_{rc}(f)$ defined in Equation C.2. The matched receiver filter also has a square root of a raised-cosine response (Equation C.6). Hence the noise bandwidth of the receiver can be determined as

$$
\begin{aligned}
B_N &= \int_0^\infty |H(f)|^2 \, df \\
&= \int_0^{(1+\alpha)/2T} |Y_{rc}(f)| \, df \\
&= \frac{1}{2T} \quad \text{(independent of } \alpha\text{)}
\end{aligned}
\tag{C.29}
$$

As defined in Equation C.1 and shown in Figure C.3, the (normalized) peak sample value at the detector of a raised-cosine pulse is 1. Using unnormalized pulses of amplitude E_s, we can determine the error rate of a binary $(+E_s, -E_s)$ raised-cosine channel as

$$
P_E = \tfrac{1}{2}\,\text{erfc}(z)
\tag{C.30}
$$

where $z^2 = (E_s/N_0)T$ and T is the duration of a signal interval.

Although Equation C.30 was derived for a raised-cosine channel, it is more general in that it is applicable to any binary system using antipodal signaling. Thus Equation C.30 is plotted in Figure 4.23 as the best performance achievable by any digital transmission system detecting one pulse at a time. (Lower error rates are possible when redundant signals or error correcting codes are used.)

The error rate performance presented in Equation C.22 for multilevel systems is based on noise power at the detector. As long as the signal bandwidth is identical for all systems, Equation C.22 is valid since the noise bandwidth of the receiver is independent of the number of levels. When the signaling rate is held constant, however, the bit rate increases with the number of levels. To compare multilevel systems on the basis of a given data rate, the signaling interval T and, hence, the noise bandwidths must be adjusted accordingly.

If T is the signaling interval for a two-level system, the signaling interval T_L for an L-level system providing the same data rate is determined as

$$
T_L = T \log_2 L
\tag{C.31}
$$

Using the noise bandwidth of a raised cosine filter in Equation C.29, we extend Equation C.22 to multilevel systems as

$$P_E = \left(\frac{1}{\log_2 L}\right)\left(\frac{L-1}{L}\right)\text{erfc}(z) \tag{C.32}$$

where

$$z = \frac{V/(L-1)}{(N_0/T_L)^{1/2}}$$

Equation C.32 can be simplified and presented in a more customary form by using the relationship that energy per symbol $E_s = E_b \log_2 L = V^2 T_L$ where E_b is the energy per bit:

$$P_E = \left(\frac{1}{\log_2 L}\right)\left(\frac{L-1}{L}\right)\text{erfc}(z) \tag{C.33}$$

where

$$z = \frac{(\log_2 L)^{1/2}}{L-1}\left(\frac{E_b}{N_0}\right)^{1/2}$$

Equation C.33 is plotted in Figure 4.26 for 2, 4, 8, and 16 levels. These curves represent the ideal relative performances of multilevel baseband systems at a constant data rate. The bandwidth requirements of the higher-level systems decrease in proportion to $\log_2 L$.

Equation C.33 and Figure 4.26 represent the performance of multilevel systems with respect to E_b/N_0 (common data rate but different bandwidths). The following relationship can be used to determine error rates with respect to signal-to-noise ratios at the decision circuit (different data rates but common bandwidth).

$$\text{SNR} = \frac{\text{signal power}}{\text{noise power}}$$

$$= \frac{(E_b)(\log_2 L)(1/T_L)}{(N_0)(1/2T_L)} \tag{C.34}$$

$$= (2)(\log_2 L)\left(\frac{E_b}{N_0}\right)$$

where $1/2T_L$ is the minimum "Nyquist" bandwidth of the signal.

The SNR obtained in Equation C.34 represents the ratio of signal power at the sample time to noise power at the detector. This ratio is sometimes referred

to as a postdetection SNR because it is the signal-to-noise ratio at the output of the signal processing circuitry.

Some communications theorists use predetection signal-to-noise ratios in determining error rates. Since predetection SNRs are measured prior to bandlimiting the noise, a noise bandwidth must be hypothesized to establish a finite noise power. Commonly, a bit rate bandwidth $(1/T)$ or a Nyquist bandwidth $(1/2T)$ is specified. The latter specification produces SNRs identical to that in Equation C.34. Exceptions occur with double-sideband modulation using coherent demodulation (e.g., 2-PSK) where the predetection SNR is 3 dB higher than the postdetection SNR. (All signal power is coherent to the demodulator carrier reference, but only half of the noise is "coherent.")

C.3.4 Partial Response Systems

The error rate equation for a $1 + D$ partial response system is obtained by incorporating the following modifications to a full response system:

1. The error distance is exactly one-half the error distance of a corresponding full response system (see Figure C.6).
2. The noise bandwidth of the receiver (assuming square root partitioning) is obtained by integrating H_{RX} of Equation C.14:

$$B_N = \int_0^{1/2T} \cos \pi f T \, df \qquad (C.35)$$

$$= 1/(\pi T)$$

Thus the noise bandwidth of a $1 + D$ partial response system is 2 dB lower than the noise bandwidth of a raised-cosine (full response) system obtained in Equation C.29. Since the error distance is reduced by 6 dB, partial response filtering incurs a net penalty of 4 dB—with respect to unfiltered signal power at the source. In terms of channel powers, the partial response system incurs a smaller penalty owing to greater spectrum truncation in the transmit filter.

The difference in channel powers of the two systems can be obtained by integrating the respective channel power spectra. If square root partitioning is used, the channel power of the partial response system is 2 dB below the channel power of the corresponding full response systems. In this case, the difference in channel powers is exactly equal to the difference in noise bandwidths because the receiver filter responses are matched to the respective channel spectra. That is, integration of channel spectra has essentially been accomplished in Equations C.29 and C.35.

When spectrum truncation loss in a transmitter is considered, a square root partitioned partial response system is only 2 dB worse than that of a corresponding full response system. Of course, the partial response system requires less bandwidth than the full response system. For completeness, the error rate

equation of a square root partitioned $1 + D$ partial response system is

$$P_E = \left(\frac{1}{\log_2 L}\right)\left(\frac{L-1}{L}\right)\text{erfc}(z) \tag{C.36}$$

where $z = (\pi/4)(E_b/N_0)^{1/2}(\log_2 L)^{1/2}/(L-1)$ and E_b is the energy per bit on the channel.

Equation C.36 is identical to Equation C.33 except for the factor $\pi/4$ resulting from the lower error distance, the lower noise bandwidth, and the lower channel power of the partial response system.

C.4 CARRIER TRANSMISSION SYSTEMS

C.4.1 Filter Design

Except for a few relatively uncommon frequency modulated systems, digitally modulated carrier systems can be designed and analyzed with baseband-equivalent channels. Carrier based filters are derived by translating the baseband filters to bandpass filters centered at the carrier frequency. The output pulse response of the channel is determined by the composite of the baseband-equivalent filters. Thus pulse shaping can be achieved by filtering the baseband signals or the modulated signals. Partitioning of the channel filter function is dependent on the application. In all cases the composite channel response is identical to that defined in Equation C.2 for raised cosine channels or Equation C.13 for partial response systems.

C.4.2 Error Rate Analysis

Error rate analyses of baseband systems can be directly applied to carrier systems—under one important condition: coherent modulation and demodulation must be used. For example, coherent demodulation of a 2-PSK signal $y(t) = \cos[\omega t + \phi(t)]$ involves implementing the following equations:

$$\begin{aligned}
\phi(t) &= \text{low pass}\{\cos[\omega t + \phi(t)] \cdot 2\cos(\omega t)\} \\
&= \text{low pass}\{[(\cos \phi(t)\cos \omega t - \sin \phi(t)\sin \omega t]2\cos \omega t\} \\
&= \text{low pass}\{\cos \phi(t)(1 + \cos 2\omega t) - \sin \phi(t)\sin 2\omega t\} \\
&= \cos \phi(t) \\
&= +1 \text{ for } \phi(t) = 0 \text{ and } -1 \text{ for } \phi(t) = \pi \tag{C.37}
\end{aligned}$$

Notice that coherent demodulation involves multiplying the received signal by a local carrier that is exactly in phase with the respective incoming signal. Hence coherent demodulation is closely related to matched filter detection as presented

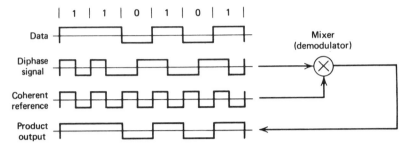

Figure C.9. Coherent demodulation (detection) of digital biphase signal.

in Equations C.26 and C.27. To complete optimum detection of a digitally modulated signal, the baseband equivalent of the receiver filter function must also be matched to the carrier envelope: $\cos \phi(t)$. When coherent demodulation (also called coherent detection) is used, the error rate performance is identical to the analogous baseband system. Hence the error rate of a coherently detected 2-PSK system is provided in Equations C.19 or C.28.

The correspondence of coherent demodulation to matched filter detection is illustrated in Figure C.9. For convenience, digital biphase (diphase) is compared to an NRZ baseband signal. The basic principle also applies to sine wave carriers at any frequency. The important point to notice in Figure C.9 is that the output of the coherent demodulator (or equivalently, the matched filter) is identical to the NRZ signal. Furthermore, the noise power density coming out of the coherent demodulator is identical to the noise power density of the baseband system. (Positive weighted white noise is statistically no different than negative weighted white noise.)

Figure C.9 demonstrates that coherently demodulated carrier-based signals produce the same signal-to-noise ratios at the detector as baseband systems—despite the fact that (double-sideband) carrier systems require twice as much bandwidth. Coherent demodulation leads to a receiver noise bandwidth equal to the baseband-equivalent noise bandwidth because only one-half of the noise power in the carrier signal bandwidth is passed by the coherent demodulator. (Noise in the carrier signal bandwidth that is out of phase with respect to the coherent reference is translated to twice the carrier frequency and therefore eliminated by the low-pass filter, $\sin \omega t \cdot \cos \omega t = \frac{1}{2} \sin 2\omega t$.)

C.4.3 QAM Error Rates

The error rate equation of a coherently detected QAM system is identical to the error rate of the corresponding multilevel baseband system applied individually to each quadrature channel. Thus the error performance of a 16-QAM system is provided in Equation C.33 for $L = 4$ levels. In a QAM system coherent demodulation causes one-half of the noise power in the carrier signal bandwidth to show up at the I-channel detector and one-half shows up at the Q-channel

detector. Of course, the total signal power is divided in two so the signal-to-noise ratio at the individual detectors is identical to the carrier signal-to-noise ratio (i.e., predetection SNR is equal to postdetection SNR). The error rate performance of QAM systems is plotted in Figure 6.20 in terms of E_b/N_0. To relate those results to signal-to-noise ratios, use Equation C.34 using L as the number of levels on each quadrature channel.

C.4.4 PSK Error Rates

The error rate of a multilevel PSK system is derived most easily by using quadrature channel representations for the signals. For example, Figure C.10 displays the regions of decision errors for a representative signal phase in an 8-PSK signal. The received signal is processed by two orthogonal phase detectors to produce quadrature signals $Y_I(t)$ and $Y_Q(t)$ in Equations 6.10 and 6.11, respectively.

If the transmitted phase is $\pi/8$ (corresponding to data values 011), a decision error results if noise causes $Y_Q(t)$ to go positive at the sample time. [$Y_Q(t)$ is positive downward to represent the Q (sine) channel as lagging the I (cosine) channel.] The normalized error distance is $\sin(\pi/8)$. A decision error also results if noise causes $-Y_Q(t)$ to exceed $Y_I(t)$, indicating the phase is greater than $\pi/4$. This latter condition can be represented as a negative value for the transformed signal $Y_B(t) = 0.707\ Y_I(t) + 0.707\ Y_Q(t)$. [$Y_B(t)$ is a projection of the received signal onto a $-\pi/4$ basis vector (see Figure 6.13)]. Examination of Figure C.10 reveals that the error distance for this second type of error is also $\sin(\pi/8)$. Since the noise variance $Y_B(t)$ is identical to the noise variance of $Y_I(t)$ (by virtue of the 0.707 multipliers), both types of errors are equally likely.

In general terms, the error distance of a PSK system with N phases is $V \cdot \sin(\pi/N)$ where V is the signal amplitude at a detector (i.e., the radius of a

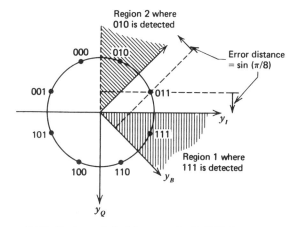

Figure C.10. Regions of decision error for 8-PSK signal at $\pi/8$ (011).

PSK signal constellation). A detection error occurs if noise of the proper polarity is present at the output of either of two phase detectors. A detection error, however, is assumed to produce only a single-bit error. The general expression for the theoretical error rate for PSK modulation is now determined by modifying Equation C.22 as

$$P_E = \left(\frac{1}{\log_2 N}\right) \text{erfc}(z)$$

where

$$z = \frac{\sin(\pi/N) \cdot V}{\sqrt{2}\sigma} \tag{C.38}$$

The signal amplitude V can be expressed as

$$V = \left[E_b(\log_2 N)\left(\frac{1}{T}\right)\right]^{1/2} \tag{C.39}$$

and the rms noise voltage σ as

$$\sigma = \left[N_0\left(\frac{1}{2T}\right)\right]^{1/2} \tag{C.40}$$

for noise in a Nyquist bandwidth.

Combining Equations C.38, C.39, and C.40 relates PSK error rates to energy per bit to noise density on the channel:

$$P_E = \left(\frac{1}{\log_2 N}\right) \text{erf}(z) \tag{C.41}$$

where

$$z = \sin\left(\frac{\pi}{N}\right)(\log_2 N)^{1/2}\left(\frac{E_b}{N_0}\right)^{1/2}$$

Equation C.41 is plotted in Figure 6.16 for PSK systems with various numbers of phases.* To determine error rates with respect to signal-power-to-noise-power ratios use the following:

$$\text{SNR} = \log_2(N)\left(\frac{E_b}{N_0}\right) \quad (N > 2) \tag{C.42}$$

*For 2-PSK systems, the error rates as specified in Equations C.38 or C.41 should be divided by 2 because only one phase detector is needed and it produces errors for one polarity of noise only.

APPENDIX D
TRAFFIC TABLES

Table D.1 is a table of maximum offered loads A for various blocking probabilities B and number of servers N. The blocking probabilities are for infinite sources with lost calls cleared (erlang-B, Equation 10.8).

Table D.2 is a table of maximum offered loads A for various blocking probabilities B, number of servers N, and finite number of sources M. The offered load A is determined as $M\rho$ where ρ is the average source activity assuming no calls are cleared. The blocking probability for finite sources is determined from Equation 10.13.

Note: Entries in the following tables were obtained from reference [15] of Chapter 9: *Telephone Traffic Theory, Tables, and Charts*, Siemens Aktiengesellschaft, Munich, 1970.

TABLE D.1 Maximum Offered Load Versus B and N[a]

N/B	0.01	0.05	0.1	0.5	1.0	2	5	10	15	20	30	40
1	.0001	.0005	.001	.005	.010	.020	.053	.111	.176	.250	.429	.667
2	.014	.032	.046	.105	.153	.223	.381	.595	.796	1.00	1.45	2.00
3	.087	.152	.194	.340	.455	.602	.899	1.27	1.60	1.93	2.63	3.48
4	.235	.362	.439	.701	.869	1.09	1.62	2.05	2.50	2.95	3.89	5.02
5	.452	.649	.762	1.13	1.36	1.66	2.22	2.88	3.45	4.01	5.10	6.60
6	.728	.996	1.15	1.62	1.91	2.28	2.96	3.76	4.44	5.11	6.51	8.19
7	1.05	1.39	1.58	2.16	2.50	2.94	3.74	4.67	5.46	6.23	7.86	9.80
8	1.42	1.83	2.05	2.73	3.13	3.63	4.54	5.60	6.50	7.37	9.21	11.4
9	1.83	2.30	2.56	3.33	3.78	4.34	5.37	6.55	7.55	8.52	10.6	13.0
10	2.26	2.80	3.09	3.96	4.46	5.08	6.22	7.51	8.62	9.68	12.0	14.7
11	2.72	3.33	3.65	4.61	5.16	5.84	7.08	8.49	9.69	10.9	13.3	16.3
12	3.21	3.88	4.23	5.28	5.88	6.61	7.95	9.47	10.8	12.0	14.7	18.0
13	3.71	4.45	4.83	5.96	6.61	7.40	8.83	10.5	11.9	13.2	16.1	19.6
14	4.24	5.03	5.45	6.66	7.35	8.20	9.73	11.5	13.0	14.4	17.5	21.2
15	4.78	5.63	6.08	7.38	8.11	9.01	10.6	12.5	14.1	15.6	18.9	22.9
16	5.34	6.25	6.72	8.10	8.88	9.83	11.5	13.5	15.2	16.8	20.3	24.5
17	5.91	6.88	7.38	8.83	9.65	10.7	12.5	14.5	16.3	18.0	21.7	26.2
18	6.50	7.52	8.05	9.58	10.4	11.5	13.4	15.5	17.4	19.2	23.1	27.8
19	7.09	8.17	8.72	10.3	11.2	12.3	14.3	16.6	18.5	20.4	24.5	29.5
20	7.70	8.83	9.41	11.1	12.0	13.2	15.2	17.6	19.6	21.6	25.9	31.2
21	8.32	9.50	10.1	11.9	12.8	14.0	16.2	18.7	20.8	22.8	27.3	32.8
22	8.95	10.2	10.8	12.6	13.7	14.9	17.1	19.7	21.9	24.1	28.7	34.5
23	9.58	10.9	11.5	13.4	14.5	15.8	18.1	20.7	23.0	25.3	30.1	36.1
24	10.2	11.6	12.2	14.2	15.3	16.6	19.0	21.8	24.2	26.5	31.6	37.8
25	10.9	12.3	13.0	16.0	16.1	17.5	20.0	22.8	25.3	27.7	33.0	39.4
26	11.5	13.0	13.7	15.8	17.0	18.4	20.9	23.9	26.4	28.9	34.4	41.1
27	12.2	13.7	14.4	16.6	17.8	19.3	21.9	24.9	27.6	30.2	35.8	42.8
28	12.9	14.4	15.2	17.4	18.6	20.2	22.9	26.0	28.7	31.4	37.2	44.4

TABLE D.1 *(Continued)*

N/B	0.01	0.05	0.1	0.5	1.0	2	5	10	15	20	30	40
29	13.6	15.1	15.9	18.2	19.5	21.0	23.8	27.1	29.9	32.6	38.6	46.1
30	14.2	15.9	16.7	19.0	20.3	21.9	24.8	28.1	31.0	33.8	40.0	47.7
31	14.9	16.6	17.4	19.9	21.2	22.8	25.8	29.2	32.1	35.1	41.5	49.4
32	15.6	17.3	18.2	20.7	22.0	23.7	26.7	30.2	33.3	36.3	42.9	51.1
33	16.3	18.1	19.0	21.5	22.9	24.6	27.7	31.3	34.4	37.5	44.3	52.7
34	17.0	18.8	19.7	22.3	23.8	25.5	28.7	32.4	35.6	38.8	45.7	54.4
35	17.8	19.6	20.5	23.2	24.6	26.4	29.7	33.4	36.7	40.0	47.1	56.0
36	18.5	20.3	21.3	24.0	25.5	27.3	30.7	34.5	37.9	41.2	48.6	57.7
37	19.2	21.1	22.1	24.8	26.4	28.3	31.6	35.6	39.0	42.4	50.0	59.4
38	19.9	21.9	22.9	25.7	27.3	29.2	32.6	36.6	40.2	43.7	51.4	61.0
39	20.6	22.6	23.7	26.5	28.1	30.1	33.6	37.7	41.3	44.9	52.8	62.7
40	21.4	23.4	24.4	27.4	29.0	31.0	34.6	38.8	42.5	46.1	54.2	64.4
41	22.1	24.2	25.2	28.2	29.9	31.9	35.6	39.9	43.6	47.4	55.7	66.0
42	22.8	25.0	26.0	29.1	30.8	32.8	36.6	40.9	44.8	48.6	57.1	67.7
43	23.6	25.7	26.8	29.9	31.7	33.8	37.6	42.0	45.9	49.9	58.5	69.3
44	24.3	26.5	27.6	30.8	32.5	34.7	38.6	43.1	47.1	51.1	59.9	71.0
45	25.1	27.3	28.4	31.7	33.4	35.6	39.6	44.2	48.2	52.3	61.3	72.7
46	25.8	28.1	29.3	32.5	34.3	36.5	40.5	45.2	49.4	53.6	62.8	74.3
47	26.6	28.9	30.1	33.4	35.2	37.5	41.5	46.3	50.6	54.8	64.2	76.0
48	27.3	29.7	30.9	34.2	36.1	38.4	42.5	47.4	51.7	56.0	65.6	77.7
49	28.1	30.5	31.7	35.1	37.0	39.3	43.5	48.5	52.9	57.3	67.0	79.3
50	28.9	31.3	32.5	36.0	37.9	40.3	44.5	49.6	54.0	58.3	68.5	81.0
51	29.6	32.1	33.3	36.9	38.8	41.2	45.5	50.6	55.2	59.7	69.9	82.7
52	30.4	32.9	34.2	37.7	39.7	42.1	46.5	51.7	56.3	61.0	71.3	84.3
53	31.2	33.7	35.0	38.6	40.6	43.1	47.5	52.8	57.5	62.2	72.7	86.0
54	31.9	34.5	35.8	39.5	41.5	44.0	48.5	53.9	58.7	63.5	74.2	87.6
55	32.7	35.3	36.6	40.4	42.4	44.9	49.5	55.0	59.8	64.7	75.6	89.3
56	33.5	36.1	37.5	41.2	43.3	45.9	50.5	56.1	61.0	65.9	77.0	91.0
57	34.3	36.9	38.3	42.1	44.2	46.8	61.5	67.1	62.1	67.7	78.4	92.6
58	35.1	37.8	39.1	43.0	45.1	47.8	62.6	68.2	63.3	68.4	79.8	94.3
59	35.8	38.6	40.0	43.9	46.0	48.7	53.6	59.3	64.5	69.7	81.3	96.0
60	36.6	39.4	40.8	44.8	46.9	49.6	54.6	60.4	65.6	70.9	82.7	97.6
61	37.4	40.2	41.6	45.6	47.9	50.6	55.6	61.5	66.8	72.1	84.1	99.3
62	38.2	41.0	42.5	46.5	48.8	51.5	56.6	62.6	68.0	73.4	85.5	101.
63	39.0	41.9	43.3	47.4	49.7	52.5	57.6	63.7	69.1	74.6	87.0	103.
64	39.8	42.7	44.2	48.3	50.6	53.4	58.6	64.8	70.3	75.9	88.4	104.
65	40.6	43.5	45.0	49.2	51.5	54.4	59.6	65.8	71.4	77.1	89.8	106.
66	41.4	44.4	45.8	50.1	52.4	55.3	60.6	66.9	72.6	78.3	91.2	108.
67	42.2	45.2	46.7	51.0	53.4	56.3	61.6	68.0	73.8	79.6	92.7	109.
68	43.0	46.0	47.5	51.9	54.3	57.2	62.6	69.1	74.9	80.8	94.1	111.
69	43.8	46.8	48.4	52.8	55.2	58.2	63.7	70.2	76.1	82.1	95.5	113.
70	44.6	47.7	49.2	53.7	56.1	59.1	64.7	71.3	77.3	83.3	96.9	114.
71	45.4	48.5	50.1	54.6	57.0	60.1	65.7	72.4	78.4	84.6	98.4	116.
72	46.2	49.4	50.9	55.5	58.0	61.0	66.7	73.5	79.6	85.8	99.8	118.
73	47.0	50.2	51.8	56.4	58.9	62.0	67.7	74.6	80.8	87.0	101.	119.
74	47.8	51.0	52.7	57.3	59.8	62.9	68.7	75.6	81.9	88.3	103.	121.

TABLE D.1 *(Continued)*

N/B	0.01	0.05	0.1	0.5	1.0	2	5	10	15	20	30	40
75	48.6	51.9	53.6	58.2	60.7	63.9	69.7	76.7	83.1	89.5	104.	123.
76	49.4	52.7	54.4	59.1	61.7	64.9	70.8	77.8	84.2	90.8	105.	124.
77	50.2	53.6	55.2	60.0	62.6	65.8	71.8	78.9	85.4	92.0	107.	126.
78	51.1	54.4	56.1	60.9	63.5	66.8	72.8	80.0	86.6	93.3	108.	128.
79	51.9	55.3	56.9	61.8	64.4	67.7	73.8	81.1	87.7	94.5	110.	129.
80	52.7	56.1	57.8	62.7	65.4	68.7	74.8	82.2	88.9	95.7	111.	131.
81	53.5	56.9	58.7	63.6	66.3	69.6	75.8	83.3	90.1	97.0	113.	133.
82	54.3	57.8	59.5	64.5	67.2	70.6	76.9	84.4	91.2	98.2	114.	134.
83	55.1	58.6	60.4	65.4	68.2	71.6	77.9	85.5	92.4	99.5	115.	136.
84	56.0	59.5	61.3	66.3	69.1	72.5	78.9	86.6	93.6	101.	117.	138.
85	56.8	60.4	62.1	67.2	70.0	73.5	79.9	87.7	94.7	102.	118.	139.
86	57.6	61.2	63.0	68.1	70.9	74.5	80.9	88.8	95.9	103.	120.	141.
87	58.4	62.1	63.9	69.0	71.9	75.4	82.0	89.9	97.1	104.	121.	143.
88	59.3	62.9	64.7	69.9	72.8	76.4	83.0	91.0	98.2	106.	123.	144.
89	60.1	63.8	65.6	70.8	73.7	77.3	84.0	92.1	99.4	107.	124.	146.
90	60.9	64.6	66.5	71.8	74.7	78.3	85.0	93.1	101.	108.	126.	148.
91	61.8	65.5	67.4	72.7	75.6	79.3	86.0	94.2	102.	109.	127.	149.
92	62.6	66.3	68.2	73.6	76.6	80.2	87.1	95.3	103.	111.	128.	151.
93	63.4	67.2	69.1	74.5	77.5	81.2	88.1	96.4	104.	112.	130.	153.
94	64.2	68.1	70.0	75.4	78.4	82.2	89.1	97.5	105.	113.	131.	154.
95	65.1	68.9	70.9	76.3	79.4	83.1	90.1	98.6	106.	114.	133.	156.
96	65.9	69.8	71.7	77.2	80.3	84.1	91.1	99.7	108.	116.	134.	158.
97	66.8	70.7	72.6	78.2	81.2	85.1	92.2	101.	109.	117.	135.	159.
98	67.6	71.5	73.5	79.1	82.2	86.0	93.2	102.	110.	118.	137.	161.
99	68.4	72.4	74.4	80.0	83.1	87.0	94.2	103.	111.	119.	138.	163.
100	69.3	73.2	75.2	80.9	84.1	88.0	95.2	104.	112.	121.	140.	164.

[a] N is the number of servers. The numerical column headings indicate blocking probability (%).

TABLE D.2 Maximum Offered Load Versus B, N, and Finite Sources M[a]

N	M	.01	.05	0.1	0.5	1.0	2	5	10	15	20	30	4
1	2	.0002	.0010	.0020	.0100	.0200	.0400	.100	.202	.307	.417	.659	.9
	3	.0002	.0008	.0015	.0075	.0151	.0303	.077	.159	.246	.341	.559	.8
	4	.0001	.0007	.0013	.0067	.0134	.0270	.069	.143	.224	.312	.519	.7
	5	.0001	.0006	.0013	.0063	.0126	.0254	.065	.136	.213	.298	.498	.7
2	3	.030	.067	.095	.212	.300	.425	.678	.980	1.23	1.47	1.97	2.
	4	.023	.052	.074	.167	.238	.342	.560	.832	1.07	1.30	1.78	2.
	5	.021	.046	.065	.149	.213	.308	.510	.767	.997	1.22	1.70	2.
	6	.019	.043	.061	.139	.200	.289	.482	.731	.955	1.18	1.65	2.
	7	.018	.041	.058	.133	.191	.277	.464	.707	.928	1.15	1.62	2.

TABLE D.2 *(Continued)*

N	M	.01	.05	0.1	0.5	1.0	2	5	10	15	20	30	40
3	4	.186	.317	.400	.685	.864	1.09	1.50	1.95	2.31	2.65	3.35	4.
	5	.148	.254	.322	.561	.715	.918	1.30	1.72	2.08	2.42	3.13	3.
	6	.131	.227	.288	.505	.648	.837	1.20	1.62	1.97	2.31	3.02	3.
	7	.122	.211	.268	.473	.609	.790	1.14	1.55	1.90	2.24	2.95	3.
	8	.116	.201	.255	.452	.583	.759	1.10	1.51	1.85	2.19	2.90	3.
	9	.111	.194	.246	.437	.565	.737	1.07	1.47	1.82	2.16	2.86	3.
	10	.108	.188	.240	.426	.551	.720	1.05	1.45	1.79	2.13	2.84	3.
	15	.100	.174	.222	.396	.514	.675	.994	1.38	1.72	2.06	2.76	3.
4	5	.500	.748	.889	1.33	1.59	1.89	2.24	2.98	3.43	3.86	4.76	5.
	6	.408	.617	.737	1.12	1.36	1.64	2.15	2.71	3.16	3.60	4.51	5.
	7	.365	.554	.665	1.02	1.24	1.52	2.01	2.56	3.02	3.46	4.38	5.
	8	.340	.517	.621	.963	1.17	1.44	1.92	2.47	2.93	3.37	4.30	5.
	9	.323	.492	.592	.922	1.13	1.39	1.86	2.41	2.87	3.31	4.24	5.
	10	.310	.474	.571	.892	1.09	1.35	1.82	2.36	2.82	3.27	4.20	5.
	15	.280	.429	.518	.816	1.00	1.25	1.71	2.24	2.70	3.14	4.08	5.
5	6	.951	1.31	1.51	2.08	2.40	2.77	3.39	4.04	4.58	5.09	6.17	7.
	7	.794	1.11	1.28	1.80	2.10	2.45	3.07	3.73	4.28	4.80	5.91	7.
	8	.716	1.00	1.16	1.66	1.94	2.29	2.90	3.56	4.12	4.65	5.77	7.
	9	.668	.940	1.09	1.56	1.84	2.18	2.78	3.45	4.01	4.55	5.68	7.
	10	.635	.896	1.04	1.50	1.77	2.11	2.71	3.37	3.94	4.48	5.61	6.
	12	.592	.839	.979	1.42	1.68	2.01	2.60	3.27	3.84	4.38	5.52	6.
	15	.556	.791	.924	1.35	1.60	1.92	2.51	3.18	3.74	4.29	5.44	6.
	20	.525	.748	.876	1.28	1.53	1.84	2.42	3.09	3.66	4.21	5.37	6.
6	7	1.51	1.97	2.21	2.90	3.26	3.69	4.38	5.12	5.73	6.32	7.59	9.
	8	1.28	1.70	1.91	2.55	2.90	3.32	4.02	4.78	5.41	6.02	7.32	8.
	9	1.17	1.55	1.76	2.37	2.71	3.12	3.82	4.59	5.24	5.85	7.17	8.
	10	1.09	1.46	1.65	2.25	2.58	2.98	3.69	4.47	5.12	5.74	7.07	8.
	15	.926	1.25	1.43	1.97	2.29	2.68	3.38	4.17	4.84	5.48	6.84	8.
	20	.865	1.17	1.34	1.87	2.17	2.56	3.26	4.05	4.72	5.37	6.74	8.
	30	.813	1.10	1.27	1.77	2.07	2.45	3.15	3.94	4.62	5.28	6.66	8.
7	8	2.15	2.70	2.98	3.76	4.17	4.63	5.39	6.20	6.89	7.56	9.01	10.
	9	1.85	2.36	2.62	3.36	3.75	4.22	5.00	5.85	6.56	7.25	8.74	10.
	10	1.70	2.17	2.42	3.13	3.52	3.99	4.78	5.64	6.37	7.07	8.58	10.
	15	1.39	1.80	2.02	2.68	3.05	3.51	4.31	5.21	5.97	6.70	8.25	10.
	20	1.28	1.67	1.88	2.52	2.88	3.33	4.14	5.05	5.81	6.56	8.13	10.
	30	1.19	1.56	1.76	2.38	2.74	3.18	3.99	4.90	5.68	6.44	8.03	9.
8	9	2.85	3.48	3.80	4.65	5.09	5.59	6.41	7.30	8.05	8.80	10.4	12.
	10	2.49	3.08	3.37	4.20	4.64	5.14	6.00	6.92	7.71	8.48	10.2	12.
	11	2.29	2.85	3.14	3.94	4.38	4.89	5.76	6.71	7.51	8.29	9.99	12.
	12	2.16	2.70	2.97	3.77	4.20	4.17	5.59	6.56	7.37	8.17	9.88	12.
	15	1.93	2.43	2.70	3.46	3.89	4.40	5.29	6.29	7.13	7.94	9.69	11.
	20	1.76	2.24	2.49	3.23	3.65	4.16	5.06	6.07	6.93	7.76	9.54	11.
	30	1.63	2.08	2.32	3.03	3.45	3.95	4.86	5.89	6.77	7.61	9.41	11.
	10	3.59	4.30	4.64	5.57	6.03	6.56	7.44	8.39	9.22	10.0	11.9	14.
	11	3.18	3.84	4.17	5.07	5.54	6.09	7.00	8.01	8.87	9.72	11.6	13.
	12	2.94	3.57	3.89	4.79	5.26	5.81	6.75	7.78	8.66	9.52	11.4	13.
	13	2.78	3.39	3.71	4.59	5.06	5.61	6.57	7.62	8.51	9.39	11.3	13.

TABLE D.2 *(Continued)*

N	M	.01	.05	0.1	0.5	1.0	2	5	10	15	20	30	40
9	14	2.66	3.26	3.57	4.44	4.91	5.47	6.43	7.50	8.41	9.29	11.2	13.
	16	2.50	3.08	3.38	4.23	4.70	5.27	6.24	7.33	8.25	9.15	11.1	13.
	18	2.39	2.95	3.25	4.09	4.56	5.13	6.11	7.21	8.15	9.06	11.0	13.
	20	2.31	2.87	3.15	3.99	4.46	5.02	6.02	7.13	8.07	8.99	11.0	13.
	30	2.12	2.64	2.91	3.73	4.19	4.76	5.77	6.90	7.87	8.81	10.8	13.
10	11	4.38	5.15	5.52	6.49	6.98	7.54	8.47	9.49	10.4	11.3	13.3	15.
	12	3.91	4.64	5.00	5.97	6.47	7.04	8.02	9.09	10.0	11.0	13.0	15.
	13	3.63	4.34	4.69	5.65	6.16	6.74	7.75	8.86	9.81	10.8	12.8	15.
	14	3.45	4.13	4.47	5.43	5.94	6.54	7.56	8.69	9.66	10.6	12.7	15.
	15	3.31	3.98	4.32	5.27	5.78	6.38	7.41	8.56	9.55	10.5	12.6	15.
	16	3.20	3.86	4.19	5.14	5.65	6.25	7.30	8.46	9.46	10.4	12.6	15.
	18	3.04	3.68	4.01	4.95	5.46	6.07	7.13	8.31	9.33	10.3	12.5	15.
	20	2.93	3.56	3.88	4.81	5.32	5.93	7.01	8.21	9.23	10.2	12.4	15.
	25	2.75	3.36	3.68	4.59	5.10	5.72	6.81	8.04	9.08	10.1	12.3	14.
	30	2.65	3.25	3.56	4.47	4.98	5.59	6.69	7.93	8.99	10.0	12.2	14.
11	12	5.19	6.01	6.41	7.44	7.95	8.53	9.50	10.6	11.6	12.5	14.7	17.
	13	4.68	5.46	5.85	6.88	7.40	8.01	9.04	10.2	11.2	12.2	14.4	17.
	14	4.37	5.13	5.51	6.54	7.07	7.69	8.76	9.94	11.0	12.0	14.2	17.
	15	4.15	4.90	5.27	6.30	6.84	7.47	8.56	9.77	10.8	11.9	14.1	17.
	16	3.99	4.72	5.09	6.12	6.66	7.30	8.40	9.63	10.7	11.7	14.0	16.
	17	3.86	4.59	4.95	5.98	6.52	7.17	8.28	9.53	10.6	11.7	14.0	16.
	18	3.76	4.48	4.84	5.86	6.41	7.06	8.18	9.44	10.5	11.6	13.9	16.
	20	3.60	4.31	4.66	5.68	6.23	6.88	8.03	9.31	10.4	11.5	13.8	16.
	25	3.36	4.04	4.39	5.40	5.95	6.62	7.79	9.10	10.2	11.3	13.7	16.
	30	3.32	3.90	4.24	5.24	5.79	6.46	7.64	8.98	10.1	11.2	13.6	16.
12	13	6.03	6.90	7.31	8.39	8.92	9.52	10.5	11.7	12.7	13.8	16.1	19.
	14	5.47	6.31	6.72	7.80	8.35	8.98	10.1	11.3	12.4	13.4	15.8	18.
	15	5.13	5.95	6.35	7.44	8.00	8.65	9.77	11.0	12.1	13.2	15.7	18.
	16	4.88	5.69	6.09	7.18	7.75	8.42	9.56	10.8	12.0	13.1	15.5	18.
	17	4.70	5.50	5.90	6.99	7.56	8.24	9.40	10.7	11.8	13.0	15.5	18.
	18	4.56	5.35	5.74	6.83	7.41	8.09	9.27	10.6	11.8	12.9	15.4	18.
	20	4.34	5.12	5.51	6.60	7.18	7.87	9.08	10.4	11.6	12.8	15.3	18.
	25	4.03	4.77	5.16	6.24	6.83	7.54	8.78	10.2	11.4	12.6	15.1	18.
	30	3.85	4.58	4.96	6.04	6.63	7.35	8.61	10.0	11.3	12.5	15.0	18.

[a] The numerical column headings indicate blocking probability (%).

GLOSSARY

Many of the following definitions have been obtained from CCITT recommendations and are so indicated. Others have been obtained from Engineering and Operations in the Bell System and are indicated by EOBS.

Abbreviated dialing. A subscriber can dial a short code that is converted in the originating exchange into a form that will establish connection with the local, national, or international number required. (CCITT)

Added channel framing. *See* Bunched frame alignment signal.

Added digit framing. *See* Distributed frame alignment signal.

Alternate mark inversion signal (AMI) (bipolar signal). A pseudoternary signal, conveying binary digits, in which successive "marks" are normally of alternative, positive, and negative, polarity but equal in amplitude, and in which "space" is of zero amplitude. (CCITT)

Alternate mark inversion violation (AMI) (bipolar violation). A "mark" that has the same polarity as the previous "mark" in the transmission of AMI signals. (CCITT)

Alternate routing. A means of selectively distributing traffic over a number of routes that ultimately lead to the same destination. (EOBS)

Amplitude distortion. Distortion of a transmission signal caused by nonuniform passband attenuation as a function of frequency.

Amplitude modulation (AM). Modulation in which the amplitude of an alternating current is the characteristic varied. (CCITT)

Antipodal signaling. A technique of encoding binary signals so that the symbol for a 1 is the exact negative of the symbol for a 0. Antipodal signaling provides optimum error performance in terms of the signal-to-noise ratio.

APSK. Amplitude and phase shift keying, a form of digital modulation using a combination of both amplitude and phase modulation.

ARQ. *See* Request repeat system.

Asynchronous transmission. A mode of communication characterized by start/stop transmissions with undefined time intervals between transmissions.

Automatic call distributor (ACD). A switching system used to distribute automatically and evenly incoming calls to a number of stations on a first-come, first-served basis. Applications include operator assistance and airline reservations.

547

Automatic number identification (ANI). The process of automatically identifying a calling number.

Availability. (1) With respect to switching systems: the number of outlets available from a particular inlet. (2) With respect to equipment in general: the percentage of time the equipment is operational.

Balanced code. A line code whose digital sum variation is finite (balanced codes have no dc component in their frequency spectrum). (CCITT)

Baseband. Literally, the frequency band of an unmodulated signal. A baseband signal is an information bearing signal that is either transmitted as is or used to modulate a carrier.

Baud. Unit of signaling speed (symbols per second).

Bipolar coding. *See* Alternate mark inversion.

Blocked calls cleared (BCC). A service discipline in which unserviceable requests are rejected by the system without service. Also called lost calls cleared. (LCC).

Blocked calls held (BCH). A service discipline in which unserviceable requests stay "in the system" without being serviced but having a portion of their desired service time elapse until service begins. Also called lost calls held. (LCH)

Blocking. The inability of the calling party to be connected to the called party because either (1) all suitable trunk paths are busy or (2) a path between a given inlet and any suitable free outlet of the switching network of a switching system is unavailable. (EOBS)

Bridged tap. An extra pair of wires connected in shunt to a main cable pair. The extra pair is normally open circuited but may be used at a future time to connect the main pair to a new customer. Short bridged taps do not affect voice frequency signals but can be extremely detrimental to higher frequency digital signals.

Bunched frame alignment signal (added channel framing). A frame alignment signal in which the signal elements occupy consecutive digit time slots. (CCITT)

Burst isochronous. A transmission process that may be used where the information bearer channel rate is higher than the input data signaling rate. The binary digits being transferred are signaled at the digit rate of the information bearer channel and the transfer is interrupted at intervals to produce the required mean data signaling rate. The interruption is always for an integral number of digit periods. (CCITT)

Busy hour (of a group of circuits, a group of switches, or an exchange, etc.). The busy hour is the uninterrupted period of 60 min of which the traffic is at the maximum. (CCITT)

C-Message weighting. Selective attenuation of voiceband noise in accordance with the subjective effects as a function of frequency.

Call congestion. Blocking probability of a trunk group.

Call forwarding. A customized feature available on computer controlled switches that allows customers to direct all incoming calls to another number.

Call waiting. Subscriber A whose termination is in the active state associated with a call to subscriber B is given a call waiting indication (CWI) that a caller subscriber C, is attempting to obtain connection. He may either (1) ignore or reject the indication and continue with the existing call, (2) terminate the existing call and answer subscriber C, or (3) hold the existing call and answer subscriber C. Thus the three party service functions are available to the extent that they are provided in the system. (CCITT)

Calling number indication (CNI). A service whereby the calling subscriber's number can be identified by means of a visual or verbal indication at the called terminal. (CCITT)

Carterfone decision. A 1968 decision by the FCC that allows telephone company customers to connect their own equipment to the network as long as it does not adversely affect the network.

CBX. Computer controlled PBX.

CCIS. Common channel interoffice signaling.

CCS. A measure of traffic intensity expressed as so many hundred call seconds per hour. 36 CCS = 1 erlang.

Central office. Usually used to refer to a local switching system that connects lines to lines and lines to trunks. It may be more generally applied to any *network switching system*. The term is sometimes used loosely to refer to a telephone company building in which a switching system is located and to include other equipment (such as transmission system terminals) that may be located in such a building. (EOBS)

Centrex service. The provision to subscribers, by means of a specially equipped public telephone exchange, of services normally available only in PABXs (e.g., internal dialing of PABX type, operators' desk, direct access to the network, direct dialing-in, transfer of calls). (CCITT)

Channel-associated signaling. A signaling method in which the signals necessary for the traffic carried by a single channel are transmitted in the channel itself or in a signaling channel permanently associated with it. (CCITT)

Channel bank. Terminal equipment for a transmission system used to multiplex individual channels using FDM or TDM techniques.

Circuit switching. The principle of establishing an end-to-end connection between users of a network. The associated facilities are dedicated to the particular connection and held for the duration of the call.

Clock. Equipment providing a time base used in a transmission system to control the timing of certain functions such as the control of the duration of signal elements, and the sampling. (CCITT)

Codec. An assembly comprising an encoder and a decoder in the same equipment. (CCITT)

Coded mark inversion (CMI). A 2-level nonreturn-to-zero code in which binary zero is coded so that both amplitude levels A_1 and A_2 are attained consecutively, each for half a unit time interval $(T/2)$. Binary one is coded so that either of the amplitude levels, A_1 and A_2, are attained alternately for one full unit time interval (T). (CCITT)

Coherent demodulation. Demodulation using a carrier reference that is synchronized in frequency and phase to the carrier used in the modulation process.

Common channel signaling. A signaling method using a link common to a number of channels for the transmission of signals necessary for the traffic by way of these channels. (CCITT)

Common control. A form of automatic control for a switching system that concentrates all control functions into one equipment shared by all connections.

Community dial office (CDO). A small, normally unattended switching system that is used in small communities and is controlled from a larger central office.

Companding. The process of compressing a signal at the source and expanding it at the destination to maintain a given end-to-end dynamic range while reducing the dynamic range between the compressor–expandor.

Concentration. The process of switching some number of lightly used channels or sources onto a smaller number of more heavily used channels.

Conference call. A call between more than two participants. (CCITT)

CPFSK. Continuous phase frequency shift keying.

Crossbar switch. A relay mechanism consisting of 10 horizontal paths and 10 or 20 vertical paths. Any horizontal path can be connected to any vertical path by means of magnets. A 2-stage operation is used to close any crosspoint. First, a selecting magnet shifts all selecting fingers in a horizontal row; then a holding magnet shifts a vertical actuating card to close the selected contacts. (EOBS)

Crosstalk. Unwanted signal transfer from one circuit into another.

Datagram. A single packet message.

Data set. *See* Modem.

Data under voice (DUV). An arrangement for transmitting 1.544-megabits per second (Mbps) pulse streams in the bandwidth available underneath the portion of the baseband used for voice channels on existing microwave systems. (EOBS)

Dataphone digital service (DDS). A service offering of the Bell System providing digital channels at 2.4, 4.8, 9.6, 56, or 1544 kbps.

dBm. Power level in decibels relative to 1 mW.

dBrnC. Power level of noise with C-message weighting expressed in decibels relative to reference noise. Reference noise power is $-90\,\text{dBm} = 10^{-12}\,\text{W}$.

dBrnC0. Noise power measured in dBrnC but referenced to the zero-level transmission level point.

DDD. Direct distance dialing.

Delta modulation. Technique for digitally encoding an analog signal by continuously measuring and transmitting only the polarity of the signal slope. The analog signal is reconstructed as a staircase approximation of the input.

Demand traffic. First attempt traffic. (Offered traffic not including retries.)

Despotic (synchronized) network (master clock). A network synchronizing arrangement is despotic when a unique master clock exists with full power of control of all other clocks. (CCITT)

DID. Direct inward dialing.

Digital multiplex equipment (muldem). Equipment for combining, by time division multiplexing (multiplexer), a defined integral number of digital input signals into a single digital signal at a defined digit rate and also for carrying out the inverse function (demultiplexer). *Note:* When both functions are combined in one equipment at the same location the abbreviation "MULDEX" may be used to describe this equipment. (CCITT)

Digital signal. A signal constrained to have a discontinuous characteristic in time and a set of permitted discrete values. (CCITT)

Digital switching. A process in which connections are established by operations on digital signals without converting them to analogue signals. (CCITT)

Direct inward dialing. Calls can be dialed from a telephone connected to the public network directly to extensions on a PABX. (CCITT)

Distributed frame alignment signal (added digit framing). A frame alignment signal in which the signal elements occupy nonconsecutive digit time slots. (CCITT)

Double-sideband modulation. A modulation technique in which a baseband signal with no dc energy directly modulates a carrier to produce both upper and lower sidebands but no carrier energy.

DS1: Digital signal-1. Primary multiplex level in North American TDM hierarchy. Ascending levels are DS2, DS3, and DS4.

DTE. Data terminal equipment.

Dual tone multifrequency (DTMF) signaling. Generic name for push-button telephone signaling equivalent to the Bell System's TOUCH-TONE®.

Dynamic range. The range of power levels (minimum to maximum) achievable by a signal or specified for equipment operation.

Echo. A wave that has been reflected or otherwise returned with sufficient magnitude and delay to be perceived in some manner as a wave distinct from that directly transmitted. *Note: Echoes* are frequently measured in decibels (dBs) relative to the directly transmitted wave.

Echo cancellor. A device that removes talker echo in the return branch of a four-wire circuit by subtracting out a delayed version of the signal transmitted in the forward path.

Echo suppressor. A voice-operated device placed in the four-wire portion of a circuit and used for inserting loss in the echo path to suppress echo. (CCITT)

Elastic store. A storage buffer designed to accept data under one clock but deliver it under another, so short-term instabilities (jitter) in either clock can be accommodated.

Electronic automatic exchange (EAX). Designation of stored program control switching machines developed by General Telephone and Electronics.

Electronic switching system (ESS). Designation of stored program control switching machines developed by AT&T.

E&M lead signaling. A specific form of interface between a switching system and a trunk in which the signaling information is transferred across the interface via 2-state voltage conditions on two leads, each with a ground return, separate from the leads used for message information. The message and signaling information are combined (and separated) by a signaling system appropriate for application to the transmission facility. The term *E&M lead signaling* is used also in some special-services applications. (EOBS)

End office. Class 5 switching office. Also referred to as a central office.

Envelope delay. Derivative of channel phase response with respect to frequency. Ideally, the phase response should be linear, indicating that all frequencies are delayed equally.

Equalization. The practice of compensating for transmission distortions with fixed or adaptive circuitry.

Erlang. A measure of traffic intensity. Basically, a measure of the utilization of a resource (e.g., the average number of busy circuits in a trunk group, or the ratio of time an individual circuit is busy).

Error-second. A measurement of system performance for digital transmission facilities. An *error-second* is a 1-sec interval during which one or more bit errors occur. (EOBS)

Exchange area. A contiguous area of service defined for administrative purposes that typically comprises an entire town or city and includes the immediate countryside and suburbs. An exchange area may have one end office or many end offices interconnected by trunks and tandem offices.

Expansion. The switching of a number of input channels onto a larger number of output channels.

Facsimile. The black and white reproduction of a document or a picture transmitted over a telephone connection. (CCITT)

Far-end crosstalk (FEXT). Unwanted energy coupled from one channel or circuit into another circuit and appearing at the far end of the transmission link.

FIFO. First-in first-out service discipline for a queue.

Final group. A trunk group that acts as a final (last-chance) route for traffic. Traffic can overflow to a *final group* from high-usage groups that are busy. Calls blocked on a *final group* are not offered to another route. (EOBS).

Flow control (in data network). The procedure for controlling the rate of transfer of packets between two nominated points in a data network, for example between a DTE and a data-switching exchange. (CCITT)

Foreign exchange circuit (FX). An extension of service from one switching office to a subscriber normally serviced by a different switching office.

Four-wire circuit. A circuit using two separate channels for each direction of transmission. When wire-line transmission is involved, each direction of transmission is provided by a separate pair of wires.

Fractional speech loss. The fraction of speech that gets clipped in a TASI system because all channels are in use when talk spurts begin.

Frame. A set of consecutive digit time slots in which the position of each digit time slot can be identified by reference to a frame alignment signal. The frame alignment signal does not necessarily occur, in whole or in part, in each frame. (CCITT)

Frame alignment. The state in which the frame of the receiving equipment is correctly phased with respect to that of the received signal. (CCITT)

Frame alignment recovery time. The time that elapses between a valid frame alignment signal being available at the receive terminal equipment and frame alignment being established. *Note:* The frame alignment recovery time includes the time required for replicated verification of the validity of the frame alignment signal. (CCITT)

Frame alignment signal. The distinctive signal used to enable frame alignment to be secured. (CCITT)

Frequency diversity. In radio systems the use of one or more back up transmitters, channels, and receivers to protect against atmospheric (multipath) fading.

Frequency-division multiplex (FDM). A method of providing a number of simultaneous channels over a common transmission path by using a different frequency band for the transmission of each channel. (EOBS)

Frequency modulation (FM). One way to modify a signal to make it "carry" information. The frequency of a carrier signal is modified in accordance with the amplitude of the information signal. (EOBS)

Frequency-shift keying (FSK). A modulation technique for transmitting digital information having two, or possibly more, discrete states. Each of the discrete states is represented by an associated frequency. The most common form is binary *FSK*, which uses two frequencies to represent the two states. (EOBS)

Full-duplex. Transmission in two directions simultaneously.

Full-echo suppressor. An echo suppressor in which the speech signals on either path control the suppression loss in the other path. (CCITT)

Glare. Simultaneous seizure of both ends of a two-way trunk.

Half-duplex transmission. Transmission in both directions but only in one direction at a time.

Half-echo suppressor. An echo suppressor in which the speech signals of one path control the suppression loss in the other path but in which this action is not reciprocal. (CCITT) Half-echo suppressors are normally used in pairs—one at each end of the long-distance circuit.

HDB3. A modified AMI (bipolar) line code in which strings of four zeros are encoded with an AMI violation in the last bit. (CCITT)

Highway. A common path or a set of parallel paths over which signals from a plurality of channels pass with separation achieved by time division. (CCITT)

Hook flash. A momentary depression of the switchhook to alert an operator or equipment, but not so long as to be interpreted as a disconnect.

Hot standby. Redundant equipment kept in an operational mode as backup for primary equipment. Usually, automatic switching to the standby equipment occurs when the primary equipment fails.

Hybrid. A device used to connect a two-wire, bidirectional circuit to a four-wire circuit.

Idle character. A control character that is sent when there is no information to be sent. (CCITT)

Impulse noise. Short-duration, high-amplitude bursts, or spikes, of noise energy, much greater than the normal peaks of message circuit noise on a transmission channel. (EOBS)

Inband signaling. Signaling transmitted within the same channel and band of frequencies used for message traffic.

Integrated digital network. A network in which connections established by digital switching are used for the transmission of digital signals. (CCITT)

Integrated services digital network. An integrated digital network in which the same digital switches and digital paths are used to establish connections for different services, for example, telephony, data, etc. (CCITT)

Intercept. Calls that, for reasons such as those listed below, cannot reach the wanted number may be intercepted and diverted to an operator, an answering machine, or a tone to give the caller appropriate information. (CCITT) (1) Change of a particular number including advice of new number. (2) Renumbering of a group of numbers or a change of dialing code. (3) Wrong information in telephone directory. (4) Dialing of an unallocated code. (5) Dialing of a number or numbers allowed by the numbering plan but not yet allocated or no longer in service. (6) Route(s) out of order. (7) Route(s) congested. (8) Subscriber's line temporarily out of order. (9) Suspension of service owing to nonpayment.

Intersymbol interference. Interference in a digital (or any time division) transmission system caused by a symbol in one signaling interval being spread out and overlapping the sample time of a symbol in another signal interval.

Isochronous. A signal is isochronous if the time interval separating any two significant instants is theoretically equal to the unit interval or to a multiple of the unit interval. *Note:* In practice, variations in the time intervals are constrained within specified limits. (CCITT).

Jitter. Short-term variations of the significant instants of a digital signal from their ideal positions in time. (CCITT)

Junctor. A connecting circuit between switching stages or frames of a single switching office. A junctor may also provide signaling supervision and talking battery for connected subscriber instruments.

Justifiable digit time slot (stuffable digit time slot). A digit time slot that may contain either an information digit or a justifying digit. (CCITT)

Justification (pulse stuffing). A process of changing the rate of a digital signal in a controlled manner so that it can accord with a rate different from its own inherent rate, usually without loss of information. (CCITT)

Justification ratio (stuffing ratio). The ratio of the actual justification rate to the maximum justification rate. (CCITT)

Justification service digits (stuffing service digits). Digits that transmit information concerning the status of the justifiable digit time slots. (CCITT)

Justifying digit (stuffing digit). A digit inserted in a justifiable digit time slot when that time slot does not contain an information digit. (CCITT)

Key telephone set. A telephone set with special buttons (keys) to provide such capabilities as switching between lines, call holding, or alerting of other telephones.

Line code. A code chosen to suit the transmission medium and giving the equivalence between a set of digits generated in a terminal or other processing equipment and the pulses chosen to represent that set of digits for line transmission. (CCITT)

Load balancing. Adjusting the assignment of very active customer lines so that all groups of customer lines in a multistage switch receive approximately the same amount of traffic.

Loading coils. Lumped element inductors inserted at periodic points in cable pairs to flatten out their voice frequency response. Although loading coils improve voice frequency transmission they severely attenuate higher frequencies as required for digital transmission.

Lost calls cleared (also loss system). A mode of operation in which blocked calls are rejected by the network and may or may not return.

Lost calls held. A mode of operation that holds blocked call requests until a channel becomes available. The portion of call that gets blocked is lost.

Master clock. A clock that generates accurate timing signals for the control of other clocks and possibly other equipments. (CCITT)

Mesochronous. Two signals are mesochronous if their corresponding significant instants occur at the same average rate. *Note:* The phase relationship between corresponding significant instants usually varies between specified limits. (CCITT)

Message switching. The practice of transporting complete messages from a source to a destination in nonreal time and without interaction between source and destination, usually in a store-and-forward fashion.

Modem. A contraction of the terms modulation and demodulation. A device used to generate "voicelike" data signals for transmission over telephone lines. A modem is referred to as a "data set" in Bell System terminology.

Modified alternate mark inversion. An AMI signal that does not strictly conform with alternate mark inversion but includes violations in accordance with a defined set of rules. Examples of such signals are HDB, B6ZS, B3ZS. (CCITT)

Muldem. (*See* Digital multiplex equipment.)

Multiframe (masterframe). A set of consecutive frames in which the position of each frame can be identified by reference to a multiframe alignment signal. The multiframe alignment signal does not necessarily occur, in whole or in part, in each multiframe. (CCITT)

Multifrequency signaling (MF). A signaling method used for interoffice applications. MF signaling uses two of six possible tones to encode 10 digits and five special auxiliary signals.

Multiplexing. The process of combining multiple signals into a single channel for transmission over common facilities.

Mutually synchronized network. A network synchronizing arrangement is mutually synchronized when each clock in the network exerts a degree of control on all others. A mutually synchronized system is democratic when all clocks in the network are of equal status and exert equal amounts of control on the others, the network operating frequency (digit rate) being the mean of the natural (uncontrolled) frequencies of the population of clocks. (CCITT)

Near-end crosstalk (NEXT). Unwanted energy coupled from a transmitter in one circuit into a receiver of another circuit at the same location (near end).

Network management. Network management is the function of supervising a communications network to ensure maximum utilization of the network under all conditions. Supervision requires monitoring, measuring, and, when necessary, action to control the flow of traffic. The objective of network management is to provide service protection and to maximize the number of paid conversations by fully utilizing available equipment and facilities during normal and abnormal periods. (CCITT)

Nonblocking. A switching network that always has a free path from any idle incoming trunk or line to any idle outgoing trunk or line.

Nonreturn to zero (NRZ). A line code that switches directly from one level to another. Each level is held for the duration of a signal interval.

Nonsynchronous network (asynchronous network). A network in which the clocks need not be synchronous or mesochronous. (CCITT)

Occupancy. The fraction of time that a circuit or an equipment unit is in use, expressed as a decimal. Numerically, it is the erlangs carried per circuit. *Occupany* typically includes both message time and call-setup time. (EOBS)

Off hook. The state or condition that a telephone receiver is requesting service- or in use. Also a supervisory signal to indicate active status of a telephone or line.

Offered traffic. It is necessary to distinguish between traffic offered and traffic carried. The traffic carried is only equal to the traffic offered if all calls are immediately handled (by the group of circuits or group of switches being measured) without any call being lost or delayed on account of congestion. The flow of traffic offered, and of traffic carried, is expressed in erlangs. The amount of traffic offered and of traffic carried is expressed in erlang-hours. (CCITT)

On hook. The inactive status of a telephone or line.

One-way trunk. A trunk circuit that can be seized at only one end.

Other common carrier (OCC). Common carriers offering services in competition with established carriers, principally AT&T long lines. (*See* Specialized common carrier.)

Out-of-band signaling. A signaling technique that uses the same path as message traffic but a portion of the channel bandwidth above or below that used for voice.

Out-of-frame alignment time. The time during which frame alignment is effectively lost. That time will include the time to detect loss of frame alignment and the alignment recovery time. (CCITT)

PABX. Private automatic branch exchange (also a PBX).

Packet-mode operation. The transmission of data by means of addressed packets whereby a transmission channel is occupied for the duration of transmission of the packet only. The channel is then available for use by packets being transferred between different data terminal equipments. (CCITT)

Pair-gain system. A subscriber transmission system that serves a number of subscribers with a smaller number of wire pairs—using concentration, multiplexing, or both.

Parity. The process of adding a redundant bit to a group of information bits to maintain either odd or even numbers of 1's in the composite group of bits. A parity error results if an odd number of 1's is detected when even parity is transmitted or vice versa.

Partial response signaling (PRS). The use of controlled intersymbol interference to increase the signaling rate in a given bandwidth.

Permanent virtual circuit. A user facility in which a permanent association exists between the DTEs that is identical to the data transfer phase of a virtual call. No call setup or clearing procedure is possible or necessary. (CCITT)

Per-trunk signaling. A method of signaling in which the signals pertaining to a particular call are transmitted over the same trunk that carries the call. Interoffice signaling other than common-channel interoffice signaling falls into this category. (EOBS)

Phase distortion. Signal distortion resulting from nonuniform delay of frequencies within the passband.

Phase reversal keying (PRK). A special case of phase shift keying involving only two phases $180°$ apart.

Phase shift keying (PSK). A form of digital modulation that uses 2^n distinct phases to represent n bits of information in each signal interval.

Plesiochronous. Two signals are plesiochronous if their corresponding significant instants occur at nominally the same rate, any variation in rate being constrained within specified limits. *Notes:* (1) Two signals having the same nominal digit rate, but not stemming from the same clock are usually plesiochronous. (2) There is no limit to the phase relationship between corresponding significant instants. (CCITT)

Polar signaling. Two-level line-coding for binary data using balanced (symmetric) positive and negative levels. Also called NRZ coding.

Power spectral density. The distribution of signal power as a function of frequency.

Private branch exchange (PBX). Switching equipment used by a company or organization to provide in-house switching and access to the public network.

Protection switching. That category of restoration in which one transmission path is substituted for another to permit maintenance operations for the protection against component failure, or remedy temporary conditions such as fading. This is intended to reflect a configuration in which M paths protect N paths on the same route. (CCITT)

Psophometric weighting. Noise weighting filter recommended by CCITT.

Pulse amplitude modulation (PAM). The process of representing a continuous analog waveform with a series of discrete time samples. The amplitudes of the samples are continuous and therefore analog in nature.

Pulse code modulation (PCM). A process in which a signal is sampled, and the magnitude of each sample with respect to a fixed reference is quantized and converted by coding to a digital signal. (CCITT)

Pulse stuffing. *See* Justification.

QPRS. Quadrature channel modulation using partial response signaling on each channel.

QPSK. Quaternary phase shift keying (4-PSK).

Quadrature amplitude modulation (QAM). Independent amplitude modulation of two orthogonal channels using the same carrier frequency.

Quantization noise. The difference between the discrete sample value represented by a digital code and the original analog sample value.

Quantizing. A process in which samples are classified into a number of adjacent intervals, each interval being represented by a single value called the quantized value. (CCITT)

Raised-cosine channel. A digital transmission channel with a particular type of pulse response that produces no intersymbol interference at the sample times of adjacent signaling intervals. The designation "raised cosine" is derived from the form of the analytical frequency spectrum (1 + cosine or cosine squared).

Reference clock. A clock of high stability and accuracy that is used to govern the frequency of a network of mutually synchronizing clocks of lower stability. The failure of such a clock does not cause loss of synchronism. (CCITT)

Reframing time. The time that elapses between a valid frame alignment signal being available at the receive terminal equipment and frame alignment being established. (CCITT)

Regeneration. The process of recognizing and reconstructing a digital signal so that the amplitude, waveform, and timing are constrained within stated limits. (CCITT)

Regenerative repeater. A device used to detect, amplify, reshape, and retransmit a digital bit stream.

Relative (power) level. The expression in transmission units of the ratio P/P_0, where P represents the power at the point concerned and P_0 the power at the transmission reference point. (CCITT)

Request repeat system (automatic repeat request: ARQ). A system employing an error-detecting code and so arranged that a signal detected as being in error automatically initiates a request for retransmission of the signal detected as being in error. (CCITT)

Return loss. The difference in dB between reflected and incident energy at a signal reflection point.

Ringback. The signaling tone returned by switching equipment to a caller indicating that a called party telephone is being alerted (ringing).

Robbed digit signaling. *See* Speech digit signaling.

Sidetone. The portion of a talker's signal that is purposely fed back to the earpiece so that the talker hears his or her own speech.

Signaling. The exchange of electrical information (other than by speech) specifically concerned with the establishment and control of connections and management in a communication network.

Significant instants. The instants at which a digital line code changes state. (The boundaries of a signal interval.)

Simplex transmission. A mode of operation involving transmission in one direction only.

Singing. An audible oscillation of a telephone circuit caused by a net amount of gain in a four-wire segment of the circuit.

Single-frequency (SF) signaling. A method of conveying dial-pulse and supervision signals from one end of a trunk or line to the other, using the presence or absence of a single specified frequency. A 2600-hertz (Hz) tone is commonly used. (EOBS)

SLIC. Subscriber loop interface circuit.

Slip. At the interface between two digital systems, the insertion or deletion of data into or from the data stream caused by an offset in clock frequencies.

Space diversity. In radio systems, the use of two receiving antennas and possibly two separate receivers to provide protection against atmosphere induced signal attenuation (fading).

Span line. A repeatered T1 line from end to end but not including channel banks.

Speech digit signaling (robbed digit signaling). Signaling in which digit time slots primarily used for the transmission of encoded speech are periodically used for signaling. (CCITT)

Speed dialing. The use of a short address code to represent an often-called telephone number. The common control computer in a PBX or end office providing such a service translates the short code into the desired number.

Specialized common carrier (SCC). A common carrier company originally offering specialized leased line services. More recently, the SCCs have been competing in established markets, including dial-up services. Also referred to as "other common carriers."

Star network. A network with a single node to which all other nodes are connected.

State store. A memory map of the connection status of a switching matrix.

STS. Space–time–space digital switching structure.

Stuffing character. A character used on isochronous transmission links to take account of differences in clock frequencies. (CCITT)

Step-by-step (SXS) system. An automatic switching system using step-by-step switches. In most such systems, a call is extended progressively, step-by-step, to the desired terminal under direct control of pulses from a customer's dial or from a sender. (EOBS)

Stored-program control (SPC). A form of switching system control in which system operations are controlled by a stored program executed by one or more processors. Operation of the system can be altered significantly by changing programs. (EOBS)

Supervisory signal. A signal used to indicate the status of a line or to control equipment on the line.

Switchhook. The hook or buttons upon which a telephone handset rests when it is not being used.

Symbol rate. The reciprocal of the unit interval in seconds. (This rate is expressed in bauds.) *Note:* This definition is the same as "modulation rate." The term "symbol rate" is preferred in the case of line transmission of digital signals. (CCITT)

Synchronous. Two signals are synchronous if their corresponding significant instants have a desired phase relationship. (CCITT)

Synchronous network. A network in which the clocks are controlled so as to run, ideally, at identical rates, or at the same mean rate with limited relative phase displacement. *Note:* Ideally the clocks are synchronous, but they may be

mesochronous in practice. By common usage such mesochronous networks are frequently described as synchronous. (CCITT)

Synchronous transmission. A mode of digital transmission in which discrete signal elements (symbols) are transmitted at a fixed and continuous rate.

Talker echo. An echo of a talker's voice that is returned to the talker. When there is delay between the original signal and the echo, the effect is disturbing, unless the echo is attenuated to a tolerable level. (EOBS)

Talkoff. An inadvertent disconnect caused by speech sounds being interpreted as in-channel control signals (disconnects).

Tandem office. In general, any intermediate switch used to establish a connection. In specific terminology, a tandem office is a switch used to interconnect end offices in an exchange area.

TASI. Time assignment speech interpolation, the practice of concentrating a group of voice signals onto a smaller group of channels by dynamically switching active voice signals to idle channels.

Ternary coding. The use of all states of a three-level code to send more than one bit of information in a single symbol. This is in contrast to bipolar coding, which uses three levels, but only one of two in any particular interval. One method of interfacing binary data to a ternary line code is to encode 4 bits with three ternary symbols (4B-3T).

Tie line. A dedicated circuit connecting two private branch exchanges.

Time congestion. The ratio of time that all facilites of a system are busy (congested). Time congestion refers to the status of the system and does not necessarily imply that blocking occurs.

Time division multiplexing (TDM). Sharing a transmission link among multiple users by assigning time intervals to individual users during which they have the entire bandwidth of a system.

Time expansion. The use of more time slots on internal links of a switch than exist on external links.

Time out. A network parameter related to an enforced event designed to occur at the conclusion of a predetermined elapsed time. (CCITT)

Traffic carried. The amount of traffic carried (by a group of circuits or a group of switches) during any period is the sum of the holding times expressed in hours. (CCITT)

Traffic engineering. A network planning activity that determines the number and type of channels or communication paths required between switching points and the call-handling capacity of the switching points. (EOBS)

Traffic flow. The traffic flow (on a group of circuits or a group of switches) equals the amount of traffic divided by the duration of the observation, provided that the period of observation and the holding times are expressed in the same time units. Traffic flow calculated in this way is expressed in erlangs. (CCITT)

Transhybrid loss. The amount of isolation (in decibels) between go and return paths on the four-wire side of a four-wire to two-wire hybrid.

Transmission level point (TLP). A specification, in decibels, of the signal power at a point in a transmission system relative to the power of the same signal at a zero transmission level point (0-TLP).

Transmission reference point (TLP). A hypothetical point used as the zero relative level point in the computation of nominal relative levels. (CCITT)

Transmultiplexer (transmux). An equipment that transforms signals derived from frequency-division-multiplex equipment (such as group or supergroup) to time-division-division-multiplexed signals having the same structure as those derived from PCM multiplex equipment (such as primary or secondary PCM multiplex signals) and vice versa. (CCITT)

Transversal equalizer. A time domain equalizer utilizing a tapped delay line and weighting coefficients at each of the taps to remove intersymbol interference.

Traveling class mark. A code that accompanies connection setup messages indicating the nature of the service request and any special provisions that may be desired.

Trunk. A circuit or channel between two switching systems.

TST. Time-space-time. Digital switching structure.

Two-wire circuit. A circuit consisting of a single pair of wires and capable of simultaneously carrying two signals in opposite directions.

Two-way trunk. A trunk circuit that can be seized at either end of the circuit.

Unipolar. A binary line code using single polarity pulses and zero voltage for the two coding levels.

Vestigial sideband transmission. A form of single-sideband transmission that includes a vestige of the deleted sideband and a small amount of carrier energy.

Virtual call (circuit). A user facility in which a call setup procedure and a call clearing procedure will determine a period of communication between two DTEs in which the user's data will be transferred in the network in the packet mode of operation. All the user's data are derived from the network in the same order in which they are received by the network. (CCITT)

Wide area telecommunications services (WATS). A service that permits customers to make (OUTWATS) or receive (INWATS or 800 Service) long-distance calls and to have them billed on a bulk basis rather than individually. *WATS* is provided within selected service areas, or bands, by means of special private-access lines, which are connected to the public switched telephone network through *WATS*-equipped central offices. A single access line permits inward or outward service but not both. (EOBS)

ANSWERS TO SELECTED PROBLEMS

1.1 2000 pWC, $2(10)^{-6}$ mWC

1.3 500 pWC

1.5 250 pW at -3 dB TLP

3.1 1 kHz, 2 kHz, 3 kHz

3.2 2.43 dB degradation

3.4 49.4 Mbps

3.6 38.5 dB

3.8 01100110

3.10 23.5 dB

3.14 (a) 13 (b) 33

(c) Sample

Sample	30.2	123.2	-2336.4	8080.9
Error	.2	$-.8$.4	.9
Noise pwr	.04	.64	.16	.81
SQR	43.6	43.7	75.3	79.1

3.17 (a) 3 dB (b) 1.76 dB

3.19 SQR increases by 3.1 dB

3.21 01100110, 01100110, 11100110, 11100110 (non-inverted)

3.23 01100001, 01110100, 01110100, 01100001, 11100001, 11110100 11110100, 11100001

4.1

	0110	1000	0100	0110	0000	0010
(a)	0$-$+0	$-$000	0+00	0$-$+0	0000	00$-$0
(d)	0$-$+0	$-$00$-$	0+00	+$-$+$-$	0$-$+0	+0$-$0

4.2 PST has 1.76 dB more crosstalk

4.4 (a) .097 sec (b) .193 sec
4.6 141 bits
4.7 3 bits
4.9 .6875
4.11 (a) 1.99 dB (b) 3.78 dB
4.14 (a) 81 (b) 19 (c) 18
4.16

	+1	−3	+1	−1	+3	+3	−3
(a)	*	−2	−2	0	+2	+6	0
(b)	*	—	—	—	—	—	—
(c)	*	*	—	—	—	—	—

5.1 250
5.2 (a) 621 (b) 768
5.3 (a) .11 (b) .25
5.5 (a) 55,296
5.7 (a) 41 (b) 10,828

6.1 (a) .5 dB degradation
6.2 new BER is $1(10)^{-8}$
6.4 1600 Hz
6.5 (a) 17% of peak signal power (b) 15.3 dB reduced error distance
6.7 (a) 3.89 dB advantage for 32-QAM (b) 2.39 dB advantage for 32-QAM
6.9 11.2 dB degradation in error distance
6.11 $1.5(10)^{-3}$
6.13 3 dB

7.1 185 bits if the receiver knows the nominal data rate.
 370 bits to accomodate peak-to-peak shifts.
7.3 ± 1.3 UI with 99% probability
7.5 $5.37(10)^{-5}$ misframes/sec (once every 5.2 hours)
7.7 25.6 dB relative to 1 UI2
7.9 32%
7.11 (a) TIE = 15.44 UI (b) MTIE = 15.44 UI

8.1 .188 dB/km
8.3 16 km
8.5 $\Delta\lambda$ at 1300 nm is 5 times large than $\Delta\lambda$ at 1550 nm.

8.7 (a) 2m for mB1P (b) m + 1 for mB1C
8.9 (a) min rate = 1.528 Mbps (b) max rate = 1.560 Mbps
8.11 (a) min rate = 44.712 Mbps (b) max rate = 44.784 Mbps
8.13 68.7 usec

10.1 (a) 40 calls/hour (b) .989 (c) 9.4% (d) .188 erlangs (e) .453
10.3 1.4 erlangs
10.5 (a) 15 ports (14 ports with finite source analysis)
 (b) B = 9%
10.7 (a) 2.2% (b) 1%
10.8 (a) B = 30% (b) add 6 channels
10.10 (a) 32 erlangs (b) 46 trunk circuits
10.13 (a) B = 10% (b) B = 0.78%
10.14 2 WATS lines
10.16 (a) 83.3% (b) .833 sec (c) 30.6% (d) .833

INDEX